Lecture Notes in Computer Science 10904

Commenced Publication in 1973
Founding and Former Series Editors:
Gerhard Goos, Juris Hartmanis, and Jan van Leeuwen

More information about this series at http://www.springer.com/series/7409

Sakae Yamamoto · Hirohiko Mori (Eds.)

Human Interface and the Management of Information

Interaction, Visualization, and Analytics

20th International Conference, HIMI 2018
Held as Part of HCI International 2018
Las Vegas, NV, USA, July 15–20, 2018
Proceedings, Part I

Springer

Editors
Sakae Yamamoto
Tokyo University of Science
Tokyo
Japan

Hirohiko Mori
Tokyo City University
Tokyo
Japan

ISSN 0302-9743   ISSN 1611-3349 (electronic)
Lecture Notes in Computer Science
ISBN 978-3-319-92042-9   ISBN 978-3-319-92043-6 (eBook)
https://doi.org/10.1007/978-3-319-92043-6

Library of Congress Control Number: 2018944382

LNCS Sublibrary: SL3 – Information Systems and Applications, incl. Internet/Web, and HCI

Printed on acid-free paper

This Springer imprint is published by the registered company Springer International Publishing AG
part of Springer Nature
The registered company address is: Gewerbestrasse 11, 6330 Cham, Switzerland

Foreword

The 20th International Conference on Human-Computer Interaction, HCI International 2018, was held in Las Vegas, NV, USA, during July 15–20, 2018. The event incorporated the 14 conferences/thematic areas listed on the following page.

A total of 4,373 individuals from academia, research institutes, industry, and governmental agencies from 76 countries submitted contributions, and 1,170 papers and 195 posters have been included in the proceedings. These contributions address the latest research and development efforts and highlight the human aspects of design and use of computing systems. The contributions thoroughly cover the entire field of human-computer interaction, addressing major advances in knowledge and effective use of computers in a variety of application areas. The volumes constituting the full set of the conference proceedings are listed in the following pages.

I would like to thank the program board chairs and the members of the program boards of all thematic areas and affiliated conferences for their contribution to the highest scientific quality and the overall success of the HCI International 2018 conference.

This conference would not have been possible without the continuous and unwavering support and advice of the founder, Conference General Chair Emeritus and Conference Scientific Advisor Prof. Gavriel Salvendy. For his outstanding efforts, I would like to express my appreciation to the communications chair and editor of *HCI International News*, Dr. Abbas Moallem.

July 2018   Constantine Stephanidis

HCI International 2018 Thematic Areas and Affiliated Conferences

Thematic areas:

- Human-Computer Interaction (HCI 2018)
- Human Interface and the Management of Information (HIMI 2018)

Affiliated conferences:

- 15th International Conference on Engineering Psychology and Cognitive Ergonomics (EPCE 2018)
- 12th International Conference on Universal Access in Human-Computer Interaction (UAHCI 2018)
- 10th International Conference on Virtual, Augmented, and Mixed Reality (VAMR 2018)
- 10th International Conference on Cross-Cultural Design (CCD 2018)
- 10th International Conference on Social Computing and Social Media (SCSM 2018)
- 12th International Conference on Augmented Cognition (AC 2018)
- 9th International Conference on Digital Human Modeling and Applications in Health, Safety, Ergonomics, and Risk Management (DHM 2018)
- 7th International Conference on Design, User Experience, and Usability (DUXU 2018)
- 6th International Conference on Distributed, Ambient, and Pervasive Interactions (DAPI 2018)
- 5th International Conference on HCI in Business, Government, and Organizations (HCIBGO)
- 5th International Conference on Learning and Collaboration Technologies (LCT 2018)
- 4th International Conference on Human Aspects of IT for the Aged Population (ITAP 2018)

Conference Proceedings Volumes Full List

http://2018.hci.international/proceedings

Human Interface and the Management of Information

Program Board Chair(s): **Sakae Yamamoto, Japan and Hirohiko Mori, Japan**

- Yumi Asahi, Japan
- Linda R. Elliott, USA
- Shin'ichi Fukuzumi, Japan
- Michitaka Hirose, Japan
- Yasushi Ikei, Japan
- Yen-Yu Kang, Taiwan
- Keiko Kasamatsu, Japan
- Daiji Kobayashi, Japan
- Kentaro Kotani, Japan
- Hiroyuki Miki, Japan
- Ryosuke Saga, Japan
- Katsunori Shimohara, Japan
- Takahito Tomoto, Japan
- Kim-Phuong L. Vu, USA
- Marcelo Wanderley, Canada
- Tomio Watanabe, Japan
- Takehiko Yamaguchi, Japan

The full list with the Program Board Chairs and the members of the Program Boards of all thematic areas and affiliated conferences is available online at:

http://www.hci.international/board-members-2018.php

HCI International 2019

The 21st International Conference on Human-Computer Interaction, HCI International 2019, will be held jointly with the affiliated conferences in Orlando, FL, USA, at Walt Disney World Swan and Dolphin Resort, July 26–31, 2019. It will cover a broad spectrum of themes related to Human-Computer Interaction, including theoretical issues, methods, tools, processes, and case studies in HCI design, as well as novel interaction techniques, interfaces, and applications. The proceedings will be published by Springer. More information will be available on the conference website: http://2019.hci.international/.

General Chair
Prof. Constantine Stephanidis
University of Crete and ICS-FORTH
Heraklion, Crete, Greece
E-mail: general_chair@hcii2019.org

http://2019.hci.international/

Contents – Part I

Information and Vision

Text and Data Mining and Analytics

Contents – Part II

Information and Learning

Information in Aviation and Transport

Intelligent Systems

Service Management

Information Visualization

VisUML: A Live UML Visualization to Help Developers in Their Programming Task

Mickaël Duruisseau[1,2(✉)], Jean-Claude Tarby[2], Xavier Le Pallec[2],
and Sébastien Gérard[1]

[1] CEA LIST, Boîte 94, 91191 Gif sur Yvette, France
mickael.duruisseau@gmail.com, sebastien.gerard@cea.fr
[2] Univ. Lille, UMR 9189 - CRIStAL, 59000 Lille, France
{jean-claude.tarby,xavier.le-pallec}@univ-lille1.fr

Abstract. Developers produce a lot of code and most of them have to merge it to what already exists. The required time to perform this programming task is thus dependent on the access speed to information about existing code. Classic IDEs allow displaying textual representation of information through features like navigation, word searching or code completion. This kind of representation is not effective to represent links between code fragments. Current graphical code representation modules in IDE are suited to apprehend the system from a global point of view. However, the cognitive integration cost of those diagrams is disproportionate related to the elementary coding task.

Our approach considers graphical representation but only with code elements that are parts of the developer's mental model during his programming task. The corresponding cognitive integration of our graphical representation is then less costly. We use UML for this representation because it is a widespread and well-known formalism. We want to show that dynamic diagrams, whose content is modified and adapted in real-time by monitoring developer's actions can be of great benefit as their contents are perfectly suited to the developer current task. With our live diagrams, we provide to developers an efficient way to navigate through textual and graphical representation.

Keywords: Human-Computer Interaction
Model Driven Engineering · Software engineering
Unified modeling language · Human-centered design

1 Introduction

Human-Computer Interaction (HCI) has significantly evolved in recent years with the appearance of mobile and tactile devices, voice and gesture recognition, augmented and virtual reality, etc. Nowadays, most of the smartphone users know how to interact with a map, using simple interactions like touch,

© Springer International Publishing AG, part of Springer Nature 2018
S. Yamamoto and H. Mori (Eds.): HIMI 2018, LNCS 10904, pp. 3–22, 2018.
https://doi.org/10.1007/978-3-319-92043-6_1

but also some more complex, like swipe or pinch. In the meantime, software practitioners still develop applications only with a keyboard and a mouse. Furthermore, "development tools are showing mainly text with (so much) obstinacy" [1] despite some improvements concerning HCI in their IDE, like syntax coloration and auto-completion. We may consider software visualization tools as an improvement of the HCI, but their place in IDE and their use remain anecdotal. Visualization tools generally help developers to understand the global architecture of the application they are working on or the impact of what they are changing. Development consists mainly in producing code but not dealing with considerations of macroscopic nature. We argue these visualization tools are not focused on the most important and elementary task: programming. We claim that a graphical representation of elements that are currently knitted by a developer may be more easily accepted. The first reason is it can quickly provide information that is less visually explicit in textual code and still relevant for coding. More specifically it may highlight the different relations between elements (structural relations or specific execution flow). The second reason is that graphical representations are more suited to mobile and tactile devices (like tablets) than textual code and so, by taking advantage of them, they can provide HCI improvements of IDE.

In this paper, we present VisUML[1,2] a tool which uses a *"live diagramming"* approach and implements this point of view of software visualization. We defined *live diagrams* as being diagrams (UML or not) that display information according to the current task of the developer. We assume that currently opened elements in an IDE refer to this task and are therefore part of developers' mental models. These diagrams are updated instantly each time the code changes. Our approach consists in reducing the number of displayed elements but also to ease the navigation between code and diagrams.

To present this approach, we first describe the scientific background on which VisUML is based, as well as our design guidelines. Second, we review works that are related to our idea. Then we explain how VisUML works, with a focus on user interactions. Finally we highlight the contributions of this tool and we discuss its features and its evolutions. At last, we conclude with a summary and perspectives.

2   Scientific Background and Design Guidelines

The psychological mechanisms related to programming received much attention during the 90s [2–4]. Notable among these was the fact that developers work in little "spurts" [5] (sprint). Green [5] mentions the notion of spurts to emphasize that programming is a series of small steps where each one refers to a mental chunk or scheme. Therefore it is logical that the developer's main concern consists in connecting the spurt result with what has been produced so far. Indeed,

---

[1] VisUML website: http://these.mickaelduruisseau.fr/VisUML/.
[2] VisUML demonstration video: https://youtu.be/buyGojmbUpQ.

developers often read and analyze what has been done in order to properly "knit" (link) what they do with the rest of code.

The development environment should therefore optimize the "reading/production" cycle. It must be adapted to the current spurt and simultaneously provides quick access to information that will help developers to link their code to the existing one [6].

Thus, it is no wonder that most of current code editors propose shortcuts to go quickly to the definition of the selected element or to list all the invocations of the selected method. However, navigation is not the only way to find relevant information. Changing visual properties of code elements is another way to highlight what can interest the developers in their knitting task; for instance, the indentation clearly shows the different control structures in which the current line of code is nested, background color variation is sometimes used to identify the different places where a variable is used, etc. Visual changes can go further with concrete transformations of shapes and concerned elements. That is the speciality of what the software community called visualization tools. These ones are of great help in order to understand the existing code; this is particularly mentioned in the SoftVis/VISSOFT conferences cycle. [1] insists on the necessity of the requisite ubiquity of visualization in development environments, even if it means rethinking them entirely. A majority of the work on software visualization results in tools that allow finding or obtaining macroscopic information.

At the opposite, our approach provides microscopic information by focusing on more specific and activity related data. We do not aim to provide information about possible impacts of each modification, nor knowing which application's parts have to be rewritten. Instead, our goal is to allow developers to have a visual support of their code. This will enable them to quickly find information about entities and their relations and will also add an easy navigation mean. In addition, this visual support could be shared with peoples that have different needs in terms of visualization. For example, a product manager will not use the same kind of representations than a developer, but the displayed entities remain identical. In the same idea, developers can have a fully detailed class diagram, whereas project leader may want to see a class diagram without attributes or operations, with a color code that indicates the code quality of each class or their modification date, number of commits, etc.

Developers constantly execute code reading operations to find information in order to allow them to modify their code. We wish to shorten these reading operations by giving a quick access to elements often used or viewed by the developers. These elements are mostly represented as entities connected by links. The textual support is not very effective to represent a system with interconnected elements; however the diagrammatic representations are much more efficient in this task [7]. Furthermore, works on the psychology of senior developers show that they have mainly problems understanding the control flow rather than the basic bricks of a language (e.g. variables names) [8]. Thus, in addition to the two aspects to be displayed (entities and links), we can add the control flow between entities that are associated to the active coding task. To ensure that the access of information through the graphical representation is as fast as possible, the reading/decoding of this representation must take as little time as possible. The cognitive fit is a main concern for our tool, and the time spent when switching between a representation to another, or when switching between tasks in general (e.g. code review vs. diagram creation, class dependencies search vs. debugging), is thus very important. This concern is the heart of the cognitive dimensions [9] and can be found as a rule *cognitive integration* in the physics of notations [10]. In our case, when switching from a textual code editor to a graphical representation, it is clearly necessary that developers keep their references. For example it is important that developers recognize the entities they have just been manipulating. Therefore the graphical representation has to be close to their mental model and display information about:

- Entities linked to the active coding task: whether they are open or not in the IDE and which element is currently active
- Neighbourhood data: accessible via variables of a method, attributes of an object, inheritance...
- Control flow: the content of a currently consulted method, or at the origin of a search or navigation.

We use the UML language for the graphical representation because it remains a language known and mastered by developers, even if according to different surveys it is not enough used in firms. This selection was made according to the principle of cognitive integration [9]: adapt to the knowledge of developers.

Forward and Dzidek [11,12] attest that two of the three most widely used UML diagrams are the class and the sequence ones. We consequently chose them in VisUML. The **class diagram** is important because developers can recognize the entities they manipulate, as well as consult other related entities thanks to the different types of links.

In order to display the control flow, especially for the body of a method, we opted for the **sequence diagram**. We assume that this diagram is a complementary visual support for developers that are working on the implementation of a method. They will be able to see all the classes that are involved in this method, and especially the different exchanges (and thus links) between them. In addition, the temporality of these exchanges is emphasized because it is represented on the *y-axis*. This correspondence between *y-axis* and *temporality* reduces subjectivity in the layout, and is therefore less subject to interpretation than communication diagram, where the placement at *x* and *y* is arbitrary. Activity diagrams and states machines can be used for the implementation of a method but their point of view (activity or state) adds a semantic gap which is likely to increase the time required for decoding (without taking into account the ordering of instructions).

The navigation between the code and these two types of representation (class and sequence diagrams) is explained in the next section. Finally, we chose to display our diagrams in web pages (of a web browser) so that we can easily connect our module into any IDE. This aspect may seem purely technical but it is not: one of the cognitive dimensions is the visibility in which the juxtaposition of two points of view is a way to easily switch from one to the other. If the IDE does not allow two large windows to be displayed next to each other (each one on a screen or both on the same screen), it is natively possible with our approach, even when using a tablet or any display device with an OS containing a web browser. However, in addition to these web pages, we also made modeling tools plugins (see Sect. 4.4) that enable live diagrams on them.

3 Related Works

In this part, we first describe works about visualization tools for code, their UML diagrams features as well as the possible interactions and navigation. Then we talk about reverse-engineering tools, especially how they generate diagrams and in what way the generated elements and source code are connected. Finally we conclude with a summary of these reviews and an opening to the presentation of our tool.

Code visualization tools usually allow to have graphical and exhaustive views of projects. These views can use 2D or 3D (for example [13] uses 2D diagrams connected in 3D and 2D diagrams overloaded with information in 3D like a city map with buildings), even in virtual reality. However, they require significant cognitive efforts (understanding the graphic representation, which is often unusual for the developers), as well as large screens or even specific equipment for VR. The advantage of this type of representation is the "macroscopic" view of projects (dependencies between packages or classes, code versioning...). The major drawback of these representations is the "off-line" aspect since they do not allow to reflect in real time the project in its current state. On the contrary, and on a "microscopic" aspect, [14] can see in real time the code of a project in a simplified way and thus making it easier to navigate in it. Unfortunately, this representation is only textual and requires time to adapt to the developer.

Since class diagrams, sequence diagrams and code, share elements, IDE and modeling tools propose more and more often a "find usages" or "find in diagram" command. These commands may be triggered in two ways. The first one is by using the top toolbar menus, but it requires to select which element must be looked for. The second is via contextual menus; in that case, the element is already selected and the menu shows only information and commands about it. One of the easiest way currently implemented in MagicDraw[3] is by adding an item in the elements' contextual menu. Thus right-clicking an element will bring up the menu and browsing it will allow users to open a related diagram. However this menu is yet too much complicated since it does not just show a list of diagrams but displays the full (UML) path to a related element. Figure 1 shows this menu.

This navigation is indeed based on the UML relations between elements while we have a user-centered approach. This is especially true for tools using EMF[4], such as Papyrus, since the relations are even more complicated and less direct for the user; for instance a class inside a sequence diagram is the type of the property

---

[3] MagicDraw: http://www.nomagic.com/products/magicdraw.html.
[4] Eclipse Modeling Framework: https://eclipse.org/modeling/emf/.

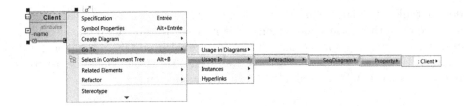

Fig. 1. MagicDraw - class diagram: "Usage In" feature requires 8 clicks

that the lifeline represents (e.g. In Fig. 1, the lifeline property has *"Client"* for type). This navigation action requires at least 8 clicks to switch from a class to any associated sequence diagram. At the opposite, in VisUML, a simple click on a method in the class diagram updates immediately the sequence diagram to display the correct method.

IntelliJ meanwhile implements a similar feature: when displaying a class diagram, it is possible for users to "Jump to Source" or "Find Usage" of any sub-elements (attributes and methods). The advantages of IntelliJ over MagicDraw is that it is directly linked to the code representation. As a result, it is possible to quickly switch from a diagram view to the code. Two visualization modes are available. The first one is classic: the diagram is displayed on a new tab (which can be shown near another tab). The second is interesting as it creates a popup window with the diagram inside. In the popup mode, nothing can be modified, but this allows a quick preview of a diagram. Despite these two modes of visualization, the work context is broken because developers must do several actions in order to generate and see any diagrams.

IntelliJ only supports class diagrams, it is therefore not possible to create the sequence diagram of a selected method. In addition, there is no navigation between diagrams (whether class diagram to class diagram, or class diagram to sequence diagram).

Figure 2 is an example of how IntelliJ shows usages of a class inside the project: an unordered list of all the occurrences of this class, without filtering options. In this example we wanted to find the class that extends *Entity*, the useful results are bordered in green, only 4 of 18 lines are relevant.

Reverse engineering is now widespread among IDE and modeling tools. It allows developers to create a graphical representation of their code, or part of code. All the tools proposing reverse engineering allow developers to produce class diagrams. However, sequence diagrams remain yet less common.

ObjectAid[5] is one of the tools (together with MaintainJ[6], VisualParadigm[7] and MagicDraw) that handle sequence diagrams. Depending on the tools, there are several ways to generate a sequence diagram. All of them (except MaintainJ) use a common menu that allows developers to choose which elements they want

---

[5] ObjectAid: http://www.objectaid.com/sequence-diagram.

[6] MaintainJ: http://maintainj.com/.

[7] VisualParadigm: https://www.visual-paradigm.com/.

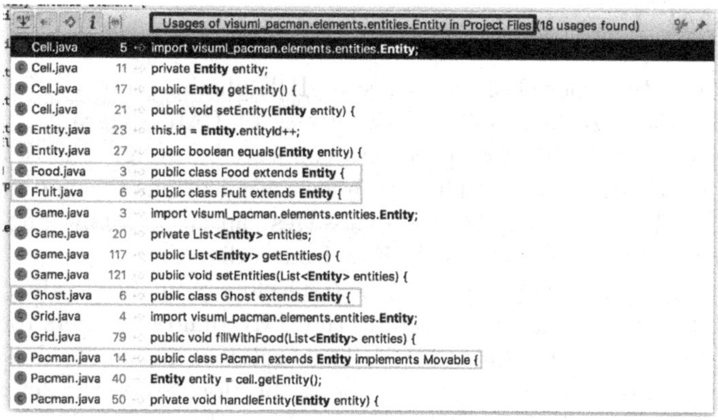

Fig. 2. IntelliJ - find usages of "Entity" (only 4 of 18 lines are relevant)

to reverse-engineer in a sequence diagram. ObjectAid and IntelliJ (only for class diagrams) also add drag & drop support, from any location in the IDE. This means that developers can add elements on diagrams easily, without having to browse the entire project. However, it is still up to them to choose which elements should appear or not. In VisualParadigm, developers have to navigate through four windows, when they want to "instant" reverse a method. For this, they must (1) select the source code folder and (2) find the correct class. Once they've found it, they must (3) select the method they want to reverse. This process is complicated as it takes at least twenty clicks.

At last, some tools are specialized in visualization after code execution, such as MaintainJ, which only works at runtime, and ObjectAid which can analyze Java stack-traces. These tools can therefore generate sequence diagrams that reflect the execution of a particular method, but not its complete representation (e.g. "*alt*" or "*loop*" fragments are missing). Although they can generate a lot of sequence diagrams which are interconnected, it is up to the developers to choose which one to display, and then navigate through them, using basic interactions, i.e. right click on a specific invocation to see its own sequence diagram.

Finally, among all the tools we have analyzed, none allows developers to have live diagrams that fit their current task. Some presented solutions propose to display diagrams at runtime, with information and values extracted from stack traces or execution, but this does not necessarily correspond to the active coding task. Moreover, interactions in these diagrams remain basic. Most of the navigation actions must be triggered with contextual menus and clicks on elements, and they simply allow developers to switch from one diagram to one of its sub-diagrams. In addition, some interactions can link the diagram with the code, but most of them use the "search" function. Overall, each navigation interaction forces users to choose from a list of elements, rather than automatically display elements that are relevant to their task.

4   VisUML Presentation

VisUML is a tool composed of two parts: an IDE plugin, presented in Sect. 4.3, and multiple visualization tools[8] of two types of UML diagrams (class and sequence diagrams). These two parts are connected through our communication bus which is named WSE. This bus allows applications to send and receive information in JSON messages. Those messages can contain any kind of information, whether it comes from the IDE or from a diagram.

As previously described, we aim to help developers in their coding task. To this end, all the information displayed on the UML diagrams refers to elements currently opened in the developer's IDE. This tool does not aim to do a full synchronization between the code and the models, but focuses on the active coding task of the developers. As a result, we use a light mechanism of synchronization, using WSE as way of communication.

Figure 3 gives an overview of the interactions between the IDE and both diagrams. In this section, we present the different parts of VisUML: the class diagram view, the sequence diagram view, the IDE plugins and to conclude, a Papyrus class diagram plugin.

The first two sub-sections are split into two parts: displayed elements and interactions, in which we present the elements that our views display, as well as the interactions that are triggered from these views. The IDE plugins part presents the interactions triggered by the IDE, as well as the messages passing through WSE. Finally, the Papyrus plugin sub-section shows that VisUML visualizations also work on modeling tools.

4.1   VisUML Class Diagram View

The UML class diagram of our tool only shows the most important part of a class diagram: the classifiers (class, interface...), their attributes and operations, as well as the links (generalization, association...) between them, and the packages. According to empirical studies on UML in industry [11,15,16], the class diagram is often used in a simple way and informally. The displayed elements and the possibilities offered by our template are thus enough for our use.

Displayed Elements. Once activated, VisUML class diagram displays all the Java elements that are currently opened in the IDE. These elements can be classes, enumerations, interfaces, ... and they may or may not be related to any other element.

In addition to these elements, VisUML also displays unopened elements that have at least one relation with an opened element. However, those elements are differentiated by their opacity, as they appear more transparent than the others.

In order to highlight the element that is active in the IDE, we change the color of the associated graphical representation to green. Moreover, any links

---

[8] As of today, VisUML visualization tools are available on web pages, Papyrus and GenMyModel plugins.

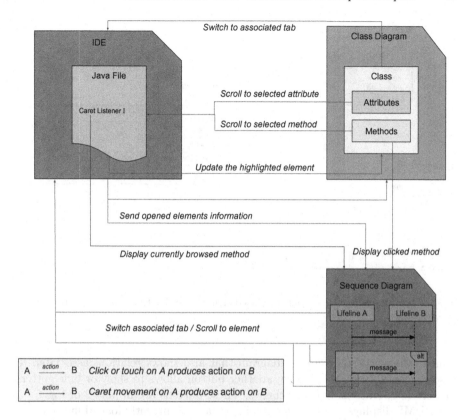

Fig. 3. Navigation interactions and resulting actions (Color figure online)

(i.e. relations) that are connected to this representation appear in bold and red. This allows users to easily detect related elements.

Figure 4 shows a class diagram with the active tab in a different color, as well as unopened related elements.

Filters. In order to allow users to limit the number of displayed elements, we added several filters to our class diagram view.

First, we added 4 visualization profiles (see part 1 of Fig. 5). Each profile has a different configuration, showing or not part of the diagram.

- **Packages only**: Display only packages, without any classifier inside
- **Classes**: Display classes, without attributes or operations
- **Public & Protected elements**: Display classes, with only public and protected attributes or methods. Private fields are hidden.
- **All details**: Default profile. Display everything.

Moreover, a simple button toggles the visibility of unopened related elements. This function aims to quickly switch between a diagram that matches exactly

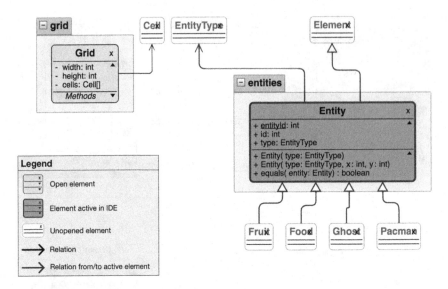

Fig. 4. Class diagram with unopened related elements and active tab highlighted (Color figure online)

opened tabs in the IDE, and a diagram that also shows relations associated to these elements. In the same idea, another button allows to show or hide getters and setters operations in every classes. These buttons are in part 2 of Fig. 5.

VisUML displays two lists of elements (that are currently loaded in the tool). These lists show respectively classifiers (class, enum, interface ...) and packages. They are displayed in part 3 of Fig. 5. On each element of either list, a checkbox allows to change the visibility of the related element. So, one can simply check or uncheck any element, which allows to quickly filter elements according to their name or package.

Finally, we chose to hide some elements that do not provide useful information to developers. For example, we hide the *java.lang.Object* element because it is inherited by all other Java classes. In the same idea, primitives types are also hidden. In the same idea, we can hide elements according to a specific framework or language. For example, when working on an Android project, we hide some Android specific elements such as *android.app.Activity*. There is currently no graphical interface to add or remove elements in this list, but this can be done easily by modifying the source code of the class diagram.

Interactions. In order to make the class diagram visualization interactive, we added interactions on each displayed element. These interactions send messages on our communication bus, and those messages are then received by any connected tools. Interactions in the class diagram view (also visible on Fig. 3) are:

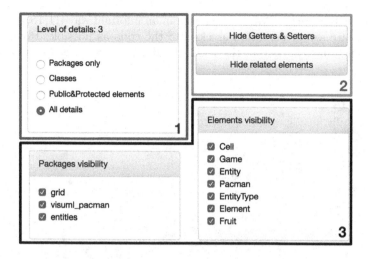

Fig. 5. VisUML filters

- A click on a classifier (besides its attributes/methods) will update the IDE by putting forward this element (changing the active tab, or opening the corresponding file).
- A click on an attribute or a method will switch the active tab to the associated file if needed, scroll the IDE to the definition of the chosen element and highlight or select it. Moreover, in the case of a method, the sequence diagram view will automatically be updated in order to display the selected method body.
- Finally, a click on the cross (X) on the top right corner of an element will delete its graphical representation and close the associated tab on the IDE.

Handling these events allows the IDE to remain entirely synchronized with our graphical representations (*"live diagram"*) when the developer navigates inside a diagram or interact with it. It is therefore possible to switch between the two representations (code and model) without losing the context of work (since navigation inside the diagram or the IDE also updates the other). For instance, the selected class (in the diagram) will always be the active tab in the IDE. With this visual aid, the developer does not need to look for the active element. Moreover, a clicked/selected attribute or method in the diagram will always be visible (i.e. the IDE's editor will scroll if needed) and highlighted in the code. Finally, the currently browsed method (in the code, according to the caret location) is emphasized in bold and blue.

4.2 VisUML Sequence Diagram View

In addition to the class diagram, we chose to implement a sequence diagram view, as explained in Sect. 2. This diagram is often used to represent control flow, here the body of a method.

Displayed Elements. UML is often used in an informal way [11, 15, 16] but we chose to create a representation as close to the code as possible. However our sequence diagram differs from the UML standard in several ways as explained below.

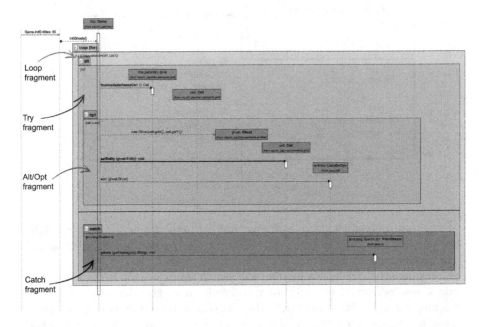

Fig. 6. Sequence diagram with colored fragments and highlighted message (Color figure online)

First, our sequence diagram acts like an UML one, but with more code specific information. For example, we create an "alt" fragment for each "try" and "catch" block and we display invocation details such as parameters (types and values). Moreover, when VisUML users let their mouse on an element, a tooltip appears with all code comments associated to this element. Finally, we added a specific color code on fragments which allows developers to easily recognize the type of a specific fragment, as well as their nesting level. We chose to be closer to the code in order to reduce the cognitive integration of the model, since our goal is to help developers in their current task but not to abstract their code. Figure 6 shows an example of colored fragments.

Interactions. As explained in the previous section, visualizations allow users to interact with the displayed elements. Possible interactions on this diagram are similar to the ones implemented on the class diagram:

– A click on a lifeline (which can be created at any time during the method) will scroll the IDE directly to the associated class, or variable assignment.

– A click on a message (link between two lifelines) or an activity (block on a lifeline), which is linked to a method invocation, will scroll the IDE to the corresponding code line.
– A click on a group, or fragment (alt, loop, ...), will highlight this group on the IDE (full selection of all the lines of this group).

In addition to these interactions, we added a caret listener in the IDE. As a result, when the developer moves the caret, we check if it is in the body of a method and send a message containing information about this method, as well as the line where the caret is located. This message is then received by the sequence diagram view, which automatically updates its display to show the graphical representation of this method. Moreover, since the message contains the location of the caret (the line in the code file), the sequence diagram knows what UML message (represented by links between Lifelines) is concerned and we highlight it by changing its color. Figure 6 shows an example of highlighted message (message *setEntity* is orange and bold).

Since the communication between the IDE and the visualizations is very fast (less than 100 ms between sending and receiving a message in normal conditions), it is possible for the users to see, in real-time, the sequence diagram lights up (highlight the related message) when the cursor moves in the code. This is especially convenient when browsing methods inside a file.

4.3 VisUML IDE Plugins

The current VisUML IDE implementation refers to an IntelliJ[9] plugin and an Eclipse[10] one. IntelliJ is a Java IDE developed by JetBrains. Android Studio[11] is based on this IDE, with specific functions for Android developers. IntelliJ uses a plugin system, allowing us to add functionality to any IntelliJ based IDE.

As mentioned, IntelliJ (as well as Eclipse) offers an API that entirely manages the interactive part of the IDE. Therefore it is possible to add listeners on any kind of event. In our case, we are mainly listening to five events:

1. A file has been opened
2. A file has been saved (or its content has been modified)
3. The user has changed the active tab
4. A tab has been closed
5. The caret has been moved

Each handled event contains parameters, primarily the name of the concerned file, which allows us to make a link between the code element inside it and the file name, as well as create a representation of that element (in which we store every needed information, such as its flags, attributes, methods, relations, etc.). Once the event has been received and the element identified, the plugin sends a specific message via WSE, to give an order to the connected applications.

---

[9] IntelliJ: https://www.jetbrains.com/idea/.
[10] Eclipse: https://eclipse.org/.
[11] Android Studio: https://developer.android.com/studio/index.html.

For events 1, 2 and 3, a *"createOrUpdateUML"* message is generated with all the information of the class (or several messages, if there are intern or anonymous classes in addition to the main class). These messages are then sent to WSE. Once received the UML class diagram (whether on the web page or on Papyrus) will create or update the UML graphical representations associated to this element.

The event 3 also sends a message of type *"highlightClass"*, containing the main class of the file in foreground on the IDE; it allows to put forward its graphical representation. In this way, the active tab of the IDE will always be highlighted on the diagrams, whether via a different background coloration, a flashing border or a zooming effect. Figure 4 shows an example of highlighted element.

The event 4 creates a *"remove"* message (or several) with the Fully Qualified Name (FQN)[12] of the element that has been closed by the user as parameter.

Finally, the event 5 sends a *"highlightMethod"* message, which is filled with the active class information, the currently browsed method and the line on which the caret is. This event is triggered by following the position of the user's caret in the code.

These five events allow the IDE and the diagram views to be synchronized at any time.

4.4   Plugins for Modeling Tools

In addition to the visualization into web pages, we also made a Papyrus plugin that listens to the same messages as these web pages, in order to create UML class or sequence diagrams. Papyrus[13] is an UML modeling tool based on Eclipse. It is developed by the Laboratory of Model Driven Engineering for Embedded Systems (LISE) which is a part of the French Alternative Energies and Atomic Energy Commission (CEA-List)[14]. Papyrus can either be used as a standalone tool or as an Eclipse plugin.

Our VisUML Papyrus plugin is still an early prototype. When connected to an IDE, the current Papyrus model shows a live representation of the opened tabs of the IDE. Several interactions have also been implemented, such as the click on a class (*switchToClass* message) and on an attribute or method (*highlightAttribute/highlightMethod*).

The advantages of Papyrus compared to a web page is that it implements a lot of useful features to UML diagrams, such as formal validation, exportation in various formats, easy refactoring, code generation, etc. Even if in our web pages the goal is to have live and transient diagrams, Papyrus allows users to store these diagrams and do UML operations on them.

In the same idea, we are developing a GenMyModel[15] plugin that works the same way. GenMyModel is an online UML editor that provides collaboration to

---

[12] A   FQN   is   an   unambiguous   name   that   specifies   an   object   (e.g. *com.myapp.model.Client*).

[13] Papyrus: https://eclipse.org/papyrus/.

[14] CEA LIST: http://www-list.cea.fr/.

[15] GenMyModel: https://www.genmymodel.com/.

its users, as well as a history mechanism. With this history, each modification is saved and can be *replayed*, helping users to understand the evolution of the UML model.

To sum up, VisUML is independent of IDE and modeling tools or UML visualizations. Any modeling tool with an open API and connectable to WSE (i.e. able to send and receive HTTP requests) can be connected to VisUML.

5 VisUML Contributions and Discussion

In this section we will describe what are the contributions of VisUML. At first, we will show that according to its technical implementation, VisUML is a flexible tool. Then we will point out that it is developer-centered, which is an important concept for such a tool, since we aim to ease the work of developers. With all these points, we partly answer to Chaudron's vision [17] (e.g. "mixing formal notations with informal notations", "higher level of integration of tools"). Indeed, we use both formal and informal representations of UML models. More, we make sure that our tools are focused on one specific aspect while being connected to each other. Finally, even if there is not yet any particular gestural interaction, our tools work on tablets and can manage touch and gestural events.

5.1 Distributed and Linked Applications

VisUML is composed of several applications that are independent and simply connected through WSE. It is therefore easy to add an application at anytime, or even replace one by another (e.g. UML visualization can be done by the GoJS[16] web pages as well as with the Papyrus plugin. With VisUML, a developer can switch at anytime between those two visualizations). Figure 7 shows how VisUML can be used in different work contexts.

In addition, because of their technologies and implementations, these applications are able to run on any platform and OS (Java works on Windows, OS X and Linux). The web pages can even work on Android and iOS, in any web browser app. Moreover, the UML class diagram view automatically switches to a specific template when it is displayed on a smartphone (with a "small" screen). This template shows only the name of the classes, but all the links between them. A simple click (or touch) on a class display all its details.

Finally, the applications are able to run on different devices (PC, tablet and smartphone) at the same time, allowing developers to use this tool in their environment without changing their habits.

VisUML is then entirely multi-platform; developers can develop on a Mac or PC, visualize their live diagrams on the same machine with a different screen, or on their tablet or another machine. They can also have a PC with their code, a laptop with Papyrus showing their UML class diagram, and a tablet with the sequence diagram on it. Thanks to WSE, each view is synchronized and interactive.

---

[16] GoJS: https://gojs.net/.

(a) 2 screens: IDE and Class Diagram   (b) 2 screens, 1 tablet: IDE, Class Diagram and Sequence Diagram

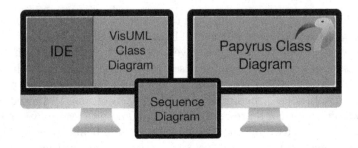

(c) 2 screens, 1 tablet: IDE, 2 Class Diagrams (VisUML and Papyrus) and Sequence Diagram

Fig. 7. Examples of VisUML contexts of work

5.2 Displayed and Highlighted Information

In addition to this liberty of devices and displays, we aim to ensure that VisUML is current-activity centred. Rather than reverse-engineering all the project to create a gigantic and unreadable class diagram, we chose to only show the opened tabs of the developer's IDE. Although this solution works and allows the user to switch easily between the two representations (code and model, since they are exactly the same), a simple UML class diagram generation from opened tabs has shown that some information were missing. Indeed, in the code, a developer can rapidly know if a class extends another class or not. However, if a diagram only shows opened Java classes, this kind of information will not be displayed.

In order to fix this lack of information, in addition to every opened elements, VisUML displays all their related elements (an element can be in relation with another if one extends or implements the other, or if there is an association between them), even if they are not opened. This allows the developer to easily navigate in the graphical representation without losing their context.

5.3 Simplify the Use of Diagrams

Another important aspect in VisUML is to reduce the number of required actions for the developers to see and navigate in their diagrams. In most of the current

IDE or modeling tools, developers have to select which classes or elements they want to reverse-engineer or model, by navigation through the list of all the possible elements of the project and checking which one they want to process.

This results in a loss of the work context of the developers. They have to stop their work, take time to select what to show, and then they can resume their task. This is a big waste of time and concentration, because they changed their context and active task (cognitive overload of short-term memory), then they will have to do some cognitive work to restart their task. Moreover, if afterwards they realize that they forgot some elements, they will have to redo all the process, which will result in another waste of time, as well as a discouragement to build UML diagrams.

With VisUML however, we want to reduce to a minimum this waste of time and the consequences. Once developers started the plugin, everything is done automatically, without any actions from them. Obviously, developers must do some actions in their code (or IDE), for example open, edit or close a file, but nothing specific to VisUML. They just have to work as usual with new live diagrams views of the code on the desired screens and devices. If they want to see an element, they just have to open it. If they open an element A that extends another element B, then B will be shown (with lower opacity), and a simple click on B will open the corresponding code in the IDE. Therefore, in addition to easily navigate between UML diagrams and the corresponding code, VisUML adds a new and efficient way of navigation inside UML diagrams; indeed, a simple click on a method in the class diagram will update the sequence diagram to display the representation of this method.

6   Discussion

In this section, we discuss some of the side effects of VisUML, as well as improvements that can be made to this tool in order to makes it even more efficient.

6.1   Malleable Environment

Our first goal with VisUML is to adapt to the current work environment of developers. Instead of making a new IDE or modeling application, we created plugins that work with commonly used IDEs. This reduce the changes needed to adopt the tool and so ease its use.

In addition, we aim to add flexibility to the environment by separating features in multiple applications that share data to each others. With VisUML, most of the data come from the code and are transmitted through our communication bus (WSE). Any compatible application can connected to WSE to listen data and send actions if needed. This On-The-Go principle allows users to configure their work environment as they wish.

Moreover, there is no limit on how many devices or applications can connect to WSE. It is possible to be in any kind of configuration. For example, a user could have VisUML views and code, another Papyrus and VisUML sequence diagram view. A third could use only its IDE while sending data to others. Then a fourth could use a tablet with GenMyModel.

On another hand, UML diagrams are not only used by developers. They are commonly used in firms by multiple peoples, all of them doing specific actions. We previously describe "profiles" in the class diagram view. The profiles we defined are developers oriented, but we thought about job oriented profiles. According to its job, a user could see only a part of a diagram, add extra information such as code quality, or even implementation of specific view, like Android layouts or SQLite databases.

As explained in Sect. 4.4, in addition to web pages, VisUML views are also available as a plugin for Papyrus (and soon GenMyModel). This allows developers to benefit from the modeler features (e.g. save, UML validation, XMI export, code generation...) without having to create the model and diagrams, as they are generated by VisUML. This aspect validates one of our goals, which was to adapt to the developers environments by implementing live diagrams on modeling tools.

6.2   Evolutions

Because our goal is to display information about the active task and not to ensure a complete synchronization, we did not add model to code transformation even if it would be easy to do it. This would allow a bidirectional modification between the model and the code, which would increase the efficiency of VisUML by reducing the number of switches between code and model.

In the same idea, many refactoring interactions are considered (some of them currently tested in a prototype) in both the class and sequence diagrams. For example, messages (links) in a sequence diagram could be moved, which implies a reorganization of the invocations order (code refactoring), as well as their belonging to a fragment or not. In the same idea, a simple deletion of the graphical representation of a link or a group could delete their corresponding element in the code. Likewise in the class diagram, developers could be able to move elements in or out packages.

Finally we also thought about adding new listeners on the IDE. For example, listeners on the code execution would enable information of debugging (breakpoints, execution errors, ...) to be displayed on diagrams. In the same idea, listeners on the syntax errors or code errors (shown in the IDE) could also be displayed on a sequence diagram (e.g. an incorrect type or a private method called in a wrong context would be highlighted).

7   Conclusion

In this paper we presented VisUML, a tool that helps developers in their coding and debugging tasks. We explained the current limitations of visualizations tools and the contributions provided by VisUML.

Whether it is to understand a project or to find interesting classes, developers spend their time looking for files, classes or methods. These subtasks disrupt developers concentration, as they force them to switch from one activity (e.g. write code) to another (e.g. find an element) and thus it breaks their context of work and overload their short-term memory. Indeed, most of the time, developers look at the code editor part of their IDE. However in order to find files they must look at menus, sub-menus and other windows, which is not efficient and results in a waste of time.

UML could help them to quickly understand a project architecture and its elements relations, since graphical representations of the code are more efficient for developers to understand all the existing relationships than text. However, modeling tools and UML itself are not generally used, or used in an informal way. Nowadays tools allow developers to reverse-engineer the entire project, which results for instance in a unreadable UML class diagram. It is also possible to create a class diagram using a subset of the project, but this requires a lot of interactions and time for developers, as they have to select which files should be considered.

Finally, we show that with our tool, developers could be able to access a view displaying live diagrams of their projects. These diagrams are designed to be easy to read as they only use information provided by the developers IDE.

We focused on two UML diagrams for the moment, but we do consider extending this set to other UML diagrams, such as Activity diagram (work in progress) or Object diagram. In the same idea, since our tool is not a modeling tool, we decided to use UML diagrams informally, and add extra information on them. For example, we thought about Android activities representations, using the XML layout, as well as the *intents* to build links between those activities. This would really be helpful for Android developers to be able to see their views and the links between them in a simple interface, without having to navigate through different files (activities and layouts files).

On the other hand, improving the visual representation of our diagrams would be interesting, but these changes are slightly outside the main domain of our work, which are interaction and navigation for the developers. Indeed, another thesis in the team [18], and in collaboration with CEA LIST, works on UML representations and semiology of graphics, which is complementary to this work.

References

1. Girba, T., Chis, A.: Pervasive software visualizations (keynote). In: Proceedings of 2015 IEEE 3rd Working Conference on Software Visualization, VISSOFT 2015, pp. 1–5, September 2015
2. Brooks, R.: Towards a theory of the cognitive processes in computer programming. Int. J. Hum.-Comput. Stud. **51**(2), 197–211 (1999)
3. Davies, S.P.: Skill levels and strategic differences in plan comprehension and implementation in programming. In: Proceedings of the Fifth Conference of the British Computer Society, Human-Computer Interaction Specialist Group on People and Computers V, pp. 487–502. Cambridge University Press, New York (1989)

4. Détienne, F.: Expert programming knowledge: a schema-based approach. In: Hoc, J.-M., Green, T.R.G., Samurcay, R., Gilmore, D. (eds.) Psychology of Programming. People and Computer Series, pp. 205–222. Academic Press (1990)

5. Olson, G.M., Sheppard, S., Soloway, E. (eds.) Empirical Studies of Programmers: Second Workshop, p. 263. Ablex Publishing Corporation, Norwood (1987)

6. Davies, S.P.: Externalising information during coding activities: effects of expertise, environment and task. In: Empirical Studies of Programmers: Fifth Workshop, pp. 42–61 (1993)

7. Larkin, J., Simon, H.A.: Why a diagram is (sometimes) worth ten thousand words. Cogn. Sci. **11**(1), 65–99 (1987)

8. Church, L., Marasoiu, M.: A fox not a hedgehog: what does PPIG know? In: 27th Annual Workshop on PPIG 2016, pp. 17–31 (2016)

9. Green, T., Petre, M.: Usability analysis of visual programming environments: a 'cognitive dimensions' framework. J. Vis. Lang. Comput. **7**(2), 131–174 (1996)

10. Moody, D.L.: The "physics" of notations: toward a scientific basis for constructingvisual notations in software engineering. IEEE Trans. Softw. Eng. **35**(6), 756–779 (2009)

11. Lethbridge, T.C., Ave, K.E.: Perceptions of software modeling: a survey of software practitioners table of contents. In: 5th Workshop From Code Centric to Model Centric: Evaluating the Effectiveness of MDD (C2M: EEMDD), pp. 1–102 (2008)

12. Dzidek, W.J., Arisholm, E., Briand, L.C.: A realistic empirical evaluation of the costs and benefits of UML in software maintenance. IEEE Trans. Softw. Eng. **34**(3), 407–432 (2008)

13. Gregorovic, L., Polasek, I.: Analysis and design of object-oriented software using multidimensional UML. In: Proceedings of the 15th International Conference on Knowledge Technologies and Data-Driven Business, pp. 47:1–47:4 (2015)

14. De Line, R., Czerwinski, M., Meyers, B., Venolia, G., Drucker, S., Robertson, G.: Code thumbnails: using spatial memory to navigate source code. In: Proceedings - IEEE Symposium on Visual Languages and Human-Centric Computing, VL/HCC 2006, pp. 11–18 (2006)

15. Chaudron, M.R., Heijstek, W., Nugroho, A.: How effective is UML modeling?: an empirical perspective on costs and benefits. Softw. Syst. Model. **11**(4), 571–580 (2012)

16. Petre, M.: UML in practice. In: Proceedings - International Conference on Software Engineering, pp. 722–731 (2013)

17. Chaudron, M.R.V., Jolak, R.: A vision on a new generation of software design environments. In: HuFaMo@ MoDELS, pp. 11–16 (2015)

18. El Ahmar, Y., Gerard, S., Dumoulin, C., Le Pallec, X.: Enhancing the communication value of UML models with graphical layers. In: Proceedings of 2015 ACM/IEEE 18th International Conference on Model Driven Engineering Languages and Systems, MODELS 2015, pp. 64–69, September 2015

Web-Based Visualization Component for Geo-Information

Ralf Gutbell[1(✉)], Lars Pandikow[2], and Arjan Kuijper[1,2]

[1] Fraunhofer Institute for Computer Graphics Research IGD, Darmstadt, Germany
`ralf.gutbell@igd.fraunhofer.de`
[2] Technische Universität Darmstadt, Darmstadt, Germany

Abstract. Three-dimensional visualization of maps is becoming an increasingly important issue on the Internet. The growing computing power of consumer devices and the establishment of new technologies like HTML5 and WebGL allow a plug-in free display of 3D geo applications directly in the browser. Existing software solutions like Google Earth or Cesium either lack the necessary customizability or fail to deliver a realistic representation of the world. In this work a browser-based visualization component for geo-information is designed and a prototype is implemented in the gaming engine *Unity3D*. *Unity3D* allows translating the implementation to JavaScript and to embed it in the browser with WebGL. A comparison of the prototype with the opensource geo-visualization framework *Cesium* shows, that while maintaining an acceptable performance an improvement of the visual quality is achieved. Another reason to use a gaming engine as platform for our streaming algorithm is that they usually feature engines for physics, audio, traffic simulations and more, which we want to use in our future work.

Keywords: Geo-information · 3D · WebGL · Unity3D · Cesium

1 Motivation

The growing computing power of PCs and the increasing public awareness of WebGL fosters the development of web-based 3D visualizations of geo-information like elevation data and 3D citymodels. The visual quality of applications like Google Earth, Google Maps, Bing Maps 3D and the 3D view of Apple's maps has become more and more realistic through the development of better photogrammetry algorithms in the past years and the increasing availability of high quality geodata also for private persons. The display of this massive and realistic geodata has become the state-of-the-art and ideas grow what kind of applications can be build on top. One application could be to use these real-world data and to interact with and modify it by using physics engines even in a web environment. This would introduce the possibility to build commercial or serious games using the real-word data without reproducing them manually

© Springer International Publishing AG, part of Springer Nature 2018
S. Yamamoto and H. Mori (Eds.): HIMI 2018, LNCS 10904, pp. 23–35, 2018.
https://doi.org/10.1007/978-3-319-92043-6_2

like it is done nowadays. Krämer and Gutbell [1] compared the available *WebGL* frameworks and their capability of rendering geodata like terrain or point clouds, but none of them include features like a physics engine out of the box.

This is why we want to integrate streaming of geodata into a gaming engine, because even a scenario of small spatial extent like a suburban area consists of too much data to be transferred initially without loosing the attention of the user. Nielson [2] showed that a latency up to 10 s is acceptable for most users. If this time is exceeded the user usually shifts his attention and leaves the application [2]. Besides the goal of our work is to bring massive geodata into an environment to display them in a browser, we want to be able to enrich them in our future work by using physics engines.

1.1 Challenge

The challenge of this work is to integrate streaming of massive elevation data into gaming engine, which is able to export this feature to a *WebGL* application. To evaluate whether our approach and implementation is successful, we compare our prototype with the popular geovisualization framework *Cesium* in Sect. 5, we define the following four criteria:

- Network traffic
- Disk space
- Memory usage
- Efficiency of rendering

In Sect. 3 we explain our approach to stream the data and go into further implementation details. In Sect. 4 how we solved the problems of visualizing huge spatial areas.

2 Related Work

Google Maps [3] is probably the best known application to view geo-information, which features the option to view the world in 3D. It efficiently loads the data and displays them quickly. Atmospheric effects and the good visual quality of the geodata, which is created with photogrammetry, guarantee for a satisfying user experience. Figure 1 shows the detailed geometry available in *Google Maps* in urban areas. There are still areas left where only 2.5D geometry is available. Mostly in rural areas, as shown in Fig. 2. A high-level API is available to integrate personal geodata and build individual applications on top of *Google Maps*'s Desktop version *Google Earth* [4].

Cesium [5] is an open-source JavaScript library focusing on rendering digital globes and geodata. *Cesium* can load all sorts of geo-information including Shape Files [6], GeoJSON [7], Imagery by using OGC conform WMS [8] services, elevation data and 3D and 2D vector data like 3D citymodels. The team around Cesium developed a set of new dataformats (QuantizedMesh [9] and *3D-Tiles* [10]) which exploit hierarchical structures to store arbitrarily big geodata. Cesium is the most flexible open-source framework to store and display individual data (Fig. 3).

Fig. 1. Urban area of San Francisco in California, USA, in *Google Maps* showing the detailed photogrammetrical created model.

Fig. 2. Rural area of Freiensteinau, Germany, with no 3D content in *Google Maps*

3   Approach

For our goal to stream geodata into a game engine, we first explain the choice to use *Unity3D* in Sect. 3.1. Then we outline how we extended the platform to be able to stream and display large elevation data in Sect. 3.2.

3.1   Why Unity3D?

Unity3D is a cross-platform game engine of *Unity Technologies*. Its first version was released in 2005 and at the time of writing this paper, the latest version is 2017.3. It consists among others of a modern rendering pipeline, an audio and physics engine and tools to develop and display animation. A large *Asset Store*

Fig. 3. Photogrammetry model of AGI headquarters using 3D Tiles in *Cesium.*

[11] offers a variety of free and commercial tools, which are easy to integrate in *Unity3D* projects. This way projects can be extended to support realistic looking clouds or specific 3D objects. The possibility to export the developed tool to *WebGL* and many other popular platforms and the native support for a physics engine made *Unity3D* the best suited framework for this work.

3.2 Streaming of Elevation Data

To render only the visible data and thus keeping the amount of loaded data minimal, we stream the elevation data. To make this possible we use a server client infrastructure. The server stores all available data and the client requests the visible tiles. Our approach bases on the Ulrich's approach described in his book *Rendering Massive Terrain using Chunked Level of Detail Control* [12].

The data is stored in a *QuadTree*, which is a hierarchical data structure. This means the world is divided symmetrically into tiles and stored in different levels of the hierarchy. Every tile in one level stores the same visual quality which is defined by the *Screen Space Error* [13], which is explained in the next subsection. The deeper the level of tile inside the hierarchy is, the smaller the spatial extent of the tile is but the more detailed information in the tile is stored. The system is depicted in Fig. 4. This way the prototype can load more detailed information close to the camera and less detailed tiles the further the distance is. To reduce the amount of data in each tile, we use the data format *QuantizedMesh*, which is widely used by applications built on top of *Cesium. QuantizedMesh* stores a *Triangulated Irregular Network* (TIN), allowing to simplify the information of the terrain, which are usually stored as regular grids. So only the important and necessary 3D coordinates are stored, produced by a simplification method based on a Triangulation algorithm for point clouds explained in *Computational geometry: algorithms and applications* [14], which triangulates and simplifies a set of point until a certain user given error threshold is reached. This error threshold will correspond to the *Screen Space Error* of each tile.

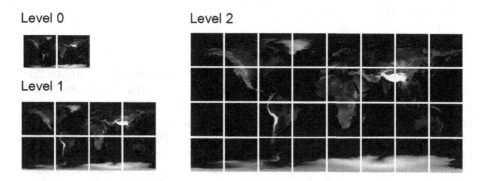

Fig. 4. The first three levels of the used QuadTree structure.

Screen-Space Error. The selection of tiles depends on their *Screen Space Error*. In *Level of Detail* (LoD) algorithms the *Screen Space Error* defines the difference of the tile in pixel between its original resolution representation and its current low resolution representation. It is a good measure to decide whether the current resolution of the tile is sufficient or a higher resolution has to be loaded. The user can define a threshold of the error in pixel and controls that way the quality of the displayed data and implicitly the resolution and amount data transferred.

High *Screen Space Error* values lead to a more inaccurate visualization and more visible changes of LoDs. Low values in contrast lead to more data transfer and more computational effort. Usually a threshold of one or two pixel is chosen. The *Screen Space Error* is mathematically defined as follows:

$$\rho = \frac{\sigma}{D}K = \frac{\sigma}{D}\frac{width_{viewport}}{2tan(\frac{fov}{2})}$$

σ is the maximum geometric error of the tile, D the shortest distance of the camera center to a point of the tile and K a scale factor, which is calculated by the size of the *Viewport* and the *Field of View*. The geometric error of a tile is the maximum deviation of the simplified mesh of a *QuadTree* tile to the original full resolution. This is calculated and stored during the creation and implicit simplication of the *QuadTree* for every tile. It is possible to approximate the distance D with a *Bounding Box* of the tile.

4   Implementation Details

Unity3D supports the development of 64-bit applications but the engine code uses 32-bit precision internally. Therefor the resulting range of values is $\pm 1.18 \times$

10^{-38} to $\pm 3.40 \times 10^{38}$, but the decimal places are truncated at the 6th or 7th position [15]. This determines how precise the world can be rendered in *Unity3D*. Out of convenience we define that one unit inside *Unity3D* represents one meter in the real world.

4.1 Enabling Double Precision

If we project the earth in relation to the coordinate system *WGS84* [16] to 2D, the width of the map is exactly the circumference of the earth (40.075 km). If we place the map in the center the distance to the edge is 20.037,5 km, which need eight decimal places. Using a 32-bit representation would mean that an object can be positioned at the edge with an accuracy of 100 Meter. Not only the placement of objects is imprecise, the control of the camera and the rendering is also affected by the limited accuracy. To eliminate this effect we define the center of the map when the system starts and use a method called *Floating Origin*, as described by Thorne [17], to guarantee the highest computational accuracy around the area of the camera. *Floating Origin* defines that the camera isn't moving in the scene, instead the scene moves around the camera.

4.2 Camera Setup

Another challenge are the great distances (*Draw Distance*) which have to be rendered. The *draw distance* can easily be several hundreds of Kilometers when the camera is placed in a high altitude looking to the horizon, leading to *Z-Fighting* [18, p. 833]. *Z-Fighting* is the flickering of a pixel because two objects "fight" for the representation in the pixel because their distance to the camera is computationally close. The computational accuracy of the camera is bound by the distance of the *Far* and *Near plane*. The larger the distance between the *Near-* and *Far Planes* the less accurate is the calculation deciding which object is represented in the pixels. Especially visualizing digital globes and the huge distances is problematic since we have to render objects very close to the camera and also objects which are many kilometers away from the camera.

To solve this three cameras are combined, which render different distance intervals to the camera. The three images of these cameras are combined from back to front and result in the final image displayed to the user viewing the entire range. Figure 5 shows the rendering when only one camera is used for the whole view range in contrast to the combined image of all three cameras of our setup.

The setup of the three cameras is shown in Table 1.

5 Results

In our problem definition in Sect. 1.1 we identified the performance criteria that we use for evaluation. These criteria have been measured for our *Unity3D* prototype and *Cesium*. To compare these two systems we defined two computer

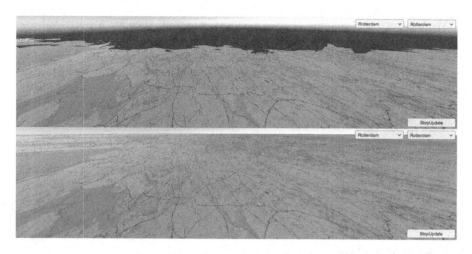

Fig. 5. Comparison of a rendering with one camera for the whole view range (upper image) and the result of our setup with three cameras rendering the final image (lower image).

Table 1. Order and the range of the Clipping-Planes of the cameras covering the earth.

Camera	Render order	Near-plane	Far-plane
NearCamera	3	0.1	2000
MiddleCamera	2	1000	1000000
FarCamera	1	1010000	100000000

Table 2. Computer systems used for evaluation.

Type	Desktop	Notebook
Operating system	Windows 10	Windows 10
CPU type	i5 6600 @ 3.3 GHz	Intel Core i7-3740@M @ 2.7 GHz
GPU type	Radeon R9 390, 4 GB	NVIDIA NVS 5200M, 1 GB
Main memory	16 GB	16 GB

setups and two use cases (see Sects. 5.1 and 5.2) used by non expert users (see Table 2).

We used data of a rural area in Hessian, Germany. The elevation data is stored as *QuantizedMesh* files and the imagery is served via a *Geoserver* [19] as *WMS* service. The details of the used data are:

- Extent: 33587,2 × 31948,8 m
- Resolution: 1 Pixel per meter
- Imagery: 255 files, size: 10,2 GB
- Elevation data: 909.233 files, size: 2,05 GB

We measure the amount of used memory when all tiles for a portrayal are loaded. The network traffic will be monitored by *Google Chrome*'s debugging tool. To measure the performance of the rendering the frames per second are logged and their average value is calculated. Another important setup is the camera and its movement. Therefor we defined two use cases described in the following Subsects. 5.1 and 5.2. To guarantee an objective comparison the following configurations of the experiments will be identical:

- Position and view direction of the camera
- Resolution
- Size of the *Viewport*
- Field of View
- Data

5.1 First Use Case

The first Use-Case is testing the performance when the systems start and a big amount of data has to be loaded. It is recorded how long it takes to load the data and how many frames per second are rendered. The camera is positioned at a low altitude and points towards the horizon. The view to the horizon forces the systems to load a high amount of visible tiles from the *QuadTree* structure. The records of this use case are listed in Table 3 and the view of both systems are shown in Fig. 6. It is noticeable that the *Cesium* renders the curvature of the earth and our approach doesn't.

Table 3. Performance measurement of the first use case

	Unity3D		Cesium	
	Desktop	Notebook	Desktop	Notebook
Average FPS	57.19	56.4	58.33	55.2
Memory	705.31 MB	705.31 MB	138.65 MB	138.65 MB
Tile requests	846	846	882	882
Received of tile data	11.7 MB	11.7 MB	10.15 MB	10.15 MB
Loading times in seconds	13.58	14.778	13.35	14.85

5.2 Second Use Case

The second use case simulates a typical user interaction of moving the camera through a scene. The camera movement is controlled by a script that is integrated into our *Unity3D* prototype and *Cesium*. This tests the impact to the performance when the movement of the camera forces the systems to load the appearing tiles.

The camera starts with a top down view of the scene. The measurement starts when all tiles of the initial camera position are loaded. Then the camera

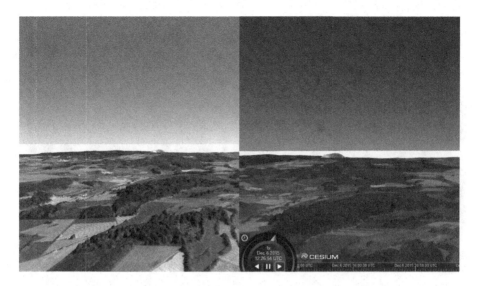

Fig. 6. Visual result for the first use case of *Unity3D* (left) and *Cesium* (right)

moves close to the surface without rotating so that the camera still looks down to the ground. After that, the camera rotates so that it looks to the horizon and starts to move for 5 s along the surface. The camera's last movement is a rotation of 180° to look to the other side of the horizon. The measurement ends, when all tiles for the new camera position and direction are loaded. Table 4 lists the results and Fig. 7 shows screenshots of both systems of this use case.

Table 4. Performance measurement of the second use case

	Unity3D		Cesium	
	Desktop	Notebook	Desktop	Notebook
Average FPS	58.81	49.7	56.87	52.3
Memory	710.58 MB	710.58 MB	261.94 MB	261.94 MB
Tile requests	1016	1016	878	878
Received of tile data	10.83 MB	10.83 MB	6.41 MB	6.41 MB
Loading times in seconds	23.35	25.08	22.946	24.85

6   Evaluation

The comparison shows that *Cesium* is more economical using the computer's resources than our *Unity3D* prototype and achieves better results in mainly all measured criteria. The network load of the *Unity3D* prototype is up to 35% higher than in Cesium, which is most likely because *Cesium* renders the earth

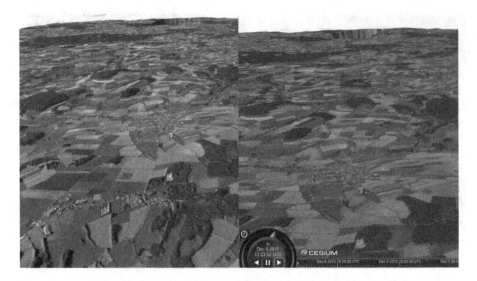

Fig. 7. Visual result for the second use case of the *Unity3D* prototype (left) and *Cesium* (right)

as an ellipse, which our prototype doesn't. This is due to *Horizon Culling*, which means the tiles below the horizon are occluded by other tiles and will not be loaded and displayed.

A big difference between both systems is the amount of loaded memory. *Cesium* uses much less memory than *Unity3D*, most likely because *Unity3D* allocates on startup a continuous block of memory, which is set manually and isn't adjusted over time. The size of allocated memory has to be chosen so that it is sufficient for all loaded objects during the runtime of the application. If it's chosen too big, the browser might crash because it isn't able to allocate the requested amount of memory. In our specific use cases a size of around 700 MB has proven to be sufficient without sacrificing too much unnecessary memory. In addition *Unity3D* automatically allocates memory for the engine code. Even though both systems aren't high end computers the framerates are more than sufficient to have an interactive experience. This allows to integrate more objects like 3D citymodels, other objects like city furniture, etc.

6.1 Comparison of Visual Quality

The visual quality is of particular interest, but an objective comparison is only possible to a limited extent. There are, however, a number of features that characterize a good visualization. An overview of the relevant visualization features are listed in Table 5.

These include especially the complexity of the lighting model and the shadow cast. The most striking difference in comparison is the illumination of the terrain. In *Cesium* only ambient lighting is used, so all surfaces are illuminated with

Table 5. Comparison of visualization features of *Cesium* and *Unity3D*

	Cesium	Unity3D
Directional lighting	Limited	Yes
Shadows	No	Yes
Atmosphere	Yes	Yes
Water surface	Yes	Limited
Clouds	No	No

Fig. 8. Comparison of *Unity3D's* (left) and *Cesium's* (right) lighting effects at sunset.

the same constant intensity. Although the aerial images already contain low illumination information in the form of shadows, the scene is flat and detailed in comparison with this lighting model. In addition, the sunshine is reflected in the lighting of the terrain. The illustrations of both systems are shown in Fig. 8. A more complex illumination is also possible in Cesium using the *QuantizedMesh* 1.0 format. For this purpose, however, the surface normals must be calculated when processing the data.

The most significant improvement is the support of shadow calculation. Shadows are an important tool for human perception, as they enable objects to be positioned more precisely in space. The absence of shadows in a virtual scene often makes them seem artificial. In 3D geo-information systems, shadows are also of great relevance for realism. Since the calculation of shadows has a great influence on the performance, these are currently only displayed in medium quality in *Unity3D*. A comparison of visualization at low sunshine is demonstrated in Fig. 8.

Both systems model the atmospheric dispersion of incoming sunrays taking into account the sun's level. An estimation which modeling rather corresponds to a realistic representation is not easily possible for the layman.

7   Conclusion

This work shows that *Unity3D* can be used to display massive amounts of geo-information in a browser. The prototype developed for this demonstrates that a visually appealing and at the same time smooth representation of a large area can be realized. The adaptive and robust design of the system allows streaming of any amount of data over the network and from the hard disk.

Within the scope of this work, a solid foundation was created, to exploit the features of a game engine in combination with real world data. For many reasons many features of *Unity3D* were not taken into account. This will be the focus of our future work, as described in Sect. 8.

However, the use of *Unity3D* can also be considered critical, since these additional features may not be needed and can only be a ballast that slows down the system. At the same time, there is a lack of functions which simplify the correct use of georeferenced data. It can be argued that the use of a smaller, more lightweight renderer with a specialized focus on geographic data can achieve similar or even better results. Ultimately, the goals of the developer and the customer's requirements are whether the use of *Unity3D* brings added value.

The choice of *Unity3D* as an engine has yet another great advantage: Since the development of an application in *Unity3D* is largely independent of the later target platform, it is effortless to run the created system natively on common operating systems such as Windows, Linux and Mac. The native execution can make better use of hardware resources, which can improve visual quality and performance.

8   Future Work

One of our future goals building on this prototype is to integrate game engine features like physics engine or traffic simulations, helping to create a more credible world. Here the use of *Unity3D* as a basis unfolds the full potential. The representation of the world can be extended, for example, by the addition of flora and fauna or the integration of a traffic simulation, thus offering the possibility to create a living world. The extensive Asset Store provides a wide range of visual and functional enhancements that can be added to the project with little effort.

Regarding the correctness of the visualized geodata between *Cesium* and our *Unity3D* prototype the missing ellipsoidal shape of the terrain in our prototype will be a focus of our future improvements. This can be realized by projecting the coordinates of the tiles onto the surface of the ellipse representing the earth.

Acknowledgments. We would like to thank Wiebke Mildes and Eva Klien for their valuable input and support.

References

1. Krämer, M., Gutbell, R.: A case study on 3D geospatial applications in the web using state-of-the-art WebGL frameworks. In: Proceedings of the 20th International Conference on 3D Web Technology Web3D 2015, pp. 189–197. ACM, New York (2015)
2. Nielson, J.: Website Response Times, December 2010. http://www.nngroup.com/articles/website-response-times
3. Google Maps. https://developers.google.com/maps/
4. Google Earth. https://developers.google.com/earth-engine/
5. Cesium - WebGL Virtual Globe and Map Engine. https://cesiumjs.org
6. Shape File Specification. https://www.esri.com/library/whitepapers/pdfs/shapefile.pdf
7. GeoJSON Specification. http://geojson.org/
8. ISO. Geographic information - web map server interface. ISO 19128:2005, International Organization for Standardization, Geneva (2005)
9. QuantizedMesh Specification. https://github.com/AnalyticalGraphicsInc/quantized-mesh
10. 3D Tiles Specification. https://github.com/AnalyticalGraphicsInc/3d-tiles
11. Unity Asset Store. https://www.assetstore.unity3d.com/
12. Ulrich, T.: Rendering massive terrains using chunked level of detail control (2002)
13. Lindstrom, P., Pascucci, V.: Visualization, of large terrains made easy. In: Proceedings of IEEE Visualization 2001, San Diego, CA, USA, 24–26 October 2001, pp. 363–371 (2001)
14. de Berg, M., Cheong, O., van Kreveld, M.J., Overmars, M.H.: Computational Geometry: Algorithms and Applications, 3rd edn. Springer, Heidelberg (2008). https://doi.org/10.1007/978-3-540-77974-2
15. IEEE Standard for Floating-Point Arithmetic. IEEE Std 754-2008
16. WGS 84: EPSG Projection - Spatial Reference. http://spatialreference.org/ref/epsg/wgs-84/
17. Thorne, C.: Using a floating origin to improve fidelity and performance of large, distributed virtual worlds. In: 4th International Conference on Cyberworlds (CW 2005), 23–25 November 2005, Singapore, pp. 263–270 (2005)
18. Möller, T., Haines, E., Hoffman, N.: Real-Time Rendering, 3rd edn. Peters (2008)
19. GeoServer - An Open Source Server for Sharing Geospatial Data. http://geoserver.org/

A System to Visualize Location Information and Relationship Integratedly for Resident-centered Community Design

Koya Kimura[1]([✉]), Yurika Shiozu[2], Kosuke Ogita[1], Ivan Tanev[1], and Katsunori Shimohara[1]

[1] Graduate School of Science and Engineering, Doshisha University, Kyoto, Japan
{kimura2013,ogita2016,itanev,kshimoha}@sil.doshisha.ac.jp
[2] Faculty of Economics, Aichi University, Nagoya, Japan
yshiozu@vega.aichi-u.ac.jp

Abstract. This research aims to realize the resident-centered community design by utilizing of information and communication technology (ICT), and create an opportunity to regain relationship within the community by visualizing media spot defined as the place where communication is active in an area. In order to visualize and analyze the location information and the relationship integratedly, we developed a system using the Web interface. As a result of visualization using the system, it became easy to guess the type of meeting and attendees, which can help analysis. On the other hand, it turned out that there is room for improvement in drawing speed and analysis efficiency.

Keywords: Resident-centered community design · Visualization
Location information · Relationship · Web application

1 Introduction

The structure of local communities in Japan has needed to change after the Tohoku Earthquake. Japan is the first country to have a super-aging society, first noted in 2007. According to the latest Annual Health, Labor and Welfare Report, on the other hand, the connections have been weak among members of the communities, but the number of people who want to help in the community has increased [2]. That is, local communities need to be vitalized for the regaining relationship.

This type of social situation has prompted the development of several approaches from the computer science field. For example, the Strategy Proposal of 2012, the Japan Science and Technology Agency includes the person-watching system for the elderly and smart city system are examples of these approaches [3]. Despite the need for these approaches in many communities, most communities cannot use them casually.

© Springer International Publishing AG, part of Springer Nature 2018
S. Yamamoto and H. Mori (Eds.): HIMI 2018, LNCS 10904, pp. 36–44, 2018.
https://doi.org/10.1007/978-3-319-92043-6_3

The research on "constructing" the system has been successful, but that on its "usage" is in the developing stage. Moreover, the introduction of the system does not always guarantee local activation, which always depends on the locals who will notice their problems and act to solve them. If residents can be aware of their invisible relationship that they are usually unconscious, and if they understand its importance, they could vitalize their community through sustaining and promoting it? That is the concept of resident-centered vitalization of a local community.

This research aims to realize the resident-centered community design by utilizing of information and communication technology (ICT). We defined "a media spot" as the place where communication is active in an area. As a place for creating and strengthening the relationship, we visualize media spot [1] and create an opportunity to regain the relationship within the community.

In recent years, SaaS (Software as a Service) type data visualization service has been prospering due to spread of cloud computing. On the other hand, there is no service for visualizing and analyzing location information and relationship integratedly. In this research, in order to visualize and analyze media spots, we developed a tool using a web interface that can visualize the location information and the relationship integratedly. We visualized location information and passing-each-other data from Bluetooth collected in the field experiment that we have been doing since 2013 using this system.

2   Resident-centered Vitalization of the Local Community

Ushino's research (1982) on local resident-based regional development explained the importance of such concept and proposed a system called "Kande System" [4]. Ushino said that after the industrialization and urbanization in the 1950s, the village communities in the rural areas were divided by the agricultural policy and then re-integrated in the 1970s to create a new regional system. The importance of local resident-based regional development has already been a significant research topic since the 1980s.

Meanwhile, Yoshizumi's case study (2013) analyzed the way for locals to develop regions sustainably and suggested the "Eco Card System" [5]. In this system, the locals are given a stamp card called "Eco Card" that promotes environmental activities, thereby creating a setup for the locals to be involved in the region. This system highlights the importance of visualizing or making the locals notice the problems for them to manage local resident-based regional development.

With the introduction of information communication technology (ICT), it is temporarily possible to solve the challenges of a local community. However, to vitalize a local community continuously, residents must solve them positively. For residents to solve the challenges of a local community by themselves, they must be conscious of these challenges. Thus, "resident-centered" development means that "residents themselves solve the challenges faced by their local community."

In this research, we aimed to establish a methodology that enables them to do so by visualization.

3   Overview of the Visualization and Analysis System

3.1   User Interface

Figure 1 shows a user interface that is actually visualized using the system. The upper left side of the screen is a map, the upper right side of the screen is the relationship (undirected graph) based on the passing-each-other data and the lower part of the screen is the query input part. We can input start time, interval to visualize, and send a query. Then, the result from the API is drawn on the upper part of the screen.

3.2   Architecture

This system consisted of a "database" in which raw data is stored, "Web API" for returning data based on data acquired from the database, and "Web front-end" for drawing data acquired from the API.

We used "MySQL," which is one of the RDBMS. A table was built for each type of data in the database.

The Web API was built using "Django" which is one of the Web frameworks described using Python. Django, like a general Web framework, can easily build Web applications by following the design pattern. In this research, we used Python-based scientific computing library, therefore we used Python web framework that has a high affinity. In addition, we used the "Django REST framework" package specialized for API creation.

We used HTML5, CSS3, Vue.js to construct the Web front-end. Vue.js is a JavaScript framework that makes it easy for browsers to update screens dynamically. We adopted this mechanism because this system mainly focuses on updating screens on the browser. We used Google Maps JavaScript API and D3.js to draw data.

The flow of the system is shown in Fig. 2. First of all, we connect to the Web front-end using a Web browser and designate the data period etc. The Web front-end builds a query to send to the API and sends it to the API Endpoint. Then, the Web front-end draws using the data returned from the API.

There are two types of APIs, one that returns location information based on a query and one that returns an undirected graph showing the relationship. The API that returns location information calls data from the database, converts it to JSON format, and returns it to the client. The API that returns the undirected graph calls the data of the passing-each-other data from Bluetooth from the database, converts it to the undirected graph, and then returns the data to the client in the JSON format. For conversion to an undirected graph, we used Python's network analysis library "NetworkX".

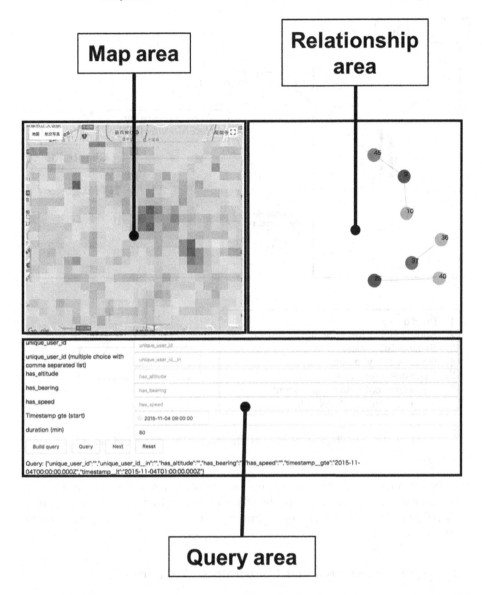

Fig. 1. User interface of the visualization and analysis system

4   Experiment Method

4.1   Overview of Field Experiment

We acquired residents' activity data with smartphones lent to them in the Makishima area in Uji City, Kyoto, Japan, in cooperation with members of the non-profit corporation Makishima Kizuna-no-Kai. Activity data means

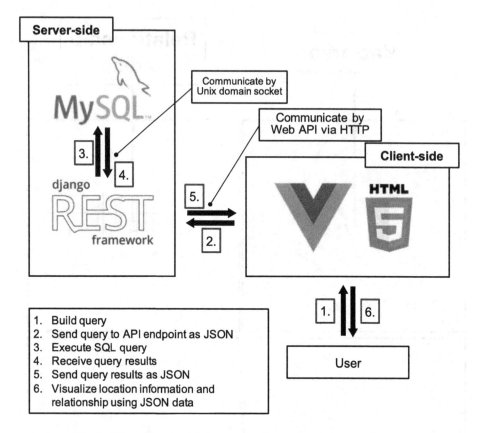

Fig. 2. Flow diagram of the system focused on server and client

"location information," "send and receive emails," "telephones reception and transmission" and "passing-each-other data from Bluetooth [6]." In this paper, we analyzed "location information" and "passing-each-other data from Bluetooth" using the system.

4.2   Experimental Area and Cooperators' Attribute

This field experiment has been conducted in the Makishima area in Uji City, Kyoto, Japan.

Uji City is located in the south of Kyoto on the south side of Kyoto City. The population of Uji City as of April 1, 2016, is about 190,000 and 15,000 (about 8%) of them live in the Makishima area. Uji City has attracted lots of attention as a residential area near Kyoto, Osaka, and Kobe since the early 1960's. As a result, residential land was developed in Uji City and the population remarkably increased. The Makshima area is one of the development areas in Uji City, and there is a densely populated area such as a housing complex. Blocks of the development area in the early 1960's are aging, but the population of the whole Makishima area is slightly increasing.

The experimental cooperators live in the Makishima area. They are members of the non-profit corporation Makishima Kizuna-no-Kai. Table 1 shows the attributes of the experimental cooperators. A lot of experimental cooperators are over 65 years old. The reason is that people who retired at mandatory age mainly join the regional development.

Table 1. Attributes of field experiment

Area	Makishima, Uji, Kyoto, Japan
Cooperators	20 to 50 people
Age	30 to 70 years old

Table 2 shows the periods of the field experiment. We instructed the experimental cooperators to use the lent smartphone at all times for the duration of the experiment. However, from the perspective of informed consent, we instructed that the experimental cooperators can switch off the smartphone when he/she does not want to inform his/her location information.

Table 2. Periods of field experiment

1st. period	Nov. 11, 2013 to Dec. 10, 2013 (30 days)
2nd. period	Feb. 11, 2015 to Mar. 27, 2015 (45 days)
3rd. period	Jul. 11, 2015 to Jan. 11, 2016 (185 days)

4.3 Experimental Installation

Table 3 shows the specification of smartphones that are used in the field experiment. Smartphones used in the first period were discredited mostly due to sluggish actions and small screen. Accordingly, we lent them stylus pens for the improvement of usability. This way partly resolves that discredit.

Following the suggestion of the first period, smartphones using the second and third periods was chosen as quick action and big screen. Some of the experimental cooperators have his/her smartphone due to the spread of smartphones compared to the first period. That discredit is significantly resolved by these external factors also.

5 Results of the Analysis and Discussion

Figure 3 shows a certain period using the system. When we watched only location information, we could know that the meeting was held at the facilities near

Table 3. Specification of smartphones

	First period	Second and third periods
Manufacture	Fujitsu	ASUS
Model number	ARROWS Kiss F-03E	ZenFone 5 A500KL
OS	Android 4.0.4	Android 4.4.2
Network career	NTT docomo	IIJ Mobile (MVNO of NTT docomo)
CPU	Qualcomm Snapdragon S4 MSM8960	Qualcomm Snapdragon 400 MSM8926
Clock frequency	1.5 GHz	1.2 GHz
Core	Dual Core	Quad Core
RAM	1 GB	2 GB
Location information	GPS	GPS and GLONASS
Bluetooth	4	4
Sensor	G-Sensor	G-Sensor/E-Compass/Proximity Light/Hall Sensor

the plotted place, but we could not know which people were actually present. However, the structure of the undirected graph shows who was present at the meeting. On the other hand, it was very difficult to estimate the type of event with only the graph structure, but it was made possible to some extent by combining with the location information.

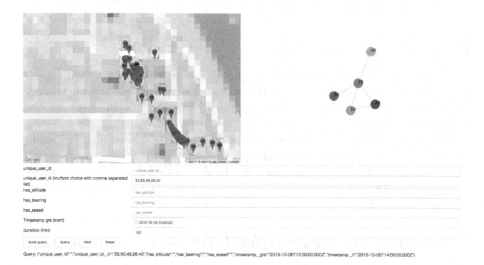

Fig. 3. A result of visualization by the system for analysis

Figures 4 and 5 show the days when the meeting was held and the holidays.
Several problems also surfaced when analyzing with the system.

First, it is known that there is a limit to the number of location information
that can be drawn at one time, and the action becomes slow when rendering the

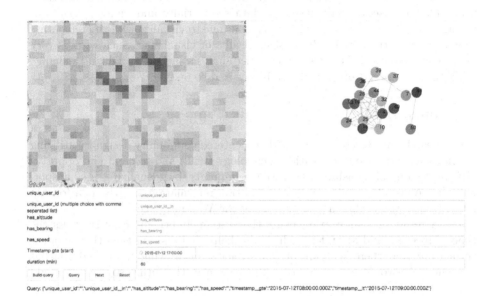

Fig. 4. A result of visualization that we can observe many edges and nodes

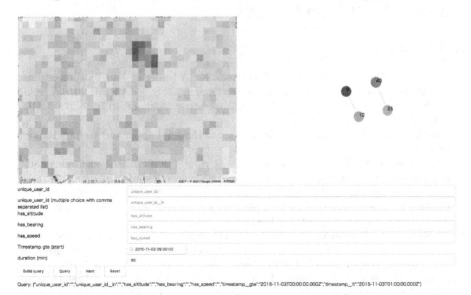

Fig. 5. A result of visualizing a holiday

raw data because the location information is dynamically rendered by the client. As a countermeasure, it is possible to summarize markers, to thin out location information, etc. However, we considered methods that do not lose meaning from a set of location information.

Next, the system visualizes location information and relationship at a certain time. That is, users should visualize data sequentially and analyze results of visualizing it. In other words, users must continue to visualize while incrementing time counter until they find interesting results of visualizing. Such work is very inefficient, therefore it is necessary to assist analysis such as visualizing a change of centrality in a certain period in advance and so on.

6 Conclusion

This research aims to realize the resident-centered community design by utilizing of information and communication technology (ICT) and to create an opportunity to regain relationship within the community by visualizing media spot. In order to visualize and analyze the location information and the relationship integratedly, we developed a system using the Web interface.

As a result of visualization using the system, it became easy to guess the type of meeting and attendees, which can help analysis. On the other hand, it turned out that there is room for improvement in drawing speed and analysis efficiency. We would like to improve the system in the future so that we could get more interesting analysis results.

Acknowledgment. This work was supported by JSPS KAKENHI Grant Numbers JP16J03602, JP16K03718, and JP17KT0086.

References

1. Kimura, K., Shiozu, Y., Tanev, I., Shimohara, K.: A leader and media spot estimation method using location information. In: Yamamoto, S. (ed.) HIMI 2016. LNCS, vol. 9735, pp. 550–559. Springer, Cham (2016). https://doi.org/10.1007/978-3-319-40397-7_53
2. Ministry of Health, Labour and Welfare: Annual Health, Labour and Welfare Report 2013–2014 (2014)
3. Japan Science and Technology Agency: Research and development on fundamental technologies of cyber physical systems and their social implementation - A case study on promoting aged people to social activities, CDS-FY2012-SP-05 (2013)
4. Ushino, T.: Comprehensive district plan by inhabitants and "Kande" system. J. Rural Plann. Assoc. 1(3), 19–29 (1982)
5. Yoshizumi, M.: A study on actively community-based environmental town planning toward sustainable communities: a case study on the eco-community program in Nishinomiya, Hyogo, Japan. J. City Plann. Inst. Jpn 48(3), 831–836 (2013)
6. Kimura, K., Shiozu, Y., Tanev, I., Shimohara, K.: Visualization of relationship between residents using passing-each-other data toward resident-centered vitalization of local community. In: Proceedings of the Second International Conference on Electronics and Software Science (ICESS 2016), pp. 122–127 (2016)

Reversible Data Visualization to Support Machine Learning

Boris Kovalerchuk[1(✉)] and Vladimir Grishin[2]

[1] Department of Computer Science, Central Washington University, Ellensburg,
WA 98922, USA
borisk@cwu.edu
[2] ViewTrend Int., 1001 Colonial Avenue SE, Palm Bay, FL 32909, USA

Abstract. An important challenge for Machine Learning (ML) methods such as the Support Vector Machine (SVM), and others, is the selection of the structure of ML models for given data. This paper shows that the abilities of the pure analytical ML methods to address this challenge are limited. It is due to the fundamental nature of the ML methods, which rely on the available training data, which can result in overgeneralized or overfitted model. In the proposed visual analytics approach, domain experts are put into the "driving seat" of the ML model development to control the model overgeneralization and overfitting. In this approach, domain experts work interactively with multidimensional data, and the ML data classification models, presented in the lossless reversible visualizations. This paper shows that it enhances the ML classification models, and decreases the use of external and irrelevant-to-the-domain assumptions in the ML models.

Keywords: Multidimensional data · Visualization · Machine learning
Classification · Reversible lossless visualization

1 Introduction

An important challenge for Machine Learning (ML) methods, such as the Support Vector Machine (SVM) and others, is the *selection and evaluation of the structure of machine learning models*. It includes (1) a class of models to be explored (linear, nonlinear with a type of non-linearity), and (2) *characteristics* of the models such as the set of possible kernels. Usually these structures are chosen empirically, based on problem knowledge and multiple runs of the algorithm with *limited data visualization*. Typically, many categories of characteristics of the ML models are external for the given task and data. They are imposed by the ML method not derived from the data. This process can lead to *inadequate models*. Such models can lack interpretation, provide wrong predictions on new unseen data, can be overfitted or overgeneralized.

In the current machine learning practice, visualization is commonly used, for illustration and explanation of the ideas, of many algorithms such as the Support Vector Machine, or the Fisher Linear Discriminant Analysis (LDA), but much less for the actual discovery of the n-D rules, models due to the difficulties to adequately represent the n-D data in 2-D.

© Springer International Publishing AG, part of Springer Nature 2018
S. Yamamoto and H. Mori (Eds.): HIMI 2018, LNCS 10904, pp. 45–59, 2018.
https://doi.org/10.1007/978-3-319-92043-6_4

Besides, some ML methods use only a part of the available information to construct the models. For instance, an ML algorithm can use only the cases from each class, which are close to the cases of the opposing class, in the training data. Such very limited usage of training dataset information can prohibit getting models that are more efficient. Visualization can help to discover situations where it can decrease the efficiency of ML models. It is illustrated in Fig. 1.

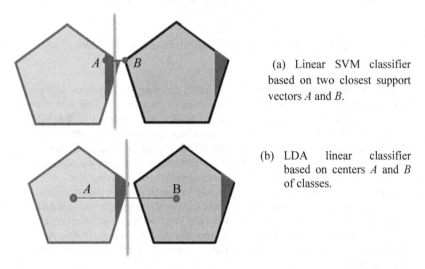

(a) Linear SVM classifier based on two closest support vectors *A* and *B*.

(b) LDA linear classifier based on centers *A* and *B* of classes.

Fig. 1. Two classes classified by linear SVM and simplified LDA (Kovalerchuk 2018). (Color figure online)

Figure 1 shows an example where SVM uses only two closest support vectors *A* and *B*, while LDA uses all training cases to construct a line that discriminates classes. In Fig. 1a we use the geometric interpretation of the linear SVM (Bennett and Campbell 2000; Bennett and Bredensteiner 2000) that shows that SVM uses the closest support vectors of the two classes.

Respectively, in Fig. 1a linear SVM uses a red line that connects *two closest support vectors* (SV) *A* and *B* from opposing classes (blue and grey areas that constitute training data *D*). This red line is a basis of the green discrimination line, which bisects the red line in the middle and is orthogonal to it.

In contrast, in Fig. 1b, the simplified Fisher LDA uses the *average points of all of the training data* of each class (points *A* and *B*), connects them with the red line. Then the orthogonal green line bisects it in the middle. The green line serves as the discrimination line. In Fig. 1, two algorithms produce different green discrimination lines, which are *error free on the training data*. However, the LDA, which used the training data more fully to build the discrimination line, has *less error* on violet *validation data* of the left class. While visualization in Fig. 1 clearly and quickly shows this, discovering it analytically would require extensive work to build both models completely, and generate quite specific validation data, which will allow detecting a lower accuracy of the SVM model.

This paper proposes a way to improve the construction of the ML models, which uses the more complete information in the visual form, visualized in 2-D to get a more accurate classification. This approach is based on the reversible lossless General Line Coordinates (GLC) (Kovalerchuk and Grishin 2017; Kovalerchuk 2018). It opens the opportunity for (1) visual discovery of linear separatability of the classes, (2) ensuring the finding of it by the classifiers, and (3) getting additional visual confidence in the linear separation by the domain users in the understandable form.

Next, the actual boundary between the classes can be *non-linear, even when the linear separation exists*, because the linear separation can *severely overgeneralize* the training data as we show below. The proposed visual analytics approach supports the estimation of the level of nonlinearity of boundaries, and respectively selecting better parameters of the non-linear ML models.

Consider this overgeneralization challenge, of the ML models, using the 4-D Iris data (Lichman 2013). These data consist of 150 cases of three classes of Iris: Setosa, Versicolor and Virginica. Each case is represented by sepal and petal length and width as a 4-D point. Figure 2 shows the examples of Iris petals of these classes. Later on, in one of our experiments, petals are modeled by ovals of respective length and width.

Setosa   Versicolor   Virginica

Fig. 2. Examples of Setosa, Versicolor and Virginica Iris petals.

Figures 3 and 4 show the examples of *logistic regression* (Big data 2018), *SVM* (FitcSVM 2018) and *Decision Tree* (Taylor 2011) classification of these classes of Iris, provided in these references as machine learning tutorials and lectures. In these figures, the results are visualized in two dimensions (petal length and width). While all of three classification models are quite accurate, on the given cases, they *very differently generalize and classify the data outside the given cases*.

Moreover, all of them *significantly overgeneralize* given 150 cases relative to human generalization that we report in the experiment section later. In fact, according these models, irises with all possible combinations of petal length and width exist. For instance, the logistic regression and the Decision Tree allow Setosa petal to be more than 4 times longer than those in the Setosa training data, and Versicolar petal to be more than 4 times shorter than in the Versicolar the training data (see Figs. 3 and 4).

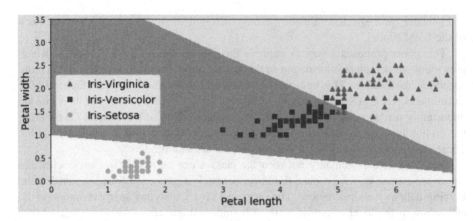

Fig. 3. Example of logistic regression classification of these classes of Iris (Big data 2018).

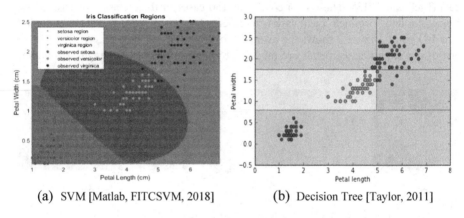

(a) SVM [Matlab, FITCSVM, 2018]   (b) Decision Tree [Taylor, 2011]

Fig. 4. Example of SVM and Decision Tree classification of three classes of Iris.

2   Approach

Historically in ML, overgeneralization was associated with the situation, where only the positive examples (example of one class) are available, which "do not provide information for preventing overgeneralization of the inferred concept." (Carbonell et al. 1983). It is contrasted with the situation when both positive and negative examples are available, where "negative examples *prevent overgeneralization* (the induced concept should never be so general as to include any of the negative examples)" (Carbonell et al. 1983). However, as we see in Figs. 3 and 4 the presence of the cases of the opposing classes prevent only such an *extreme overgeneralization* as the direct overlap of classes.

A computational approach to control overgeneralization is proposed in (Pham and Triantaphyllou 2008) based on partitioning the space of the training data according to the data density. The expansion of the training data area is made proportional to the density of points in the area measured by *homogeneity degree* HD defined in that paper. The HD approach assumes that areas that have higher HD can be expanded more up to the point where expanded areas of opposing classes reach each other. The application of this approach to Iris data will definitely make more conservative generalization than shown in Figs. 3 and 4. However, it still can be overgeneralization as we discuss below.

The major drawbacks of this approach are: (1) the density of training data in each area is low in a high dimensional space; (2) the higher density in the area does not justify larger expansion of this area.

This heuristic rule is not derived from the given task, but it is an *external hypothesis*. It is also not a type of question that can be answered by the domain expert easily to clarify the relation of the larger expansion hypothesis to the domain task.

Our approach is different. It is putting the domain expert into the *"driving seat"* of the ML model development. The premise is that interaction of the domain experts with the visualization of the n-D data and the alternative discrimination functions, models in these visualizations will enable the making of better ML classification models, and a decrease in the use of external and irrelevant-to-the-domain assumptions in the ML models.

The feasibility of this approach is partially supported by our previous studies where the domain expert (radiologist) was able to identify deficiencies of the rules discovered by ML algorithms, when these rules were presented in the understandable and visual forms (Kovalerchuk et al. 2000, 2012).

Selecting and evaluating the class of ML models requires identifying:

- the *type of the ML model* to be discovered – linear or non-linear,
- the *form of non-linearity,* and
- the level of generalization *confidence* of the discovered ML model.

The model can be overfitted, overgeneralized, or a right one from the viewpoint of the domain expert. Here we explore the abilities of the visual approach, combined with analytical means, to test and increase the confidence in the ML model including finding the areas of high confidence.

Assume that we discovered a linear model M, which separates the given training and validation data with 100% accuracy. Can we assign this model M the highest confidence, and apply it to predict the class of *new unseen data* with high confidence? How to check that we can use the model M, with high confidence for such new data? Another situation happens when a linear model M separates the training and validation data with, say, 60% accuracy. How to modify M to be able to apply it to predict the class of new unseen data with high confidence?

The uncertainty in both situations is coming from the fact that typically with high-dimensional data we do not know for sure *how representative* given training and validation data, that are used to build model M, for prediction on the new unseen data. In this situation, making an assumption on the probability distribution of high-dimensional data outside of the training data rarely can be of high confidence. As a result, we cannot assign a reliable probabilistic level of confidence to the predictions outside of the training data.

Moreover, there are situations where some unseen data do not belong to any of the training data class and their classification *must be refused* by a classifier. As we have seen in Figs. 3 and 4 it is not done for Iris data by all three ML algorithms.

Another example is classification of letters. Let letters *A* and *B* represented as 16-D data (Lichman 2013) be classified successfully by some ML model that does not refuse to classify any case. Thus, any letter, encoded as a 16-D point, will be classified as *A* or *B* which is a vast overgeneralization.

Thus, the fundamental task is developing methods to test and increase confidence in the ML models. While the most efficient approach for this is *adding new training data and attributes*, however, it is not possible, in many real world situations.

An alternative idea is adding other types of *additional information*. The examples are below. The expert can bring the information that data must be within hyperspheres, centered about the given templates, or the expert may select and analyze the n-D points close to the border between the classes, and/or far away from the training data, and tell that those points must belong to other classes than the model *M* suggests. In the experiment section we explore this alternative idea.

The generation of new data can be done analytically without visualization or interactively with visualization. The analytical way can be quite challenging to implement. Let we want to generate n-D points that are close to the classification line. This task is *ill-posed mathematically* without setting constrains around that line.

Next, we need to deal with the infinite number of n-D points in the area limited by a threshold that also needs to be set up. Therefore, we need to limit the number of points that will be randomly generated. Those assumptions are difficult to set up and justify formally and rigorously. In contrast, the human expert can see and select such points visually if an appropriate visualization of the n-D data and n-D classifier are provided. This will support the finding of the anomaly and the generation of the confidence/non-confidence evaluation of the ML model faster.

3 Experiments

This section presents experiments conducted to check the proposed approach, to see how the domain experts can limit the overgeneralization of ML models, using visualization. It is conducted as a series of experiments with the participants (computer science students).

Each participant worked with the visualization plots, assigning the level of human confidence between 0 and 10, for classification of the different cases and models. The data used in these experiments were selected in a way to ensure that students can serve as reasonable "experts" in classifying these data.

Figure 5 shows data of the two classes, where two disjoint areas represent class 1. In experiment 1 (Fig. 5, left), participants evaluate their confidence in classification of the red point to class 1 or class 2 using the confidence scale from 0 to 10 with 10 indicating the max of confidence.

Fig. 5. Setting of experiments 1 (left) and 2 (right) (Color figure online)

In experiment 2 (Fig. 5, right), participants evaluate their confidence in classification of the blue point to class 1 or 2 using the same confidence scale. These experiments can clarify the level of human confidence in classification of highly contested points.

Figure 6 shows the setting of experiment 3, where participants evaluate their confidence in alternative separation lines between the two classes using the same confidence scale from 0 to 10. The first alternative is a linear discrimination line with a small margin and the second one is a non-linear discrimination line. This experiment can clarify the human preference between simpler, but less accurate line on the left vs. the more complex, and likely more accurate discrimination line on the right.

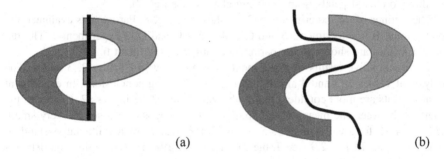

(a)   (b)

Fig. 6. Setting of experiment 3 to evaluate the two alternative separation lines between the classes.

Table 1 presents the quantitative result of these experiments, with 7 participants (Computer Science students). In the experiments 1 and 2, participants are highly uncertain in classifying the red and blue points into either class. The confidence is just about 50:50.

Table 1. Results of experiments 1–3.

Class	Experiment 1		Experiment 2		Experiment 3		
	Mean	Stdev	Mean	Stdev	Alternative	Mean	Stdev
Class 1	4.14	1.81	5.14	1.81	(a)	5.00	2.20
Class 2	6.86	1.25	6.29	1.48	(b)	7.14	2.17

The experiment 3 shows that participants prefer a more accurate non-linear discrimination line (b), while the automatic ML algorithms may stop by finding the linear discrimination line in the alternative (a), because many ML algorithms, in the attempt to balance simplicity and accuracy, assume that simpler lines are more robust. While this general assumption is valid, it is not task specific, and for the given data, it may or may not be applicable.

We conducted five more experiments, with 11 participants (computer science students). In these experiments, we used the same Iris data from the UCL Machine Learning repository (Lichman 2013).

In experiments 4–6, the participants evaluated their confidence in the petal classification of the two petals (denoted as A and B) into class 1 or 2. An oval defined by its length and width represents each petal. In the experiments 5–8, participants evaluated petals represented by 2-D points (x, y) where x is the length and y is the width of the petal, in the 2-D Cartesian coordinates. Thus, in these experiments participants observe only the parameters of the petals (not the actual petals simulated as ovals in experiment 4).

The participants marked their confidence in the same confidence scale from 0 to 10. The petals A and B have been selected in the middle between the two classes in the contested "grey" area where the intuitive classification is difficult. The additional information in experiment 4 is given as ovals, which represent the centers of classes and the two closest petals from opposing classes (see Fig. 7).

The settings of the experiments 5–7 are shown in Fig. 8. Participants evaluated the points A and B in experiments 5 and 6, and points C and D in experiment 7. The red and black lines are shown to the participants only in experiment 6.

The quantitative results of experiments 4–8 are shows in Tables 2, 3 and 4. The analysis of Tables 2, 3 and 4 allows making the following conclusions. In experiment 4, participants are more confident that both ovals A and B are in class 2. However, the difference between the means of the confidences for opposing classes are very small: 5.27 vs. 4.91 for oval A and 5.45 vs. 4.93 for oval B with quite large standard deviations from 1.91 to 2.9 (see Table 2). In other words, in average the confidence is about 50:50 in experiment 4.

In experiment 5, participants are more confident in classifying point (oval) A into class 1 and point (oval) to B into class 2: 7.82 vs. 2.27 for A and 6.36 vs. 3.45 for point (oval) B standard with standard deviations from 1.4 to 2.45.

Thus, this result is quite different from the result of experiment 4. In experiment 5, point A is classified into the different class (class 1) with high confidence (7.82 vs. 4.91 for class 1 in experiment 4). In experiment 5, point B is classified into the same class as in experiment 4, but with slightly higher confidence (6.36 vs. 5.45).

The important difference between these experiments is that in experiment 4 participants observe *actual ovals*, but in experiment 5 they observe only 2-D points that represent *parameters* of ovals (length and width) in 2-D Cartesian Coordinates. The other difference is that in experiment 4 participants observe only 6 ovals, but in experiment 5 they observe ten times more (about 60) 2-D points that represent 102 ovals. With all these differences, the conclusion is that the use of *parameters* of actual objects gives *more confidence* than the use of *actual objects* in classifying difficult cases from the "grey" border area between classes.

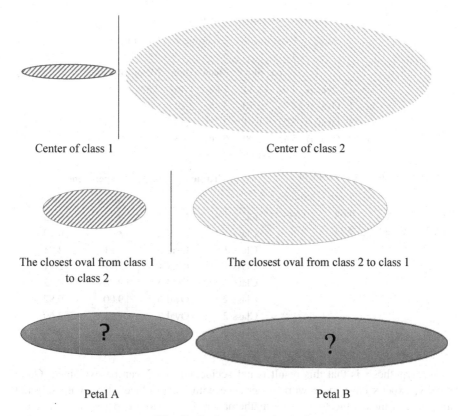

Center of class 1   Center of class 2

The closest oval from class 1   The closest oval from class 2 to class 1
        to class 2

Petal A   Petal B

Fig. 7. Setting of experiment 4

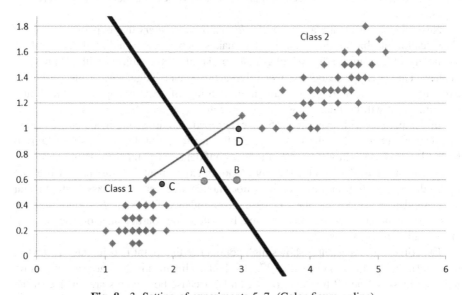

Fig. 8. 3. Setting of experiments 5–7. (Color figure online)

Table 2. Results of experiments 4–6

		Experiment 4		Experiment 5		Experiment 6	
		Mean	Stdev	Mean	Stdev	Mean	Stdev
Class 1	Oval A	4.91	2.31	7.82	1.40	7.82	2.95
Class 1	Oval B	4.73	2.70	3.45	2.19		
Class 2	Oval A	5.27	1.91	2.27	1.48		
Class 2	Oval B	5.45	2.90	6.36	2.46	5.82	3.59

Table 3. Results of experiment 7

		Mean	Stdev
Class 1	Oval C	9.64	0.64
Class 2	Oval D	9.27	1.71

Table 4. Results of experiment 8.

		Mean	Stdev
Class 1	Oval 1	9.73	0.45
Class 1	Oval 2	8.91	0.79
Class 1	Oval 3	6.91	2.81
Class 1	Oval 4	5.91	2.91
Class 2	Oval 5	9.78	0.42
Class 2	Oval 6	9.00	0.82
Class 2	Oval 7	7.11	2.69
Class 2	Oval 8	6.22	2.7

Our hypothesis is that this result is not accidental, but it can be explained. Quite typically, experts have more work experience with actual objects than with selected parameters of those objects in some mathematical form. More experience with actual objects can lead to more justified judgements on classification of objects by observing the actual objects. In actual objects, experts can observe selected parameters differently, e.g., holistically.

Next, actual objects are a source of *additional parameters* that experts can use for classification of objects. Pure intuitively it makes more sense to tell 50:50 (i.e., refuse to classify objects in the "grey" area) than to classify those objects with a high confidence. This reasoning suggests that relying on human judgement in the setting of experiment 4 is *more reliable* than in the setting of experiment 5. In the future experiments it will be interesting to test this conclusion on other objects.

In comparison with experiment 5, in experiment 6, adding the black classification line does not change the relatively large confidence for point A (78.2%, stdev 2.95) to be in class 1, but slightly decreases the low confidence in point B to be in class 2, to 5.82 (with standard deviation 3.59) from 6.36, in experiment 5. Thus, presence of the tip (in the form of the classification line) does not change much the classification result and confidence. Moreover, the decrease in confidence for point B shows that the closeness of point B to the classification line alerts the participants on the risk the classification of B to class 2.

Experiment 7 consistently shows high confidence that point C is in class 1 (9.64%, std. 0.64); and point D is class 2 (9.27, std. 1.71). This means that the participants do not limit classes 1 and 2 by their convex hulls, because both points are outside of the respective convex hulls.

The design of experiment 8 is shown in Fig. 9, where participants assigned their confidence values to ovals 1–4 to class 1 and ovals 4–8 to class 2. Experiment 8 consistently shows decreasing confidence from inner ovals to outer ovals for both classes: from high confidence 9.73 with stdev 0.45 (class 1) and 9,78 with stdev 0.42 (class 2) to low confidence 5.91 with stdev 2.91 (class 1) and 6.22 with stdev 6.22 (class 2) of the outer ovals. For the outer ovals these numbers are close to the results of experiment 4 (5.27 for oval A and 5.45 for oval B) and less close to experiment 5 (7.82 for point A and 6.36 for point B and even further from experiment 6 for point A (7.82, stdev 2.95), but closer for point B (5.82, stdev 3.59).

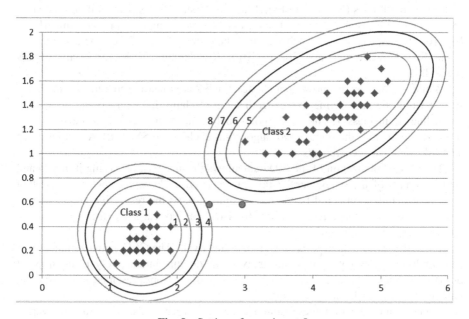

Fig. 9. Setting of experiment 8.

The experiments presented in this section show that participants can control overgeneralization of data in a variety of visualization settings. All cases that are far away from training data were classified with low confidence, in essence, refused to be classified. This is in a sharp contrast with automatic ML models shown in Figs. 3 and 4, where no case was refused.

4   Reversible Data Visualization Method to Support Machine Learning

While the experiments in the previous section were done on 2-D data, in this section, we show the use of the lossless visualization of 4-D data to support finding a Machine Learning classification model.

The *lossless visualization approach for multidimensional data* that we use below is based on the concept of the *General Line Coordinates* (GLC) and one of its specific forms denoted as *Parametrized Shifted Paired Coordinates* (PSPC) (Kovalerchuk 2018; Kovalerchuk and Grishin 2017). GLCs allow visualizing n-D data with full *preservation* of n-D information in 2-D. In this sense, GLC is *lossless* and *reversible* allowing restoring all n-D information from 2-D visualization of an n-D point presented in 2-D as a graph.

Figure 10 shows Iris data of classes 1 and 2 using all 4 attributes x_1–x_4 represented in PSPC in 2-D. In PSPC each 4-D point $\mathbf{x} = (x_1, x_2, x_3, x_4)$ is represented as a line (arrow) from point (x_1, x_2) in Cartesian coordinates (X_1, X_2) to point (x_3, x_4) in Cartesian coordinates (X_3, X_4). In Fig. 10a, coordinates (X_3, X_4) are shifted relative to (X_1, X_2) in such way that the center of class 1 became a single 2-D point. Similarly, in Fig. 10b this shift is done in such way that the center of class 2 became a single 2-D point. See (Kovalerchuk and Grishin 2017; Kovalerchuk 2018) for details of this method.

While Fig. 10 clearly shows that classes 1 and 2 are linearly separated by the black line, it would be an extreme overgeneralization to claim that every point, above that black line, is in class 1, and every point, below it, is in class 2. However, as we discussed in Sect. 1, the popular ML algorithms such as the SVM, LDA and Decision Trees do such overgeneralization.

The GLC visualization of the n-D data allows the analysis, observing it in 2-D and setting control of the overgeneralization, using *implicit domain knowledge*. The first layer of generalization could be the convex hull, around the respective graphs of n-D points of classes (simple arrows in PSPC for 4-D data). For class one, the convex hull is shown in Fig. 10a in green.

This figure shows the next layer as two dotted ovals. The GLC visualization system can produce it automatically or an analyst can do this interactively using implicit domain knowledge how far the unseen cases can differ from presented in the training data at the different levels of confidence.

Generalization beyond these dotted ovals, to larger ovals, will lead to overlap classes, as the visualization clearly shows. Thus, the analyst needs either to stop the generalization at these ovals, or change the shape of the border (the model of the border).

The change of the model would mean expanding ovals only in the areas, where they do not touch each other. It is difficult to justify such change of the model. We simply do not have any information, to justify this change of the model, beyond our desire to avoid overlap, in the area where the ovals already touch each other.

For instance, for (2), it does not allow the Iris parameters x_i to be equal to zero. In contrast, the decision tree in Fig. 4b overgeneralizes, to such unrealistic Iris cases, as well as to the cases with the petal length 4 times greater than in the training data.

The representation of n-D data as the 2-D *GLC graphs* such as shown in Fig. 10 has some similarity with the *manifold* approach. Commonly, the manifold approach uses a graph to find a surface in a subspace of much smaller dimension than the original n-D space (Gorban et al. 2007; McQueen et al. 2016). This graph captures the distances between the nearest given n-D points to construct the surface. While this is valuable information, it is only partial information about the relations between the n-D points. It

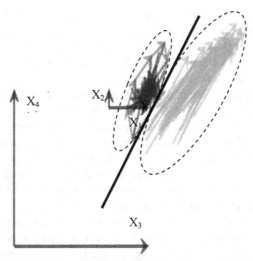

(a) Iris data in PSPC anchored in class 1. The class 1 convex hull is in green.

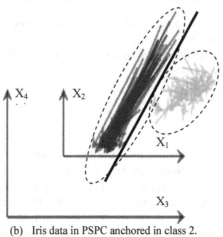

(b) Iris data in PSPC anchored in class 2.

Fig. 10. 4-D data reversible visualization to identify ML model parameters. (Color figure online)

can be insufficient in discovering the patterns in the n-D data. Thus, this method is lossy; it preserves only a part of all the n-D information in a lower-dimension subspace. The manifolds are defined on n-D points not on 2-D graphs.

In contrast, the GLC graphs preserve all the information about the n-D points. While this is a difference between GLC graphs and manifolds, the commonalities between them are in shrinking the dimensions of the multidimensional data. The manifold can be expanded on the manifold surface, but not outside of this surface. This is like the expansion of the GLC graphs within the convex hulls, as shown in Fig. 10. Thus, manifolds control the overgeneralization, by allowing the data of the classes to be only on the surface of the manifold.

A similar control for the GLC graphs can be derived by analyzing the properties of these graphs within the oval. For instance, in Fig. 10a arrows of the yellow class have a dominant direction and in Fig. 10b, the arrows of the red class have a dominant direction. This means that they occupy a fraction of the respective n-D areas.

This fraction of the area can be described mathematically similarly, to the manifold, e.g., by constructing a function of the two new attributes, angle and length of the arrow. In Fig. 10, the center of the oval that contains the GLC-graphs represents some n-D point. The n-D points, which are located around that n-D point, have their graphs within the oval. Thus, similar GLC graphs visualize similar n-D points. For hyper-cubes, it is proved mathematically in (Kovalerchuk 2018).

The advantage of the GLC graph approach is that the domain experts can be in the "driving seat" in analyzing and constructing the ML models controlling the generalization. In contrast, for the manifolds it is not clear how the domain experts can be in the same "driving seat" in constructing and limiting the manifolds. Moreover, the manifold can be higher dimension than 3, i.e., without natural visualization.

The restricted generalization in Fig. 10 illustrates the capabilities of the GLC visualization approach:

(1) Building interactively a *more accurate border between the classes* to avoid overgeneralization, and
(2) Getting a *better-explained description of the classes* that avoids the confusing description of the classes.

5   Conclusion

The proposed vision of the domain expert in the "driving seat" of model development depends on the success of the two important steps: (1) visualization of the data of the classes, making the classes separable in the visualization, and (2) abilities of the domain expert to generalize the data in these visualizations for constructing and correcting the models. This paper shows the feasibility of both steps. More examples of the success of step (1) are presented in (Kovalerchuk and Grishin 2017; Kovalerchuk 2018; Kovalerchuk and Gharawi 2018). Can we ensure that it will always be successful? The answer is the same no, as for the analytical ML models. For some data, successful ML models do not exist. If accurate enough analytical ML exists for the given data, then the chances that the visualization model will exist will also be higher.

In general, visualizations of the n-D points, as GLC graphs in 2-D, allow the observation of all of their values in all the subspaces, including the overlap in each subspace. This supports the discovery of the efficient ML models including the support vectors in the SVM and the subspaces, where the data classes are separable.

Discovering visually just one such subspace is sufficient for solving the ML problem for the given data. Such effective tools include permuting and reversing the coordinates, in combination with the analytical search, for the efficient classification rules, and the visualization of them in GLC.

Holistic shapes of graphs (Grishin and Soula 2003) allow their comparison to be more effective in the selection of subspaces. Besides that, the graphs give plenty of information, about the relationship between the parameters, in the subspaces and between the subspaces. These graphs are represented, in the different GLCs for the n-D data visualization, giving additional information for the mutual properties of data classes, relative to the linear and non-linear separation.

The future studies are expanding the class of ML models, which can benefit from the GLC visualization of the multidimensional data, and the ML models.

References

Bennett, K.P., Campbell, C.: Support vector machines: hype or hallelujah? ACM SIGKDD Explor. Newsl. **2**(2), 1–13 (2000)

Bennett, K.P., Bredensteiner, E.J.: Duality and geometry in SVM classifiers. In: ICML, pp. 57–64, 29 June 2000

Big Data and Machine Learning (2018). http://www.cnblogs.com/luweiseu/p/7826679.html

FITCSVM, Mathworks (2018). https://www.mathworks.com/help/stats/fitcsvm.html?s_tid=gn_loc_drop

Carbonell, J.G., Michalski, R.S., Mitchell, T.M.: An overview of machine learning. In: Michalski, R.S., Carbonell, J.G., Mitchell, T.M. (eds.) Machine Learning. SYMBOLIC, vol. I, pp. 3–23. Springer, Heidelberg (1983). https://doi.org/10.1007/978-3-662-12405-5_1

Gorban, A.N., Kégl, B., Wunsch, D.C., Zinovyev, A. (eds.): Principal Manifolds for Data Visualisation and Dimension Reduction. LNCSE, vol. 58. Springer, Heidelberg (2007). https://doi.org/10.1007/978-3-540-73750-6. ISBN 978-3-540-73749-0

Grishin, V., Soula, A.: Pictorial analysis: a multi-resolution data visualization for monitoring and diagnosis of complex systems. Int. J. Inf. Sci. **152**, 1–24 (2003)

Kovalerchuk, B., Vityaev, E., Ruiz, J.: Consistent knowledge discovery in medical diagnosis. IEEE Eng. Med. Biol. **19**(4), 26–37 (2000)

Kovalerchuk, B., Delizy, F., Riggs, L., Vityaev, E.: Visual data mining and discovery with binarized vectors. In: Holmes, D.E., Jain, L.C. (eds.) Data Mining: Foundations and Intelligent Paradigms. ISRL, vol. 24, pp. 135–156. Springer, Heidelberg (2012). https://doi.org/10.1007/978-3-642-23241-1_7

Kovalerchuk, B., Grishin, V.: Adjustable general line coordinates for visual knowledge discovery in n-D data. Inf. Vis. (2017). https://doi.org/10.1177/1473871617715860

Kovalerchuk, B.: Visual Knowledge Discovery and Machine Learning. Springer, Cham (2018). https://doi.org/10.1007/978-3-319-73040-0

Kovalerchuk, B., Gharawi, A.: Decreasing occlusion in interactive visual knowledge discovery. In: Human-Computer Interaction International Conference, Las Vegas (2018, in print)

Lichman, M.: UCI machine learning repository. University of California, School of Information and Computer Science, Irvine, CA (2013). http://archive.ics.uci.edu/ml

McQueen, J., Meila, M., VanderPlas, J., Zhang, Z.: megaman: manifold learning with millions of points (2016). https://arxiv.org/abs/1603.02763v1

Taylor, J.: STAT 2002, Data Mining, Stanford (2011). http://statweb.stanford.edu/~jtaylo/courses/stats202/trees.html

Pham, H.N.A., Triantaphyllou, E.: The impact of overfitting and overgeneralization on the classification accuracy in data mining. In: Maimon, O., Rokach, L. (eds.) Soft Computing for Knowledge Discovery and Data Mining, pp. 391–431. Springer, Boston (2008). https://doi.org/10.1007/978-0-387-69935-6_16

Segmented Time-Series Plot: A New Design Technique for Visualization of Industrial Data

Tian Lei, Nan Ni[✉], Ken Chen, and Xin He

School of Mechanical Science and Engineering, Huazhong University of Science and Technology, Wuhan, China
nanni@hust.edu.cn

Abstract. Time-series plots have been widely used in the fields of data analysis and data mining because of its good visual characteristics. However, when researching and analyzing the massive data formed in industrial, some short-comings of the traditional time-series plot make the visualization of big data ineffective, which is not conducive to data analysis and mining.

In this paper, the traditional time-series graph is improved and a segmented time-series plot that can be used for massive industrial data analysis is proposed. In addition, this paper describes in detail the steps of making the segmented time-series plot. The method can reduce the information overload and the interface issues by limiting the amount of information presented.

Keywords: Data visualization · Design technique · Industrial data
Segmentation method

1 Background

In recent years, data visualization has been widely used and put into function due to its convenience, comprehensibility and accuracy in industry. For example, through the integration of images, 3D animation and computer-controlled technology with the solid model, the visualization technology of operation simulation visualizes the state of the equipment so that the manager has a specific concept of the equipment as well as its position, shape and all other parameters [1]. However, how to display the industrial data better to the user has been a new topic in industry, with the further application of information visualization in industry, especially in the face of massive, widely distributed, complex, fast processed and uneven evaluated industry data [2].

2 Status of Industry Data Visualization

Most industrial data are time series which are collected at different points of time and reflect the change status of things or phenomena. Therefore, the time-series plot is usually used for visualization. Time-series plot is also called transition diagram that describes the variable in relation with time. Scatter plot, bar chart and broken line chart can all be used to visualize time-series data. For example, the graph below shows China's GDP and GDP growth rate from 1985 to 2015 in relation with time [3]. This

© Springer International Publishing AG, part of Springer Nature 2018
S. Yamamoto and H. Mori (Eds.): HIMI 2018, LNCS 10904, pp. 60–69, 2018.
https://doi.org/10.1007/978-3-319-92043-6_5

figure shows the data clearly to readers. However, different challenges appear when the data visualization in industry is concerned compared with Internet data visualization (Fig. 1).

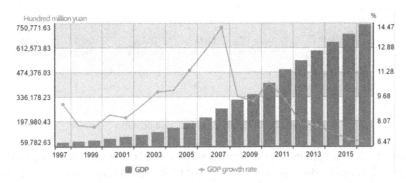

Fig. 1. China's GDP and GDP growth rate from 1985 to 2015

The first problem that needs to be solved is that the huge amount of data leads to a failure of effective and immediate display in the monitoring points. As industrial data comes primarily from various sensors and a common device may have been equipped with hundreds of sensors that update at a very high frequency, the amount of data in industry is much larger compared with traditional Internet data. How can so many categories of data be arranged and selected on a narrow screen?

The second problem is how to balance 'big' with 'small'. The problem is divided into two aspects, one of which is how the whole and the part can be effectively combined. The staff should pay attention to not only the overall trend of the data but also the abnormal values. In the face of such nested data, the staff needs to grasp the overall situation while capturing some local changes quickly. The other aspect is how to balance local and details. In a display of local data, we also want to get some specific details including related data, historical data, abnormal data, data trends, data forecasting, etc. And this requires designers to grasp the relationship between local and detail display.

The last problem lies in the translation efficiency from data into effective information for the searching and delivery of users. It has been common in Internet companies, such as Taobao and Jingdong from China, to use big data to analyze their users' habits and hobbies to find their favorite topics and push the relevant products [4]. However, this practice with big data is rarely utilized in industry.

These issues can lead to poor readability and low efficiency of the graphics, thus leading to poor user experience (Fig. 2).

Based on these problems, different scholars have made their own attempts to visualize time-series data. The main goal of visualizing time-series data is to reduce the dimensions and reduce the noise interference [5]. Among the various solutions, visualization method based on segmentation is frequently used for its advantages in data compression and noise filtering. Li proposed that the data should be sorted and classified according to certain requirements and arranged according to a certain order so as to facilitate people to analyze and study the problems [6].

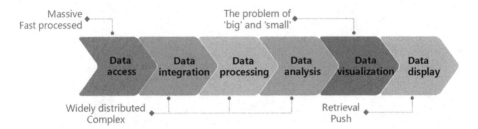

Fig. 2. Difficulties in data visualization

According to the differences in segmentation methods, segmentation-based visualization method can be divided into PAA [7] and PLR [8]. PAA approximates the entire sequence with the average of each segment by equally dividing the time series. PLR approximates the original time series with a number of straight line segments that are adjacent to each other whose interval are not necessary equal. The PAA method is rarely applied because it uses only equal division without considering the actual situation of the time-series plot, thus failing in retain the change trends of the original data.

Tian et al. described the superiority of segmentation-based visualization and improved the segmentation of time-series data at important points based on PLR theory. This method well retains the trend of the original data and clearly shows the abnormal and important points [9]. Similarly, Yu et al. also proposed a new feature-point-based segmentation method that can accurately represent time-series data, which preserves the main state features of the time series [10]. However, the PLR method does not guarantee that there will be only one basic trend in each segment, thus focusing too much on local details while ignoring the overall features.

In order to make up for the shortcomings of PLR, this study proposes a more effective method – segmented time-series plot – to achieve the visualization of time-series plot from the perspective of massive data display and balance between "big" and "small".

3   Industrial Data Visualization Based on Segmented Time Series

This method proposes a segment processing to the traditional time-series plot based on the existing situation of presenting industrial data with a time-series plot. The specific steps are as follows.

(1) The data is segmented with PLR as follows (R is the preset distance threshold);

- Take the start and end point as the initial segment point;
- Find the point that keep the largest distance from the segment line which should be greater than R;

- Take the points that meet the previous conditions as new segment points;
- If there is any point in the segments that has a distance greater than R, go to step 2. If not, the segmentation is then finished [11].

(2) Process each segment into a separate time series. That is, to extract the data from each segment to remake a time-series plot, the height of which is constant.

(3) Distinguish each time-series plot according to the design elements such as graphics, color, text, etc. The way we recommend here is to differentiate each time-series plot with different color backgrounds. But pay attention that the background should not be so fancy that user's attention is transferred;

(4) Connect the remade time-series plot in chronological order. Starting from the second segment, the time-series plot is shifted longitudinally according to the positions of each segment point until the initial point of the second segment coincides with the end point of the first segment. And repeat this step until the initial point of all the segments coincides with the end of the previous segments. Gestalt psychology suggests that people tend to percept a part as the whole. According to the continuity principle in perceptual organization, these sections are relatively easy to be perceived as a whole because they are connected at segment points [12].

(5) Conduct interaction design for the connected time-series plot. Human-computer interaction technology in information visualization can be summarized into five main categories: Dynamic Filtering Technology, Overview + Detail Technology, Pan + Zoom Technology, Focus + Context Technology, Multi-view Correlation Technology [13].

4 Case Analysis

4.1 Data Source

The data presented comes from experiments on a milling machine under various operating conditions. There are 16 sets of experimental data with differences in the three independent variables. Three different types of sensors (sonic sensors, vibration sensors and current sensors) are utilized for data acquisition. The lateral wear is not measured constantly but at intervals.

The data is organized in a Matlab structure shown in Table 1.

Table 1. Data structure

Field name	Description
case	Case number (1–16)
run	Counter for experimental runs in each case
time	Duration of experiment (restarts for each case)
DOC	Depth of cut (does not vary for each case)
feed	Feed (does not vary for each case)
material	Material (does not vary for each case)
smcDC	DC spindle motor current

Take case 1 (depth of cut is 1.5 mm, feed is 0.5 mm, material A) as an example. The current curve of the DC spindle motor in relation with time is shown in Fig. 3.

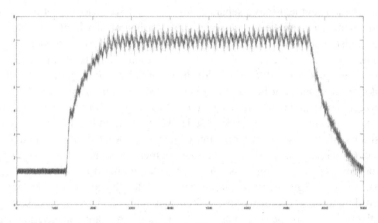

Fig. 3. The DC spindle motor of case 1

4.2   Visualization Process

Time-Series Plot Segmentation
Firstly, start point A and end point B are connected as shown in Fig. 4.

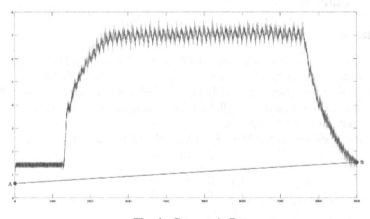

Fig. 4. Connect A, B

With R preset to 1.4, find the farthest point C from the line AB and tell whether the distance from point C to line AB is greater than R. If yes, point C will be chosen as a new segment point and connected with A and B respectively as shown in Fig. 5. If the distance is less than R, R should be properly adjusted. The value of R is determined by

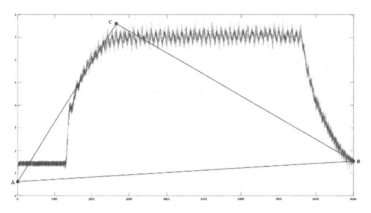

Fig. 5. Connect A, C and B, C

the degree of subdivision required and the ordinate of the time-series plot, which is proportional to the degree of subdivision.

Repeat this step until the farthest distance from all points to the segment line in each segment is less than R. Then the segmentation is finished. The time-series plot is finally divided into four sections, namely a, b, c and d, as shown in Fig. 6.

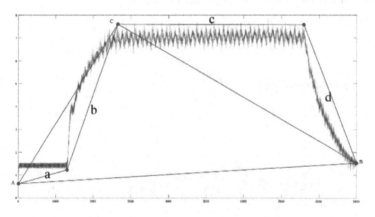

Fig. 6. Segmentation is completed

Independent Remake

First, extract each segment's data and make it into a time-series plot while ensuring that the height of each segment is approximately the same. It should be noted in this step that the same method is used to visualize each segment of the data. We still use the line chart here as shown in Fig. 7.

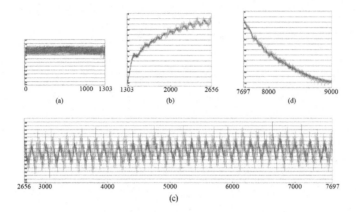

Fig. 7. Independent remake

Visual Distinction

At this step, the background of the line chart for each segment is reset. According to the range of ordinate values, the background color of the line chart with smaller ordinate value is replaced by the light blue while the darker blue is applied in line charts with larger ordinate value, as shown in Fig. 8. It should be noted that the background color and the polyline color should have a greater distinction.

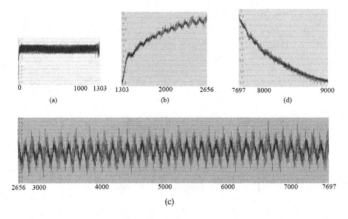

Fig. 8. Visual distinction. (Color figure online)

Connection

Connect the remade line charts together in the original chronological order. The initial point of the second segment is coincident with the end point of the first segment. And this step is repeated until the initial point of all the segments coincide with the end point of previous segments. The result is shown in Fig. 9.

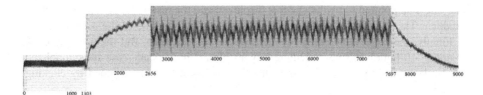

Fig. 9. Connection

Add Interactive Action

Then the time-series plot is basically completed and will be displayed as the main part in the interface of Data Viewer (DV) software. DV is a data visualization tool for data analysts and their superiors. The design of DV should meet three requirements of users: First, it should support switching among multiple variables and timelines and be able to explore and navigate data interactively. Second, it should be scalable to handle large amounts of time-series data. Third, it should be better readable (Fig. 10).

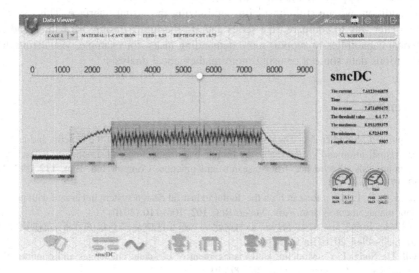

Fig. 10. Main interface of DV

The segmented time-series plot is displayed in the central part of the interface.

Limiting the amount of information presented can alleviate information overload and interface issues both overall and in part. Interaction is added here.

First, when clicking on a segment of the segmented time-series plot, the user can read the threshold lines and special values carefully by enlarging the selected area while compressing other areas. This method not only shows all the segments but also ensures a complete temporal context.

Second, the DV also provides a standard slider assembly and a timeline. The time slider divides the data in the time-series plot into two types: regular data and selected data. The data pointed to by the slider is displayed and the rest of data except the outlier is hidden. The specifics of this area are shown on the right side of the screen.

Finally, the page shows only one case in the DV interface. But the user can click the upper left button to switch. Different types of data in the same case are symbolized and displayed at the bottom of the page. The data type can be switched by clicking buttons.

At the same time, DV also provides search capabilities. But the search interface provides the ability to find data instead of discovering information. It supports keyword searches on time-series data, such as time and values.

5 Summary

In this paper, we introduce the segmented time-series plot that visualize the massive time-series data from industry. The segmented time-series plot improves the traditional time-series plot, which shows the details of the data in a multi-level way while clearly showing the trends in the data. Furthermore, this method also provides interactive techniques to support the industrial data analysis in a visual way.

In addition, we are still considering optimizing the visual interface. This can provide a better way to compare time-series data. In the future design, we will extend the analytical capabilities to support more demanding tasks, for example, visualizing data from various data sources or using more mathematical analysis.

References

1. Akpan, I.J., Shanker, M.: The confirmed realities and myths about the benefits and costs of 3D visualization and virtual reality in discrete event modeling and simulation: a descriptive meta-analysis of evidence from research and practice. Comput. Ind. Eng. **112**, 197–211 (2017)
2. Jiang, X.F., Li, J.Y.: Research on the digital industrial design system integrated with product data management system. Adv. Mater. Res. **102**, 106–110 (2010)
3. 王玫弘: 新常态下中美综合经济实力对比——基于国内生产总值的分析. 美国研究 **30** (05), 31–49+6 2016. (in Chinese)
4. Gil, D., Song, I.-Y.: Modeling and management of big data: challenges and opportunities. Future Gener. Comput. Syst. **63**, 96–99 (2016)
5. Lu, Z., Zhangm, Q.: Clustering by data competition. Sci. China (Inf. Sci.) **56**(01), 65–77 (2013)
6. 李晓明: 采用Excel制作数据分段统计表的研究. 电脑知识与技术 **10**(23), 5572–5579 (2014). (in Chinese)
7. Lin, W.-M., Gow, H.-J., Tsay, M.-T.: A partition approach algorithm for nonconvex economic dispatch. Int. J. Electr. Power Energy Syst. **29**(5), 432–438 (2006)
8. Stoica, P., Soderstrom, T., Ahlen, A., Solbrand, G.: On the convergence of pseudo-linear regression algorithms. Int. J. Control **41**(6), 1429–1444 (1985)
9. 田野, 张忠能: 改进的基于重要点的时间序列数据分段方法. 微型电脑应用 **28**(2), 48–51 (2012). (in Chinese)
10. 喻高瞻, 彭宏, 胡劲松, 等: 时间序列数据的分段线性表示. 计算机应用与软件 **24**(12), 17–18 (2007). (in Chinese)

11. 孙焕良,邱邦华,魏渊华. 一种优化的自底向上时间序列分段算法. 沈阳建筑大学学报(自然科学版) (06), 1049–1052 (2007). (in Chinese). Assessed 12 Oct 2017

12. Evergreen, S., Metzner, C.: Design principles for data visualization in evaluation. New Dir. Eval. **2013**(140), 5–20 (2013)

13. Cockburn, A., Karlson, A., Bederson, B.B.: A review of overview + detail, zooming, and focus + context interfaces. ACM Comput. Surv. (CSUR) **41**(1), 2 (2009)

Research on the Fuzziness in the Design of Big Data Visualization

Tian Lei, Qiumeng Zhu[(✉)], Nan Ni, and Xin He

School of Mechanical Science and Engineering,
Huazhong University of Science and Technology, Wuhan, China
`qiumeng_zhu@hust.edu.cn`

Abstract. In consecution to use and process information immediately, the relationship among a huge number of information is necessary to be read and understand. Information visualization as an effective method to optimize this process, using the charts to help people comprehend and process information intuitively and quickly. The accuracy of the information in the visualization chart is based on the readability and integrity of the information transition, once the chart does not meet this requirement, the accuracy of the information will be greatly reduced, and even may be misunderstood or cannot obtain the problem of information.

This paper will analyze and deduce the causes of ambiguous in the information visualization from the aspects of ambiguity definition and fuzziness experimental research. To solve this problem, the investigation collects 30 samples based on five complex information visualization charts, we will use infographic as the research object to explore the impact of fuzziness on the user in the visualization process and explore the causes and mechanisms of this effect by quantitative experiments.

Keywords: Information visualization · Fuzziness

1 Introduction

In the time of information explosion, people need to quickly handle the mass of information and understand the complex relevance among them. Information visualization as an effective method to optimize this process, using the charts to help people comprehend and process information intuitively and quickly. There is no doubt that the information visualization will promote the information resource development and utilization. If users can quickly and accurately comprehend all the information contained in the infographic, then the method of visualizing these information is effective and accurate. Otherwise, if users misunderstand the information or cannot get all the information content of the problem, so this appearance means that the method of visualizing the information remains fuzziness.

2 Literature Review

In contemporary times, there are many problems with the process of visualization due to huge and complex data. For example, improper or unreasonable key information extraction may result in chaotic and unsystematic problems in the information

© Springer International Publishing AG, part of Springer Nature 2018
S. Yamamoto and H. Mori (Eds.): HIMI 2018, LNCS 10904, pp. 70–77, 2018.
https://doi.org/10.1007/978-3-319-92043-6_6

visualization; inconsistent or irregular characterization of information types may result in that users are unable to quickly and accurately understand the information in a short time. When the origin comprehensible information becomes incomprehensible, the fuzziness of the visualization occurs. In the Webster's Dictionary, the "understanding" means that "the ability to distinguish the nature of things, or to grasp the inherent behavior of things or results" [1]. If users cannot distinguish the nature of things or cannot understand the internal relations, the thing like "incomprehension", "misunderstand" will occur. This is called information fuzziness.

Fuzziness is a very common topic. The problem of fuzziness exits in many fields. For example, in the linguistics, a kind of words like "many" "little" "possible" "probably" is full of fuzziness. However, there is no common acceptable definition until now, fuzziness is an objective attribute without a certain boundary in the amount. In the Oxford Advanced Learner's Dictionary, fuzziness has two meanings: one is the quality of being not clear in shape or sound and the other one is the fact of being confused and not expressed clearly [2]. Nevertheless, this does not affect scholars' enthusiasm for the study of ambiguity. For instance, scholars use mathematical methods to study the fuzziness - fuzzy mathematics. Its essence is to discretize the associated problem, in consecution to study and deal with the phenomenon of fuzziness [3]; scholars also use psychological methods to study the fuzziness - fuzzy psychology. That is, in a case of uncertain decision-making, subconscious psychology as the main tone, to make a subconscious psychological choice [4].

In the branches of design science, fuzziness is a widespread phenomenon, because the design itself is full of numerous emotional information. On the base of the emotional information, how to propose a more user-friendly design, or to evaluate these designs, is a very important part of the design process. For example, KY Chanb et al. used the method of intelligent fuzzy regression to study the nonlinear fuzzy relation between emotional response and design variables, and they used it as a tool to obtain more reliable information about customer demand [5]; Wang established an evaluation model in an inaccurate, non-deterministic product design environment, which based on the integrity of the user preference information [6]. These studies are interesting, but they are fragmented and not systematic. In the design of information visualization, there are similar problems in theory and practice.

In this paper, we will use infographic as the research object to explore the impact of fuzziness on the user in the visualization process and explore the causes and mechanisms of this effect by quantitative experiments.

3 Investigation of Fuzziness

The aim of this investigation is to understand the mechanism of the generation about fuzziness in the cognize space during the process of information visualization, and to find a method of avoiding the fuzziness.

The research is divided into two part:

(1) The evaluate to the subjects' cognition of infographic;
(2) The experimenter evaluate the subjects' cognition degree.

3.1   The Experiment About the Subjects' Cognition on Infographic

For information visualization of simple data, such as line chart, pie chart, it is less chance that the fuzziness happens. Since the data is simple, the relationship of data is simple, and the way of information visualization is simple. But for more complex data, namely the data itself is complex, since the relationship is complicated, the kinds of graphics, symbols needed are complex, it is easy to produce cognitive fuzziness generally. According to the relation of the data in this study, the main methods of complex information visualization are divided into four categories, shown in Table 1.

Table 1. The classification of infographic

Relationship type	Members
Logic relationship	Venn diagram
	Clustermap
Hierarchal relationship	Tree diagram
	Circle packing
	Sunburst
	Mosaic plot/cube tree map
Circulation relationship	Chord diagram
	Network diagram
	Sankey diagram
	Flow chart
Date comparison relationship	Bullet chart
	Scatter plot matrix
	Radar chart
	Boxplot

Logical relationship charts and contrast data chart are learned in the Chinese education, based on the degree of complexity, five methods of information visualization are selected from other kinds of charts in the hierarchical and circulation relation. They are chord diagram method, Sankey diagram method, circle packing method, sunburst method and tree diagram method. The independent variable is the five degree of the information visualization degree. The control variable of this study is the content of the information and the color of the chart. The dependent variable as the degree of familiarity, the degree of complexity, the degree of intact, the degree of consecution, the degree of access to information, the degree of color influence, the degree of language influence, the degree of readable. The corresponding problem are shown in Table 2. While the material of experiment are shown in Fig. 1.

The experiment subjects are thirty university students aged between 18 and 25, 15 subjects are men and 15 subjects are women, among them, 25 subjects are undergraduate students and 5 are postgraduate students. The major of 10 subjects are associated with design engineering and rest 20 subjects' major are not related to design engineering.

Table 2. Experimental questions

No.	Questions
Q1	Are you familiar with this kind of charts?
Q2	Do you think the chart is complicated?
Q3	Can you understand the complete information expressed in this chart?
Q4	Do you think the chart is organized?
Q5	Do you think the chart is useful for getting data?
Q6	How much do you think the color of the chart affecting reading?
Q7	How much do you think the language has caused you to read?
Q8	How was your reading experience on the chart?

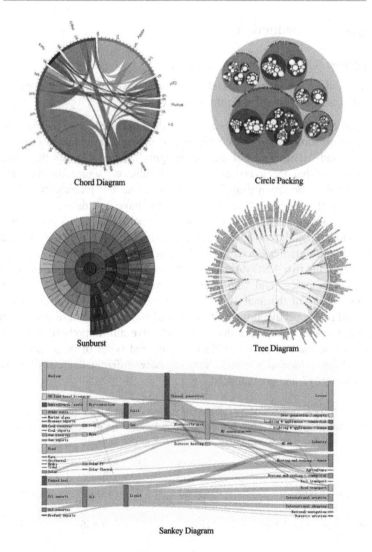

Fig. 1. Experimental materials (Chord Diagram, Circle Packing, Sunburst, Tree Diagram, Sankey Diagram)

The experiment steps are divided into two parts:

(1) The subjects read the prepared experiment material.
(2) The subjects mark the experiment material based on the five points scale method.

3.2   The Evaluate to the Cognition Degree of Subjects

The experimenter interview each subjects after the questionnaire finished. Then according to the experimental material of the entire information content, the experimenter judge the subjects' understanding of the experimental materials and mark the degree of subjects' understanding by using five points scale method.

4   Analysis and Deduction

The investigation result reveals:

(1) The method of information visualization does have an impact on the subjects' accurate understanding of the information content.

We conducted an ANOVA analysis of the data obtained from the three subjects "readability of information graphics", "degree of information obtained from information graphics" and "mastery of information integrity", Table 3 shows the results.

Table 3 shows that the average variance of subjects' scores on "the degree of readability of the information graphics" is 15.001, and the probability value of the corresponding null hypothesis "the subjects do not have any difference on the readability of the information graphics" is 0.000, less than the critical value of $\alpha = 0.01$, which shows that the null hypothesis is not established, that is, the degree of readability of these information graphics does exists the differences.

The results of the two ANOVA analyzes, "the degree of information obtained from the information graphics" and "the degree of mastery of information integrity" are similar to the results of the analysis of "the degree of readability of the information graphics". And the respective probability values of the null hypothesis are less than the critical value $\alpha = 0.01$, indicating that their respective null hypothesis does not hold. Which means that the two cases "subjects in different types of information obtained different levels of information", and "subjects have different degrees of mastery of information integrity" are exists indeed.

Table 3. The ANOVA analysis of the "degree of subjects' readable of infographic", "degree of subjects' information obtained from infographic" and "mastery of degree of experimenters' information integrity"

	Sum of squares	df	Mean square	F	Sig.
The degree of readability of infographic	60.004	4	15.001	20.959	0.000
The degree of access to information from infographic	18.624	4	4.656	4.405	0.002
The degree of mastery of infographic	22.131	4	5.533	8.267	0.000

Based on the results of ANOVA analysis of the three data types, concluded that same information through different visualization methods, the users have cognitive differences in terms of readability, comprehension, and integrity.

(2) The better visualization chart organized, the less ambiguous of the subjects incorrect interpretation on the chart of the ambiguous.

The bivariate correlation analysis are applied on the scores of "the degree of organization of information graph" and the scores of "the degree of readability of information graph", "the degree of information obtained from information graph" and "the degree of mastery of information integrity", finishing results are shown in Table 4.

From Table 4, it shows that there are significant correlation between "the degree of organization of information graph" and "the degree of readability of information graph", "the degree of mastery of information integrity" on the level of $\alpha = 0.01$, which corresponds to the correlation coefficient of 0.619 and 0.431 respectively. This shows that the higher degree of organization of the information graph, the more conducive to the participants to read the information, the subjects are also easier to think they have completely understand the information in the graph.

The null hypothesis that the correlation exists between "the degree of the organization of information graph" and "the degree of information obtained from the information graph" is 0.37, which is greater than the critical value of $\alpha = 0.05$, indicating that there is no correlation between them. That is, the degree of participants get information from the infographic has no relevance to the degree of the organization of graph. This is a bit different from our usual perception. In most cases, we think that the well-organized information may be more beneficial to users to obtain information. The unusual result may be related to the user's cognitive style, which users may extract information according to their own mode of thinking or focus, rather than read all the information step by step like a scanner.

Table 4. The partial correlations are on the subjects' marks on "degree of subjects' readable of infographic", "degree of subjects' information obtained from infographic" and "subjects' mastery degree of information integrity"

		The degree of readability of infographic	The degree of information obtained from infographic	The degree of mastery of infographic
The degree of organization of infographic	Pearson correlation	0.619	−0.075	0.431
	Sig. (2-tailed)	0.000	0.370	0.000

(3) The subjects are more familiar to the theme of the information, the less likely they have ambiguity in understanding.

We conducted a bivariate correlation analysis of "subjects' familiarity with the information involved in the infographic" and "the degree of mastery of information integrity". Table 5 shows the results that there is a significant correlation between "subjects' familiarity with the information involved in the infographic" and "the degree of mastery of information integrity" at the level of $\alpha = 0.01$, the corresponding correlation coefficient is 0.601. This result is consistent with common experience. For example, from the time-share charts of stocks, we can see that subjects with relevant experience can get the meanings of curves, bars and other representations in the graph in a relatively short period of time. And the whole meanings can also be obtained in a relatively short period of time. This is because subjects are more familiar with the information content, it is easier to understand the infographic.

Table 5. The correlation analysis on "degree of subjectes' familiarity to involved infographic information" and "the degree of mastery of information integrity"

		The degree of mastery of infographic
The degree of familiarity with the information involved in infographic	Pearson correlation	0.601
	Sig. (2-tailed)	0.000

(4) The color collocation of the infographic and reading disorder caused by the language are not correlated with whether the subjects grasp of the information completely.

We tested participants' scores on the influence of "chart color on reading" and "the chart language", respectively. Bivariate correlation is conducted with those two scores with the score of "subject's degree of mastery of information integrity", the finishing results in Table 6. From Table 6, we can see that, the probability values of the null hypothesis that "the influence of chart color on reading" and "the reading disorder caused by language of chart" are related to "subjects' degree of mastery of information integrity" are greater than the critical value of $\alpha = 0.05$. It indicates that the fuzziness of reading caused by the color of the chart or graphic language does not correlate with completely mastered the information.

This conclusion seems inconsistent with our daily experience. For example, when we use a pie chart to represent the consumption of a physical quantity, we might give the consumed part a black color and the remaining part a white color. However, because subjects have different semantic relations to black and white in their daily life, some people may think that white is consumed, black persists, while others are opposite. In this case, it is difficult to judge whether all users have an accurate understanding of this information.

Therefore, we suppose that the relationship between the color of the infographic and the reading disorder caused by the language and the subjects' complete mastery of the information may also be influenced by the user's background knowledge and educational level. If participants have a high level of education or a background in design, they may self-correct some of the information in the cognitive system to reduce the ambiguity of the individual's understanding of the infographic. In this experiment, one third of the participants had a design knowledge background and all of the participants were undergraduate students from Huazhong University of Science and Technology (China Top 10).

Table 6. The correlation analysis is done to "the reading impact degree of color of chart" and "the reading barrier caused by language of chart" with "subjects' mastery degree of information integrity" respectively.

		The degree of mastery of infographic
The reading impact degree of color of chart	Pearson correlation	−0.111
	Sig. (2-tailed)	0.176
The reading barrier caused by language of chart	Pearson correlation	0.059
	Sig. (2-tailed)	0.475

5 Conclusion

Based on the experimental study and analysis of the differences in user perception space under different visualization methods, this research finds out the method of information visualization, the information graph's own consecution degree, information content and the subjects' familiarity is the main cause of ambiguity in information visualization. And the language, color and collocation of the information graph do not affect the subjects producing ambiguity of information content in sensation space.

Acknowledgments. We would like to express my gratitude to all those who helped me during the writing of this thesis. We gratefully acknowledge the help of the 30 subjects who participated in the experiment, who has offered us valuable data in the academic studies.

We also owe a special debt of gratitude to all the companions in lab who have always been giving us valuable suggestions and critiques which are of help and importance in making the thesis a reality.

Last but not the least, my gratitude also extends to my family who have been assisting, supporting and caring for me all of my life.

References

1. Merriam-Webster's Learner's Dictionary. http://www.learnersdictionary.com/
2. Oxford Advanced Learner's Dictionary. http://www.oxfordlearnersdictionaries.com/
3. Haoliang, Z.: Basic theory and application of fuzzy mathematics. Mine Constr. Technol. (Z1), 70–80+96 (1994)
4. Tongtao, Z.: Fuzzy language and fuzzy psychology. J. Xiamen Liniversity **4**, 140–146 (1987)
5. Chanb, K.Y., Kwonga, C.K., Dillonb, T.S., Funga, K.Y.: An intelligent fuzzy regression approach for affective product design that captures nonlinearity and fuzziness. J. Eng. Des. **22** (8), 523–542 (2011)
6. Wang, J.: Int. J. Prod. Res. **35**(4), 995–1010 (1997). 16 p.

Interactive Point System Supporting Point Classification and Spatial Visualization

Boyang Liu[1]([✉]), Soh Masuko[2], and Jiro Tanaka[1]

[1] Waseda University, Fukuoka, Japan
waseda-liuboyang@moegi.waseda.jp
[2] Rakuten Institute of Technology, Rakuten, Inc., Tokyo, Japan

Abstract. Point system is structured marketing strategy offered by retailers to motivate customers to keep buying goods or paying for the services. However, current point system is not enough for reflecting where points come from. In this paper, concept of point classification is put forward. Points are divided into different categories based on source. We introduce mission into point system. Mission content is designed to guide consumption. In our system, point, mission and virtual pet will be spatially visualized using AR technique. The state of virtual pet depends on the evaluation for users. Users need to adjust themselves to keep their pets in a good state. Users can manipulate on the GUI or use gestures to interact with system.

Keywords: Interactive system · Spatial visualization
Gesture interaction · Value creation · Gamification

1 Introduction

Point system is structured marketing strategy offered by retailers to encourage consumers to keep buying goods or paying for the services. According to the spending in consumption, retailers give consumers a certain amount of points as reward. These points can be exchanged into goods or service [1].

On May 1, 1981 American Airlines launched the first loyalty marketing program of the modern era which was called a Advantage frequent flyer program [2]. This revolutionary program is thought to be the first one to reward frequent fliers with miles that can be accumulated and redeemed for free travel. Many travel providers and airlines saw the incredible value in offering customers an incentive to use a company exclusively and be rewarded for their loyalty. After the success of this program, dozens of travel industry companies launched similar programs within a few years. This program is also considered as the initial modern consumer reward program.

In early part of 2010, Card Linked Offers appears as a new loyalty marketing technique for retailers, brands, and financial institutions, stemming from a rise

© Springer International Publishing AG, part of Springer Nature 2018
S. Yamamoto and H. Mori (Eds.): HIMI 2018, LNCS 10904, pp. 78–89, 2018.
https://doi.org/10.1007/978-3-319-92043-6_7

in popularity of both coupons and mobile payment [7]. Loyalty cards are a form of tracking and recording technology that enables retailers to collect data about their customers' demographic and purchase behaviors [3,8]. As recompense for recording data from consumers, customers can receive loyalty points which can be redeemed for exclusive discounts and rewards [9]. Many vendors apply loyalty systems to collect customer-specific data that may be exploited for many reasons, e.g., price discrimination and direct marketing [4,10]. There are some findings clearly show that there is significant evidence of the effect of all loyalty programs on building and maintaining customer retention [5].

For the purpose of inspiring customer loyalty, some ideas of game-based marketing are proposed to engage customers [6,12]. People play games everywhere. Frequent Flyer Programs (FFPs) and other loyalty systems prove that gaming and marketing can fuse together perfectly. Modern FFPs use a number of gaming elements to engender loyalty, including point accumulation, level climbing, rewards and challenges [6].

2 Problem

The first noticeable problem is that all the points are treated the same way in current systems. In spite that points come from a certain amount of consumption and can be used as electronic money in payment, it is not appropriate to focus only on the number of points and ignore the difference of their source.

Point system is a model of transaction. After purchase, customers can get point as reward. However, it is not advisable to get points from spending without considering the value created in consumption. Current systems with single criteria need to be improved to make consumption comprehensive.

Users usually get point information from textual description and cannot keep the impression or even ignore these textual description. Point is one type of immediate feedback mechanism to reward users who meet requirements. However, the limitation of static display makes it not appealing.

3 Goal and Approach

The goal of research is to create an interactive point system with multi-value criteria that supports point classification and spatial visualization.

The source of points is used to classify points. Points obtained from different kinds of goods are considered different. Mission is put forward in the system to introduce multi-value criteria into consumption. We will combine the theme such as healthy diet, environmental protection into content design. If users succeed in completing these valuable tasks, level will be upgraded to reward their efforts. Compared to the current systems that only reward consumption, mission is a complete change in shopping. By doing that, we aim to arouse public attention on the value creation in consumption.

For the limitation of displaying the information in textual form, we propose a new approach – spatial visualization [13]. Spatial visualization can display the traditional textual information in the form of spatial model. In spatial visualization, points are represented as visualized objects. They are visualized as different objects according to the source. The number of points will also be emphasized through the visual feedback mechanism. Mission is visualized as the star in game scene. The content of mission will be given in textual form on the screen after users click the star. Inspired by tamagotchi [11] and game-based design [6], a virtual pet is visualized in our system. Our system measures users' state from two aspects – point and pet. Based on spatial visualization, we provide graphical user interfaces (GUI) for users. In our system, users can not only see the visualization, but also interact with the system by clicking on screen or using gesture as input. In the system, users can get mission instruction and complete it. If mission is done, dynamic animation will be displayed to give visual feedback to users and level of pet will be upgraded. Our system will measure result of points and pet's health state and show the information to the user through the GUI. The state of pet includes level and two indicators. The number of completed mission is reflected by the level of pet. Different level corresponds to different stage of pet. Pet evolves if the level goes up. Two indicators reflect the number of points and variety of points. The two indicators will be combined to measure the health state of pet. If the pet is not in good state, users need to adjust themselves to help the pet recovers. If the pet is healthy, users are expected to continue their efforts.

4 Our System

Our system classifies points by the source and visualizes them as virtual objects. It is different from current point systems which accept a single input (money) to generate a single output (point). Mission is designed as input of our system to remind users of multi-value in consumption instead of only money. We expand the feedback mechanism from single output to two kinds of output – point and virtual pet. Our system presents feedback by spatial visualization instead of the traditional two-dimensional textual information. The way of interaction is expanded from only operation on two-dimensional screen to the spatial gesture interaction.

4.1 Point Classification

Quantity, time limitation and use range of points are taken into consideration in current point system. However, the source of point is an important feature indicating the difference between points. In this paper, we propose the idea of point classification. We classify points based on where they are collected from. Points are from various source such as food, book or flight. In our system, there several types of points indicating the source of points. Points from same source are classified into the same category. By using AR, users are reminded of the source of points and can know the number of each category in a more intuitive way.

4.2   Multi-value Creation

Point system is a system rewarding points to encourage spending. Rewarding points for spending is a way to promote consumption and retain customers. It is economy-oriented. However, such design ignores other important factors in the consumption. Many value factors are critical for making decision, such as health diet and environmental protection. These factors are not taken into consideration in current design. Therefore, we import the mission into our system to remind users of multi-value which cannot be reflected with only points [15].

In the system, mission will be given during shopping. Users can get point from a certain amount of consumption and get feedback from virtual pet after the completion of mission. It motivates users to take full account of the value of consumption.

4.3   Interactive System with Spatial Visualization

The point system is actually a reward system based on gamification theory. Point is a mechanism for immediate feedback and tracking progress. In the current system, customers are informed of point information in the textual form. The problem is that this immediate feedback cannot be fully expressed by static textual description. It prevents point systems from showing appealing reward in a dynamic way.

Point. Point is visualized in our system. The source of points is reflected by the objects which the points are visualized as. One object represents one category of points. Coin models are placed next to the object that represents the source of the points. Users can get a rough idea of how many points there are by spatial visualization. Users can know the specific number of each category after clicking the coin model. The spatial visualization of points is shown as Fig. 1.

Virtual Pet. The spatial visualization of virtual pet is shown as Fig. 2. In the picture, state of pet is different. In subgraph (a), two indicators are high and the pet is active without illness. In subgraph (b), two indicators are normal and the pet looks calm. In subgraph (c), blue indicator is low and the pet looks unhappy. In subgraph (d), two indicators are low and the pet is sick. The characters are colorful and simplistically designed creatures based on animals and people. These pets look like common animals. At the same time, they can make movements like human beings. Users can decide pet's name according to their preference. After the pet interface is opened, the pet will appear on the screen. Its name and level is displayed at the bottom-right corner of the pet. Information about the pet's state is displayed in the top-right conner of the screen. There are two indicators to determine how healthy and happy the pet is. Each indicator is a measure of consumption. If user gets many points, the green indicator will be high. If user gets many kinds of points, the blue indicator will be high. These indicators have a direct relationship with the user's consumption. The two indicators are

Fig. 1. Spatial visualization of three kinds of points in the system.

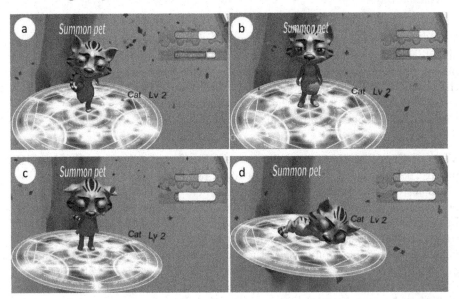

Fig. 2. Spatial visualization of virtual pet with different expression. Pet is energetic in subgraph (a). Pet looks calm in subgraph (b). Pet is unhappy in subgraph (c). Pet is sleeping in subgraph (d).

combined to evaluate the health of the pet. The expression of pet is influenced by the health of pet. If both indicators are high, it will be active. If two indicators is normal on average, it will look fine. If one indicator is not good, pet's expression will be unhappy. If both indicators are not good, pet may be ill. The pet goes through several distinct stages of development throughout its life cycle. Each stage lasts a period of time, depending on the level of pet. The level depends on mission completion. If mission is done, pet will get experience value. When the threshold is met, level will be upgraded. After reaching a certain level, the pet reaches a new stage and its appearance changes, which is the evolution of pet. The body shape of the pet varies depending on how many points there are. By introducing the level and two indicators to evaluate the health state of virtual pet, user's consumption are evaluated and the result is shown to user by spatial visualization.

Mission. Mission is an important method in our system to introduce the multi-value criteria (Fig. 3). In our system, the content of the mission is displayed in a game-like environment. The process of obtaining mission is designed to be a dynamic animation. Users can know the content of mission from reading text on the screen. After completing the mission, the system will play fireworks animation to create joyful environment in which users can feel successful.

Fig. 3. Spatial visualization of mission.

Usage. When users go shopping, they can start point system and enter mission interface. There is one box in the center of the screen. The white clouds float in the air. After users click the box, it will emit light and the box will open.

After the animation stops, a star will rise into the air. After users click the star, it will blow up and users can see several patterns on the ground. If users click the pattern, they can view the content of mission displayed on the GUI (Fig. 3). If users complete the mission, it will be confirmed by the system after the consumption. If it is confirmed, the pet will receive the experience value and it will be recorded.

The number, variety of points and the completion of mission will affect pet's health state. On the pet interface, virtual pet will appear after some animations. The pets life cycle stages are baby, child, teen, and adult. There is a state bar at the top right corner of the screen indicating how healthy and happy the pet is. In state bar, there are two different indicators – Hunger and Happy.

The higher each indicator is, the better the pet's state is. Hunger is related to the number of points earned by the user. After the user gets points, Hunger indicator will rise. Happy is related to the variety of points. If user gets points from various source, the Happy indicator will be high. Filling up the Hunger can be achieved by obtaining points. Filling up the Happy can be achieved by getting points from buying different goods. The two indicators will be considered together to assess the health state of virtual pet. Virtual pet will make different action with different expression according to the health state of virtual pet (Fig. 2) and the expression of pet will change.

Level is used to describe the pet's current life stage. The pet's experience value increases after the user completes mission. After the experience value reaches threshold, the pet will level up. When it reaches a certain level, the pet will go to new life stage, which is considered as evolution.

The results, including points, mission and virtual pet, are presented to the user via spatial visualization.

Spatial Gesture. Users can interact with the system based on the GUI provided by the system. When users interact with the system, they can operate on two-dimensional screen. However, this makes the virtual world created by AR cannot be well integrated with the real world. This reduces the users' interest in the system to some extent. In order to improve the interactive experience of the point system, we consider replacing the traditional two-dimensional interaction with the spatial gesture interaction. We use leap motion to capture the user's real-world hands movement and map it into the virtual world. Users can interact with system using gesture (Fig. 4).

When the camera scans the surrounding scenes, the features of point card are extracted and compared with the recorded features. After that, the coordinate system of real world and screen coordinate will establish mapping. User's hand movement is obtained by leap motion and it is shown in the real-time video captured by webcam. Users can see the real-world movement of their hands on the screen. The virtual objects are superimposed to real world in the real-time video. Therefore, users can adjust the position and movement of hand to touch and interact the virtual objects as if they were objects in real world.

Fig. 4. Spatial gesture interaction in the system.

5   Implementation

5.1   Development Environment

The hardware devices used for the development of the prototype system include a laptop, a webcam, a smartphone and leap motion. Windows 10 Home Edition is installed in the laptop. The processor is Intel(R) Core(TM) i7-6500U CPU @2.50 GHz 2.59 GHz. The RAM is 8.00 GB. Webcam is connected to the laptop to capture video and send video stream to the laptop for processing. The development software is Unity 2017.2.0f3(64-bit), a cross-platform game engine. Unity 3D is used to develop and render three-dimensional system. Vuforia SDK is used as a foundation for AR implementation, which uses computer vision technology to recognize and track image targets and simple 3D objects in real-time. After recognizing the image on the smartphone, virtual objects created in Unity 3D are superimposed over the image. After that, users can see the spatial objects and interact with them. The Leap Motion controller is a small USB peripheral device placed on a physical desktop, facing upward. Using two monochromatic IR cameras and three infrared LEDs, the device observes a roughly hemispherical area, to a distance of about 1 meter. The LEDs generate IR light and the cameras generate reflected data. It is sent via a USB cable to the laptop and analyzed by the Leap Motion software. Leap Motion is used to get the high-precision gesture information. With the information provided by leap motion, spatial interaction can be detected and sent to laptop as input data.

5.2   Main Work

In our system, the image of point card are used as the target of recognition. Image database is created and the image of point card is uploaded to the database. The image will be analyzed by the algorithm of image recognition. The database containing the image and Vuforia SDK are imported into unity. We connect the webcam to the laptop and configure the webcam information in unity. We install the leap motion software on the laptop, download and import unity core assets into unity.

In this research, we propose mission into our system. We expand the feedback from only points to point and virtual pet. Spatial visualization is used to give visual feedback instead of textual information. To make our point and level meaningful, we try to combine game with point system and design how users interact with our system. We define game rules to connect the point system with new design.

Specifically, Unity project is bulit and spatial models of different objects representing different points are created in the project. Coin model is set up next to visualized points to inform users of number information. Point models are set to receive and react to user operation. Point, level and some other information are recorded in our system. Data transmission and interface jump are processed to keep system consistent.

Virtual pet spatial model are set in the center of point card. The changes in the shape of pet are controlled to reflect the change in the number of points. Pet is given a variety of animations in reaction to user operation. We change the pet's state dynamically and visualize the information in state bar.

We design the way to get mission and presented it in the form of animation in reaction to user actions. We render the scene in real time to provide a precise visual effect. We create a gamified environment with light, sounds and dynamic particle effects. Multi-value are imported into mission content and mission is visualized on GUI. Animations are designed and points are rewarded after the completion of mission.

We define some spatial gesture in the system to interact with spatial objects. Leap motion is used to capture the user's real-world hands movement and the movement is mapped into the virtual world spatial gesture. Input data is processed and feedback is given via visualization.

6   Related Work

Our research is a new exploration of point system. Although there are no research similar to our system, some theoretical researches are the basis of our system. Zichermann and Linder considered harnessing the power of games to create extraordinary customer engagement with Game-Based Marketing [6]. They thought the most powerful way to create and engage a vibrant community is game mechanics – points, levels, badges, challenges, rewards and leaderboards.

The research conducted by Choi and Kim investigated why people continue to play certain online games [14]. Their results shows that people continue to play online games if they have optimal experiences and personal interaction can be facilitated by providing appropriate goals, operators and feedback.

The research conducted by Neal et al. proposed the idea that value drives loyalty. They thought buyers who are considering a purchase in a particular product or service category scan their product/service options and develop a consideration set, in which they develop a hierarchy of products based on their assessment of value.

The paper written by Annika Hupfeld et al. explicated how people's everyday shopping practices and orientations are shaped by loyalty scheme and contribute to the creation of value through personal data. Although our system emphasizes value creation, the methods are quite different.

7   Discussion

We propose a new interactive point system based on current point system. Compared to the current system, our system has the following advantages:

1. Point classification
2. Spatial visualization
   (a) Point
   (b) Virtual pet
3. Multi-value criteria

Current point systems only focus on the time limitation and application conditions. Our system explores the value of point. In our system, point is different from electronic money. We focus on the source of points because it reflects their preferences in the past. We propose a new method to classify points in our system. With new point classification, our system can have a positive impact on evaluating consumption.

We propose the spatial visualization method which is not available in the present system. Users can know information more directly and it is easier for them to retain the impression. The user's behavior is also measured and presented to the user by spatial visualization of virtual pet. It is a unique feature in our system. The current system affects the user's consumption behavior through rewarding points. However, in our system, multi-value criteria are proposed. We introduce mission into our system. Our system assists users to asses their shopping from many aspects, which can help them make comprehensive decision. The value of consumption is explored and created in our system.

In addition, the attraction and interaction between users and point system is greatly enhanced in our system. We use the game design in our system. We expand the feedback from only points to points and virtual pet. Points are used for rewarding spending. Virtual pet is created for giving the intuitive feedback to the user. The user can adjust his shopping habit to help the pet in a good state. User can have a more enjoyable experience via interaction with the game-based system.

Based on the spatial visualization, we expand the interaction on two-dimensional screen to the three-dimensional interaction. The three-dimensional interaction allows users to interact with the system in a more natural way, thereby enhancing the sense of immersion in the game-based system.

However, there are disadvantages in our system. Spatial visualization system requires a large number of predefined models. Building different models for various consumption is a complex and huge project even if consumption is classified. As the number of points increases, the cost and difficulty of spatial visualization increases.

8   Conclusion and Future Work

In this paper, a new interactive point system is introduced. The whole system is designed and implemented based on game design. We propose a new idea of point classification, paying attention to the source of points. Points are visualized to facilitate information browsing and virtual pet is visualized to indicate the user's consumption. Mission is given to instruct users to make a comprehensive thinking during consumption. The purpose of our system is to help users make better choices. User can obtain points from consumption and get feedback from pet after completing the mission given by the system. Depending on the number of points and the variety of points, the state of the virtual pet will change. Users need to complete mission with value creation to care their pets.

In the current system, interaction is limited between users and the system. We consider combining the SNS features into our system to enhance interaction between users. The mission in the system is given randomly. We consider customizing mission for each user based on their personal information.

References

1. Liu, B., Tanaka, J.: AR-based point system for game-like shopping experience. In: International Conference on E-Business and Applications, pp. 41–45. ACM (2018)
2. Kotler, P.: According to Kotler: The World's Foremost Authority on Marketing Answers Your Questions. AMACOM Division of American Management Association, New York (2005)
3. Hupfeld, A., Speed, C.: Getting something for nothing?: a user-centric perspective on loyalty card schemes. In: Proceedings of the 2017 CHI Conference on Human Factors in Computing Systems, pp. 4443–4453. ACM (2017)
4. Enzmann, M., Schneider, M.: Improving customer retention in e-commerce through a secure and privacy-enhanced loyalty system. Inf. Syst. Front. 7(4), 359–370 (2005)
5. Magatef, S.G., Tomalieh, E.F.: The impact of customer loyalty programs on customer retention. Int. J. Bus. Soc. Sci. 6(8), 78–93 (2015)
6. Zichermann, G., Linder, J.: Game-Based Marketing: Inspire Customer Loyalty Through Rewards, Challenges, and Contests. Wiley, New York (2010)
7. Passingham, J.: Grocery retailing and the loyalty card. Int. J. Market Res. 40(1), 55 (1998)
8. Lacey, R., Sneath, J.Z.: Customer loyalty programs: are they fair to consumers? J. Consum. Market. 23(7), 458–464 (2006)
9. Mauri, C.: Card loyalty. A new emerging issue in grocery retailing. J. Retail. Consum. Serv. 10(1), 13–25 (2003)
10. Felgate, M., Fearne, A., Di Falco, S., et al.: Using supermarket loyalty card data to analyse the impact of promotions. Int. J. Market Res. 54(2), 221–240 (2012)

11. Wikipedia Contributors: Tamagotchi. Wikipedia, The Free Encyclopedia, 12 Dec. 2017. Web. 16 Jan. 2018
12. Huotari, K., Hamari, J.: Defining gamification: a service marketing perspective. In: Proceeding of the 16th International Academic MindTrek Conference, pp. 17–22. ACM (2012)
13. Chubb, C.E., Dosher, B.A., Lu, Z.L.E., et al.: Human Information Processing: Vision, Memory, and Attention. American Psychological Association, Washington, DC (2013)
14. Choi, D., Kim, J.: Why people continue to play online games: in search of critical design factors to increase customer loyalty to online contents. CyberPsychol. Behav. **7**(1), 11–24 (2004)
15. Neal, W.D.: Satisfaction is nice, but value drives loyalty. Market. Res. **11**(1), 20 (1999)

A Topological Approach to Representational Data Models

Emilie Purvine[1]([✉]), Sinan Aksoy[2], Cliff Joslyn[1], Kathleen Nowak[2], Brenda Praggastis[1], and Michael Robinson[3]

[1] Pacific Northwest National Laboratory, Seattle, WA 98109, USA
{emilie.purvine,cliff.joslyn,brenda.praggastis}@pnnl.gov
[2] Pacific Northwest National Laboratory, Richland, WA 99354, USA
{sinan.aksoy,kathleen.nowak}@pnnl.gov
[3] American University, Washington, DC, USA
michaelr@american.edu

Abstract. As data accumulate faster and bigger, building representational models has turned into an art form. Despite sharing common data types, each scientific discipline often takes a different approach. In this work, we propose representational models grounded in the mathematics of algebraic topology to understand foundational data types. We present hypergraphs for multi-relational data, point clouds for vector data, and sheaf models when both data types are present and interrelated. These three models use similar principles from algebraic topology and provide a domain-agnostic framework. We will discuss each method, provide references to their foundational mathematical papers, and give examples of their use.

Keywords: Relational data · Vector data · Hypergraph models
Topological data models · Data-agnostic models

1 Introduction

The potential for drowning in data has become ubiquitous across all scientific domains. Instruments are built to collect massive amounts of data from anything we can get our hands on. When that does not suffice, those instruments are refined to reduce error rates and collect even more data. As data collection methods evolve so must the models used to study the systems. The way data are represented is important to how it is interacted with, visualized, and understood.

In many application domains, researchers have developed highly specialized methods to represent and build models for their specific data (e.g., signature-based malware detection in cyber security or bottom-up mass-spectrometry-based protein identification processes in proteomics). We are not advocating for ignoring or dismissing these targeted models. Rather, we offer three examples of topological methods that are broadly applicable across many domains and may allow researchers to see a new side of their data. Methods built over many years

© Springer International Publishing AG, part of Springer Nature 2018
S. Yamamoto and H. Mori (Eds.): HIMI 2018, LNCS 10904, pp. 90–109, 2018.
https://doi.org/10.1007/978-3-319-92043-6_8

within specific domains are well understood and yield predictable results. We propose these more generic topological methods to reveal hidden structure and strengthen prior hypotheses or build new ones.

2 Hypergraphs for Relational Data

Data that describe relationships are pervasive. The famous Enron email data set describes relationships between people via being on the same email [1]; protein interaction data help to understand what happens when proteins encounter each other in a cell [2]; bibliometrics tracks coauthorship relationships [3].

As one kind of relational mathematical model, graphs are popular in data science, typically used where there are sparse connections among a large collection of entities. However, graphs code pairwise associations between entities. Thus they should be seen as a special case of multi-way associations among an arbitrary number of entities, which are encoded by hypergraphs. In fact, some domain scientists have argued recently that hypergraphs are more appropriate [4–6]. As hypergraph-structured data are more general than graph-structured data, hypergraphs formally generalize graphs as mathematical objects. Therefore, we can conceive of hypergraphs literally as "multidimensional graphs." And, while hypergraph methods and applications are relatively rare, hypergraph-structured data are relatively ubiquitous, for example, whenever information presents naturally as set-valued, tabular, or bipartite data.

Hypergraphs typically have been studied from the network science (i.e., graph) perspective or in connection to each application separately. Our group is pursuing a hybrid graph and topological treatment. This allows the generalization from graphs to hypergraphs in a canonical, principled manner rather than for each application, which accounts for the high-dimensional structure contained in the multi-way relationships while remaining in agrement with network science and graph theory. We note that in network science, authors occasionally use the word *topology* to refer to graph structure or connectivity, e.g., [7]. Rather, we mean the mathematical domains of topology [8] and algebraic topology [9].

2.1 Definitions

Hypergraphs are generalizations of graphs [10]. Formally, a hypergraph $\mathcal{H}$ is a collection of subsets $\mathcal{E}$, called *edges*, of a set of elements V, called *vertices*, and we write $\mathcal{H} = (V, \mathcal{E})$. We sometimes refer to edges in an hypergraph as *hyperedges* to distinguish them from graph edges. Consequently, whereas a graph edge, (v, w), consists of precisely two vertices, a hyperedge, $f \in \mathcal{E}$, can contain any number of vertices, $f \subseteq V$. The number of edges a vertex belongs to is its *vertex degree*. Conversely, the number of vertices contained within an edge is its *edge cardinality*. A *k-uniform* hypergraph is a hypergraph where all edges have cardinality k. In particular, a 2-uniform hypergraph is just a graph.

Research on hypergraphs, particularly in the mathematics literature, mostly is limited to studying k-uniform hypergraphs. For instance, much work on the

spectral theory of hypergraphs [11,12], hypergraph coloring [13], extremal problems [14], and hypergraph transversals [15] considers the k-uniform case. However, as real hypergraph data are often non-uniform, we require tools to analyze,
interact with, and understand such data.

A fundamental notion underlying a multitude of graph analytics tools is *graph
distance*. Oft-studied metrics such as diameter, average shortest path length, and
centrality measures are all based on measuring how "far" vertices in an network
are from each other, defined as the length of the shortest walk connecting them.
Beyond distances and diameters, walks in a graph can be used to make sense
of large data through the concept of a random walk. For example, an efficient
implementation of PageRank, one of the algorithms Google uses to rank web-
pages in search results, relies heavily on random walks in graphs [16]. However,
extending the concept of a walk to general hypergraphs introduces subtleties
that do not arise in the case of graphs.

In a graph, a *walk of length k* is a sequence of vertices $v_0, v_1, \ldots, v_k$, each of
which is adjacent to the next via an edge. Because two adjacent vertices belong
to exactly one edge in a graph (and two incident edges intersect at exactly one
vertex), we can equivalently describe a walk as a sequence of incident edges or
adjacent vertices, i.e.,

$$\underbrace{v_0, v_1}_{\text{adjacent}}, \ldots, \underbrace{v_{k-1}, v_k}_{\text{adjacent}} \quad \longleftrightarrow \quad \underbrace{e_1}_{\{v_0, v_1\}}, \ldots, \underbrace{e_k}_{\{v_{k-1}, v_k\}}.$$

A third equivalent way to represent a walk is the alternating vertex-edge form:
$v_0, e_1, v_1, e_2, \ldots, e_k, v_k$. When vertices are not allowed to repeat, we call the walk
a *path*. These three equivalent definitions of a graph path also happen to coincide
with the topological definition of a path, i.e., a continuous map from the interval
$[0, 1] \subset \mathbb{R}$ to the graph thought of as a topological space.

In contrast, this three-way equivalence does not hold for hypergraphs: two
hypergraph edges can intersect at any number of vertices, and two vertices can
belong to any number of shared edges. This simple observation, along with the
goal to keep the correspondence with the topological definition of a path, motivates two walk concepts for hypergraphs that are dual but nonetheless different:
walks on the vertex level (consisting of successively adjacent vertices) and walks
on the edge level (consisting of successively intersecting edges). For ease of presentation, we will focus on edge-level walks, keeping in mind there is a dual
concept for vertex-based walks. We define an *s-walk* on a hypergraph, where s
controls the size of the edge intersection, as follows:

Definition 1. *For $s \geq 1$, an s-walk of length k between edges $f, g \in \mathcal{E}$ is a
sequence of edges, $f = e_0, e_1, \ldots, e_k = g$, where for $i = 1, \ldots, k$ and $I_i = e_{i-1} \cap e_i$,
we have $|I_i| \geq s$ and $e_{i-1} \neq e_i$.*

Observe that a 1-walk on a graph corresponds to the usual graph walk,
whereas s-walks for $s \geq 2$ are only possible for hypergraphs. We note that
several notions closely related to s-walks have previously appeared in the literature. In [17], Lu and Peng consider walks in k-uniform hypergraphs to build an

s-Laplacian matrix of a hypergraph. Wang and Lee [18] consider edge intersection properties in walks unrelated to the size of the intersections. Perhaps most pertinent to this investigation, Bermond et al. [19] introduce and analyze *s-line graphs* of hypergraphs, which are graphs derived from hypergaphs by representing each hyperedge as a vertex, and link two such vertices if their corresponding hyperedges intersect in at least s vertices. Our proposed s-walks on hypergraphs correspond precisely to graph walks on these s-line graphs. Hence s-line graphs may be used as auxiliary graphs for computing s-walk-based metrics.

A number of basic, yet important, properties of walks in graphs immediately extend to s-walks on hypergraphs. Consequently, just as the length of the shortest walk defines a bona fide distance metric for graphs, s-distance, defined as follows, serves as a distance metric in hypergraphs.

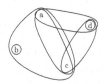

Fig. 1. An illustration of a hypergraph with edges $\{a, b, c\}$, $\{a, c\}$, $\{a, d\}$, $\{c, d\}$, $\{b\}$, and $\{d\}$. This hypergraph has 1-diameter of 3 (which is achieved by the 1-walk b, abc, ad, d), average 1-distance of 1.5, infinite 2-diameter, and 3-diameter of 0.

Definition 2. *Let $\mathcal{H} = (V, \mathcal{E})$ be a hypergraph and $\mathcal{E}_s$ denote the set of hyperedges in $\mathcal{H}$ with at least s vertices. Define the distance function $d_s : \mathcal{E}_s \times \mathcal{E}_s \to \mathbb{Z}_{\geq 0}$*

$$d_s(f, g) = \begin{cases} length\ of\ shortest\ s\text{-}walk & if\ s\text{-}walk\ between\ f,\ g\ exists \\ \infty & otherwise \end{cases}.$$

From this distance metric, a number of graph-theoretic concepts generalize naturally, For example, s-diameter (resp. average s-distance) of a hypergraph is the maximum (resp. average) s-distance between all pairs of hyperedges with at least s edges. A hypergraph is s-connected if its s-diameter is finite. We illustrate these with an example in Fig. 1.

From the perspective of data analytics, s-walk-based metrics may uncover higher-order interactions among the entities that are evident neither from classical hypergraph walks which don't control for intersection size nor the raw relational data itself. For example, a user-group hypergraph exhibiting lower average 1-distance than 4-distance suggests groups are more closely related via sparse overlappings of their members. In general, depending on the hyperdata in question, it may be the case that higher-cardinality intersections signify stronger ties between groups, or only intersections within a certain size range are meaningful due to noise or other artifacts in the data. In such cases, s-walk-based metrics provide a framework that not only recovers classical hypergraph and graph walks (by taking $s = 1$), but also allows the flexibility to filter and control for higher-dimensional interactions as well.

Hypergraph Visualization and Sensemaking. Understanding hypergraph data can be significantly more difficult than understanding graph data. Adequate hypergraph visualizations are lacking when data get large, and more intricate interactions inevitably lead to more complex data analysis and interpretation. However, making the move to use hypergraphs as representational models should allow researchers to learn more about their complex relational data. Sensemaking in graph data has been an active research area in recent years [20–22]. Before that large graph visualization became prominent [23,24]. If researchers in application domains focus more on hypergraphs over graphs, we expect that work on sensemaking and visualization for graphs can extend into hypergraphs. For example, Hoff et al. observe that although their latent space approach to social network analysis and visualization is focused on binary relations (i.e., graphs), the work can extend to more general relational data [25].

2.2   Example: Enron Email Data

Email or other mass communications are a good example of data that naturally fit into a hypergraph structure. Originally released in 2004 by the Federal Energy Regulatory Commission during its investigation, the Enron email data set is a highly cited and studied corpus of emails from the inboxes of roughly 150 Enron employees [1]. While there has been considerable research on classification and clustering based on the email text [1,26,27], our focus is on the hypergraph structure of user interactions implied by the set of emails.

To date, the Enron corpus has been studied overwhelmingly with graphs. Properties such as the distance between two users or the diameter of the entire graph can shed light on the extent to which knowledge is shared or disseminated within the company. A graph can be built from these data by letting the vertices be the set of users, or email addresses, in the data. Then, an edge is created between user1 and user2 if there is an email from user1 to user2 or vice versa (see graph type 1 in Fig. 2). This graph can help understand how information flows through the group of users and has been well studied in the literature [28–31]. Another potential graph model considers the set of recipients for each email, if there are multiple, and adds all possible edges between the users in that set (see graph type 2 in Fig. 2). Rather than flow of information, this models shared knowledge. Connected users have some common knowledge by virtue of receiving the same email.

This second representation lends itself most naturally to a hypergraph structure. Rather than connecting all pairs of users that receive the same email, we create a single hyperedge for each email. The vertices in that hyperedge are all of the recipients of that email (see hypergraph in Fig. 2). To build this hypergraph, we used a data set hosted by KONECT [32] that provides a set of sender, recipient, timestamp rows. We collected all recipients that occur for any given ⟨sender, timestamp⟩ pair and put them into a single hyperedge. Admittedly this assumes that only one email was sent from the given sender at the given time. Although that may not always be true, the incidents of email collision are probably rare.

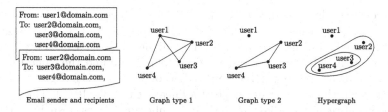

From: user1@domain.com
To: user2@domain.com,
    user3@domain.com,
    user4@domain.com

From: user2@domain.com
To: user3@domain.com,
    user4@domain.com,

Email sender and recipients   Graph type 1   Graph type 2   Hypergraph

Fig. 2. Example creation of two types of graphs and one hypergraph from an email. Notice that graph type 2 would be the same if only the first email were seen, while the hypergraph representation is able to disambiguate the two emails.

It is not our goal to do a comprehensive study of the Enron email hypergraph. Instead, we aim to show one example, s-diameter, where the hypergraph can be more informative to a user than a traditional graph. Rather than considering the Enron hypergraph as one single system, we break the data into two-week subsets. Figure 3 shows the vertex and hyperedge counts over time. The maximum number of hyperedges (emails) occurs in the hypergraph representing the two-week period October 19-November 2, 2001, which coincides with the beginning of the Securities and Exchange Commission inquiry into Enron finances. We will study the evolution of the s-diameter over time for various values of s and how that evolution correlates with the vertex and hyperedge counts.

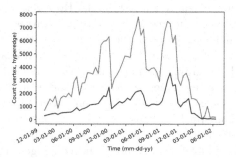

Fig. 3. Vertex (red) and hyperedge (blue) counts in the two-week Enron hypergraphs. (Color figure online)

Notably, it is not difficult to show that if the 1-diameter of a hypergraph is d, then the diameter of the associated graph is either $d-1$, d or $d+1$. Specifically, if we consider the graph where each hyperedge is replaced with graph edges for all pairs of vertices within that hyperedge, then its graph diameter would be roughly equal to the 1-diameter of the hypergraph. The question is: is there added value to calculating the s diameter for any value of $s > 1$? Do we learn any new information that the 1-diameter could not tell us? In the case of the Enron email hypergraph, the answer appearss to be: *yes*.

We consider the s-diameter of only the largest s-connected component of each hypergraph (rather than saying that s-diameter is ∞ if the hypergraph is not s-connected). In Fig. 4(a), we show the 1-diameter of the Enron hypergraph over time. Although there are two maxima roughly around the same times as two of the peaks in the vertex and hyperedge counts, there is a slight downward or flat trend in the 1-diameter (slope of a linear fit is -3×10^{-8}). The Pearson correlation between the 1-diameter and the number of vertices (resp. hyperedges) is 0.34 (resp. 0.35). If this were a graph, we would be forced to stop here and perhaps conclude that the diameter, or largest path distance between two users, of the Enron graph is only slightly correlated to the number of users or emails, i.e., connection distance does not seem to grow as the company grows. However, because of the richer structure contained in the hypergraph, we are able to look further at s-diameter for $s > 1$. When we do this, it becomes clear that the larger values of s change our perception of the distances in the system. As s grows so does the correlation between s-diameter and both vertex and hyperedge count. Figure 4(c) shows how the correlation between the s-diameter and the vertex (resp. hyperedge) counts increases as s increases, reaching a maximum at $s = 14$ (resp. $s = 12$) before decreasing slightly. In Fig. 4(b), we show the 12- and 14-diameters over time. The resemblance of these to the corresponding vertex and hyperedge counts in Fig. 3 is clearly much higher than the 1-diameter, as expected by a correlation of nearly 0.9. This indicates that as we force the intersection size (s) to be larger, we must travel farther to form a path between two users. This, by itself, is unsurprising as putting more constraints on the intersections could never require shorter paths. However, what is surprising is that paths exist at all when intersections are required to be large.

To validate that high correlation of the s-diameter with vertex and hyperedge counts for large values of s does not stem from random chance, we created a sequence of random hypergraphs using a Chung-Lu model [33]. This sequence of random hypergraphs has the same vertex degree and edge size distributions (in expectation) as the sequence of two-week Enron hypergraphs. In these random hypergraphs, the 1-diameter has slight negative correlation with the vertex and hyperedge counts (-0.34 and -0.24 respectively), but by $s = 4$, there are no s-paths. In other words, all intersections between hyperedges in the random model are of size 3, at most. This means that the existence of s-paths for large s in the Enron hypergraphs along with the high correlation with vertex and edge size is significant. There is more value in considering the Enron data as a hypergraph than as a graph.

3   Topological Analysis of Vector Data

We pivot now from complex relational data to high-dimensional numeric data. This type of data, which can be thought of as a high-dimensional point cloud, is common and often studied under the lens of dimensionality reduction tools, such as principle component analysis (PCA) or nonnegative matrix factorization (NMF). However, these techniques can introduce loss of information and may

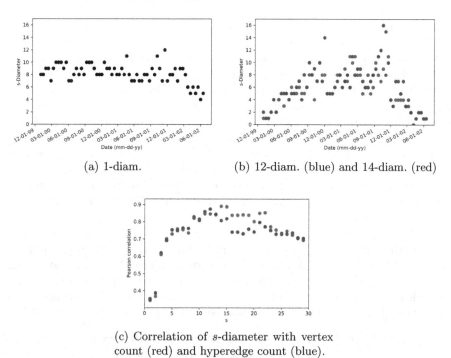

(a) 1-diam.   (b) 12-diam. (blue) and 14-diam. (red)

(c) Correlation of s-diameter with vertex
count (red) and hyperedge count (blue).

Fig. 4. (a), (b) s-diameters versus time for the two-week Enron hypergraphs. (c) Correlation of s-diameter versus s. (Color figure online)

make attribution of anomalies difficult. Instead, our group, and others, have worked with data's intrinsic high dimensionality using persistent homology, a technique within topological data analysis (TDA). Persistent homology skirts around the curse of high computational times that often come along with high-dimensional data by considering only distances between points rather than the ambient dimensionality.

3.1   Definitions

Topology is the study of geometric properties that are unaffected by bending or stretching a space (such properties are called *invariants*). TDA approaches identify changes in a system of interest by tracking topology that describes the system. Because topological properties are invariant under scaling, they ignore noise and only identify true fundamental changes to the underlying system.

Most TDA methods operate on (topological spaces derived from) numerical point clouds, so the first step must be to construct vectors $x \in \mathbb{R}^n$ that represent the system state. In some cases, data come in the form of vectors natively. For example, if the system is a biological sample, the measured data could be protein abundance values, i.e., an ordered list of measured values for each protein found in the sample. This list becomes the vector representing the biological sample state. However, there are other systems where a numerical vector is not the measured data. Consider a cyber network where the data describing the system is collected by a *packet analyzer* (also called a network analyzer or packet sniffer). This computer program or piece of hardware can intercept traffic as it passes over a digital network or part of a network. As data streams flow across the network, the sniffer captures each packet and generates a corresponding log. The information recorded in each log can vary with the packet analyzer, but it almost always contains time intercepted, source Internet Protocol (IP) address, destination IP address, source port, destination port, protocol, number of packets, and number of bytes. Thus, given a set of flow logs, some natural quantitative statistics to compute include number of source/destination IPs, number of packets, number of bytes, number of (source IP, destination IP) pairs, and number of protocols.

In this work, we concentrate on time series data, but other organizing principles, such as geographic area or biological family, can be used to group data. To vectorize time series data, there are two main approaches. The first, which can be thought of as a *feature vectorization*, is to construct a vector of statistics for each time window, from t to $t + \epsilon$. For example, for a cyber network given all network flow logs in the time window we could construct a vector $x_t \in \mathbb{R}^4$ as

$$x_t = (\# \text{ of source IPs seen}, \# \text{ of destination IPs seen},$$
$$\text{total } \# \text{ of packets sent}, \text{total } \# \text{ of bytes sent}).$$

Then, a collection of time windows yields a set of feature vectors; a point cloud. An example of this is featured in Sect. 3.2.

Another way to vectorize a time series is to use a *Takens embedding*. Given a time series $X(t)$, fix a lag parameter $\eta > 0$, and a dimension $m \in \mathbb{Z}_{>0}$. The Takens embedding of the time series then is a lift to the map

$$t \to x_t = (X(t), X(t - \eta), X(t - 2\eta), \ldots, X(t - (m-1)\eta)).$$

In other words, the time series sequence $X(0), \ldots, X(N)$ becomes a set of m-dimensional points $x_{(m-1)\eta}, \ldots, x_N$, or another point cloud. Takens proved that under the correct choice of m and η, this embedding represents the underlying dynamics of the dynamical system [34][1]. In particular, if the time series is periodic, the Takens embedding will contain a cycle (see example in Sect. 3.3).

Both vectorization methods for time series provide mathematical abstractions, or representations, for the data that are domain agnostic. Specifically, once the transformation to a point cloud has been made, the analysis proceeds only based on those values not on the original domain of the data.

The main tool in TDA for analyzing a point cloud is persistent homology, a topological invariant that distinguishes spatially robust topological features from smaller ones more likely to be noise. Broadly speaking, persistent homology takes a point cloud; constructs a sequence of topological spaces over the

---

[1] There are ways to intelligently choose m and η. For example, Khasawneh and Munch in [35] use false nearest neighbors to choose m and the first zero of the autocorrelation function to choose η.

point cloud by considering a filtration of spatial resolutions; and computes a set of 2-dimensional points, called a *persistence diagram*, where each point (b, d) delineates the birth and death of an important topological feature in the filtration sequence of topological spaces. Shorter-lived features can be thought of as noise, while longer-lived ones represent the more robust features in the space, representative of the *true* shape of the underlying space.

Given a point cloud, $V \subset \mathbb{R}^n$, the two natural topological filtrations that can be constructed are the Čech filtration and Vietoris-Rips filtration. For both, one starts by placing a ball of some small radius ϵ around each data point. The *Čech complex*, C_ϵ, is constructed by taking all subsets of points whose associated balls have a nonempty intersection. On the other hand, the *Vietoris-Rips complex*, R_ϵ, is constructed by taking all subsets of points whose associated balls have nonempty *pairwise* intersections. The two constructions are demonstrated geometrically in Fig. 5, where two-element sets are shown with a line connecting the two elements, three-element sets are filled-in triangles, and four-element sets are filled-in tetrahedra. Notice that the Vietoris-Rips complex has a filled-in triangle connecting the three points on the right, whereas the Čech complex has an open triangle. This is because the corresponding circles overlap pairwise but not all together. The triangle and tetrahedron on the left are present in both because of the respective 3- and 4-way intersections. A filtration is formed by increasing ϵ and growing the complexes accordingly.

Given a filtration, the persistence diagram can be computed for topological features of different dimensions (see Fig. 6). Features in dimension 0 are connected components; in dimension 1 they are loops (e.g.,the open triangle or longer cycles); in dimension 2, (not pictured), they are voids such as a hollow tetrahedron. Topological features in dimensions larger than two become difficult to conceptualize. For a rigorous discussion of the persistent homology methodology, the interested reader is directed to [36–38].

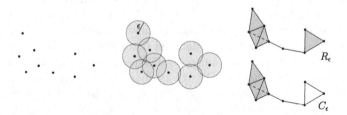

Fig. 5. A point cloud (left) can be completed to a Čech complex C_ϵ (lower right) or to a Vietoris-Rips complex R_ϵ (upper right) based on a resolution parameter ϵ (center).

To topologically compare two point clouds using persistent homology, their persistence diagrams are compared using a distance metric. The most intuitive and widely used persistence diagram distance metric is the Wasserstein distance. Roughly speaking, this metric details the minimum movement required to move points in one persistence diagram to those in another diagram where points are

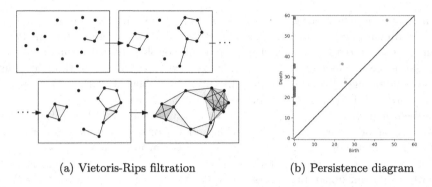

(a) Vietoris-Rips filtration   (b) Persistence diagram

Fig. 6. Example Vietoris-Rips filtration and corresponding persistence diagram for a point cloud. Blue points in the persistence diagram (and the red square) correspond to 0-dimensional features, and orange points correspond to 1-dimensional features. Note that there is always one feature in dimension 0 with infinite death time (the red square) corresponding to the one connected component that does not die once fully connected. (Color figure online)

allowed to move to the diagonal (where birth = death) if no better choice exists. The exact formulation of the Wasserstein distance is not required to understand the concept, so we refer the reader to Sect. 3 in [39].

To use persistent homology for anomaly detection, assume we are given a system baseline representing normal behavior (if this is not the case, a baseline can be learned from an initial time period within the data). All data are broken into overlapping widows, and a numerical vector is associated to each window using one of the procedures already described. Collecting the vectors from the baseline windows provides a point cloud, $\mathcal{B}$, which represents normal system behavior. Computing persistent homology on $\mathcal{B}$ provides a reference barcode, C. Then, for each new window, construct its numerical vector x using the same procedure. Compute the persistence barcode, C', for $\mathcal{B} \cup \{x\}$ and the distance between C and C'. This distance affords a measure of how much the system changed by adding this one new window. For a detailed description of this anomaly detection procedure, see [36]. One advantage to this technique is an interpretation via the persistence stability theorem [40,41]. In short, this theorem says that if the distance between the barcodes is large, then there must be a large distance in the point clouds. Because of how we created our two point clouds, this means that a large distance in the barcodes can only come from x being far away from the baseline point cloud.

Interpretability and Interaction. This method of transforming data from its raw form into a high-dimensional point cloud may seem to be such an abstraction that interpretability is lost. However, the opposite seems to be true. The topological summary through the lens of persistent homology simplifies the data by cutting through the noise (by way of the persistence stability theorems) to find the most robust features [42]. Additionally, keeping the data in its natural high-dimensional setting rather than using dimension reduction techniques allows for root cause analysis once anomaly scores have been calculated.

While not discussed herein, Mapper [43] is another tool worth mentioning because of its wide use to visualize and interpret high-dimensional data. Mapper uses local clustering behavior to reconstruct global data shape. It has been used in a variety of domains to provide new interpretations of data sets and discover previously unknown clustering behavior [44]. A "users guide" for TDA is provided in [45] that offers more detail into practical uses, interpretations, and specific implementations of both persistent homology and Mapper.

3.2 Cyber Kill Chain

Here, we provide an example of using our TDA anomaly detection algorithm to identify malicious behavior in a cyber network. It is not our goal to describe the cyber background in detail. Rather, we hope to show the method's flexibility. For more detail on cyber kill chains and attacks, see [46]. The data for our case study were simulated by a cyber security team at Pacific Northwest National Laboratory. The experimental network consisted of 16 virtual machines: 15 Linux workstations and one web server. Each workstation ran a Markov chain script to simulate browsing the external web and the corporate network site. The entire experiment ran in four phases for a total of 52 min:

Phase 1: 15 min of baseline network traffic
Phase 2: Network reconnaissance using nmap port scans
Phase 3: Normal network traffic to simulate an attacker "planning"
Phase 4: Remote buffer overflow, gain privileged shell, change root password.

We ran our persistent homology anomaly detection algorithm with two different feature vectorization strategies. For the first, we constructed a vector of length 9 with the following entries:

- # distinct source IPs (SIP)
- # distinct destination IPs (DIP)
- # distinct (SIP, DIP) pairs
- # distinct protocols
- total # bytes
- total # packets
- # distinct (SIP, Source Port) pairs
- # distinct (DIP, Dest Port) pairs
- # records

Then, each entry in the vector was normalized by the window length (60 s in our case). The second scheme was similar to the first. Slightly different statistics were targeted to specific protocols and normalized by the number of the records in each window. For each protocol $p \in \{$ICMP, TCP, UDP$\}$, let $|p|$ be the number of records in the window with protocol p. Then, the vector contains:

- $|p|/$(total number of records)
- (number of packets in protocol p)$/|p|$
- (number of bytes in protocol p)$/|p|$

- (bytes in protocol p)/(packets in protocol p)
- $|p|$/(bytes in p/packets in p)

Figure 7 shows the anomaly scores for each window. We see that both have clear spikes, but they pick up on different phases of the attack. The first vectorization (Fig. 7(a)) measures the average statistical properties for each record. Notice that this flags anomalies during the later phase of the scenario. Specifically, it picks up on the remote buffer overflow exploit when more data are being sent per record than one would expect. Likewise, the second vectorization (Fig. 7(b)) measures the average statistical properties per unit time. This flags the reconnaissance phase because a port scan is generating many more records than usual.

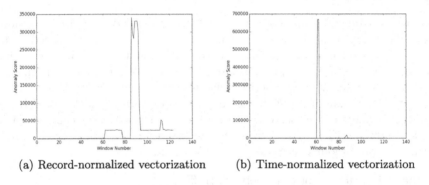

(a) Record-normalized vectorization   (b) Time-normalized vectorization

Fig. 7. Anomaly scores (y axis) over time (x axis) for our cyber kill chain data set.

3.3   Power Grid Demand Data

Operating a power grid is a careful balance between generating enough power to satisfy consumer demand without generating too much power that cannot be stored and is therefore wasted. To solve this problem of *unit commitment* [47], operators must constantly forecast the demand and increase or decrease production to match the forecasted data. Understanding the cyclic patterns in demand can be crucial to the demand forecasting problem.

The sample data used in this example was generated from a 240-bus model of the California Independent System Operator (CAISO) electric power market within the Western Interconnection grid. This model was developed in [48] as an extension to earlier 225-bus [49] and 179-bus [50] CAISO models. Using data derived from published CAISO transmission studies and Western Electricity Coordinating Council's Transmission Expansion Planning Policy Committee to seed the model, they formed resource data for system conditions modeled as the year 2004. Others have used this model and data to study transmission planning, e.g., [51]. In this work, we only use the hourly demand profiles for one of the available 21 sub-regions of CAISO to show how the Takens embedding can help discover cyclic patterns.

In Fig. 8(a), we show sample demand for a two-week interval (Sunday–Saturday) in one region of the CAISO. Notice that Sunday has the lowest demand followed by Saturday and then Friday, while Monday–Thursday have similar demand. A two-dimensional ($m = 2$) Takens embedding for just Monday–Thursday, with lag $\eta = 5$ (the first approximate zero of the autocorrelation function), is shown in Fig. 9(a) alongside its persistence diagram in Fig. 9(b)[2]. Points close to the diagonal represent short-lived noisy topological features, whereas the single point farther away from the diagonal signals that there is a one-dimensional topological feature (a loop) in the Takens embedding. The presence of this loop tells us there is periodicity to the data, while the *maximum persistence*, the maximum $(b - d)$, gives us an idea of how much this is real periodicity versus just due to noise.

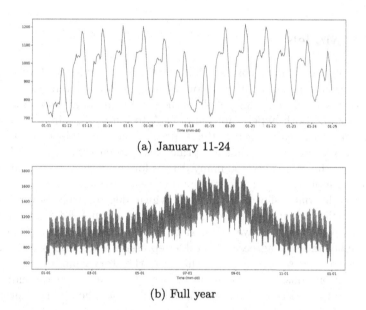

(a) January 11-24

(b) Full year

Fig. 8. Modeled demand in the "20MEXICO" region of the CAISO for two-weeks and the full year of sample generated data. The y axis is in kilowatt hours (kwH).

If we instead take a year-long view, as shown in Fig. 8(b), it becomes clear that there is still some periodicity within the weeks, but the demand rises and falls with the seasons. The Takens embedding of the full year will not yield any interesting topological features because of the seasonal variability in behavior. Still, local Takens embeddings can be used to discover and classify different types of local periodic behavior.

---

[2] The persistence diagram is a 2D representation of the barcode. A bar with birth = b and death = d is plotted at the point (b, d). Because death occurs after birth, all points will be above the $y = x$ diagonal.

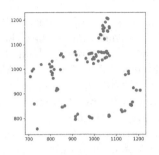
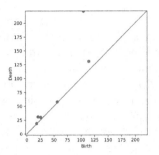

(a) Two-dimensional Takens embedding for January 12-15

(b) Persistence diagram for Takens embedding

Fig. 9. Two-dimensional Takens embedding and corresponding persistence diagram.

4   Sheaf Models

Data may also come in a combination of relational tables and vectors. For example protein expression information within a protein interaction network, or employee rank and salary for people in the Enron email database. Sheaves are a way to combine both hypergraphs and TDA to infer global information from the local data. Whereas typically this type of data is combined in an ad hoc manner, sheaves provide a canonical principled inference method [52,53]. The basics of sheaf theory require too many details for this survey. Thus, we provide only a toy example here and refer the curious reader to [52].

Consider the data shown in Table 1, which depicts what several people reported to police when questioned as witnesses to a crime. Each observation is local to the individual, but we ask a global question: How many perpetrators might there be, and who agrees? This table represents a scenario where each row corresponds to an *observation*, or, in this case, the story from a single witness. Each observation consists of one or more data *fields*, each corresponding to a column in the table. Each column specifies a data *type*, the observed property.

Table 1. Witness descriptions of crime scene details (perpetrator and incident time).

Row	Height	Weight	Time
Alice	(null)	170 lbs	(null)
Bob	5'8"	170 lbs	(null)
Charlie	(null)	250 lbs	5:00 pm
Debbie	5'8"	(null)	5:30 pm

A variety of distinct hypergraphs can be constructed from this kind of tabular data. For this example, it is most convenient to consider the hypergraph $\mathcal{T}$ in which each row corresponds to a vertex, and each edge corresponds to a set of rows with at least one nonempty column in common [54]. For Table 1, this hypergraph is shown in Fig. 10(a). Notice in particular that each vertex is itself a 1-edge. The presence of the 3-edge indicates that three witnesses (Alice, Bob, and Charlie) reported the criminal's weight. Although Debbie did not report the weight, the 2-edges indicate that, like Bob, she reported the height, and, like Charlie, she reported the time of the incident.

Formally, let $T = (t_{i,j})$ be a data table with m rows and n columns. Notationally, we will let $i = 1, \ldots, m$ index rows and $j = 1, \ldots, n$ index columns. Then, $T = (V, \mathcal{E})$ is a hypergraph whose vertices $V = \{1, \ldots, m\}$, and $e = \{i_1, \ldots, i_k\} \subset V$ is an edge if there is a j such that $t_{i,j}$ is nonempty for all $i \in e$.

Observe that this hypergraph only represents which fields (columns) of the data table are shared among observations (rows), but it does not account for the data types of the entries nor the entries themselves. This is remedied by building a *sheaf* structure on the hypergraph. As a technical matter, sheaves are not built on hypergraphs, rather on topological spaces. However, because this hypergraph has the property that all subsets of a given edge are also edges themselves, there is a natural way to associate a topological space (the Alexandrov topology).

Thus, we can represent the entirety of the data by constructing a sheaf in the following way [55]. Notice that each edge in the hypergraph implicitly specifies a set of columns – ones that are nonempty in the rows corresponding to the edge. For the 1-edges (which are also the rows), these are merely the nonempty columns in each row. Hence, to each edge of the hypergraph, we associate the set of all possible data vectors indexed over the nonempty columns corresponding to that edge. Each of these sets of data vectors are called *stalks* over the edges.

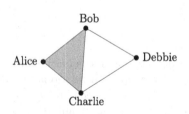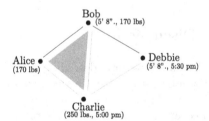

(a) Hypergraph associated to the data in Table 1.

(b) A graphical representation of the consistent data in Table 1.

Fig. 10. Underlying hypergraph and consistent sections for data in Table 1.

Given this construction, the data table entries are elements of the stalks over the 1-edges. However, what about the stalks for the other edges? If two edges are adjacent, they share at least one column in the data table. Thus, their stalks contain a common subspace, which means those corresponding entries are comparable. If those rows agree upon projecting to the common subspace, then this projection is what should be chosen as the element of the stalk. On the other hand, if the data elements are not in agreement on this common subspace, this implies that the rows themselves are in conflict. If each column is additionally associated to a metric, a numerical value can be assigned to the disagreement. The maximum amount of disagreement over the entire hypergraph is called the *consistency radius* [52] and it can provide evidence for determining the root cause of disagreements, e.g., noise, equipment malfunction, or a faulty underlying hypergraph model. Additionally the consistency radius can serve as a basis for imputing corrections to noisy data.

A collection of edges and elements from the stalks over those edges that agree upon projection is called a *section*. Larger sections, consisting of more rows, represent a more self-consistent data set, while smaller sections, those where adding any other row causes a conflict, are less so. In the specific Table 1 example, those edges in which there is agreement are shown in Fig. 10(b), while those in disagreement are grayed out. The largest section consists of the triple Alice, Bob, and Debbie because Alice and Bob agree that the criminal weighed 170 lbs., and Bob and Debbie both said that he or she was 5'8". On the other hand, Charlie (alone) also is a section, but it cannot be extended to a larger section without disagreement. In fact, his disagreements might indicate that Charlie witnessed an entirely different criminal than Alice, Bob, and Debbie.

The Local-to-Global Promise of Sheaf Theory. Since its origin dates back to the 1930s and 40s, sheaf theory is not a new mathematical discipline. However, using sheaves in data analysis is rather new. Sheaf theory is about local-to-global inference involving a collection of overlapping and interacting local data sources to infer a global state. Thus, as data sources have become more universal, the need to synthesize disparate sensors around a common theme is more important.

5   Discussion

In this paper, we have described three related models resting in the mathematics of topology for representing and exploring data: hypergraphs for relational data, point clouds for vector data, and sheaves when both are present and related. Moreover, these mathematical structures each incorporate domain-agnostic analysis tools that are able to discover insights into the data that domain-specific tools may miss through the use of higher-order structures. First, topologically inspired paths in hypergraphs see chains of highly intersecting groups that can shed light on how knowledge is shared through a network. Next, persistent homology applied to point clouds uses topological structure to detect anomalies and differentiate between noise and periodicity. Finally, sheaf theory provides a local-to-global inference mechanism. As the Internet of things grows and sensors become both smaller and more ubiquitous, a principled and general technique, like sheaf theory, will be necessary to make sense of it all. Topology, on its own, may not be able to draw declarative conclusions or confirm phenomenological hypotheses. However, these methods can provide a different kind of feature for more robust classification algorithms.

Acknowledgements. We wish to thank Prof. Francisco Munoz, University of Chile, for providing the sample power grid data studied in Sect. 3.3, and Will Hutton, Pacific Northwest National Laboratory, and his team for providing the data described in Sect. 3.2. This work was supported in part by (a) the Applied Mathematics Program of the Office of Advanced Scientific Computing Research within the Office of Science of the U.S. Department of Energy (DOE) through the Multifaceted Mathematics for Complex Energy Systems (M2ACS) project, (b) the Asymmetric Resilient Cybersecurity Initiative at Pacific Northwest National Laboratory, and (c) the High Performance Data Analytics program at the Pacific Northwest National Laboratory (PNNL). PNNL is operated by Battelle for the United States Department of Energy under Contract DE-AC05-76RL01830.

References

1. Klimt, B., Yang, Y.: The enron corpus: a new dataset for email classification research. In: Boulicaut, J.-F., Esposito, F., Giannotti, F., Pedreschi, D. (eds.) ECML 2004. LNCS (LNAI), vol. 3201, pp. 217–226. Springer, Heidelberg (2004). https://doi.org/10.1007/978-3-540-30115-8_22
2. Pržulj, N.: Protein-protein interactions: making sense of networks via graph-theoretic modeling. BioEssays **33**(2), 115–123 (2011)
3. Newman, M.E.J.: Coauthorship networks and patterns of scientific collaboration. Proc. Nat. Acad. Sci. **101**(Suppl. 1), 5200–5205 (2004)
4. Silva, J., Willett, R.: Hypergraph-based anomaly detection of high-dimensional co-occurrences. IEEE Trans. Pattern Anal. Mach. Intell. **31**, 563–569 (2009)
5. Guzzo, A., Pugliese, A., Rullo, A., Saccá, D., Piccolo, A.: Malevolent activity detection with hypergraph-based models. IEEE Trans. Knowl. Data Eng. **29**, 1115–1128 (2017)
6. Hwang, T., Tian, Z., Kuangy, R., Kocher, J.P.: Learning on weighted hypergraphs to integrate protein interactions and gene expressions for cancer outcome prediction. In: International Conference on Data Mining (2008)
7. Winterbach, W., Mieghem, P.V., Reinders, M., Wang, H., de Ridder, D.: Topology of molecular interaction networks. BMC Syst. Biol. **7**(1), 90 (2013)
8. Munkres, J.R.: Topology. Prentice Hall Incorporated, Upper Saddle River (2000)
9. Hatcher, A.: Algebraic Topology. Cambridge University Press, Cambridge (2002)
10. Berge, C.: Hypergraphs: Combinatorics of Finite Sets. North Holland, Amsterdam (1989)
11. Chung, F.: The laplacian of a hypergraph. Expanding graphs (DIMACS series), pp. 21–36 (1993)
12. Cooper, J., Dutle, A.: Spectra of uniform hypergraphs. Linear Algebra Appl. **436**(9), 3268–3292 (2012)
13. Krivelevich, M., Sudakov, B.: Approximate coloring of uniform hypergraphs. J. Algorithms **49**(1), 2–12 (2003)
14. Rödl, V., Skokan, J.: Regularity lemma for k-uniform hypergraphs. Random Struct. Algorithms **25**(1), 1–42 (2004)
15. Alon, N.: Transversal numbers of uniform hypergraphs. Graphs Comb. **6**(1), 1–4 (1990)

16. Sarma, A.D., Molla, A.R., Pandurangan, G., Upfal, E.: Fast distributed pagerank computation. Theoret. Comput. Sci. **561**, 113–121 (2015)
17. Lu, L., Peng, X.: High-ordered random walks and generalized laplacians on hypergraphs. In: Frieze, A., Horn, P., Prałat, P. (eds.) WAW 2011. LNCS, vol. 6732, pp. 14–25. Springer, Heidelberg (2011). https://doi.org/10.1007/978-3-642-21286-4_2
18. Wang, J., Lee, T.T.: Paths and cycles of hypergraphs. Sci. China, Ser. A Math. **42**(1), 1–12 (1999)
19. Bermond, J.C., Heydemann, M.C., Sotteau, D.: Line graphs of hypergraphs I. Discret. Math. **18**(3), 235–241 (1977)
20. Pienta, R., Abello, J., Kahng, M., Chau, D.H.: Scalable graph exploration and visualization: sensemaking challenges and opportunities. In: 2015 International Conference on Big Data and Smart Computing (BigComp), pp. 271–278. IEEE (2015)
21. Chau, D.H., Kittur, A., Hong, J.I., Faloutsos, C.: Apolo: interactive large graph sensemaking by combining machine learning and visualization. In: Proceedings of the 17th ACM SIGKDD International Conference on Knowledge Discovery and Data Mining, pp. 739–742. ACM (2011)
22. Chau, D.H.P.: Data mining meets HCI: making sense of large graphs. Ph.D. thesis, Carnegie Mellon University (2012)
23. Van Ham, F., Perer, A.: "Search, show context, expand on demand": supporting large graph exploration with degree-of-interest. IEEE Trans. Vis. Comput. Graph. **15**(6), 953–960 (2009)
24. Herman, I., Melançon, G., Marshall, M.S.: Graph visualization and navigation in information visualization: a survey. IEEE Trans. Vis. Comput. Graph. **6**(1), 24–43 (2000)
25. Hoff, P.D., Raftery, A.E., Handcock, M.S.: Latent space approaches to social network analysis. J. Am. Stat. Assoc. **97**(460), 1090–1098 (2002)
26. Bader, B.W., Berry, M.W., Browne, M.: Discussion tracking in enron email using PARAFAC. In: Berry, M.W., Castellanos, M. (eds.) Survey of Text Mining II. Springer, London (2008). https://doi.org/10.1007/978-1-84800-046-9_8
27. Decherchi, S., Tacconi, S., Redi, J., Leoncini, A., Sangiacomo, F., Zunino, R.: Text clustering for digital forensics analysis. In: Herrero, Á., Gastaldo, P., Zunino, R., Corchado, E. (eds.) Computational Intelligence in Security for Information Systems, pp. 29–36. Springer, Heidelberg (2009). https://doi.org/10.1007/978-3-642-04091-7_4
28. Diesner, J., Carley, K.M.: Exploration of communication networks from the enron email corpus. In: Proceedings of Workshop on Link Analysis, Counterterrorism and Security, SIAM International Conference on Data Mining 2005, pp. 3–14 (2005)
29. Leskovec, J., Lang, K.J., Dasgupta, A., Mahoney, M.W.: Community structure in large networks: natural cluster sizes and the absence of large well-defined clusters. Internet Math. **6**(1), 20–123 (2010)
30. Chapanond, A., Krishnamoorthy, M.S., Yener, B.: Graph theoretic and spectral analysis of enron email data. Comput. Math. Organ. Theory **11**(3), 265–281 (2005)
31. Priebe, C.E., Conroy, J.M., Marchette, D.J., Park, Y.: Scan statistics on enron graphs. Comput. Math. Organ. Theory **11**(3), 229–247 (2005)
32. KONECT: Enron Network Dataset, April 2017. http://konect.uni-koblenz.de/networks/enron
33. Aksoy, S.G., Kolda, T.G., Pinar, A.: Measuring and modeling bipartite graphs with community structure. J. Complex Netw. **5**, 581–603 (2017)

34. Takens, F.: Detecting strange attractors in turbulence. In: Rand, D., Young, L.S. (eds.) Dynamical Systems and Turbulence. Springer, Heidelberg (1981). https://doi.org/10.1007/BFb0091924

35. Khasawneh, F.A., Munch, E.: Chatter detection in turning using persistent homology. Mech. Syst. Sig. Process. **70–71**, 527–541 (2016)

36. Bruillard, P., Nowak, K., Purvine, E.: Anomaly detection using persistent homology. In: Cybersecurity Symposium 2016. IEEE (2016)

37. Edelsbrunner, H., Harer, J.: Persistent homology-a survey. In: Surveys on Discrete and Computational Geometry: Twenty Years Later. AMS (2007)

38. Ghrist, R.: Barcodes: the persistent topology of data. Bull. AMS **45**(1), 61–75 (2008)

39. Cohen-Steiner, D., Edelsbrunner, H., Harer, J., Mileyko, Y.: Lipschitz functions have L p-stable persistence. Found. Comput. Math. **10**(2), 127–139 (2010)

40. Chazal, F., Cohen-Steiner, D., Glisse, M., Guibas, L.J., Oudot, S.Y.: Proximity of persistence modules and their diagrams. In: Proceedings of the Twenty-fifth Annual Symposium on Computational Geometry, SCG 2009, pp. 237–246. ACM, New York (2009)

41. Chazal, F., de Silva, V., Oudot, S.: Persistence stability for geometric complexes. Geom. Dedicata **173**(1), 193–214 (2014)

42. Chazal, F.: High-Dimensional Topological Data Analysis. CRC Press, Boca Raton (2017)

43. Singh, G., Mémoli, F., Carlsson, G.E.: Topological methods for the analysis of high dimensional data sets and 3D object recognition. In: SPBG, pp. 91–100 (2007)

44. Lum, P.Y., Singh, G., Lehman, A., Ishkanov, T., Vejdemo-Johansson, M., Alagappan, M., Carlsson, J., Carlsson, G.: Extracting insights from the shape of complex data using topology. Sci. Rep. **3**, srep01236 (2013)

45. Munch, E.: A user's guide to topological data analysis. J. Learn. Anal. **4**, 47–61 (2017)

46. Korolov, M., Myers, L.: What is the cyber kill chain? Why it's not always the right approach to cyber attacks. CSO Online, November 2017

47. Padhy, N.P.: Unit commitment-a bibliographical survey. IEEE Trans. Power Syst. **19**(2), 1196–1205 (2004)

48. Price, J.E., Goodin, J.: Reduced network modeling of WECC as a market design prototype. In: Power and Energy Society General Meeting, pp. 1–6. IEEE (2011)

49. Yu, N.P., Liu, C.C., Price, J.: Evaluation of market rules using a multi-agent system method. IEEE Trans. Power Syst. **25**(1), 470–479 (2010)

50. Jung, J., Liu, C.C., Tanimoto, S., Vital, V.: Adaptation in load sheddding under vulnerable operating conditions. IEEE Trans. Power Syst. **17**(4), 1199–1205 (2002)

51. Munoz, F.D., Hobbs, B.F., Ho, J.L., Kasina, S.: An engineering-economic approach to transmission planning under market and regulatory uncertainties: WECC case study. IEEE Trans. Power Syst. **29**(1), 307–317 (2014)

52. Robinson, M.: Sheaves are the canonical datastructure for information integration. Inf. Fusion **36**, 208–224 (2017)

53. Joslyn, C.A., Hogan, E.A., Robinson, M.: Towards a topological framework for integrating semantic information sources. In: Semantic Technology for Intelligence, Defense and Security (2014)

54. Dowker, C.: Homology groups of relations. Ann. Math. **56**, 84–95 (1952)

55. Robinson, M.: Sheaf and duality methods for analyzing multi-model systems. In: Pesenson, I., Gia, Q.L., Mayeli, A., Mhaskar, H., Zhou, D.X. (eds.) Novel Methods in Harmonic Analysis. Birkhäuser (2017, in press)

Trade-Off Between Mental Map and Aesthetic Criteria in Simulated Annealing Based Graph Layout Algorithms

Armin Jörg Slopek[1(✉)], Carsten Winkelholz[1], and Margaret Varga[2,3]

[1] Department of Ergonomics and Human-Systems Engineering,
Fraunhofer Institute for Communication, Information Processing and Ergonomics,
Fraunhoferstr. 20, 53343 Wachtberg, Germany
`armin.slopek@fkie.fraunhofer.de`
[2] Seetru Ltd., Albion Dockside Works, Bristol BS1 6UT, UK
[3] Department of Zoology, University of Oxford, Oxford OX1 3SY, UK

Abstract. Dynamic graph visualization is a key component of interactive graph visualization systems. Whenever a user applies filters or a graph is modified by other reasons, a new visualization of the modified graph should support the user's Mental Map of the previous visualization to facilitate fast reorientation in the new drawing. There exist specialized graph layout algorithms which adopt the concept of Mental Map preservation to create recognizable layouts for similar graphs. In this work we used Simulated Annealing algorithms to calculate layouts which fulfill aesthetic and Mental Map requirements simultaneously. We investigated criteria of both types and conducted an experiment to examine the competition and trade-off between aesthetics and mental map preservation. Our findings show that even without explicitly optimizing Mental Map criteria, recognition can be supported by simply using the previous layout as a starting point, rather than a new layout with randomly allocated vertices. This results in better aesthetic quality as well as lower algorithm runtime. Another finding is that a simple weighted sum between aesthetic and the Mental Map may not be as effective as one might expect, especially if the weight assigned to the Mental Map is higher than the weight for aesthetics. Finally, we propose approaches for changing other aspects of the Simulated Annealing algorithm to obtain better graph layouts.

1 Introduction

The visualization of data is an important aspect of today's business and research applications. Datasets can typically be viewed as an information network – for example, when entries within a relational database reference other entries, the whole database can be represented as a graph of vertices and edges.

Graph layout algorithms provide the means to visualize such networks or graphs by assigning positions to vertices, mostly in two or three dimensional

© Springer International Publishing AG, part of Springer Nature 2018
S. Yamamoto and H. Mori (Eds.): HIMI 2018, LNCS 10904, pp. 110–125, 2018.
https://doi.org/10.1007/978-3-319-92043-6_9

space. Some algorithms may also assign coordinates to edge intersections or a specific trajectory for each edge, however such algorithms are outside the scope of this work.

Changes which may trigger the redrawing of an existing graph can have multiple causes. The dataset which is represented by the graph may have been subject to change, e.g. when new entries were added or deleted or references have been adjusted. Furthermore, the user may set (or unset) filter options in an interactive visualization system or view data in another context, which also leads to a redrawing of a graph.

It is believed that, once a human generates a Mental Map of a graph layout, it is best to maintain the Mental Map for subsequent layouts [1,4,8]. This enables humans to reorient themselves in the new updated graph visualization and perceive changes in the graph more quickly. This can be achieved by applying specialized *layout transition algorithms*, which consider a preexisting layout rather than computing a completely new layout, cf. *layout initialization algorithm*. However, too much emphasis on maintaining the Mental Map can lead to unaesthetic layouts, which may severely hamper readability and perception of the graph by humans.

We have investigated criteria for: (1) improving the readability of a graph layout (aesthetic criteria), and (2) maintaining the Mental Map. We use the Simulated Annealing (SA) algorithm [2,4], a metaheuristic approximation technique to explore the trade-off in terms of aesthetic cost, Mental Map cost and algorithm runtime between the two types of criteria.

Although work has been conducted on the effectiveness of layout transition algorithms compared with layout initialization algorithms [1,8,10], there has been no evaluation and comparison of the trade-off between aesthetic and Mental Map criteria in a cost function-based metaheuristic algorithm such as Simulated Annealing. We adapt the SA-based algorithm proposed by Lee et al. [4], whose cost function encompasses five aesthetic and six Mental Map related criteria. For any input, the cost function evaluates all criteria and returns the sum. We modified that behavior to sum up the results for both categories and then return a *weighted sum* instead. Hereby we want to gain new insights into how the results of SA-based graph layout algorithms emerge and how to possibly improve human assimilation and understanding.

This paper is structured as follows: In section two, we go over preliminary definitions and survey optimization criteria for aesthetics and Mental Map preservation. Then we present our methodology in section three and evaluate the results of our work in section four. Finally we conclude by summing up our findings and proposing new approaches for further research in this area in section five.

2   Preliminaries

In this section we describe and define the terminology used throughout this paper and give an overview of the algorithm proposed by Lee et al. [4], which is the subject of our investigation.

To visualize a Graph $G = (V, E)$, where V is a set of vertices and $E \subseteq V \times V$ is a set of edges, we first need to calculate a layout $L(G)$ for that graph. It provides coordinates for each vertex in the given space – typically two or three dimensions. This is depicted by the function $pos : V \to \mathbb{R}^n$, where $n \geq 1$ denotes the number of dimensions. The edges $e \in E$ can be assumed to be straight lines connecting their incident vertices and thus be represented as a vector in n-dimensional space. If edges are to be represented by more complex figures, such as a sequence of straight lines or splines (Bézier curves), the respective parameters also have to be provided by the layout (position of edge bends, control points for Bézier curves etc.). Sometimes it may also be desirable to provide dedicated positions for placing vertex and edge labels.

Then, in a next step, a layout $L(G)$ can be used as a skeleton to create the actual visualization. Geometric shapes such as circles or polygons are placed at the vertices' reference coordinates and adjacent vertices are connected using a visual edge representation as discussed above. While the graph G is a logical view on the data, $L(G)$ logically represents the visualization.

For each graph there is an infinite number of layouts, as even slightly moving one of the vertices results in a different layout. However, the solution space can be reduced by imposing conditions, such as a finite area (or volume for three-dimensional layouts) where vertices can be placed. Further restrictions may be implied by drawing conventions [3]:

- Straight-line drawing: Vertices are connected by straight lines.
- Poly-line drawing: A superset of straight-line drawings, where vertices are connected by a sequence of straight lines.
- Octilinear drawing: A subset of poly-line drawings, where two segments must enclose an angle α where α is an integer multiple of $45°$. Segments may only run horizontally, vertically or at a $45°$ angle in-between.
- Orthogonal drawing: A subset of octilinear drawings, where the angle α between two segments must be an integer multiple of $90°$. Segments may run only horizontally and vertically.
- Grid drawing: Vertices and intersections of edges must be placed at integer coordinates.
- Upward/Downward drawing: For directed acyclic graphs (DAGs) only, placing vertices and edges in such a way that edges only run in vertically non-decreasing (upward) or non-increasing (downward) directions.

Nevertheless, even after restricting the space for possible layouts, it may still be (infinitely) large. For a given graph, not all of these layouts are equally well suited for visualization. Thus, finding (an approximation of) the best layout according to some criteria is an *optimization problem*.

We define two different classes of algorithms which provide graph layouts. Layout initialization algorithms (Function 1) take as input a graph $G = (V, E)$ and provide a layout $L(G)$ as output. Layout transition algorithms (Function 2) on the other hand need two graphs $G_1 = (V_1, E_1)$ and $G_2 = (V_2, E_2)$ and a layout $L(G_1)$ as input to calculate $L(G_2)$. Additionally, the algorithms may require further algorithm-specific parameters, e.g. electrical charge, spring length and stiffness for spring algorithms.

$$initialize : G \rightarrow L \qquad (1)$$

$$transition : G \times G \times L \rightarrow L \qquad (2)$$

Layout initialization algorithms generate graph layouts with respect to aesthetic criteria which are intended to support readability of the resulting visualization. Aesthetic criteria include, but are not limited to: Minimizing edge intersections (crossings), minimizing the number of edge bends, minimizing the variance of edge lengths (i.e. provide a layout where all edges are of roughly the same length), maximize the minimum angle between two adjacent edges (i.e. two edges incident to the same vertex) [3,8]. Minimizing the number of edge intersections and bends are believed to be by far the most important criteria to be optimized [6,7]. It is however not possible to optimize all criteria at once. Some may even contradict each another, for example when an edge intersection could be avoided by detouring one of the intersecting edges, thus adding to its length and number of edge bends.

While it is convenient that this kind of algorithm provides layouts without any further information but the graph to be visualized – and possibly algorithm-specific parameters – this is also one of its drawbacks. Solving the graph layout optimization problem is usually done iteratively, starting at a random point with a randomly generated layout which is then adjusted in each iteration until the algorithm halts. Therefore, once a graph has been altered and a new visualization is deemed necessary, it may look completely different from the previous one. This implies two problems. As the user's Mental Map [5] is destroyed, he must spend time to reorient himself. But even when the user generated a new Mental Map, it may still be hard to identify the changes, i.e. spot newly added or removed vertices and edges.

In order to mitigate the destruction of the Mental Map and the need for reorientation, a second class of algorithms – layout transition algorithms – were proposed. Besides the graph to be visualized, they take another graph and its layout as input. The resulting layout is optimized for both aesthetic appearance – as in a layout initialization algorithm – but also for Mental Map preservation (recognition). To fulfill that task, the algorithm needs the preexisting graph and its layout. As the Mental Map criteria are optimized *in addition* to aesthetic criteria, even more competition among the criteria arises. Especially the two groups of criteria – aesthetic and Mental Map – contradict each other [8,9]. The objective of our work is therefore to investigate this competition.

Let $G_1 = (V_1, E_1)$ be the preexisting graph, $L_1 = L(G_1)$ its layout, $G_2 = (V_2, E_2)$ the new graph and $L_2 = L(G_2)$ the new layout, i.e. the transition algorithm's output. Criteria for Mental Map preservation are defined on the set of common vertices $V_{common} = V_1 \cap V_2$ and common edges $E_{common} = E_1 \cap E_2$. Mental map preserving criteria include [4]:

– Relative vertex positions: Let $v_1, v_2 \in V_{common}$. If v_1 was placed to the left/right of v_2 in the preceding layout (L_1), it should also be placed to the left/right in the new layout (L_2). The same condition should hold for placing vertices above/below each another.

- Average relative distance: Let $p_{i,old}$ and $p_{i,new}$ be the positions of vertex $v_i \in V_{common}$ in the old and new layout respectively. The average euclidean distance between these positions $d(p_{i,old}, p_{i,new})$ for all common vertices should be as low as possible, as moving them too far from their original position destroys the Mental Map.
- Nearest neighbor between: In addition to the above assumptions, let $p_{j,new}$ be the position of vertex $v_j \in V_2, j \neq i$. The distance $d(p_{i,old}, p_{i,new})$ should be smaller than any $d(p_{i,old}, p_{j,new})$. In other words, each vertex should be its own nearest neighbor and no other vertex should be positioned closer to its original position.
- Nearest neighbor within: Let $v_i, v_j \in V_{common}$ and $nn_L : V_{common} \rightarrow V_{common}$ determine a vertex' nearest neighbor in terms of euclidean distance the respective layout. Then, the following condition should hold: $\forall v \in V_{common} :$ $nn_{L_1}(v) = nn_{L_2}(v)$. The nearest neighbor relations should be retained.
- Let $dir_L : E_{common} \rightarrow \{N, NW, W, SW, S, SE, E, NE\}$ determine an edge's direction in either layout. The directions of common edges should be maintained: $\forall e \in E_{common} : dir_{L_1}(e) = dir_{L_2}(e)$.

The algorithm proposed by Lee et al. [4] uses a cost function (Function 3) to evaluate a given layout. It is derived from the cost function of Davidson and Harel's algorithm [2]. Therefore, it can be used in a SA-based algorithm for layout transition and also layout initialization, when the Mental Map related criteria are "deactivated" (do not contribute to the cost function). The function encompasses five aesthetic and six Mental Map criteria, including some of the criteria described in this section.

$$cost(L(G_1), L(G_2)) = aesthetic(L(G_2)) + mm(L(G_1), L(G_2)) \qquad (3)$$

For the purpose of our work, we modified the cost function (Function 4) to be parametrized with a mental map preservation factor (mmp-factor). For $mmp = 0.5$ this resembles the original cost function by Lee et al. [4], but effectively returns half the cost. This may influence the SA algorithm's probability to accept an inferior neighbor layout L', which depends on the *absolute* cost difference divided by the current temperature (Eq. 5). This effect can be compensated for by using $0.5 \cdot T_0$ as the initial temperature, where T_0 is the initial temperature in the original algorithm.

$$
\begin{aligned}
cost_{mmp}(L(G_1), L(G_2)) = (1 - mmp) \cdot aesthetic(L(G_2)) \\
+ mmp \cdot mm(L(G_1), L(G_2)), 0 \leq mmp \leq 1
\end{aligned} \qquad (4)
$$

$$P_T(\text{accept } L'|cost(L') > cost(L)) = exp(\frac{cost(L') - cost(L)}{T}) \qquad (5)$$

3   Methodology

The evaluation of the trade-off between Mental Map preservation and optimizing aesthetics is achieved by calculating several graph layouts, using Lee et al.'s algorithm [4], with our modified cost function (Function 4) as described in Sect. 2. Then, for each layout we separately evaluate the aesthetic and Mental Map cost as well as the time (in seconds) it took the Simulated Annealing algorithm to calculate the result.

3.1   Random Graph Generation and Modification

We use six different graph structures (Fig. 1) as a basis for the layouts. They are graphs of roughly the same size in terms of the number of vertices $|V|$, but with different levels of connectivity, i.e. different $|V|$-to-$|E|$ ratios. This ensures a diverse basis of graph structures for the layout calculations, which is important as the different $|V|$-to-$|E|$ ratios lead to varying cost function results. For example, the graph in Fig. 1f will likely have higher aesthetic cost than the graph in Fig. 1d, since it contains considerably more edges which may cause more edge intersections.

For each graph, ten series of random modifications $S_{i,j} = (G_1, \ldots, G_{10}), 1 \leq i \leq 6, 1 \leq j \leq 10$ were created, where G_1 is one of the initial graph structures and each consecutive G_k is determined by the function $modify : G \to G$ (Function 6). Except for the tree structure (Fig. 1c), all initial structures are always the same. However, the ten tree modification series $S_{3,j}$ start with different initial trees, as they can be randomly generated.

$$G_{k+1} = (V_{k+1}, E_{k+1}) = modify(G_k), 1 < k \leq 10 \qquad (6)$$

The modification scheme, which infers G_{k+1} from G_k works as follows: First, it is decided whether G_k shall be modified by inserting or deleting vertices, the former occurs with probability $p_{insert} = 0.7$ and the latter with $p_{delete} = 0.3$. The probabilities were chosen because adding vertices is the more interesting use case for layout transition, as removing elements from the graph can be visualized by fading their visual representations. We further decided to *either* delete *or* insert elements from/to the graph, since performing both operations at the same time can be emulated by applying the *modify* function consecutively.

In the case of a delete operation, up to three vertices are randomly selected and together with their incident edges removed from G_k. The probabilities to select either one, two or three vertices are uniformly distributed ($p(x = X) = \frac{1}{3}$). When vertices are added to the graph, the same probabilities apply to add either one, two or three vertices. If V_k is not empty, each of the new vertices is connected to one $v \in V_k$. Hereby G_{k+1} is guaranteed to be connected if G_k was connected and G_{k+1} is a tree if G_k was a tree. The latter aspect is important for the graph series $S_{3,j}$, which has a tree as initial graph structure and all subsequent G_ks should remain a tree. To maintain comparability with the series of the other five initial graph structures, this scheme was applied for all random modifications.

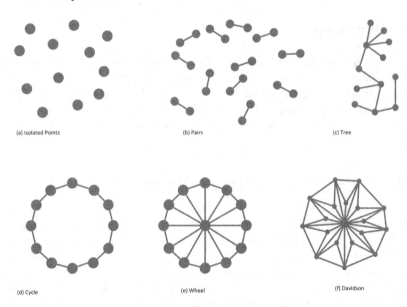

Fig. 1. The initial graph structures

Of course, the initial graph structures and the way they are modified cover only a tiny fraction of all possible graphs and graph visualizations. We aimed to emulate a small database system, where the vertices are table entries and edges foreign keys which hold a reference to other entries. Since the time for layout calculation depends heavily on the graph size ($|V|$), we chose relatively small initial graph structures and accordingly, relatively small changes for each step of the *modify* function.

3.2 Evaluation

For each of the dynamic graph series $S_{i,j} = (G_1, \ldots, G_{10})$, six series of layouts $(L(G_1), \ldots, L(G_{10}))$ were calculated. Five of them are the result of Lee et al.'s algorithm [4] using our modified cost function (Function 4) with $mmp = 0, 0.25, 0.5, 0.75$ and 1. However the first layout $L(G_1)$ in each series is an initialized layout, where from a random layout as starting point, only aesthetic cost are optimized. The sixth series serves as a control group where each $L(G_k)$ is an initialized layout. The only difference between layout initialization and layout transition with $mmp = 0$ is that the former uses a random layout as starting point and the latter assigns each $v \in V_{common}$ its previous position and random positions for all remaining (new) vertices.

Then, aesthetic cost, Mental Map cost and algorithm runtime for each layout are separately evaluated. The cost are compared by dividing the transitioned layouts' cost by the initialized layouts' cost to determine the relative quality of both layout transition and initialization (Functions 7 and 8). Thus, a quotient greater than 1.0 indicates the layout initialization algorithm obtained a better result, a quotient lower than 1.0 on the other hand implies layout transition generated a better result. We also investigate the quotient of layout transition runtime divided by layout initialization runtime.

$$Q_{aesthetic}(L_{transitioned}, L_{initialized}) = \frac{aesthetic(L_{transitioned})}{aesthetic(L_{initialized})} \tag{7}$$

$$Q_{mm}(L_{transitioned}, L_{initialized}) = \frac{mm(L_{transitioned})}{mm(L_{initialized})} \tag{8}$$

3.3   Limitations

The aesthetic part of Lee et al.'s algorithm [4], which is based on the work of Davidson and Harel [2], requires four parameters which provide relative weights between the aesthetic criteria. Finding an optimal set of parameters is not a trivial task. Parameters which are suitable to create a layout for one graph may be completely improper when visualizing another graph. We decided to find parameters which draw the Davidson graph (Fig. 1f) nicely and apply them to each of the layout calculations. Therefore, some of the resulting graph layouts do not look nicely in terms of human readability. However, this approach ensures comparability between our results. It is furthermore easy to see that not all aesthetic criteria actually lead to a nice looking graph layout. For instance, equal edge lengths and evenly spread out vertices may be desirable to draw trees (Fig. 1b, c) but are entirely negligible for graph structures as presented in Fig. 1e, f.

During a short preliminary experiment we found the parameters $\lambda_1 = 0.2, \lambda_2 = 0, \lambda_3 = 1$ and $\lambda_5 = 200$ to produce layouts for the Davidson graph similar to the one presented in Fig. 1f. The parameter λ_4 depends on λ_5 and the minimal vertex-to-edge distance. Setting $\lambda_2 = 0$ effectively turns off the Borderlines criterion, which ensures vertices do not come too close to any of the borders of the specified rectangle, in which vertices may be placed. Our preliminary experiment showed that using this criterion prevented the layout algorithm from *efficiently* using the available space, which lead to otherwise not nice-looking layouts, e.g. by inserting unnecessary edge intersections or generating an overall cluttered view of the graph. The values for λ_1 and λ_3 for node distribution and equal edge lengths respectively are relatively low, as they do not contribute to a nice drawing of the Davidson graph.

The rectangular area in which vertices are placed is a 1000×1000 raster. Initial temperature for the Simulated Annealing algorithm was 10^5 and a geometric temperature reduction schedule with $\gamma = 0.75$ and the polynomial to determine the stage size was $p(n) = 30 \cdot n$ [4], where n is the number of vertices. The algorithm halts after the same layout occurred for three consecutive stages.

All results presented in Sect. 4 must be interpreted in the context of the parameters used for the algorithm and also the initial graph structures and random modification scheme. Furthermore, as we investigate the Simulated Annealing algorithm's ability to optimize aesthetic and Mental Map related criteria by using a white-box approach, the results do not necessarily have to reflect a human's perception. This means, it is has to be verified to which extent the cost functions for aesthetics and Mental Map reflect a human's perception. Finding a good set of parameters is a complex task which greatly depends on the graph to be visualized and is outside the scope of our work.

4   Results

4.1   Hypotheses

Before we started the layout calculation and evaluation, we have come up with hypotheses regarding the experiment's outcome.

1. Except for 0% mental map preservation factor (mmp-factor), the initialized layouts are expected to have less aesthetic cost, since the layout initialization algorithm solely optimizes this type of criteria. However, when using a mmp-factor of 0%, the layout transition algorithm will also only optimize aesthetics. It is thus difficult to make a prediction for this case, but possibly the layout transition algorithm's results have even less aesthetic cost, since the algorithm starts from an already good layout rather than random positions.
2. The aesthetic quotient $Q_{aesthetic}$ will negatively correlate with the mmp-factor.
3. Transitioned layouts are expected to always have less Mental Map cost than the initialized layouts. Even with 0% mmp-factor, the randomness is eliminated and instead, vertices are assigned their previous locations as a starting point for subsequent layout calculations.
4. The Mental Map quotient Q_{mm} will positively correlate with the mmp-factor.
5. For 0% mmp-factor we expect the runtime for layout initialization and transition to be equal as in this case the algorithms are equivalent.
6. For 0% < mmp-factor < 100% the runtime of layout transition is likely higher than for layout initialization. This is because in each iteration, more criteria need to be evaluated by the cost function, thus prolonging the time to complete one iteration. Furthermore, there it may take more iterations until a trade-off between all criteria is found.

4.2   Development of Aesthetic and Mental Map Criteria

Each line in Table 1 represents the arithmetic mean values of the aesthetic quotients $Q_{aesthetic}$ (Function 7). For mmp-factor 100% all aesthetic quotients are in the order of magnitude 10^{11} to 10^{15} (Table 1). Also, at a first glance, the aesthetics of the Davidson and Wheel graphs degrade a lot when using the layout transition algorithm, even if the mmp-factor is 0%. However, when looking at the variance in Table 2 it is evident that further investigation of the parameters is needed.

Comparing the mean aesthetic quotients of Table 1 to their variance in Table 2, shows that all aesthetic quotients which are considerably larger than

Table 1. Experimental results – arithmetic mean values of $Q_{aesthetic}$

mmp-factor	0%	25%	50%	75%	100%
Davidson	1.87×10^{10}	1.85×10^9	6.01×10^{11}	1.05×10^8	1.40×10^{11}
Isolated	1.18	3.21	6.56	4.11	5.39×10^{12}
Pairs	1.18	1.12	1.20	1.51	9.75×10^{14}
Ring	1.11	1.13	1.18	1.46	1.37×10^{15}
Wheel	9.08×10^{12}	1.89×10^{11}	5.66×10^{11}	1.22	1.10×10^{15}
Tree	1.07	3.77	4.99	6.23	3.28×10^{15}

Table 2. Experimental results – variance of $Q_{aesthetic}$

mmp-factor	0%	25%	50%	75%	100%
Davidson	1.44×10^{21}	1.37×10^{19}	3.25×10^{24}	9.84×10^{16}	2.47×10^{22}
Isolated	4.71×10^{-2}	8.16	1.12×10^2	1.05×10^1	1.40×10^{25}
Pairs	1.61×10^{-2}	3.10×10^{-2}	1.56×10^{-2}	7.26×10^{-2}	6.29×10^{29}
Ring	3.79×10^{-2}	2.80×10^{-2}	2.49×10^{-2}	4.08×10^{-2}	7.30×10^{29}
Wheel	5.13×10^{26}	3.23×10^{23}	2.88×10^{24}	3.25×10^{-1}	6.48×10^{29}
Tree	3.98×10^{-2}	3.51×10^1	9.13×10^1	1.47×10^2	6.14×10^{30}

one have an even higher variance, roughly the mean value squared. For further investigation, the mean aesthetic quotient for the ten graph series $S_{6,j}$ (i.e. the Davidson graph series) are detailed in Table 3. The bottom row of Table 3 indicates the probability that a given transitioned layout has less aesthetic cost than

Table 3. Experimental results – all aesthetic quotients $Q_{aesthetic}$ of the Davidson graph

mmp-factor	0%	25%	50%	75%	100%
1	0.339	2.05	0.481	0.245	2.83
2	7.98×10^{10}	0.0182	0.118	1.05×10^9	6.83×10^9
3	1.77×10^{-12}	0.0404	0.0273	0.227	1.15×10^1
4	0.370	9.08×10^9	0.0833	0.00768	3.30
5	0.364	0.399	6.01×10^{12}	0.171	4.90×10^{10}
6	1.08×10^{11}	0.757	0.533	1.58×10^1	4.37×10^{11}
7	0.799	0.912	1.01	0.832	2.99×10^{11}
8	0.369	9.45×10^9	0.545	0.808	2.27×10^{11}
9	0.543	0.659	0.939	0.588	3.25×10^{11}
10	0.896	0.767	0.992	0.527	5.68×10^{10}
$p(Q_{aesthetic} < 1)$	0.8	0.7	0.8	0.8	0

its newly initialized counterpart. This probability is inferred from the experimental results. The finding shows that – except for 100% mmp-factor – the layout transition algorithm was able to produce layouts for the Davidson graph series with better aesthetic cost compared to the initialization algorithm.

The aesthetic quotients of the graphs Isolated Points, Pairs, Ring and Tree – which are graphs without runaway values – are within the same order of magnitude and plotted in Fig. 2a. The quotients of the Davidson and Wheel graphs are also included in this Figure, however the values differ from Table 1, since runaway values had to be excluded to be able to plot the quotients.

The Ring and Pairs graphs have equivalent aesthetic quotients. Up to 50% mmp-factor they are slightly above 1 with an increase to 1.5 if using 75% mmp-factor. The Isolated Points graphs' aesthetic quotient rises noticably as the mmp-factor increases. Yet the transitioned and initialized layout look equally well in terms of human readability. That is due to the lack of edges in this graph: Effectively, only the vertex distribution criterion dictates the aesthetic cost, as the contribution of uniform edge length, edge crossings and vertex-edge distances are close to zero. The Tree graphs' aesthetic quotient scales almost linearly with the mmp-factor.

All Mental Map quotients Q_{mm} are plotted in Fig. 2b. With 0% mmp-factor the results show large differences between the graph structures, since the Mental Map quotients are widely spread. The minimum improvement was about 5% cost reduction for the Isolated Points graph and up to 40% less Mental Map cost for the Davidson graph.

All graphs' Mental Map quotients scale with the mmp-factor, with a large drop from 75% to 100% weight. Interestingly, the Pairs and Wheel graphs have the exact same results for all five mmp-factor.

At 25% mmp-factor all but the Davidson graph Mental Map quotients drop. The most significant drop occurs for the Isolated Points graph, which had the highest quotient at 0% mmp-factor and is now on par with the Davidson graph's quotient (both 0.6). Between 25% and 75% mmp-factor the Pairs, Ring Wheel and Tree graphs' Mental Map quotient decrease almost linearly. The Davidson graph has a noticeable drop at 50% mmp-factor but remains stable until the next step.

To further evaluate the overall quality of layout transition vs. initialization, we merge the results of aesthetic and Mental Map analysis. Therefore, $Q_{aesthetic}$'s and Q_{mm}'s median values are multiplied and the product is plotted in Fig. 3a. The idea behind this approach is that, for instance, five times lower Mental Map cost even out five times higher aesthetic cost. Except for an offset, the trajectories are the same when compared to the aesthetic quotients (Fig. 2a). This implies that, whatever the actual mmp-factor may be, aesthetics dictate the overall quality of a layout.

However, layout transition performs better in most cases as the overall quality is slightly better for the Pairs, Ring and Tree graphs and considerably better for the Davidson and Wheel graphs. Only the Isolated Points graph has a better overall quality when using layout initialization.

4.3   Algorithm Runtime

The runtime quotients which compare the runtime of layout transition to layout initialization are illustrated in Fig. 3b. As a basis for the runtime analysis we chose the actual runtimes measured in seconds rather than iterations of the Simulated Annealing algorithm. This is because just analyzing the iterations would not account for runtime differences per iteration. This is however an important aspect, because some of the additional Mental Map criteria require $\mathcal{O}(|V|^2)$ time in each iteration to be evaluated. Hence, measuring the relative runtime in terms of seconds instead of iterations allows for a more realistic assessment of layout transition's feasibility for interactive graph visualization systems.

Since all quotients compare the respective runtime of layout transition to the same runtime of layout initialization, one can compare the quotients for each graph. For instance, the Davidson graph's $Q_{runtime}$ for 100% mmp-factor is 3

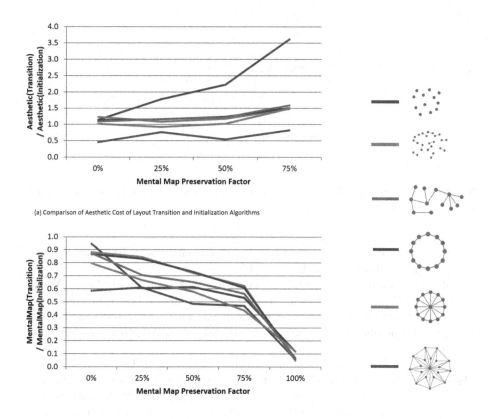

(a) Comparison of Aesthetic Cost of Layout Transition and Initialization Algorithms

(b) Comparison of Mental Map Cost of Layout Transition and Initialization Algorithms

Fig. 2. Experimental results – evolution of $Q_{aesthetic}$'s mean value (top) and Q_{mm}'s mean value (bottom)

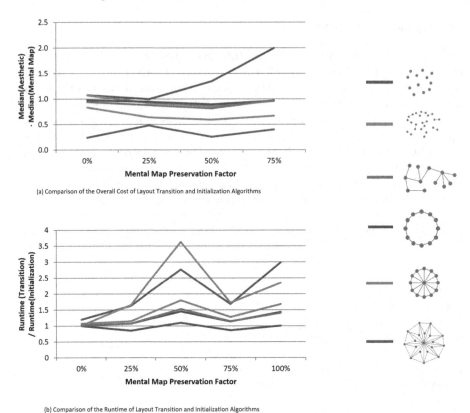

(a) Comparison of the Overall Cost of Layout Transition and Initialization Algorithms

(b) Comparison of the Runtime of Layout Transition and Initialization Algorithms

Fig. 3. Experimental results – evolution of Q_{total}'s median value (top) and the runtime comparison (bottom)

and for 0% roughly 1.2 (Fig. 3b). Thus, layout transition with 100% mmp-factor takes about 2.5 times longer than with 0%.

All quotients share a w-like trajectory with a peak at 50% mmp-factor. While the Pair and Ring graphs had similar aesthetic and Mental Map quotients, their runtime quotients differ a lot: The Pair graph has (for the most part) the highest relative runtime and the Ring graph the second lowest (together with the Tree graph).

Setting the mmp-factor to 50% – which is equivalent to Lee et al.'s original algorithm – yields the overall highest relative runtime for all graphs. The fact that with 100% mmp-factor runtimes are always higher than for 0%, shows that Mental Map criteria take more time to converge than aesthetic criteria.

4.4 Result Interpretation

The first hypothesis, which states that initialized layouts have better aesthetic cost than their transitioned counterparts for mmp-factor > 0% is true. When using 0% mmp-factor, i.e. using the initialization algorithm with a non-random starting point, aesthetic cost improved in some cases, but for the most part aesthetic cost were slightly increased. On the other hand, even though Mental Map related criteria were not explicitly optimized in this case, the cost in this category improved between 5% and 40%. This shows that when using a previous layout rather than a randomized one, existing layout initialization algorithms are also capable of implicitly maintaining the Mental Map.

However, in some instances of the Davidson and Wheel graphs, aesthetic cost of transitioned layout increased more than 10^{10} times, even when only aesthetic cost were optimized. Due to the limited scope of our experiment it is not feasible to make a general statement as to why this occurs.

When increasing the mmp-factor, the quotient $Q_{aesthetic}$ increases a bit, but remains relatively stable, hence the results confirm the second hypothesis. When optimizing only Mental Map criteria and ignoring aesthetics, i.e. setting mmp-factor to 100%, all transitioned layouts had more than 10^{10} times higher aesthetic cost than initialized layouts. When only Mental Map cost dictate the cost function result, the newly added vertices can be arbitrarily placed. This is because new vertices are only associated to aesthetic cost and thus, moving them to an arbitrary position does not change the cost function result. On the other hand, moving an old vertex inevitably increases the cost function, since any movement is directly translated to aesthetic cost. In summary, the modification of the cost function proposed in this paper (Function 4) is not suitable for high degrees of Mental Map preservation (mmp-factor $> 0\%$). An alternative approach would be to apply the mmp-factor only to old vertices, but not to new ones.

Hypotheses three and four could be confirmed, as all transitioned layouts had lower Mental Map cost compared to initialized layouts and with rising mmp-factor, the relative cost were further reduced. When analyzing the algorithm runtime, hypothesis five can be partially confirmed. With 0% mmp-factor, runtimes of layout transition and initialization were mostly equivalent, thus confirming hypothesis five. However, it took about 1.2 times longer when transitioning layouts based on the Davidson graph.

When the mmp-factor is set between 0% and 100%, layout transition takes longer than initialization, the only exception being graphs based on the Isolated Points structure. At 100% mmp-factor relative runtimes vary between one (initialization and transition take equal time) for the Isolated Points graph structure and three for the Davidson graph structure. From this it could be inferred that Mental Map criteria take longer to converge than aesthetic criteria. However, we believe that the prolonged runtime is another result of entirely neglecting aesthetic cost of new vertices. In the first stages, the underlying Simulated Annealing algorithm has high acceptance rates of non-improving layouts. In this case, such non-improving layouts are layouts, where old vertices have been displaced from their original location. This displacement then provides the potential for reducing cost in later iterations, as the old vertices are moved back to their original locations. Furthermore, the displacement per iteration is higher at the beginning, since vertices move less in later iterations. These unnecessary steps –

displacing old vertices and then moving them back to their original locations – may be considered a crucial factor for the prolonged runtime.

5   Conclusions and Future Work

In this work we have examined how aesthetic and Mental Map criteria compete and influence one another. To achieve this goal, formal definitions to distinguish between layout initialization and transition algorithms were introduced and optimization criteria for aesthetic and Mental Map preservation surveyed.

The main contribution is an experiment based on the layout transition algorithm proposed by Lee et al. [4]. A modification of the original cost function allows to weight the aesthetic and Mental Map criteria within the cost function, making it more flexible and allowing for a greater insight in how the Simulated Annealing algorithm's results emerge. The experiment compares the aesthetic and Mental Map cost of transitioned layouts to the according cost of initialized layouts.

Three major aspects can be derived from the experiment. Firstly, an explicit optimization of Mental Map criteria is not necessary in order to improve Mental Map cost. When calculating a layout for a modified graph, rather than using a random layout in the first iteration of the Simulated Annealing procedure, the positions of old vertices can be used. By using the previous layout as a starting point for all subsequent calculations, the Mental Map is implicitly maintained and the associated cost improve by 5% to 40%, depending on the concrete graph structure. Thus, existing layout initialization algorithms may suffice when they do not start with a random layout.

Furthermore, weighing aesthetic and Mental Map criteria equally – as in the original algorithm by Lee et al. [4] – has the highest runtime. When focusing on either aesthetic or Mental Map, the relative runtime between transition and initialization reduces compared to mmp-factor 50%. This is because the competition between the two types of criteria is reduced, as one type of criteria dominates the cost function.

Lastly, the extreme increase in aesthetic cost when only optimizing Mental Map cost but ignoring aesthetics shows a disadvantage of our modification to the cost function. The more focus is on Mental Map preservation, i.e. if the mental map preservation factor (mmp-factor) is greater than 50%, the placement of new vertices contributes less to the cost function result, since new vertices only have aesthetic but no Mental Map related cost. Thus, if a weighted sum-based cost function is desired, the weight should only apply to costs caused by old vertices.

The focus of our work lies on the Simulated Annealing algorithm's ability to handle the trade-off and competition between aesthetic and Mental Map criteria. Therefore, further research should study how the cost function results correlate with human perception of aesthetics and Mental Map preservation and if so, which weight(s) is/are optimal.

Further research can be conducted with alternative approaches for weighing the Mental Map. For instance, transitioned layouts may look nicer when applying the weight factor between aesthetics and Mental Map only to old vertices and take the new vertices' aesthetic cost fully into account to achieve optimal placing of new vertices while preserving the Mental Map criteria for old ones.

The effects of changes to other aspects of the Simulated Annealing algorithm are also subject to further investigation. It may be possible to obtain better results when adjusting the probability to select a certain vertex to be moved, e.g. by first deciding whether to move an old or new vertex and then selecting a random vertex from the given group. Then, when a graph with 99 vertices receives one new vertex, its probability to be moved in each iteration is 50% rather than 1%, possibly allowing the algorithm to place the new vertex in an optimal position with respect to the old vertices, while not moving the old vertices too much around and thus, destroying the Mental Map.

References

1. Bridgeman, S., Tamassia, R.: A user study in similarity measures for graph drawing. In: Marks, J. (ed.) GD 2000. LNCS, vol. 1984, pp. 19–30. Springer, Heidelberg (2001). https://doi.org/10.1007/3-540-44541-2_3
2. Davidson, R., Harel, D.: Drawing graphs nicely using simulated annealing. ACM Trans. Graph. **15**(4), 301–331 (1996)
3. Di Battista, G., Eades, P., Tamassia, R., Tollis, I.G.: Graph Drawing: Algorithms for the Visualization of Graphs. Prentice Hall, Englewood Cliffs (1999)
4. Lee, Y.-Y., Lin, C.-C., Yen, H.-C.: Mental map preserving graph drawing using simulated annealing. In: Proceedings of the 2006 Asia-Pacific Symposium on Information Visualisation, APVis 2006, Darlinghurst, Australia, vol. 60, pp. 179–188. Australian Computer Society Inc. (2006)
5. Misue, K., Eades, P., Lai, W., Sugiyama, K.: Layout adjustment and the mental map. J. Vis. Lang. Comput. **6**, 183–210 (1995)
6. Purchase, H.C., Cohen, R.F., James, M.I.: An experimental study of the basis for graph drawing algorithms. J. Exp. Algorithmics **2** (1997)
7. Purchase, H.: Which aesthetic has the greatest effect on human understanding? In: DiBattista, G. (ed.) GD 1997. LNCS, vol. 1353, pp. 248–261. Springer, Heidelberg (1997). https://doi.org/10.1007/3-540-63938-1_67
8. Purchase, H.C., Hoggan, E., Görg, C.: How important is the "mental map"? – an empirical investigation of a dynamic graph layout algorithm. In: Kaufmann, M., Wagner, D. (eds.) GD 2006. LNCS, vol. 4372, pp. 184–195. Springer, Heidelberg (2007). https://doi.org/10.1007/978-3-540-70904-6_19
9. Purchase, H.C., Samra, A.: Extremes are better: investigating mental map preservation in dynamic graphs. In: Stapleton, G., Howse, J., Lee, J. (eds.) Diagrams 2008. LNCS (LNAI), vol. 5223, pp. 60–73. Springer, Heidelberg (2008). https://doi.org/10.1007/978-3-540-87730-1_9
10. Saffrey, P., Purchase, H.C.: The "mental map" versus "static aesthetic" compromise in dynamic graphs: a user study. In: Proceedings of the Ninth Conference on Australasian User Interface, AUIC 2008, Darlinghurst, Australia, vol. 76, pp. 85–93. Australian Computer Society Inc. (2008)

Analysis of Factor of Scoring of Japanese Professional Football League

Taiju Suda[(✉)] and Yumi Asahi[(✉)]

School of Information and Telecommunication Engineering,
Department of Management System Engineering, Tokai University,
Tokyo, Japan
5bjml221@mail.u-tokai.ac.jp, asahi@tsc.u-tokai.ac.jp

Abstract. In "The Japan Professional Football League (J League)", the number of customers is increasing every year since 2011. The Japanese football team participates in the Russian World Cup in 2018. Therefore, J league market is expected to become more active. In this research, we analyze the score trend of the league with the aim of proposing tactics and training for the J League team. In each piece of data, position information obtained by dividing a field into an X-axis and a Y-axis are given. Therefore, in the data to use, this research pay attention to the data on the play involved in the score and the position where the play started. In this study, first, cluster analysis is performed to classify the start position of play involved in scores. After that, factor analysis and covariance structure analysis are carried out, and the play highly relevant to the score is discovered. Before analysis, data cleaning is carried out so that similar variables did not exhibit a strong correlation. First, the start position of the play involved in scores is classified by cluster analysis. From the score data, play related to the score was extracted and classified. After this, good results have been obtained with clusters showing mainly attacks from their own field. Therefore, in Japanese professional football, it can be predicted that there is some tendency in score from own field. Next, factor analysis/covariance structure analysis is performed on each cluster, and tactics related to scores are discovered. Factor analysis was conducted and latent variables related to the score were extracted. Define that latent variable as a score-related tactic and analyze the relationship between different tactics using covariance structure analysis. Those with low relevance are considered independent tactics. From the analysis results, in J League found that "side attack" and "pass to empty space" are strongly related to the score. Also, on the left side of the field, the score tendency using "dribbling" was weak. Japanese players, this can be expected to be related to having few players using left foot more than the right. Therefore, it is possible to propose "training of side attacker with excellent physical strength and speed" and "counterattack/strengthen side attack". Furthermore, we found out that it is a task to put emphasis on cultivating left-handed players. This analysis focused on the attack from the own field. The future task is to analyze the attack pattern from the enemy field and judge whether it is haste from the length of attack time.

Keywords: Sports marketing · Data visualization · Cluster analysis
Factor analysis · Covariance structure analysis

© Springer International Publishing AG, part of Springer Nature 2018
S. Yamamoto and H. Mori (Eds.): HIMI 2018, LNCS 10904, pp. 126–135, 2018.
https://doi.org/10.1007/978-3-319-92043-6_10

1 Introduction

In this study, we analyze "The Japan Professional Soccer League (J League)". There are 54 clubs in the J League. J League has separated 3 leagues. The top league of them is called J1-League. After Round robin tournament which plays games at home & away is carry out, a club having most points in the league win. Currently, 18 clubs belong to the J1-League.

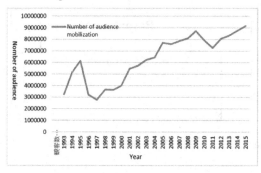

Fig. 1. Trends in number of attendants (upper-left)

Table 1. Start point percentage of scoring (upper-right) play

	U raw a Red Diamonds (%)	Yokohama F-M arinos (%)
From the spiled ball	11.8	7.5
From the dribbling	5.9	7.5
From the long pass	3.9	1.9
From the short pass	29.4	5.7
From the through-pass	5.9	13.2
From the cross	15.7	17
From the set-piece	19.6	32.1
From the penalty kick	2.0	3.8
Others	5.9	11.3
Total	100	100

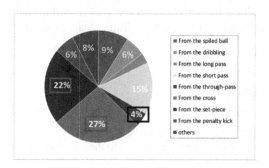

Fig. 2. Scoring trend of Japan national team (left)

In the J League, since 2011, the number of audience mobilization is increasing year by year (Fig. 1). Also, from past data, we can see that the number of audience mobilization increased in the following year of carried out World Cup. From participating in the Russian World Cup in 2018 of Japanese national team, it can predict that the J League market will be further activated.

Before the analysis, we compared the percentage of the play just before all score of "Urawa Red Diamonds" and "Yokohama F-Marinos" (Table 1). There were several contrasting results between both teams. From this result, it can be predicted that each team belonging to J League has a different score tendency. On the other hand, it can be seen that the Japanese national team has a strong tendency of a score from "through-pass (22%)" and "cross (24%)" (Fig. 2).

Based on these current situations, this study's purpose is proposing how to cultivate player and develop team tactics, by analyzing tendency of the play related to scoring in J League and finding the characteristic play of Japanese soccer.

2   Data Definition and Cleaning

Data handled in this analysis is category data which recorded all play history for the last 45 games of the league of 2016 season. All of them have been given the information of location obtained by dividing the field into X-axis & Y-axis (Fig. 3).

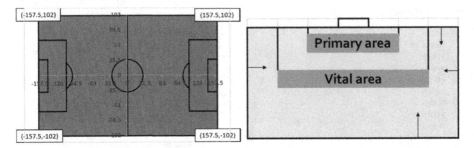

Fig. 3. The coordinates of the field   **Fig. 4.** Area name near the goal

Before analyzing the data, "Vital area" and "Primary area" are set in the penalty area. "Vital area" means the area in front of the penalty area, the play in this area is easy to score. "Primary area" indicates a position further closer to the goal than the vital area. In other words, they are areas where make easily chances. Table 2 shows mainly used variables.

Table 2. Variables mainly used

Ball_X	The ball X coordinate of the corresponding play
Ball_Y	The ball Y coordinate of the corresponding play
Attack start history No	Attack history No (Not change from the start to the end of attack)
Goal	Scored goal without own goal
Assist	Set to 1 if the play before the shot with the goal was intentional
Dribbling	Dribbling (assumed that the first touch of dribbling is not a trap)
Pass	Play which is passed a ball to a team mate. (Including through-pass, back-pass and center)
Center	Including set-piece like a corner kick
Through-pass	Through-pass
Back-pass	Back-pass
Enter in penalty area	The following play is with in the penalty area and the own team's play. (Not applicable when playing in the penalty area already)
Enter in aside of the penalty area	The same idea as "Enter in penalty area"
Enter the area 30 m from goal	The following play is with in 30 m from the enemy's goal line
Enter in "Vital area"	Refer to Fig. 4
Enter in "Primary area"	Refer to Fig. 4

The penalty area, Vital area, and Primary area are including in the area of 30 m from goal.

If "Enter in Vital area" have been "1", "Enter the area 30 m from goal" was same ("1").

If they had same "Attack start history No", and any of them (ex. "Enter in ~ ") had been inputted "1", changed the value which is farther from enemy goal, "1" to "0".

Through-pass, Back-pass, and Center are kinds of a pass. If "Through-pass" have been "1", "Pass" was same ("1"). If they had same "Attack start history No", and any of them (ex. " ~ -pass") had been inputted "1", changed the valuable, "1" to "0".

3   Cluster Analysis

The start position of the play which involved in a score is segmented by cluster analysis. At first, the first play data reached to score were extracted from all the play data. After that, cluster analysis carried out, using the information location variables (gave coordinate information by X & Y axis) of extracted data. Ward's method was used in clustering and square Euclidean distance was used in the measurement of distance (Figs. 5 and 6).

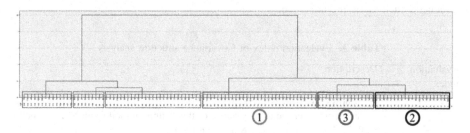

Fig. 5. Dendrogram

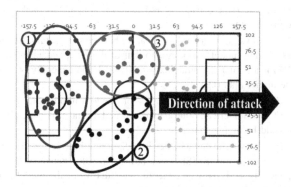

Fig. 6. Plot of "Starting point of scoring" play"

As the result of the cluster analysis, the point of first play related score was divided into 6 clusters. In this study, we used 3 clusters which had shown good results. They were mainly from own-field.

From this result, we analyzed the tendency of play related to the score from own-field in the J League.

4   Factor Analysis and Covariance Structure Analysis

Find tendencies of strategy for scoring using Factor analysis, after that the covariance structure and analyzed the relationship. We restored play-data related to scoring from the variable, "Attack_start_history_No" of all starting point data which were contained three clusters. Factor analysis was performed used each play data, and the extracted latent variables defined as "tactics involved in scoring". In factor analysis, Maximum likelihood method was used. Factors were extracted using Gutman-Kaiser criteria. The Gutman-Kaiser criterion is criteria that adopt only factors whose eigenvalues are 1 or higher (Table 3).

After that, we carried out the covariance structure and analyzed the relationship between "tactics". Those with low associations between "tactics (latent variables)" are considered as independent tactics, and the play data (explanatory variables) have influenced each tactic are analyzed.

Table 3. Evaluation index of Covariance structure analysis

Evaluation index	Description
GFI	GFI shows the fit of the model It can see whether the total variance of the saturation model can be explained well by the variance of the estimation model It is a numerical value between 0 and 1, and it is generally said that GFI should be higher than 0.9
AGFI	AGFI is GFI which takes a degree of freedom into account. AGFI is also a numerical value between 0 and 1, and it is generally said that AGFI should be higher than 0.9
NFI	NFI shows how much the deviation between the saturation model and the estimation model is superior to that of the independent model NFI is a numerical value between 0 and 1 and generally said that NFI should be higher than 0.95
CFI	CFI is NFI which takes a degree of freedom into account CFI is a numerical value between 0 and 1, and it is generally said that CFI should be higher than 0.95
RMSEA	RMSEA is Indicates whether it is divergent between model distribution and true distribution. RMSEA is a numerical value between 0 and 1, and it is generally said that it is generally said that fit of the model is good when RMSEA is lower than 0.05

4.1   Cluster1 (from Near Own Goal)

By using Gutman-Kaiser criteria, up to the third factor was adopted as latent factors. From the constituent elements of each latent factor, they were defined as tactics. The tactics affecting the play started from Cluster 1 to score were found to be "side attack", "Breakthrough a wing by dribbling" and "Final ball to the vital area". Covariance structure analysis was carried out with reference to this result (Figs. 7 and 8).

Table 4. Eigenvalue of cluster1 (upper-left)

Factor	Total	Accumulation (%)
1	**1.952**	**27.884**
2	**1.393**	**47.781**
3	**1.291**	**66.222**
4	0.845	87.288
5	0.654	87.634
6	0.458	94.170
7	0.408	100.000

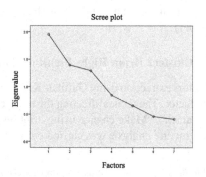

Scree plot

Fig. 7. Scree plot of cluster1 (upper-right)

Table 5. Factor structure of cluster1 (left)

Factor matrix	Factor 1	Factor 2	Factor 3
Center	**0.762**	−0.025	−0.003
Enter in "Primary area"	**0.739**	0.021	−0.063
Assist	**0.564**	−0.083	**0.418**
Dribbling	−0.005	**0.993**	**0.113**
Enter in "Penalty area"	−0.016	**0.352**	−0.063
Enter in "Vital area"	−0.064	0.096	**0.702**
Through-pass	0.042	−0.062	**0.327**

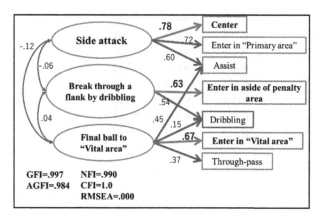

Fig. 8. Covariance structure model of cluster1

Elliptical objects are latent variables which had been defined as "tactics". Rectangular objects are endogenous variables that explain a latent variable. They are the ones had been recorded during data investigation. Numerical values on the arrows from latent variables to endogenous variables indicate the degree of explanation for each endogenous variable to latent variables. In this covariance structure analysis, "AGFI", "CFI" and "RMSEA" showed ideal values (Tables 4 and 5).

From the results, it was found that in the play from this cluster, "Side attack using dribbling and pass" has a strong relevance to the scoring. It could be judged that "center from the flank" and "breakthrough the wing using dribbling" being performed frequently.

4.2   Cluster2 (from Right Flank of Own-Field Near Halfway Line)

By factors extraction using Gutman-Kaiser criteria, up to the third factor was adopted as a latent factor. The tactics affecting the score by the attack from Cluster 2 were found to be "side attack", "Make chance using dribbling" and "Final ball to the vital area". Covariance structure analysis was carried out with reference to this result (Figs. 9 and 10).

Table 6. Eigenvalue of cluster2 (upper-left)

Factor	Total	Accumulation (%)
1	**2.002**	**28.593**
2	**1.369**	**48.149**
3	**1.120**	**64.154**
4	0.998	78.416
5	0.658	87.812
6	0.448	94.212
7	0.405	100.000

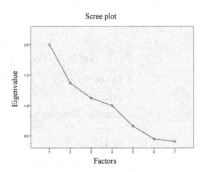

Scree plot

Fig. 9. Scree plot of cluster2 (upper-right)

Table 7. Factor structure of cluster2 (left)

Factor matrix	Factor 1	Factor 2	Factor 3
Center	**0.798**	0.033	0.001
Enter in "Primary area"	**0.729**	0.011	0.074
Assist	**0.494**	−0.092	**0.728**
Pass	**−0.105**	**−0.131**	0.074
Dribbling	−0.066	**0.989**	**0.126**
Enter in "Vital area"	−0.091	**0.168**	**0.389**
Through-pass	0.017	−0.078	**0.440**

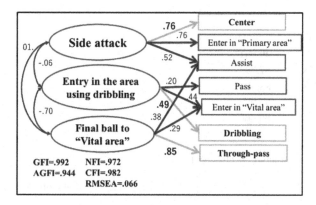

Fig. 10. Covariance structure model of cluster2

In this covariance structure analysis, "AGFI", "CFI" showed ideal values, but "RMSEA" showed the fit of the model was a little bad.

According to the result of this cluster, same as cluster1, it was found that "Side attack using dribbling and pass" have a strong relevance to the scoring.

Furthermore, it could see that "Dribbling" or "Through-pass" were frequently used in scoring play from this cluster. Therefore, it seems that there is a tendency of scoring play from the right wing near half-way-line, "attacks targeting spaces between or behind enemies" (Tables 6 and 7).

4.3 Cluster1 (from Left Flank of Own-Field Near Halfway Line)

By factors extraction using Gutman-Kaiser criteria, up to the second factor was adopted as a latent factor. The tactics affecting the score by the attack from Cluster 3 were found to be "side attack" and "Final ball to the vital area". Covariance structure analysis was carried out with reference to this result.

In this covariance structure analysis, "CFI" showed ideal values, but "AGFI" was a little acceptable in a fit of the model. In addition, "RMSEA" showed the fit of the model was bad. From the observed evaluation index, it can be said that the reliability of this model is not high (Figs. 11 and 12).

According to the result of this cluster, same as other clusters, it was found that "Side attack using pass" has a strong relevance to the score.

On the other hand, a relationship of scoring with "Side attack using dribbling" was not found. The fit of the model is not good in this cluster, but at least it can seem that "attack from left-wing near half-way-line hardly used dribbling" (Tables 8 and 9).

Table 8. Eigenvalue of cluster3 (upper-left)

Factor	Total	Accumulation (%)
1	**2.045**	**40.901**
2	**1.345**	**67.793**
3	0.780	83.391
4	0.445	92.284
5	0.386	100.000

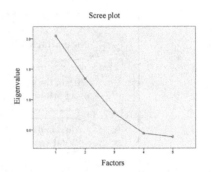

Scree plot

Table 9. Factor structure of cluster3 (left)

Factor matrix	Factor 1	Factor 2
Enter in "Primary area"	**0.736**	0.095
Center	**0.684**	−0.026
Assist	**0.551**	**0.713**
Enter in "Vital area"	**−0.109**	**0.52**
Through-pass	0.091	**0.51**

Fig. 11. Scree plot of cluster3 (upper-right)

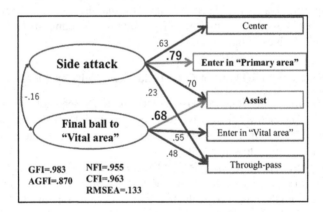

Fig. 12. Covariance structure model of cluster3

5 Discussion/Summary

In recent years of J League, it was found that attacks from own-field which aimed at vacant spaces (ex. a space between enemy defenses) much easy to score. From this result, we can propose "develop flank-man superior in physical strength/speed" for future player development. Furthermore, it thought that "tactics for using space or counter-attack" can be proposed in the strategy for getting points from own-field.

In addition, despite the results that "pass" and "dribble" are affecting scores in Cluster 1, it did not show any influence of tactics using "dribble" in Cluster 3. Furthermore, it is considered that the reason why the RMSEA value of Cluster 3 got larger was "the classification sample or play-type (variable) was less than other clusters". From these interpretations, it supposed that J-League has a problem of "Japanese players have few left-footed players".

Future analysis tasks are "Analysis of the tendency of score play starting from the enemy-field" and "Analysis of time to score until play start position to goal". Further-more, tally "the number of left-footed players in J League" which had sum-marized in a summary, and want to explore the influence on scoring.

References

1. Bollen, K.A.: Structural Equations with Latent Variables. Wiley, New York (1989)
2. Toyoda, H.: Covariance structure analysis [AMOS] in Japan, pp. 2–49. Tokyo publishing (2007)
3. Structural equation modeling Review of model construction Katsuyoshi Konnno (Shizuoka University of Science and Technology). http://www.mizumot.com/method/2012-05_Konno. pdf. Accessed 19 Dec 2017
4. Fundamental and practical analysis of covariance structure analysis - Fundamentals - Hiroshi Kano (Graduate School of Human Sciences, Osaka University) https://ssjda.iss.u-tokyo.ac.jp/ seminar2002_1.pdf. Accessed 19 Dec 2017
5. Hatta, T., Ito, Y., Kawakami, A.: On Japanese dominant handedness and dominant hands. (Graduate School of Environmental Studies-Nagoya University, Naruto University of Education). https://www.jstage.jst.go.jp/article/pamjaep/44/0/44_82/_pdf/-char/ja. Accessed 19 Dec 2017
6. J League customer mobilization data. http://footballgeist.com/audience. Accessed 19 Dec 2017
7. Football LAB League summary (2016). http://www.football-lab.jp/summary/team_ranking/ j1/?year=2016&data=goal. Accessed 2 Dec 2017

Analysis of Trends of Purchasers
of Motorcycles in Latin America

Rintaro Tanabe[1][(✉)] and Yumi Asahi[2]

[1] School of Information and Telecommunication Engineering, Course of
Information Telecommunication Engineering, Tokai University, Tokyo, Japan
5bjm1214@mail.u-tokai.ac.jp
[2] School of Information and Telecommunication Engineering, Department of
Management System Engineering, Tokai University, Tokyo, Japan
asahi@tsc.u-tokai.ac.jp

Abstract. The Latin American economy experienced the currency crisis and
the associated confusion from the early 1990s through the early 2000s. Since
2003, rapid economic growth has been achieved. As a result, in Latin America
"A" country, the impact of external demand led to the expansion of the consumer finance market. Furthermore, financial services expanded due to income
disparity correction policy implemented from 2003 to 2010. By these, purchasers due to loans increased of motorcycle and automobile, but on the other
hand rate of loans outstanding increased. In this research, we look for factors of
loans outstanding from customer data. The data used in this study is customer
data of anonymized motorcycles in Latin America "A" country from September
2010 to June 2012. From the usage data it turns out that the proportion of loans
standing is high. Therefore, it is necessary to extract variables that are factors of
loans outstanding. From there, it is necessary to grasp the characteristics of loans
outstanding. The analysis flow is data cleaning, basic aggregation, grouping of
data, variable extraction, binomial logistic regression analysis. Data is organized
by data cleaning. Data was grouped by income amount by grouping of data
Basic aggregation allows to determine the characteristics of the data. Next, we
extract the variables that cause the factor of loans outstanding by AUC. Finally,
binomial logistic regression analysis finds out how the variables extracted by
AUC affect loans outstanding. In addition, analysis results and Beforehand
studies have considered that specific variables greatly affect loans outstanding.
Therefore, this studies deeply dig up that variable. Based on the results of the
analysis, we explore the tendency of loans outstanding.

Keywords: Motorcycle · Customer data · Loans outstanding within 18 months

1 Introduction

The Latin American economy experienced the currency crisis and its associated confusion from the early 1990s through the early 2000s, so the GDP growth rate in 1990–
2002 was sluggish [1]. However, vigorous capital investment is expanded because of
the global economic expansion and the rise in primary commodity prices, expansion of
exports and inflow of investment funds since the 2000s. In addition, it has achieved

© Springer International Publishing AG, part of Springer Nature 2018
S. Yamamoto and H. Mori (Eds.): HIMI 2018, LNCS 10904, pp. 136–144, 2018.
https://doi.org/10.1007/978-3-319-92043-6_11

rapid economic growth due to the expansion of personal consumption since 2003. Therefore, In the Latin American "A" country, the expansion of exports and the influence of external demand due to the rise in primary commodity prices led to the expansion of the consumer finance market. In addition, the financial service expanded to the people who cannot take out a loan due to raising the poor to the middle class by income disparity correction policy implemented from 2003 to 2010 [2].

From the above, although the purchases increased by loans of motorcycles and cars [3], there were many customers who did not understand the contract contents of loans [4], and the rate of bad debts due to excessive debt consumption has raised [2].

In this research, we look for factors of bad debt from customer data.

2   The Data Overview

The data used in this study is anonymized customer data of motorcycles in Latin America "A" country from September 2010 to June 2012. The data is composed of the score of the credit agency A, the score of the credit agency B, the history of the tax, payment, the working year, the married/unmarried, the sex, the working state, the resident state, the main income, the side income, the dealer assessment, the division number, the down payment, the loan, interest amount, occupation, academic background, house type, region, product type, displacement, size, 6 months Bad, 12 months Bad, 18 months Bad, and so on.

(6 months Bad, 12-month Bad, and 18-month Bad mean that if it doesn't reach the price that the customer has to pay until the limits after purchasing, they will be checked. Therefore, a person who is eligible for Bad for 6 months Bad will be checked for 12-month Bad, and18-month Bad. It has never returned).

The number of customers was 14,304 in Latin America "A" country.

3   The Research Purpose

From the definition in the usage data summary, 18 months Bad customers are the most number in 6, 12, and 18 months Bad. From Fig. 1, it can be assumed the serious situation because about 20% of customers of all data are with Bad for 18 months. Due to the usage data contains many data items, it is necessary to extract variables that become a core of Bad for 18 months. In this study, we extract the variables which influence precisely 18 months bad AUC and analyze the influence of data extracted by logistic regression analysis. Based on the results, we grasp the characteristics of 18 months Bad customers. AUC is defined as the region below the ROC curve and it is an index for measuring the accuracy of the model. Therefore, AUC can be judged that the larger the numerical value of the region, the higher the accuracy.

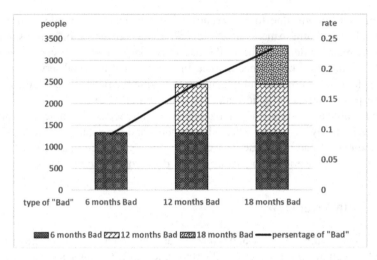

Fig. 1. Per 6,12,18 months "Bad" transition of proportion and total

4 The Analysis

The flow of analysis is performed by following procedures; data cleaning, basic tabulation, grouping, AUC, and logistic regression analysis.

At first, we supplement missing data and remove with data cleaning. Next, the data trend is grasped by basic aggregation. Then we divide into some group by the amount of main income. In AUC, the variable affecting 18 months Bad is extracted by the area under the ROC curve. Finally, we examine how the variable extracted by AUC has an impact on 18 months Bad.

4.1 The Data Cleaning

At first, we complement the missing data. As there was no blank data on the score of credit agency B, we complemented the score of credit agency A and its main income by using the score of credit agency B. In addition, we removed the interest, the down payment, the borrowing money, the age, the blank data because of the variable after complementing the credit agency A and its income, and the customer data which is impossible to calculate. As the result, the number of customer data is 13,217.

We will explain the calculation method about the score of credit agency A. We use the score of credit agency B of the customer who has a blank in the score of credit agency A and calculate the average the score of both credit agency A and B. After that, we input the calculated the score of credit agency A in the blank data. Then we supplement the main income with the score of credit agency A as well.

4.2 The Basic Aggregate

According to Fig. 2, the percentage of Bad in the Midwest, Northeast, and Northern is high. In addition, the average of main income is lower in the regions with the higher rate of Bad. Next, we look at the trends in educational backgrounds and types of occupations that are relevant to "Bad" customer's main income by region. View from Figs. 3 and 4. As a result, the same tendency was seen in all areas. The proportion of "Salary earners" in the classification of occupation and The proportion of "Graduated from Educational background 3" in the academic record was found to be large.

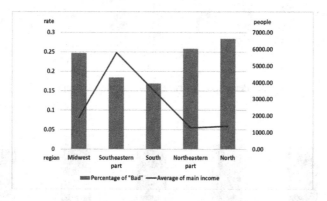

Fig. 2. By region percentage of main income and "Bad" ratio

In addition, in the prior research, the score of credit agency A and the score of credit agency B was grouping done. The ratio of 18 months Bad was calculated using them, and as a result, Fig. 5 was listed. According to the score of the credit agency A, the lower the score value, the clearly the higher the proportion of 18 months Bad. Looking at the score of credit agency B, the lower the percentage of the score, the higher the proportion of 18 months Bad is, but the groups 0, 1, 2 are not cleaning.

4.3   The Grouping

It can be seen that there is a clear difference between the ratio of Bad in the   north and the south. Moreover, it is considered that the bad factor is influenced by the main income, so we categorize groups by the main income for each revenue amount. The grouping criterion is classified by income which is published by the Ministry of Economy, Trade and Industry.

The A/B stratum get 7,475 or more a per month, the C stratum gets 1,734 to less than 7,435 a per month, D stratum gets less than 1,085 to 1,734 a per month, and E stratum gets less than 1,085 a per month.

Therefore, the A/B stratum is an affluent class, the C stratum is an intermediate class, and the D/E stratum is a poor class. We classified based on this criterion. From Table 3, it can be seen that the proportion of "Bad" in poor D and E stratum is high.

There is also a reason for summarizing the A/B stratum. There is also a reason for summarizing the A/B stratum. According to Fig. 6 the A/B stratum is a small part of the whole, and even this data is small number value [3] (Table 1).

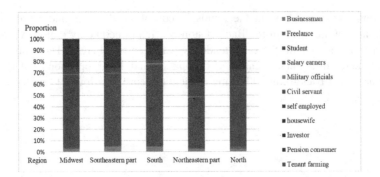

Fig. 3. Percentage of type of occupation by region

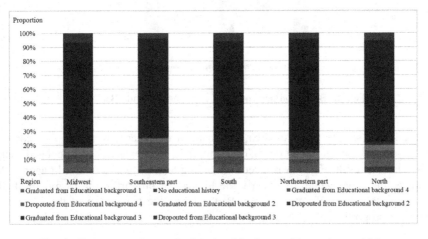

Fig. 4. Percentage of type of educational background by region variable.

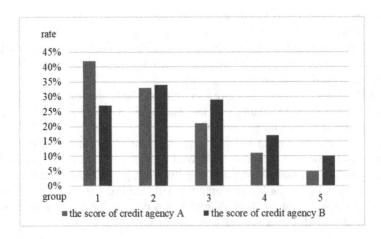

Fig. 5. Percentage of "18 months Bad" by group of credit agency

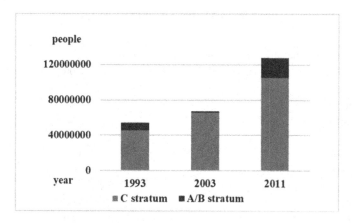

Fig. 6. Population transition by stratum

Table 1. Number of data by income stratum and percentage of 18 months Bad

	A/B stratum	C stratum	D stratum	E stratum
The number of data	147	2425	5516	5216
18 months "Bad"	13%	15%	27%	22%

4.4 The Extracting Variables

As the data of 18 months after cleaning is an objective variable and other variables are explanatory variables, we analyze the explanatory variables by logistic regression analysis and obtain the predicted probability. After that, we obtain the area below the ROC curve based on the explanatory variables. As a result, the presence/absence of negative information in the customer list, the number of inquiries to the customer list, the age, the value of real estate, the rate of list price, the borrowing money, the interest amount, the score of credit agency A, the score of credit agency B, product A, product B, D stratum were adopted. Also, from the results of AUC, the score of credit agency A is 0.688, and the score of credit agency B is 0.594. From here the credit agency A's score is strong relationship to 18 months Bad, indicating higher credibility (Table 2).

4.5 Logistic Regression Analysis

We analyze the logistic regression analysis by using the adapted variables in 4.4. The logistics regression analysis is a way to predict occurrence probability. Based on the analysis, we judge the occurrence probability of Bad customers. According to Fig. 2, the variable whose Exp (B) value is 1 or more is the number of inquiries to the customer list, the ratio of the list price, the product B, and D stratum. The customers who apply to these variables tend to be "Bad" easily. And also, the variable whose Exp (B) value is less than 1 is the presence or absence of negative information of the customer list, and product A. The customers who apply to these variables might not tend to be "Bad" easily. In addition, the higher the score of the data, the easier to be

Table 2. Analysis result of ROC curve

Variables adopted	Area	Significance probability
With or without of negative information on customer list	.554	.000
Number of Inquiries to customer lists	.553	.000
Age	.566	.000
Real estate value	.572	.000
Ratio to list price	.567	.000
Debt	.578	.000
Interest amount	.570	.000
Credit agency A score	.688	.000
Credit agency B score	.594	.000
Product A	.564	.000
Product B	.578	.000
D stratum	.552	.000

Table 3. Analysis result of logistic regression analysis

Variables	Exp(B)	Significance probability
With or without of negative information on customer list	.831	.009
Number of inquiries to customer lists	1.089	.000
Ratio to list price	1.012	.000
Debt	1.000	.000
Interest amount	.999	.001
Credit agency A score	.996	.000
Credit agency B score	.999	.000
Product A	.760	.008
Product B	1.399	.000
D stratum	1.188	.002

"Bad" because the debt is numerical data, and the value of Exp (B) is more than one point. The interest amount, the score of the credit institution A, and the score of the credit institution B are numerical data as well. The lower the score of the data, the easier to be "Bad" (B) is less than one point.

5 Consideration

Based on this analysis result, we can mainly consider two points. One is comparison between product A and product B. Since purchasers of product B are likely to become 18 months Bad customers, it is considered that it is necessary to review customer data of purchasers of product B. Buyer of product A is difficult to become 18 months Bad customer. Therefore, it can be considered to expand the range of purchasers of product A.

Next, it is about factors that make it easier for D stratum customers to become 18 months Bad customers. It is considered to be due to the expansion of financial services as well. Also, the lowest E stratum is hard to become 18 months Bad customer because it can be thought that a loan was not originally constructed.

6 The Future Tasks

In this analysis, we identified the variables which have impact on 18 months Bad. However, the payment collection rate of the customers who are 6 months Bad and 12 months Bad is lower than the 18 months Bad because of the number of times of payment. Therefore, we analyze same things in 6 months Bad and 12 months Bad and extract the influenced variables. Moreover, we reanalyze from the adapted variables respectively. For example, using the decision tree analysis with this analysis method, we search for the customers who will be "Bad", especially which variables are particularly applicable. And also, when we conduct the analysis, we analyze both cases that with the score of the credit institution "A" and "B" and not using them respectively. From the result, we can also judge whether a new credit risk model is necessary. We propose new evaluation method of credit risk model by these results.

In addition, the scores of the credit agency A are the most effective evaluation. The factors include basic aggregation, AUC results, and prior research. The content constituting the score of the credit organization A is unclear. Therefore, it is necessary to grasp the characteristics of the score of the credit agency A.

Appendix

We will explain AUC. AUC is the name of the lower area of the ROC curve as shown in Fig. 7. AUC is an index for evaluating the strength of the relationship between the analysis target and other variables. As an evaluation criterion, if AUC is 1, there is a certain relationship. If AUC is 0.5, there is no relation. The closer the AUC is to 1.0, the more the relationship is strong. As practical examples, there are the following two. "Performance evaluation of tests for diagnosing healthy group/disease group", "Examination of predictive factors of early prognosis after organ transplantation".

Next, the binomial logistic regression analysis will be explained. Introduce variables to be analyzed as categories, not numerical values you want to predict. It found whether it will be 18 months Bad or not. Factors can be found, you can predict the outstanding probability and the payment completion probability. Since the odds ratio is 1.0 or more, it can be said that it affects the analyzed variables. Examples of use include creation of credit score models and medical sites.

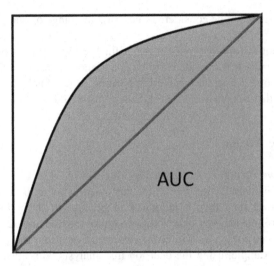

Fig. 7. The ROC curve

References

1. Cabinet Office Policy Division (Economic and Fiscal Analysis): "Trend of the global economy 2009 I", Emerging economies the impact of the financial crisis and future prospects, vol. 2, no. 3 (2009)
2. Latin America, "A" country's property and casualty insurance market, Sompo Japan Research Report, pp. 59–60. www.sjnk-ri.co.jp/issue/quarterly/data/qt64_3.pdf. Accessed 18 Dec 2017
3. METI: Commerce white paper 2012, Trends in the global economy, vol. 1, no. 6 (2012)
4. The emerging middle class of A country to push forward with loan consumption. Nikkei Business business.nikkeibp.co.jp/article/world/20110526/220227/. Accessed 20 Dec 2017

Factor Analysis of the Batting Average

Hiroki Yamato[1]([⊠]) and Yumi Asahi[2]

[1] School of Information and Communication Studies Department of
Management Systems Engineering, Tokai University, Tokyo, Japan
5bjml124@mail.u-tokai.ac.jp
[2] School of Information and Telecommunication Engineering, Department of
Management System Engineering, Tokai University, Tokyo, Japan
asahi@tsc.u-tokai.ac.jp

Abstract. This study is factor analysis of the batting average in the professional baseball in Japan. We analyze the factor influencing the batting average using the Japanese professional baseball data. There is no established method to ensure a good shot at Japanese baseball. Based on the results, clarify factors that prevent pitchers from hitting hits and factors that batters increase hits. And establish baseball teaching methods based on the clarified factor. Finally, we aim to improve the level of the professional baseball world of Japan.

The data used are the one-ball data in the regular season of Japanese professional baseball in 2015 and 2016. One-ball data is data every time a pitcher throws one ball to a batter. This time, we used only the data of the battle of right-handed pitcher and right-handed batter. The reason for limiting the data is that it is judged that it is easier to extract the characteristics of the factor when narrowing down the conditions.

In this research, factor analysis is performed first, and covariance structure analysis is performed based on extracted factors. Factor analysis extracts pitcher and batters how to approach the ball. In the covariance structure analysis, we analyze how the extracted factor affects variables.

The result of the factor analysis is that the pitcher can extract four factors, the batter can extract two factors. We named the extracted pitcher's factors "throw down low", "throw falling balls", "throw balls to escape outside", "attack in-course". We named the extracted batter's factors "upper swing", "down swing". When covariance structure analysis was performed using the result of the factor analysis, three models could be created. The three models can know how each factor influences hits, outs, batting average. From the results of these models, upper swing had a positive influence on hits, and it turned out that it had a bad influence on outs. It also proved to have a positive effect on latent variable batting time consisting of hits and outs. In summary, it turns out that doing an upper swing has a good influence on increasing the batting average.

From the analysis result, it turned out that the upper swing is important for improving the batting average. The future task can be to analyze also combinations other than right-handed pitcher versus right-handed batter who could not be done this time. In addition, we clarify the explanatory variable which has the most influence on improving batting average among latent variable upper swing.

Keywords: Sports marketing · Factor analysis

© Springer International Publishing AG, part of Springer Nature 2018
S. Yamamoto and H. Mori (Eds.): HIMI 2018, LNCS 10904, pp. 145–155, 2018.
https://doi.org/10.1007/978-3-319-92043-6_12

1   Introduction

Baseball is one of the ball games that hits a ball with a bat. It is a popular sport in Japan, the USA, Cuba etc. especially major league in America is considered to be the best league of the baseball world. The reason is that the league's economic scale is the biggest in the world. According to Fig. 1, we can see the average salary of MLB players is about 11 times that of NPB players in 2015. MLB players can get high salary. Therefore, it is easy for players to gather from all over the world and the number of player increase. Actually, NPB has 11 teams and MLB holds 30 teams. As a result, the competition of higher level is born, and powerful players are born. From Japan, sending out major players who can succeed with major like Ichiro. Japanese baseball league which Ichiro is belongs to is also high level. Besides Ichiro, Japanese baseball league has sent top players such as Hiroki Kuroda and Hideki Matsui to the major league. However, the level of Japanese professional league is starting to be thought to be getting lower in recent years. The reason is that the number of athletes trying to challenge Major League every year is decreasing from Fig. 2 and the Japanese major league challenge record that eventually lasted 22 years broke. Pitchers are active with Yu Darvish and Masahiro Tanaka and others, but no one is active in the fielder. Fielders from Japan may be disappeared in major league this year. It was third in WBC in 2017, but it seems to be struggling compared to when we were able to win successive. Therefore, raising the level of professional baseball in Japan is an urgent task and we examined the method for doing so. The problem of Japanese baseball world is various instruct method. There are too many methods, which is confusing. For example, strike to a ball from above, strongly swing the bat, etc. but it is a way to increase strong hitting and not a way to increase hits. In Japan, the method to increase hits is not clarified now. Therefore, this study is clarified factor of the increasing hits and batting average in the professional baseball in Japan and propose an instruct method to increase batting average and number of hits to improve batting record.

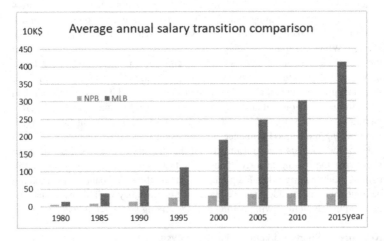

Fig. 1. Average annual salary transition comparison

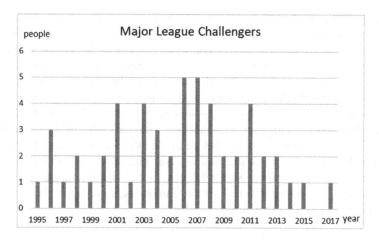

Fig. 2. Number of major league challengers

2 Detail of Used Data

We used the one-ball data in the regular season of Japanese professional baseball in 2015 and 2016. It was provided by Data Stadium. One-ball data is data every time a pitcher throws one ball to a batter. This time, used variables in this data are summarized in Table 1 below.

Table 1. Detail of used data

Details of data	
Flying distanceX	Numerical data from 0 to 280
Flying distanceY	Numerical data from 0 to 280
Hitting segment	1 when it was fly, 2 when it was a grounder, 3 when it was a liner, 4 if batter missed it
Swing	1 or 0 qualitative data
Pitching courseX	Numerical data from 0 to 200
Pitching courseY	Numerical data from 0 to 250
Breaking ball	1 for a curve, 2 "for a cutter, 3 for a shoot 4 for a shinker, 5 "for a straight 6 for a slider, 7 for a split 8 for a Off-speed pitch, 9 for a special ball
One-bound ball	1 or 0 qualitative data
Ball speed	Numerical data from 82 to 165
Batting result classification	1 is when it was swing and miss, 2 is when it was looking, 3 is when it was hit 4 is when it was out 5 is when it was home run, 6 is when it was foul, 7 is when it was bunt

Flying distance coordinates X and Y in Fig. 3 are the upper left corner of the figure above as the origin. For example, in the case of a catcher fly, the distance coordinate X is 40 and the distance of the distance Y is 240. Swing is that whether the batter swung or not. Hitting segment is batting segment. For example, fly or grounder or liner. The pitching coordinates X and Y in Fig. 4 represent the height as the Y -axis and width of the zone as the X-axis. Breaking ball is what kind balls threw. For example, slider, off-speed pitch, etc. One-bound ball is that whether the ball has bounced one or not. Ball speed is that speed of the thrown ball. Flying distanceX, flying distanceY, pitching courseX, pitching courseY and ball speed are quantity data. Swing, Result of batting, one-bound ball and breaking ball are qualitative data. The scale of them is different. Therefore, we need to align the scales before starting the analysis. This time, we try to do factor analysis and covariance structure analysis. The purpose of factor analysis is to extract elements of batter and pitcher. At the same time as converting qualitative data to quantitative data, we convert it so that features can be easily grasped by factor analysis. Details of conversion are summarized in Table 2.

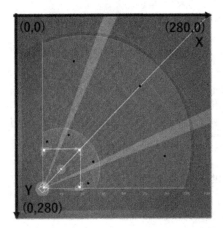

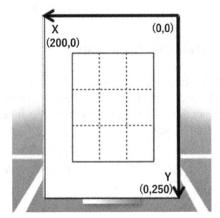

Fig. 3. Flying distance coordinates   **Fig. 4.** Pitching coordinates

Flying distanceX and Flying distanceY are very hard to use. Therefore, we redefine flying distanceX2 and flying distanceY2 from them. They adjusted so that the origin overlaps the home base. We drew a line by dividing places other than the foul area into three at 30 degrees each like Fig. 5. These lines are the line of $Y = \sqrt{3}x$ and line of $Y = 1/\sqrt{3}x$. ① which the place surrounded by Y-axis and the line of $Y = Y\sqrt{3}x$ is named left direction. ② which the place surrounded by the line of X-axis and the line of $Y = 1/\sqrt{3}x$ is named right direction. When it is caught in these zones, it is set to 1, and when it is not caught 0. In this manner, the coordinate data is converted into qualitative data. We also redefine variable distance in this data. The distance data can be obtained using the by how far the origin is from these flying distance X2 and flying distance Y2. The formula is distance $= \sqrt{(\text{flying distance x2}^2 + \text{flying distance y2}^2)}$. We redefine short distance and long distance from this distance. First, draw a line to classify the distance. The black line in the figure is a line representing the part where the distance from the origin is 42, the red line is the line showing the distance 162 like

Fig. 6. We redefine the inner side of the black line as a short distance and the outer side of the red line as a long distance. Set it to 1 when the hit ball flew into the defined place and 0 otherwise. From batting tendency, we redefine fly tendency or grounder tendency. Fly tendency is set to 1 when the ball fly and 0 otherwise. Grounder tendency is 1 when the ball rolls, and 0 otherwise.

Table 2. Change of variable

The original variable	New variable
Flying distanceX, flying distanceY	Flying distanceX2, flying distanceY2
Flying distanceX2, flying distanceY2	Left direction, right direction, distance
Distance	Short distance, long distance
Hitting segment	Fly, grounder
Batting result classification	Hitout
Pitching courseX, pitching courseY	High, low, inside, outside
Speed of ball	Fast ball, slow ball
Breaking ball	Bending ball, Falling ball, shinker

Next, we redefine pitching coordinates X and Y as inside, outside, higher and lower. In order to classify inside and outside, the zone was divided into three zones and a line was drawn like Fig. 7. It drew a line so that the proportion of thrown balls was equal for each zone. Similarly, classify high and low line to be drawn like Fig. 8. Inside is 1 when the ball is thrown into ① in Fig. 1, and 0 otherwise. Outside is 1 when the ball is thrown into ② in Fig. 1, and 0 otherwise. Higher is 1 when the ball is thrown into ① in Fig. 2, and 0 otherwise. Lower is 1 when the ball thrown into ② in Fig. 2,

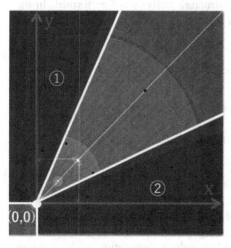

Fig. 5. Left and right direction

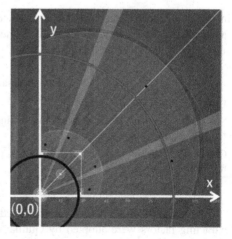

Fig. 6. Short or long distance (Color figure online)

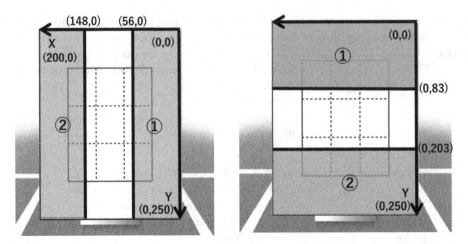

Fig. 7. Inside or outside course   **Fig. 8.** Higher or lower course

and 0 otherwise. We redefine Slider and Cutter as bending ball, and Split and Off-speed pitch as falling ball. We redefine speed of the ball as fast-ball or slow-ball. Divide the ball speed into three so that the pitching proportion is equal. Among them, we name fast or slow ones as fast-ball and slow-ball. Fast-ball classifies more than 143 km/h. Slow-ball classifies less than 132 km/h. We redefine batting result classification to hits and outs. Variable out is set to 1 when batting result classification is 5. In the same way, variable hit is set to 1 when batting result classification is 4. This is the end of organizing the data.

3 Factor Analysis

In this study, factor analysis and covariance structure analysis are performed. In the factor analysis, we extracted factor of pitcher and factor of batter. Usage variables are summarized in the Table 3 below.

Table 3. Usage variables

Usage variables (batter)
Swing, fly, grounder, long distance, short distance, left direction, right direction
Usage variable (pitcher)
One-bound ball, lower, higher, inside, outside, falling ball, bending ball, shinker, fast-ball, slow - ball

Maximum likelihood method was used for factor analysis method. Criteria for extracting factors were adopted by Gutman-Kaiser criteria. The Gutman-Kaiser criterion is a concept that adopts only factors whose eigenvalues are 1 or higher. Therefore, the yellow part of the table is adopted Considering the accumulation of the variance from Tables 4 and 5, it is understood that two factors of the batter explain the whole 69% of the total and four factors of the pitcher explain the whole 63% of the total.

Table 4. Extraction result (batter)

Factor	Sum	Variance (%)	Accumulation (%)
1	3.201	45.725	45.725
2	1.664	23.766	69.491
3	0.918	13.109	82.6
4	0.545	7.78	90.38
5	0.375	5.355	95.735
6	0.229	3.277	99.012
7	0.069	0.988	100

Table 5. Extraction result (pitcher)

Factor	Sum	Variance (%)	Accumulation (%)
1	2.387	23.872	23.872
2	1.571	15.706	39.578
3	1.357	13.574	53.152
4	1.076	10.763	63.915
5	0.837	8.368	72.283
6	0.694	6.936	79.219
7	0.643	6.434	85.653
8	0.585	5.851	91.504
9	0.508	5.08	96.584
10	0.342	3.415	100

In the batter's factor, two factors could be extracted as a result of using seven variables. In the pitcher's factor, four factors could be extracted as a result of using ten variables. Each factor matrix is shown in the following Tables 6 and 7.

Table 6. Factor matrix (batter)

Factor matrix	Factor 1	Factor 2
Fly tendency	0.928	−0.141
Short distance	−0.809	−0.549
Swing	0.638	0.462
Right direction	0.572	
Long distance	0.483	
Grounder tendency		0.97
Left direction	0.148	0.571

Table 7. Factor matrix (pitcher)

Factor matrix	Factor 1	Factor 2	Factor 3	Factor 4
Fast-ball	−0.932	−0.137	−0.152	
Slow -ball	0.564			
Bending ball	0.482		−0.267	−0.279
Lower		0.881		
Higher		−0.567		0.1
One-bound ball		0.378	0.145	
Falling ball		0.216	0.975	
Inside		−0.11		0.559
Outside		0.248	−0.144	−0.464
Shinker				0.452

We named factor 1 on the Table 6 as upper swing, factor 2 on the Table 6 as down swing, factor 1 on the Table 7 as a ball to escape outside, factor 2 on the Table 7 as a throw ball to a lower, factor 3 on the Table 7 as a falling ball, factor 4 on the Table 7 as an attack inside. Upper swing chose the name mainly from the fly tendency and right direction. Since the ball is struck from the bottom, the tip of the bat is liable to be delayed. In other words, upper swing is that because the bat is detouring, the point of swing is likely to be delayed. Down swing chose the name mainly from the grounder tendency and left direction. It is the swing to the point in the shortest way to the point, so the point is hard to delay. Ball to escape outside chose the name mainly from the bending ball and ball-speed.

4 Covariance Structure Analysis

In this analysis, we can see how factors extracted by factor analysis affect different variables. This time, we analyze the relationship between the factors and variable hit and out. Maximum likelihood method was also used for covariance structure analysis. The analysis results are summarized in the following Figs. 9 and 10.

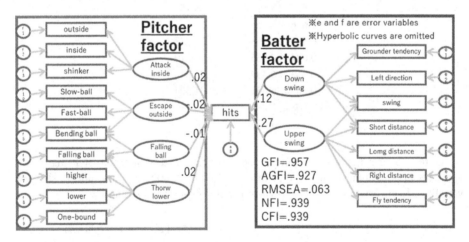

Fig. 9. Factor analysis of hits

An elliptical one is called a latent variable, and a rectangular one is called an observation variable. The latent variable refers to upper swing or down swing, and observation variable refers to left direction, right direction and so on. Latent variables refer to virtual variables such as upper swing or lower swing, and observation variables are variables of actual data such as left direction and right direction. GFI can see whether the total variance of the saturation model can be explained well by the variance of the estimation model. In other words, GFI shows the fit of the model. GFI is a numerical value between 0 and 1, and it is generally said that GFI should be higher than 0.9. NFI can see how much the deviation between the saturation model and the

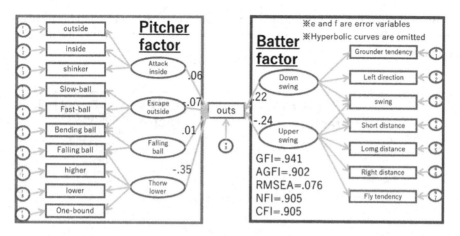

Fig. 10. Factor analysis of outs

estimation model is superior to that of the independent model. NFI is a numerical value between 0 and 1, and it is generally said that NFI should be higher than 0.95. However, GFI and NFI tend to become easier to rise as the model becomes more complicated. This is because the degree of freedom decreases as the model becomes more complicated. Therefore, we use AGFI and CFI in addition to GFI and NFI to judge whether the model is good or bad. AGFI is GFI which takes the degree of freedom into account. AGFI is also a numerical value between 0 and 1, and it is generally said that AGFI should be higher than 0.9. CFI is NFI which takes the degree of freedom into account. CFI is a numerical value between 0 and 1, and it is generally said that CFI should be higher than 0.95. RMSEA is Indicates whether it is divergent between model distribution and true distribution. RMSEA is a numerical value between 0 and 1, and it is generally said that RMSEA should be lower than 0.05. It is said to be a bad model when it exceeds 0.1.

By evaluating Figs. 9 and 10 from the above five observation points, it is understood that values other than RMSEA are satisfied. In other words, you can see that the fit of the model is good.

Arrows indicate the degree of influence between variables in Figs. 9 and 10. The degree of influence is numeric of value between −1 and 1.

It is understood that both the upper swing and the down swing have influenced on hits from Fig. 9. When comparing the two factors, we can see that the influence given by the upper swing a little is great.

It is understood that the upper swing has a negative influence and the lower throw gives a positive influence outs from Fig. 10.

Create latent variable batting rate from observation variables hits and outs. Finally, we analyze the relationship between the factors and this variable. The analysis results are summarized in the following figure.

By evaluating Fig. 11 from the five observation points, it is understood that values other than RMSEA are satisfied. In other words, we can see that the fit of the model is also good.

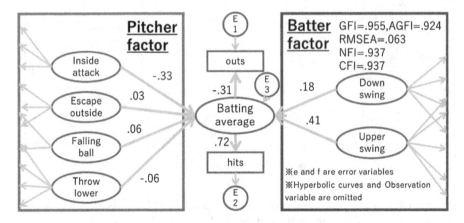

Fig. 11. Factor analysis of batting average

It is understood that the inside attack has a negative influence and the upper swing gives a positive influence outs from Fig. 11.

5   Discussion/Summary

We analyzed the factor analysis of the batting average. The result proved that upper swing is important to raise batting average. We verify the result from the professional baseball season data of 2017. The target is twenty-nine right-handed batters who reached the Stipulated At-bat in Japanese professional baseball. In the case of right pitcher vs. right batter, the batting average was compared with the fly batter and other players. A fly batter refers to a player whose fly rate exceeds the sum of the grounder rate and the liner rate. First, a t-test was performed, but significance was not noticed. Next, we decided to compare using the box plot diagram. The results are summarized in Fig. 12 below.

Box whiskers can see the maximum value, the minimum value, the proportion of the interquartile range. The red line is average of batting average. We can see no difference of average of batting average among fly batter and others. From this figure.

It can be seen that the range from the maximum value to the second quartile of the batting the average of fly batter is narrow. In other words, the fly batter shows that there are many batters with high batting rate compared with other batters.

The future task can be to analyze also combinations other than right-handed pitcher versus right-handed batter who could not be done this time. In addition, we clarify the explanatory variable which has the most influence on improving batting average among

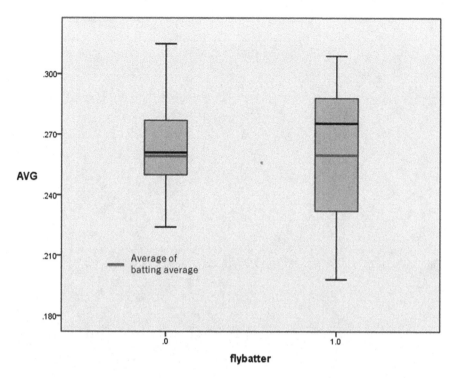

Fig. 12. Comparing batting average fly batter and others (Color figure online)

latent variable upper swing by changing the direction of the arrow of the model in the figure.

References

1. Toyoda, H.: Covariance structure analysis [Amos] -Structural equation modeling (2011)
2. Toyoda, H.: Covariance structure analysis [primer] -Structural equation modeling (2006)
3. Oshio, A.: Psychological and survey data analysis by SPSS and Amos (2005)
4. Bollen, K.A.: Structural Equations with Latent Variables. Wiley, Hoboken (1989)
5. Konno, K.: Structural Equation Modeling - Model Construction Reexamination. (Shizuoka Physics University). http://www.mizumot.com/method/2012-05_Konno.pdf
6. Kano, H.: Fundamentals and actuality of covariance structure analysis—Fundamentals. (Graduate School of Human Studies, Osaka University). http://csrda.iss.u-tokyo.ac.jp/seminar2002_1.pdf. Accessed 19 Feb 2018
7. Moriyasu, Y.: Covariance structure analysis Psychological data analysis exercise. (Graduate School of Engineering). http://cogpsy.educ.kyoto-u.ac.jp/personal/Kusumi/datasem07/moriyasu.pdf. Accessed 19 Feb 2018
8. 1.02 Essence of Baseball. http://1point02.jp/op/index.aspx. Accessed 19 Feb 2018
9. Statistical data of Japan and the world. https://toukeidata.com/index.html. Accessed 19 Feb 2018

Multimodal Interaction

Classification Method of Rubbing Haptic Information Using Convolutional Neural Network

Shotaro Agatsuma[1]([✉]), Shinji Nakagawa[1], Tomoyoshi Ono[1], Satoshi Saga[2], Simona Vasilache[1], and Shin Takahashi[1]

[1] University of Tsukuba, 1-1-1 Tennodai, Tsukuba city, Japan
`agatsuma@iplab.cs.tsukuba.ac.jp`, {`s1620660,s1620611`}`@u.tsukuba.ac.jp`,
{`simona,shin`}`@cs.tsukuba.ac.jp`
[2] University of Kumamoto, Kurokami 2-39-1, Chuo-ku, Kumamoto city, Japan
`saga@saga-lab.org`

Abstract. In previous research, we proposed a method to collect accelerations in daily haptic behaviors using a ZigBee-based microcomputer. However, the method for classifying the collected data was not sufficiently implemented. We therefore propose applying collected data to classify rubbing haptic information. In this paper, we implemented a classification approach for haptic information collected by our method. We used a convolutional neural network (CNN) to classify the information. We performed a classification experiment in which the CNN classified 18 types of information, 93.2% on average. We also performed an experiment to classify rubbed objects in real-time. The CNN was able to classify five types of objects, about 67.7% on average.

Keywords: Zigbee-based microcomputer · Haptic information
Convolutional neural network

1 Introduction

In the field of tactile display, many researchers are developing systems that employ recorded vibration as a tactile signal, and these systems present high-quality tactile sensations. To enhance this type of display method, it is necessary to collect and classify recorded vibration during haptic behaviors. Consequently, various studies of the collection and classification of recorded vibration in haptic behaviors have been performed [1,2]. However, most of these studies have collected tactile information under limited experimental environments using devices with many sensors. Therefore, it is difficult to collect haptic information outside of the experiment environment, for example in daily behavior.

In a previous research, we proposed a solution to the problem of collecting haptic information without complicated devices [3]. Using this approach, we collected only acceleration data as haptic information. Using a ZigBee-based

© Springer International Publishing AG, part of Springer Nature 2018
S. Yamamoto and H. Mori (Eds.): HIMI 2018, LNCS 10904, pp. 159–167, 2018.
https://doi.org/10.1007/978-3-319-92043-6_13

microcomputer with an accelerometer, we collected haptic information more easily than by conventional research methods. We previously reported that our method can collect acceleration data from rubbing haptic behaviors on a table. We also proposed applying collected data to classify haptic information. Our previously implementation used Support Vector Machine (SVM) and classified eight types of data with an accuracy of about 80%. However, if we change machine learning method, there is a possibility of getting higher accuracy.

Therefore, we implemented a method in the current study to classify haptic information collected by our method using machine learning. We attached a ZigBee-based wireless microcomputer with an accelerometer to the experimenter's finger or pen while they traced various objects, and we collected accelerations during haptic behaviors on a table. As a machine learning method, we used convolutional neural network (CNN) to classify the haptic information with high precision. We succeeded in classifying 18 types of data with an accuracy of about 93%. We also performed an experiment to classify rubbed objects in real-time. The CNN was able to classify five types of objects, about 67.7% on average.

2   A Method of Collecting and Classifying Haptic Information

In this study, we demonstrate our method of collecting and classifying haptic information. Acceleration in haptic behaviors is collected by a ZigBee-based microcomputer and then classified by machine learning using CNN. In the following sections, this method will be described in detail.

2.1   ZigBee-Based Microcomputer with Acceleration Sensor

Many studies have used mechanical vibrations in the band of 0–1 kHz frequency band as haptic information for haptic display, and their implementations have produced highly realistic information, such as the TECHTILE toolkit [4]. Therefore, we also collected mechanical vibration information within 0–1 kHz as haptic information. For that purpose, our implementation was based on a small ZigBee-based microcomputer equipped with an accelerometer capable of collecting vibration information in the 0–1 kHz range.

In this study, we used a ready-made device, TWE-Lite-2525A (Mono Wireless Inc. [5]), shown in Fig. 1. This device consists of a 3-axis accelerometer and a ZigBee wireless communication module with a small battery cell as a power supply. The device was 25 mm square and its weight was 6.5g; it is capable of wireless communication using ZigBee, such that it can be installed and operate on various objects. To save power, we added a sleep mode to the device, during which it measures acceleration at low frequencies until the measured acceleration exceeds the set threshold, and then begins measurement at high frequencies after the threshold is exceeded.

Fig. 1. TWE-Lite-2525A, Mono Wireless Inc. [5]. This sensor node includes a 3-axis accelerometer, ZigBee transfer module, and battery cell (CR2032).

2.2   Data Transmission

To classify haptic information by machine learning, the collected acceleration data was transmitted to a personal computer (PC) by wireless communication using ZigBee. The ADXL345 accelerometer installed on the device can measure vibrations at 800 Hz. However, I2C communication between the accelerometer and the ZigBee wireless module, as well as ZigBee communication between the wireless module and the computer, must be taken into consideration; thus, the actual value was considerably lower than 800 Hz. The measurement rate of the microcomputer was 33 Hz using the program as written at the time of purchase; this rate is insufficient for the transmission of haptic information.

Therefore, we adopted a method to pack 10 sequentially measured values into one packet. We implemented this method as a new program in a microcomputer to facilitate the measurement of 3-axis acceleration at about 330 Hz.

2.3   Classification by Machine Learning

In this subsection, we describe a method to classify collected data by machine learning. In a previous study [3], we classified data by machine learning using a support vector machine (SVM). In the current study, to improve the number of classes, we employed a CNN-based machine learning method. We used a total of 13 CNN layers for data classification. The configuration is shown in Fig. 2.

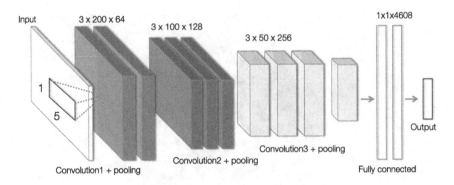

Fig. 2. Composition of CNN.

As the input data to this CNN, we used clusters of 200 consecutive points (each point includes 3-axis data: x, y, z), which were randomly extracted from the acceleration data. This method increase the number of learning data and improve generalization performance of the model. The output data of this CNN was the probability of the type of haptic information input.

We used TensorFlow [6], a machine learning library provided by Google, to build the CNN. We built the CNN with reference to the model of VGG [7] which is a typical CNN composition; we reduced the amount of data to be convoluted by one convolution layer, and increased the convolution layer to improve the accuracy of classification. We used the ReLU function [8] as the activation function and a convolution filter of size 1×5. We doubled the number of filters for each pooling to improve generalization. In the pooling layer, maximum pooling was performed at 1×2, and data was compressed to half so that CNN would be robust against shape of data. In addition, overfitting was suppressed using the Batch Normalization algorithm [9] after calculating the activation function in each convolution layer and in the fully connected layers.

3   Classification Experiment

We collected and classified haptic information using the method described in the previous section and verified the accuracy of the classification of this collected information. In addition, we performed an experiment to classify rubbed objects in real-time.

3.1   Classification of 18 Types of Haptic Information

We performed an experiment to classify 18 types of haptic information. We will describe this experiment in detail below.

Settings. The haptic information was collected by the experimenter. Attaching the ZigBee-based microcomputer to their finger or a pen, the experimenter traced the surfaces of objects with various textures, thus collecting many types of haptic information. This collection process is shown in Fig. 3. The textures used in this experiment are shown in Fig. 4.

Fig. 3. Collection of tactile data. The microcomputer was put into the 3D printed case, attached to experimenter's finger or pen and operated to collect haptic data.

Each texture to be examined was a plate-like object of about 100 mm in length and width. In Fig. 4, carpet1, carpet2, carpet3 were pieces of carpet made with different materials; sponge-g and sponge-y were the front and back of a household sponge. Objects with a texture similar to that of sponge-b were made of Styrofoam; stonetile1, stonetile2, and stonetile3 were stone tiles made with materials of different texture; whitetile1, whitetile2, and whitetile3 were white flooring tiles with different textures; woodtile1, woodtile2, and woodtile3 were wooden plates with different textures; sandpaper40, sandpaper80 were sandpapers with different roughness.

Haptic information was collected by rubbing these objects with a finger or pen at an almost constant speed of 400 mm/s. At the surface of each object, acceleration data were collected during three minutes of back-and-forth rubbing movement. By performing this operation three times per object, nine minutes of acceleration data were collected per object. As shown in Fig. 4, 10 types of objects were used in this experiment. We rubbed each object with a finger and rubbed each object except "sandpaper40"and "sandpaper80" with a pen. We collected a total of 18 types of data (10 types of finger data×8 types of pen data). We could not collect accurate data from sandpaper rubbed by pen, because the data collection process damaged the tip of the pen; therefore, we discarded these data.

We evaluated classification by machine learning using the CNN. The CNN input included 3×200 values of acceleration data, and the output was a probability of 1×18, the probability of an object having been rubbed to create the input data. During classification, to confirm the generalization performance of the model created by this CNN, we divided all data by 10 and performed 10-fold cross-validation.

Fig. 4. Textures. These are plate-like object with length 70–100 mm and width 100–130 mm.

Result. We describe the accuracy of the data classification experiment in Table 1, which shows the confusion matrix obtained by classifying 18 types of data. We obtained an accuracy of 93.2% on average, indicating that we succeeded in classifying 18 types of information with high precision.

As shown in the confusion matrix, we obtained an accuracy exceeding 80% for all results. Therefore, in this experiment, we found that critical overfitting did not occur. However, this result does not mean that there was no overfitting. The best result shown in the confusion matrix was woodtile-pen, at 97.2%; however the worst result was sandpaper40, at 83.1%. There was a difference of 14.1% between the pen and finger results, and a small amount of overfitting occurred.

For sandpaper40, the correct result was 83.1%, and the incorrect result was 17.9%. The largest of the incorrect results for sandpaper80 was 6%. Among the incorrect results for sandpaper80, sandpaper40 was incorrectly classified at 7.8%, the largest value among incorrect results. From these results, we conclude that it was relatively difficult to classify fine textures.

3.2 Object Classification in Real-Time

We performed an experiment to classify rubbed objects in real-time. We classified five types of rubbed object using the CNN model. We will describe this experiment in detail below.

Settings. An experimenter collected haptic information by attaching the ZigBee-based microcomputer to their finger and rubbing the surfaces of objects with various textures. The textures used in this experiment were carpet1, carpet2, carpet3, sponge-g, and sponge-y (Fig. 4). Haptic information was collected by rubbing these objects with a finger or pen at an almost constant speed of 400 mm/s. At the surface of each object, acceleration data were collected during five seconds of back-and-forth rubbing movement. We performed this operation

Table 1. Confusion matrix for the classification of 18 types of data.

	A	B	C	D	E	F	G	H	I	J	K	L	M	N	O	P	Q	R
A: carpet1	0.947	0	0	0	0.021	0	0	0	0.011	0	0	0	0.011	0	0	0	0.011	0
B: carpet1-pen	0	0.954	0	0.034	0	0	0	0	0	0	0	0	0.011	0	0	0	0	0
C: carpet2	0	0	0.892	0	0.039	0	0.01	0.02	0	0.02	0.01	0	0	0	0.01	0	0	0
D: carpet2-pen	0	0.01	0	0.951	0	0.01	0	0	0	0	0	0.01	0	0.01	0	0.01	0	0
E: carpet3	0.031	0	0.041	0	0.907	0	0	0	0	0	0.021	0	0	0	0	0	0	0
F: carpet3-pen	0	0	0.011	0.021	0	0.947	0	0	0	0	0	0	0.011	0	0	0.011	0	0
G: sandpaper40	0	0	0	0	0	0	0.831	0.06	0	0.048	0	0	0.048	0	0.012	0	0	0
H: sandpaper80	0	0	0.019	0	0	0	0.078	0.893	0	0	0	0.01	0	0	0	0	0	0
I: sponge-g	0	0	0	0.009	0	0	0	0	0.92	0.009	0	0.054	0	0	0	0.009	0	0
J: sponge-g-pen	0.02	0	0	0	0	0.01	0.029	0	0	0.922	0.02	0	0	0	0	0	0	0
K: sponge-y	0.01	0	0.029	0	0	0	0.01	0.01	0	0	0.941	0	0	0	0	0	0	0
L: sponge-y-pen	0	0.008	0	0	0	0	0	0	0	0	0	0.967	0.008	0.008	0	0	0	0.008
M: stonetile	0	0	0	0	0	0	0.009	0	0	0	0.019	0.019	0.925	0	0.028	0	0	0
N: stonetile-pen	0	0	0	0	0	0	0	0	0	0	0	0.009	0.009	0.972	0	0.009	0	0
O: whitetile	0.01	0	0	0	0	0	0.01	0.01	0	0	0	0	0.01	0	0.923	0	0.038	0
P: whitetile-pen	0	0	0	0	0	0	0	0	0	0	0	0	0	0.01	0	0.948	0	0.042
Q: woodtile	0	0	0	0	0	0	0	0	0	0.009	0	0	0.009	0	0.009	0	0.972	0
R: woodtile-pen	0	0	0	0	0	0	0	0	0	0	0	0	0	0.009	0	0.019	0	0.972

15 times per object each day, on six consecutive days. We collected a total of 90 observations per object.

We collected new data for this experiment because we had obtained less than 10% accuracy on average in the nine-minute data collected in Sect. 3.1. We determined that the rubbing method (e.g., finger angle, rubbing speed) changed slightly each time data was collected during that experiment. When we collected nine-minutes data, we collected data only three times per objects. Therefore, when the nine-minute data was used, CNN model learned only three types of data per object; thus, the CNN model did not attain adequate generalization performance. To solve this problem, we used six-days data. If we use six-days data representing 90 times collections per object, CNN model can learned 90 types of data per object and get better generalization performance. Therefore, CNN model using six-days data was better than using nine-minutes data.

We created the CNN model to identify objects by rubbing. The CNN input was 3×200 acceleration measurements, and the output was a probability of 1×5, the probability of an object having been rubbed to create the input data. We used 80% of all data as training data and the remaining data as test data. We used the test data to confirm the progress of CNN learning.

Using this CNN model, we classified objects rubbed by an experimenter in real-time. The instant he collected data, we inputted data and classified. He collected a total of 20 5-s observations as new data. Using this method, we classified each input value 10 times; that is, we conducted 200 classification experiments per object. The output of this CNN was the object from which the haptic information had been derived.

Result. Table 2 shows the accuracy of the experimental results. These accuracies represent averages of 20 classifications. Table 3 shows the confusion matrix for the classification of five objects.

Table 2. Accuracy of classification of haptic data collected by touching five objects.

carpet1	carpet2	carpet3	sponge-g	sponge-y
67.5%	71.0%	41.5%	83.5%	75.0%

Table 3. Confusion matrix for the classification of haptic data collected by touching five objects.

	A	B	C	D	E
A: carpet1	0.955	0.045	0	0	0
B: carpet2	0.053	0.789	0.105	0.053	0
C: carpet3	0	0.095	0.714	0.143	0.048
E: sponge-g	0	0	0	1	0
F: sponge-y	0	0	0	0	1

In Table 2, we obtained an accuracy of 67.7% on average. The best accuracy was 83.5% (sponge-g), and the worst was 41.5% (carpet3). There was a large difference between the best and the worst accuracies. In Table 3, the accuracy for detecting carpet3 was 71.4% and this represented the least accurate result, which is consistent with the trend shown in Table 2. As shown in Table 3, the accuracies for detecting sponge-g and sponge-y were 100%. Because these values are very high and there was a large difference between the best and worst accuracy results, we conclude that overfitting occurred.

The insufficient generalization performance of the CNN model may explain these results. To solve this problem in the future, we will design the CNN model using more observations than used in this experiment. In addition, we will review the CNN settings to improve the results.

4   Conclusion and Future Work

In this research, we implemented a method in the current study to classify haptic information collected by our method using CNN.

Using a compact wireless microcomputer with an acceleration sensor applied to the tip of a finger or ballpoint pen, an experimenter rubbed various objects at approximately 400 mm/s, collecting 18 types of haptic information. We then classified these data, obtaining an accuracy of 93.2%. We concluded that we were able to classify haptic information with high accuracy. In addition, we performed an experiment in which haptic data from five types of touched objects were classified using a CNN model, and obtained an accuracy of 67.2% on average. We determined that the rubbing method (e.g., finger angle, rubbing speed) may change slightly each time data is collected.

In the future, we will review our CNN settings to explore the possibility of improving classification accuracy, for example, by changing the number of convolution layers. In addition, we will apply our method in different settings.

References

1. Strese, M., Boeck, Y., Steinbach, E.: Content-based surface material retrieval. In: 2017 IEEE World Haptics Conference (WHC), Fürstenfeldbruck (Munich), Germany, pp. 352–357 (2017)
2. Abdulali, A., Jeon, S.: Data-driven modeling of anisotropic haptic textures: data segmentation and interpolation. In: Bello, F., Kajimoto, H., Visell, Y. (eds.) Euro-Haptics 2016. LNCS, vol. 9775, pp. 228–239. Springer, Cham (2016). https://doi.org/10.1007/978-3-319-42324-1_23
3. Saga, S., Nakagawa, M., Ono, T., Pan, Z., Zhang, J.: Daily haptic information collection system using zigbee microcontrollers. In: Technical Meeting on "Perception Information", vol. 2017, pp. 11–14. IEEE Japan (2017). (in Japanese)
4. Minamizawa, K., Kakehi, Y., Nakatani, M., Mihara, S., Tachi, S.: TECHTILE toolkit: a prototyping tool for designing haptic media. In: ACM SIGGRAPH 2012 Emerging Technologies. ACM (2012)
5. Mono Wireless Inc. TWE-Lite-2525A. https://mono-wireless.com/jp/products/TWE-Lite-2525A
6. Google Inc., Tensorflow. https://www.tensorflow.org/
7. Simonyan, K., Zisserman, A.: Very deep convolutional networks for large-scale image recognition. arXiv:1409.1556 (2014)
8. Glorot, X., Bordes, A., Bengio, Y.: Deep sparse rectifier neural networks. In: Proceedings of the Fourteenth International Conference on Artificial Intelligence and Statistics, pp. 315–323 (2011)
9. Ioffe, S., Szegedy, C.: Batch normalization: accelerating deep network training by reducing internal covariate shift. In: International Conference on Machine Learning, pp. 448–456 (2015)

Haptic Interface Technologies Using Perceptual Illusions

Tomohiro Amemiya[(⊠)] [iD]

NTT Communication Science Laboratories, 3-1 Morinosato-Wakamiya,
Atsugi-shi, Kanagawa 243-0198, Japan
amemiya.tomohiro@ieee.org

Abstract. With virtual reality now accessible to anyone through high-end consumer headsets and input devices, researchers are seeking cost-effective designs based on human perceptual properties for virtual reality interfaces. The author has been studying a sensory-illusion-based approach to designing human-computer interface technologies. This paper overviews how we are using this approach to develop force displays that elicit illusory continuous force sensations by presenting asymmetric vibrations and kinesthetic displays based on a cross-modal effect among visual, auditory, and tactile cues of self-motion.

Keywords: Haptics · Somatosensation · Sensory illusion

1 Introduction

After the several decades since the emergence of the virtual reality (VR) concept, the cost of the hardware for an immersive VR experience has drastically decreased [17]. Today, VR is accessible to everyone thanks to inexpensive high-end consumer-friendly head-mounted displays and input devices. However, a limited number of haptic interface displays, such as tactile or force feedback displays, have been adopted in VR systems compared with the audiovisual ones. This is mainly due to the technical difficulty in reproducing a haptic or somatosensory experience with inexpensive haptic displays.

On the other hand, innovative information displays can be designed that consider the characteristics of human perception and optimize information technologies for it. The perceptual limits on the human sensory system have often been considered to determine guidelines for designing audiovisual displays. Examples of these guidelines are video frame rates that produce smooth motion and *perceptual coding* in audio data compression for natural sound. In addition, human sensory illusions are often utilized to invent innovative audiovisual displays. Thus, some haptic interface displays can be built on the basis of the sensory-illusion-based approach without incurring much cost. This paper introduces a novel approach that exploits the nonlinearity of human perception and sensory integration for developing somatosensory and kinesthetic displays.

© Springer International Publishing AG, part of Springer Nature 2018
S. Yamamoto and H. Mori (Eds.): HIMI 2018, LNCS 10904, pp. 168–174, 2018.
https://doi.org/10.1007/978-3-319-92043-6_14

2 Haptic Displays Using Perceptual Illusions

2.1 Directed Force Perception by an Asymmetric Oscillation

Over the past two decades, a great number of force displays have been developed and studied. Most of them are grounded force displays, such as PHANToM and SPIDAR, which use mechanical linkages to establish a fulcrum relative to the ground. The fulcrum (grounding support) is required for grounded displays because of the action-reaction principle. However, since mobile devices lack a fulcrum, most conventional force display systems for mobile devices can produce neither a constant nor translational force; that is, they can generate only short-term rotational force (e.g., using the gyro effect [24] or angular momentum change [23]). Thus, in mobile devices, the haptic cues are usually limited to simple vibrotactile ones.

The author and his colleagues have succeeded in creating a force sensation of being pulled or pushed with various kinds of mobile apparatuses. The display, called Buru-Navi® [1, 2, 9], creates both a constant- and translational-force sensation by utilizing the nonlinear characteristics of human perception. The trick is to use different acceleration patterns for two directions to create a perceived force imbalance. A brief and strong force is generated in a desired direction, while a weaker one is generated over a longer period of time in the direction opposite to the desired one. Although the temporal average of the net force is physically zero (e.g., the average of the forces in each direction are the same), people who hold a device vibrating by the acceleration patterns feel as if they are being pulled to one direction because the amplitude of the weaker force is adjusted to be below a sensory threshold.

Over the past ten years, the author has been refining a method to create a sensory illusion of being pulled with a slider-crank mechanism [1–5] or spring-cam mechanism [6], and, during that time, he has developed various prototypes of ungrounded force displays as shown in Fig. 1. The author has succeeded in reducing the size and weight of the force display remarkably by using a linear electromagnetic actuator [7, 9]. The prototype was designed to be pinched by the fingers because the author focused on the finger pad, which is one of the most sensitive body surfaces. In the human finger pad, there are four major tactile mechanoreceptors, which are Pacinian corpuscles, Merkel disks, Meissner corpuscles, and Ruffini endings. The Pacinian corpuscles are sensitive to vibrations with high temporal frequencies from 100 to 300 Hz in the normal direction [13], but they seem to have nearly the same sensitivity to sliding directions tangential to the skin [22]. In contrast, SA-I and RA-I fibers, whose signals are thought to come mainly from the Merkel disks and Meissner corpuscles, respectively, are sensitive to vibrations with lower frequencies (less than 100 Hz), and some of them can clearly code the sliding or tangential force direction [18]. The SA-II fiber innervating Ruffini endings is also sensitive to skin stretch, particularly to tangential forces generated in the skin. Therefore, to create an illusory force sensation of being pulled, the author designed an asymmetrical oscillation pattern that contains the frequency components stimulating these receptors (i.e., less than 100 Hz) and contains the asymmetric magnitude which exceeds the thresholds of shearing displacement in one direction but not in the other.

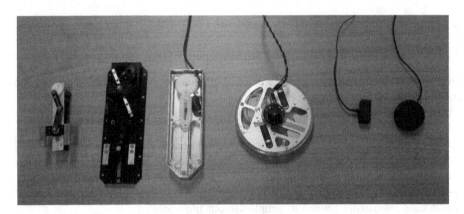

Fig. 1. Prototypes of haptic displays generating an asymmetric oscillation to induce a sensation of being pulled or pushed. The second rightmost one is $18 \times 18 \times 37$ mm (1 DoF) and the rightmost one is $\varphi 40 \times 17$ mm (2 DoF).

Our previous studies on force direction discrimination of the asymmetric oscillation confirmed that almost all the participants felt a clear illusory force of being pulled or pushed persistently and that they were able to distinguish the force direction correctly.

2.2   Wayfinding by an Illusory Force

An illusory force of being pulled as introduced above has the potential to be applied for pedestrian navigation without visual or auditory information, as if someone is being led by the hand. Furthermore, Buru-Navi® can be used by people with visual impairments to provide directional cues for wayfinding. Our previous study with a fire department showed that 91% of participants with visual impairment were able to walk safely along a predefined route without any prior training [3]. In a fire emergency, smoke hinders the visual field of people at the scene. Thus, Buru-Navi® will be useful for sighted people as well as for people with visual impairments in such a situation.

After the experiment, we raised the following question: What is the most effective and efficient way for users to understand the force direction in the application? One may think that it would be helpful to always update the force direction and maintain the same direction in global coordinates depending on the orientation of the force display, much like a compass needle points to the direction of the magnetic north. It is unclear whether the active exploration of the direction of an illusory force by hand or arm movement improves the perception of the force direction. This is because some studies have reported that tactile processing is suppressed by hand movement [15], while others have reported that active touch sensing facilitates tactile performance [18]. We have shown that active manual movement in both the rotational and translational directions enhances the precise perception of the direction of an illusory force created by Buru-Navi® [10], which suggests that the active exploration of force direction by moving the arm or hand is a good strategy for understanding the direction in a pedestrian navigation application.

3 Haptic Displays Inducing Self-Motion Perception

Self-motion, one of the most frequent movements in daily life, is experienced when we walk or ride in a vehicle. In VR theme parks or VR amusement centers, some types of self-motion can be expressed while the user sits on a chair-like vehicle, such as driving simulators or motion seats. This section introduces three studies on perceptual illusions of self-motion using haptic displays with participants seated on a chair-like vehicle.

3.1 Change in Velocity Perception of Self-Motion

In chair-like vehicles, velocity information is detected by means of visual sensory cues, and acceleration and angular acceleration information is detected by means of mechanical ones, such as vestibular and tactile sensations. However, it is unclear whether the tactile information integrated with visual information alters the velocity perception of self-motion, because some similarities exist between the perceptual and neural mechanisms when motion stimuli of vision and touch are processed.

To investigate this, we examined whether the forward velocity of self-motion is altered by applying rapid tactile flows using a vibratotactile array on a seat pan. In the study, participants viewed optical flows while gazing at a fixation cross and sitting on a tactile stimulator on the seat pan. The brief tactile motion stimulus consisted of four successive rows of 200-ms vibration with a frequency of 50 Hz, and the inter-stimulus onset between the tactile rows was varied to change the velocity of the tactile motion. The experimental results showed that the forward velocity of self-motion is significantly overestimated for rapid tactile flows and underestimated for slow ones, compared with only optical flow or non-motion vibrotactile stimulation conditions. Furthermore, temporal tactile rhythm patterns (i.e., a train of taps) did not affect the perceived velocity of self-motion as much as tactile flow stimuli, especially when the inter-stimulus onset interval was appropriate for eliciting a clear sensation of tactile apparent motion, which indicates the importance of the spatiotemporal feature of tactile motion stimuli in modulating the velocity [8].

3.2 Change in Topographic Surface Perception

It has been reported that we often misperceive a surface topography simply based on visual cues at "magnetic hills", where a slight downhill slope appears to be an uphill one due to the surroundings. We have investigated multimodal perception of a topographic surface induced by visual and body-tilt stimuli [12].

One remarkable study on haptic perception has shown that humans more strongly perceive the shape of an object during active touch from the force profile applied to the finger than from the position profile of the finger [21]. This implies that the shape perception could be induced by local changes in topographic information without vertical movement, although the shape perception by the finger and by the body will differ. To verify the hypothesis that shape perception could be induced by body tilt, we constructed an experimental system using a motion chair with two DOF in roll and pitch rotations and conducted a user study to classify the perceived shape based on visual and vestibular cues. Experimental results show that the vestibular shape cue contributed to making the shape perception larger than the visual one [12]. This result suggests that concave and convex surfaces can be expressed with only two-DOF rotating motion of the body without the participant's moving in the vertical direction while sitting on a chair-like vehicle.

3.3   Change in Boundary of Peripersonal Space

We have found that vibration on the foot sole induces a clear walking sensation even when a person is seated on a chair. Furthermore, we found that the magnitude of the subjective sensation of walking affected the objective scores, such as the reaction times of a tactile detection task [11].

Previous studies have shown that a moving sound that gives an impression of a sound source approaching the body boosts tactile reaction times when it is presented close to the stimulated body part, that is, within and not outside the PPS [14]. For instance, the PPS representation of the chest expands in the direction of walking [20]. Based on these results, we presented several vibration patterns on the soles of the feet of seated participants to evoke a sensation of pseudo-walking and examined the change in reaction times to detect a vibrotactile stimulus on the chest while the participants listened to a looming sound approaching their body, which was taken as a behavioral proxy for the PPS boundary [11]. Results revealed that a cyclic vibration consisting of low-pass-filtered walking sounds presented at the soles that clearly evoked a sensation of walking decreased the reaction times, indicating that the PPS boundary was expanded forward by inducing a sensation of walking (Fig. 2).

Fig. 2. Sole vibration to evoke a sensation of pseudo-walking expands the boundary of peripersonal space. The seated participant received a vibration pattern on the sole. Tactile reaction times on the chest when listing to a looming sound approaching the body decreased when the vibration patterns were rated high for pseudo-walking were applied, indicating that the boundary was expanded forward.

4 Conclusion

This paper introduced haptic displays based on perceptual illusions and multisensory stimuli. The sense of touch is very powerful for presenting a feeling of the existence of objects. Thus, it has been thought that there are few perceptual illusions in the haptic modality. However, several perceptual illusions in the haptic modality have been reported [16], and some have been implemented for novel information displays. This trend will continue because vision and touch work together to create a richer experience. Future information displays will ultimately utilize not only human perceptual aspects but also human perceptual flaws, such as sensory and perceptual illusions.

References

1. Amemiya, T., Ando, H., Maeda, T.: Virtual force display: direction guidance using asymmetric acceleration via periodic translational motion. In: Proceedings of World Haptics Conference 2005, pp. 619–622 (2005)
2. Amemiya, T: Haptic interface using sensory illusion. In: Tutorial in IEEE Virtual Reality 2008, Integration of Haptics in Virtual Environments: from Perception to Rendering, Reno, NV (2008)
3. Amemiya, T., Sugiyama, H.: Orienting kinesthetically: a haptic handheld wayfinder for people with visual impairments. ACM Trans. Access. Comput. 3(2), 1–23 (2010). Article 6
4. Amemiya, T., Maeda, T.: Asymmetric oscillation distorts the perceived heaviness of handheld objects. IEEE Trans. Haptics 1(1), 9–18 (2008)
5. Amemiya, T., Maeda, T.: NOBUNAGA: multicylinder-like pulse generator for kinesthetic illusion of being pulled smoothly. In: Ferre, M. (ed.) EuroHaptics 2008. LNCS, vol. 5024, pp. 580–585. Springer, Heidelberg (2008). https://doi.org/10.1007/978-3-540-69057-3_75
6. Amemiya, T., Ando, H., Maeda, T.: Hand-held force display with spring-cam mechanism for generating asymmetric acceleration. In: Proceedings of World Haptics Conference 2007, Tsukuba, Japan, pp. 572–573 (2007)
7. Amemiya, T., Gomi, H.: Distinct pseudo-attraction force sensation by a thumb-sized vibrator that oscillates asymmetrically. In: Auvray, M. (ed.) EUROHAPTICS 2014. LNCS, vol. 8619, pp. 88–95. Springer, Heidelberg (2014). https://doi.org/10.1007/978-3-662-44196-1_12
8. Amemiya, T., Hirota, K., Ikei, Y.: Tactile apparent motion on the torso modulates perceived forward self-motion velocity. IEEE Trans. Haptics 9(4), 474–482 (2016)
9. Amemiya, T.: Perceptual illusions for multisensory displays. Invited talk. In: Proceedings of the 22nd International Display Workshops (IDW 2015), Otsu, Japan, vol. 22, pp. 1276–1279 (2015)
10. Amemiya, T., Gomi, H.: Active Manual Movement Improves Directional Perception of Illusory Force. IEEE Trans. Haptics 9(4), 465–473 (2016)
11. Amemiya, T., Ikei, Y., Hirota, K., Kitazaki, M.: Vibration on the Soles of the Feet Evoking a Sensation of Walking Expands Peripersonal Space. In: Proceedings of IEEE World Haptics 2017, Munich, Germany, pp. 234–239 (2017)
12. Amemiya, T., Hirota, K., Ikei, Y.: Topographic surface perception modulated by pitch rotation of motion chair. In: Proceedings of 18th International Conference on Human-Computer Interaction (HCI International 2016), Toronto, Canada, pp. 483–490 (2016)

13. Bolanowski Jr., S.J., Gescheider, G.A., Verrillo, R.T., Checkosky, C.M.: Four channels mediate the mechanical aspects of touch. J. Acoust. Soc. Am. **84**(5), 680–694 (1988)
14. Canzoneri, E., Magosso, E., Serino, A.: Dynamic sounds capture the boundaries of peripersonal space representation in humans. PLoS ONE **7**(9), e44306 (2012)
15. Chapman, C.E., Bushnell, M., Miron, D., Duncan, G., Lund, J.: Sensory perception during movement in man. Exp. Brain Res. **68**(3), 516–524 (1987)
16. Hayward, V.: A brief taxonomy of tactile illusions and demonstrations that can be done in a hardware store. Brain Res. Bull. **75**(6), 742–752 (2008)
17. Hirose, M.: The second generation virtual reality technology. Keynote/invited Speech. In: Proceedings of 16th International Conference on Virtual Systems and Multimedia (VSMM 2010), Seoul, Korea (2010)
18. Lederman, S.J., Klatzky, R.L.: Hand movements: a window into haptic object recognition. Cogn. Psychol. **19**(3), 342–368 (1987)
19. Maeno, T., Kobayashi, K., Yamazaki, N.: Relationship between the structure of human finger tissue and the location of tactile receptors. Bull. JSME Int. J. **41**, 94–100 (1998)
20. Noel, J.-P., Grivaz, P., Marmaroli, P., Lissek, H., Blanke, O., Serino, A.: Full body action remapping of peripersonal space: the case of walking. Neuropsychologia **70**, 375–384 (2015)
21. Robles-De-La-Torre, G., Hayward, V.: Force can overcome object geometry in the perception of shape through active touch. Nature **412**(6845), 445–448 (2001)
22. Srinivasan, M.A., Whitehouse, J.M., Lamotte, R.H.: Tactile detection of slip: surface microgeometry and peripheral neural codes. J. Neurophysiol. **63**(6), 323–332 (1990)
23. Tanaka, Y., Masataka, S., Yuka, K., Fukui, Y., Yamashita, J., Nakamura, N.: Mobile torque display and haptic characteristics of human palm. In: Proceedings of ICAT 2001, pp. 115–120 (2001)
24. Yano, H., Yoshie, M., Iwata, H.: Development of a non-grounded haptic interface using the gyro effect. In: Proceedings of HAPTICS 2003, pp. 32–39 (2003)

Assessing Multimodal Interactions
with Mixed-Initiative Teams

Daniel Barber[✉]

University of Central Florida, Institute for Simulation and Training, Orlando, FL
32826, USA
dbarber@ist.ucf.edu

Abstract. The state-of-the-art in robotics is advancing to support the warfighters' ability to project force and increase their reach across a variety of future missions. Seamless integration of robots with the warfighter will require advancing interfaces from teleoperation to collaboration. The current approach to meeting this requirement is to include human-to-human communication capabilities in tomorrow's robots using multimodal communication. Though advanced, today's robots do not yet come close to supporting teaming in dismounted military operations, and therefore simulation is required for developers to assess multimodal interfaces in complex multi-tasking scenarios. This paper describes existing and future simulations to support assessment of multimodal human-robot interaction in dismounted soldier-robot teams.

Keywords: Multimodal interfaces · Human-robot interaction · Simulation
Tactile displays

1 Introduction

A desire to support the warfighters' ability to project force and increase their reach across a variety of future operations has resulted in a concerted push to advance the state-of-the-art in robotics. In current ground operations, robots are remote controlled assets supporting tasks where it is infeasible or unsafe for personnel to go (e.g. disposal of improvised explosive devices). These systems do not function collaboratively with human counterparts, requiring additional labor to not only manage the robot, but also provide force protection, which may instead create additional workload and reduce the controller's ability to perform secondary tasks, [1–3]. Although modernizations have taken place to make interfaces in dismounted applications more lightweight and portable, current interfaces for teleoperation focus on a one-to-one relationship where an operator observers sensor feeds (e.g. video) and manipulates hand controls, keeping their gaze heads-down, Fig. 1, [4].

Revolutionizing collaboration with robots will require a leap forward in their autonomy and equal development of robust interfaces. The current approach to meeting this goal is to design interfaces that model how human teammates interact today. Enabling a Soldier to use what is already familiar to them, such as speech, gestures, and vocabulary, will facilitate a seamless integration of robot counterparts. Building robot teammates that embed familiar communication methods will reduce the need for

© Springer International Publishing AG, part of Springer Nature 2018
S. Yamamoto and H. Mori (Eds.): HIMI 2018, LNCS 10904, pp. 175–184, 2018.
https://doi.org/10.1007/978-3-319-92043-6_15

Fig. 1. Interface for teleoperation of a PacBot 310 robot, [5]

training to allow Soldiers to take advantage of these new assets, and lower demands on the Soldier. Incorporating these types of interactions will lead to the creation of collaborative mixed-initiative teams where Soldiers and robots take on different roles at different times to optimize the team's ability to accomplish mission objectives, [6].

2 Multimodal Interaction

Developing advanced interfaces for human-robot collaboration that are modeled after human-to-human interactions will inherently require multimodal support. Throughout the literature, six common themes in multimodal communication efforts emerge: meaning, context, natural, efficiency, effectiveness, and flexibility. Numerous authors use multimodal communication to strive for meaning and context [7–9], more complex conveyance of information over multiple modes compared to single mode [10], and delivery of ideas redundantly (back up signals) and non-redundantly (multiple messages) [11, 12]. Ultimately, multimodal communication supports multiple levels of complexity [10].

In an effort to scope research efforts within the context of dismounted human-robot interaction (HRI), Lackey, et al. operationally defined multimodal communication as "the exchange of information through a flexible selection of explicit and implicit modalities that enables interactions and influences behavior, thoughts, and emotions," [13]. Leveraging this definition, explicit communication types from the literature for investigation within an HRI multimodal interface emerge and include speech, auditory cues, visual signals, and visual and tactile displays.

Future multimodal interfaces must support some or all of these explicit methods of communication to enable assessment of mixed-initiative team interactions. For example, to take full advantage of the auditory modality, interfaces must include functionality for both speech-to-text (STT), text-to-speech (TTS), natural language understanding (NLU), and other sound effects. Gestures are a common and natural form of communication among humans within the visual modality, and as such, robots must classify them. In addition to traditional visual displays (e.g. tablets), robots could also deliver their own gestures from manipulators and other body movements. Finally, an emerging field of research showing potential benefits is tactile displays. Tactile displays exist in many commercial-off-the-self products such as cell phones and smart watches that emit haptic cues for calls and text messages. In respect to dismounted operations, researchers using tactile belts have demonstrated improved navigation performance and wearer's ability to classify up to two-word phrases approaching the complexity of speech, [14, 15].

In an attempt to bring these individual technologies together, Barber et al. developed and assessed a prototype multimodal interface as part of the Robotics Collaborative Technology Alliance (RCTA), [16–18]. The RCTA's Multimodal Interface (MMI) supports multiple modalities for transactional communication with a robot teammate. With voice data captured on a Bluetooth headset, the Microsoft Speech Platform SDK version 11 classified speech commands to text, that were then converted into robot instructions using a natural language understanding module, [19]. For visual signals, a custom gesture glove captured arm and hand movement using an inertial-measurement unit, which a statistical model classified into gestures. Previous efforts using this glove have shown a capacity to classify 21 unique arm and hand signals, many from the Army Field Manual for Visual Signals, [20]. For robot-to-human communication, the MMI supports TTS, auditory cues, and a visual display. The MMI visual display contains all current mission information from the robot, including a semantic map, live video-feed, current command, and status, Fig. 2.

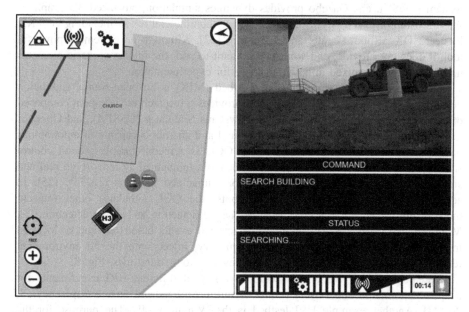

Fig. 2. Multimodal Interface (MMI) visual display. Display supports three primary areas, semantic map (left), video feed (top right), and command/status information (bottom right).

Through this combination of interfaces, users were able to give a complex speech command such as "screen the back of the building," receive confirmation, observe the robot execute, and receive feedback of mission completion without the need to look heads down at the visual display, [21]. Results from a field study revealed participants liked the ability to use multiple modalities and the interface form factor, but requested modifications to iconography, more intuitive gesture commands, and increased transparency into robot logic, [16]. Although successful in demonstrating a baseline level of human-robot interaction and multimodal assessment in a dismounted application, the types of mixed-initiative teaming available to researchers is limited to the capabilities of today's robots. The state-of-the-art in robot autonomy still does not come close to supporting human-teaming concepts such as back-up behaviors, shared mental models, and information prioritization. Without robots able to adequately perform in larger mixed-team scenarios and roles, they cannot drive adaptive multimodal interfaces with manipulation of modalities, information chunking, transparency, and dynamic reporting frequency for maintaining team situation awareness and performance.

3   Simulating Mixed-Initiative Teaming

There are several challenges when attempting to identify simulation environments for HRI experiments. For a given simulation to support mixed-initiative assessment, it must include tasks for participants to perform with robotic assets, however the majority of these environments focus on the engineering aspects of the robotic and not human interaction. For example, Gazebo is a robot simulation tool supporting Robot Operating System (ROS) users. Gazebo provides dynamics simulation, advanced 3D graphics, sensors with noise, and physics models of many commercial robots including the PR2, Pioneer2 DX, iRobot Create, and TurtleBot, [22]. Although an extremely powerful tool, it does not provide, and would be difficult to add, the human scenario elements needed to simulate mixed-initiative teaming for experimentation.

The Mixed-Initiative Experimental Testbed (MIX) is an open-source simulation designed up front for HRI, [23, 24]. MIX provides a research environment composed of two main applications: the Unmanned Systems Simulator (USSIM), and Operator Control Unit (OCU). USSIM simulates ground and air robots capable of autonomous navigation within a 3D environment. The OCU is a reconfigurable ground control station interface capable of managing one or more unmanned systems (both real and simulated) using the Joint Architecture for Unmanned Systems (JAUS), [25]. In addition to command and control of robots, the OCU simulates other relevant theoretically-driven mission tasks such as change detection and signal detection, [3]. Moreover, MIX generates a multitude of scripted events based on time or location triggers including: display visuals using imagery, injects into the 3d environment returned over robot video feeds, and audio events from sound files, Fig. 3.

Researchers using MIX have setup a variety of different HRI experiments for adaptive automation, supervisory control of multiple robots, and agent transparency, [3, 26, 27]. Another example HRI testbed is the Wingman SIL. The purpose for the Wingman program is to provide robotic technological advances and experimentation to increase the autonomous capabilities of mounted and unmanned combat support

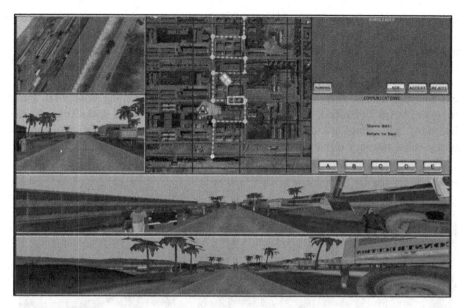

Fig. 3. MIX Testbed Operator Control Unit (OCU) simulation. OCU includes video feeds from multiple robots (i.e. air, ground) and a 360 degree indirect vision display, map for route/mission planning, and dialogs for interaction with automated agents and command, [26]. Users configure OCU layout, graphics, content, and scenario using XML.

vehicles, [28]. The Wingman SIL includes Warfighter Machine Interfaces (WMI) for a Mobility Operator, Vehicle Commander, and Robotic Gunner. Combined together, this environment supports simulation of team combat exercises. Although capable of facilitating many research efforts, systems like MIX and Wingman SIL are focused on manned (i.e. in-vehicle) missions, (e.g. supervisory control for intelligence, reconnaissance, and surveillance), and are therefore not designed for dismounted teaming studies. Without extensive modification, they are not capable of modeling scenarios where researchers can have participants act as a squad with a robot, communicating with speech and gestures similar to interactions previously described using the RCTA MMI.

The Enhanced Dynamic Geo-Social Environment (EDGE) is a multiplayer, scalable, online training environment for first responders. Developed by the US Army Research Lab (ARL), Human Research Engineering Directorate (HRED), Simulation and training Technology Center (STTC) in partnership with TRADOC G2 and the Department of Homeland Security (DHS), EDGE is a government owned platform built using the Unreal Game Engine 4, [29, 30]. Designed for extension to other applications, researchers at the University of Central Florida working with ARL leveraged EDGE to support simulation of robots in dismounted operations. Under this effort, the RCTA MMI (henceforth referred to as MMI) was modified to communicate with a modified version of EDGE called the Visualization Testbed (VTB) that included a simulated robot. This virtual robot was capable behaviors emulating the semantic navigation capabilities driven from speech commands previously demonstrated with real platforms by Barber et al., [16, 19, 21]. For example, a user issues a speech command such as "go to the north side of the bridge," and receives acknowledgement and task status as robot executes the mission. In addition to integration with VTB, the MMI was further extended to support simulation of other content and tasks. These modifications enable using time or location-based events to trigger updates to the MMI text, images, color scheme, and map independent of content coming from VTB. Furthermore, when combined, the simulations support theory-based tasks (e.g. signal detection) and multi-tasking similar to the MIX testbed, but in dismounted human-robot teaming scenarios, Fig. 4.

Fig. 4. VTB and MMI Simulation for dismounted human-robot teaming. VTB generates virtual world images, content, and simulated robot, with the MMI (overlaid top-middle) supporting interactions with the robot and other simulated communications and tasks. Characters moving in the environment support a signal detection task.

In addition to support for different interaction types, any assessment in dismounted scenarios also requires an ability for researchers to model relevant missions. One of the most frequent operations a Soldier may perform in a team is cordon and search. Cordon and search is complex in nature, with reconnaissance, enemy isolation and capture, and weapons and materials seizures, [31]. This combination of tasks makes cordon and search ideal for multimodal HRI experimentation. The VTB/MMI environment is capable of supporting cordon and search and other dismounted missions for human-in-the-loop studies. Using VTB to simulate robot teammates and an outer cordon task from the Soldier's perspective and the MMI for robot reports and commands, one can explore a variety of use-cases. In a recent example of this, in an effort to investigate adaptive multimodal communication, researchers conducted an experiment to assess recall of different robot reports using single (visual, auditory) and dual (visual and auditory) communication modalities under different environmental demands for a cordon and search mission, [32]. Although promising, there were limits to representing a dis-mounted mission, in that the task was performed on a desktop workstation, which does not provide the type of immersion or demands that may be needed to translate findings to the real world. Moreover, without the ability to provide some semblance of robot presence with participants, researchers cannot study implicit communication (e.g. social distance) or anthropomorphic affects in HRI.

4 Virtual Reality for Dismounted HRI

With recent advancements in commercial-off-the-shelf virtual reality (VR) displays, the cost associated with immersing someone in a virtual world is dramatically reduced; making incorporation of VR into human-in-the-loop experiments approachable to researchers. The HTC VIVE VR system is an example of this, with a cost of $600 and direct integration support for multiple game engines, including Unreal Engine 4 (UE4), [33, 34]. Using VR, one can address the gaps associated with the desktop-based VTB/MMI simulation for enhanced empirical validity and to cover a broader range of research. To meet this goal, this paper proposes a new simulation platform called VRMIX, which combines the UE4-based VTB simulation, MMI, and HTC VIVE to produce an immersive virtual world for exploration of multimodal interactions with mixed-initiative teams. With direct support for the HTC VIVE, developers can update VTB cordon and search scenarios for use cases where participants are "physically" present with characters and robots in the scene. The MMI, previously integrated with VTB for sharing of data, only requires integration of any visual display elements within VTB, as speech and audio modalities are supported with existing hardware (e.g. microphone, speakers), Fig. 5.

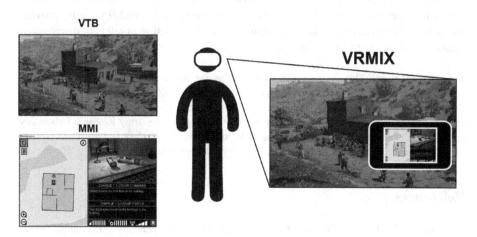

Fig. 5. VRMIX concept. Through the combination of the UE4-based VTB simulation and MMI software (left), and virtual reality headset, participants are immersed within a dismounted mission where they may interact using speech, gestures, and visual display in game (right).

In order for users to access the visual display of the MMI within VR, VTB is modified to perform screen captures (i.e. frame grabbing) of the actual MMI display software and render it in game. Thus, users are able to perform a visual search as they would in the real world by looking around with their head, command with speech and gestures, receive auditory cues, TTS, and simulated radio chatter, and access a visual display within a complex multi-tasking scenario. Furthermore, the tactile modality can be supported using the haptic channel of the HTC VIVE controllers or with integration of a tactile display. Moreover, the first-person nature of VR lends itself well to the egocentric spatial characteristics of tactile belts, [14]. Thus, VRMIX has the potential to provide a means of investigating all modalities in a laboratory setting with maximum fidelity and experimental control.

5 Conclusion

The goal for this paper is to discuss robotics research and technologies for advancing human robot collaboration, and what is needed to assess these future mixed-initiative teams. There is a clear demand to improve soldier-robot teaming established in congressional mandate and Department of Defense (DoD) funded research programs, [17, 35]. However, the state of the art in artificial intelligence and the cost to emulate the complexity of real-world mission scenarios (e.g. cordon and search) requires researchers to rely heavily on simulation. Simulation provides a means to explore future robot capabilities, keep costs low, and enable experimental control for assessment of multimodal communication. Many robotics simulation environments focus on simulation of sensors and physics for robotics development, with few platforms supporting human robot interaction. In order to drive future requirements, interface capabilities, and understand the human factors of multimodal communication, a new simulation environment called VRMIX was presented. VRMIX will provide researchers the necessary tools to assess multimodal interaction in relevant dismounted military missions with robot capabilities yet to come.

Acknowledgement. This research was sponsored by the Army Research Laboratory and was accomplished under Cooperative Agreement Number W911NF-10-2-0016. The views and conclusions contained in this document are those of the author's and should not be interpreted as representing the official policies, either expressed or implied, of the Army Research Laboratory or the U.S. Government. The U.S. Government is authorized to reproduce and distribute reprints for Government purposes notwithstanding any copyright notation herein.

References

1. Amazon.: HTC VIVE Virtual Reality System (HTC), 08 February 2018. https://www.amazon.com/HTC-VIVE-Virtual-Reality-System-pc/dp/B00VF5NT4I?th=1. Accessed 02 Aug 2018
2. Barber, D.J., Leontyev, S., Sun, B., Davis, L., Nicholson, D., Chen, J.Y.: The mixed initiative experimental (MIX) Testbed for collaborative human robot interactions. In: Army Science Conference. DTIC, Orlando (2008)

3. Barber, D.J., Reinerman-Jones, L.E., Matthews, G.: Toward a tactile language for human-robot interaction: two studies of tacton learning performance. Hum. Factors 57(3), 471–490 (2014). https://doi.org/10.1177/0018720814548063

4. Barber, D., Abich IV, J., Phillips, E., Talone, A., Jentsch, F., Hill, S.: Field assessment of multimodal communication for dismounted human-robot teams. In: The Proceedings of the Human Factors and Ergonomics Society Annual Meeting, Los Angeles, CA, vol. 59, pp. 921–925. SAGE Publications (2015)

5. Barber, D., Carter, A., Harris, J., Reinerman-Jones, L.: Feasibility of wearable fitness trackers for adapting multimodal communication. In: Yamamoto, S. (ed.) HIMI 2017. LNCS, vol. 10273, pp. 504–516. Springer, Cham (2017). https://doi.org/10.1007/978-3-319-58521-5_39

6. Barber, D., Howard, T., Walter, T.: A multimodal interface for real-time soldier-robot teaming. In: SPIE Defense, Security, and Sensing - Unmanned Systems Technology, Baltimore, Maryland USA (2016)

7. Barber, D., Lackey, S., Reinerman-Jones, L., Hudson, I.: Visual and tactile interfaces for bi-directional human robot communication. In: SPIE Defense, Security, and Sensing - Unmanned Systems Technology. Baltimore, Maryland USA (2013)

8. Bischoff, R., Graefe, V.: Dependable multimodal communication and interaction with robotic assistants. In: 11th IEEE International Workshop on Robot and Human Interactive Communication, pp. 300–305. IEEE (2002)

9. Chen, J.Y., Barnes, M.J., Qu, Z.: RoboLeader: an agent for supervisory control of multiple robots. In: Proceedings of the 5th ACM/IEEE international conference on Human-robot interaction (HRI 2010), pp. 81–82 (2010)

10. Chen, J., Joyner, C.: Concurrent performance in gunner's and robotic tasks and effects of cueing in a simulated multi-tasking environment. In: Proceedings of the Human Factors and Ergonomics Society 52nd Annual Meeting, pp. 237–241 (2009)

11. Childers, M., Lennon, C., Bodt, B., Pusey, J., Hill, S., Camden, R., Navarro, S.: US army research laboratory (ARL) robotics collaborative technology alliance 2014 capstone experiment. Army Research Laboratory, Aberdeen Proving Ground (2016)

12. Cosenzo, K., Chen, J., Reinerman-Jones, L., Barnes, M., Nicholson, D.: Adaptive automation effects on operator performance during a reconnaissance mission with an unmanned ground vehicle. In: Proceedings of the Human Factors and Ergonomics Society 54th Annual Meeting, Los Angeles, CA, pp. 2135–2139 (2010)

13. Elliot, L.R., Duistermaat, M., Redden, E., Van Erp, J.: Multimodal Guidance for Land Navigation. U.S. Army Research Laboratory, Aberdeen Proving Ground (2007)

14. Endeavor Robotics. (2018). Endeavor Robotics Products (uPOINT). (Endeavor Robotics). http://endeavorrobotics.com/products. Accessed 02 May 2018

15. EPIC.: Setting up UE4 to work with SteamVR, 08 February 2018 (EPIC). https://docs.unrealengine.com/latest/INT/Platforms/SteamVR/QuickStart/2/. Accessed 02 Aug 2018

16. Glass, D.R.: Taking Training to the EDGE, 14 March 2014 (Orlando Marketing & PR Firm Capital Communications). http://www.teamorlando.org/taking-training-to-the-edge/. Accessed 02 Feb 2018

17. Griffith, T., Ablanedo, J., Dwyer, T.: Leveraging a Virtual Environment to Prepare for School Shootings. In: Lackey, S., Chen, J. (eds.) VAMR 2017. LNCS, vol. 10280, pp. 325–338. Springer, Cham (2017). https://doi.org/10.1007/978-3-319-57987-0_26

18. Hearst, M., Allen, J., Guinn, C., Horvitz, E.: Mixed-initiative interaction: trends and controversies. IEEE Intell. Syst. 14, 14–23 (1999)

19. Kvale, K., Wrakagoda, N., Knudsen, J.: Speech centric multimodal interfaces for mobile communication. Telektronikk 2, 104–117 (2003)

20. Laboratory, U.A., Schaefer, K.E., Brewer, R.W., Pursel, R.E., Zimmermann, A., Cerame, E., Briggs, K.: Outcomes from the first wingman software-in-the-loop integration event: January 2017. US Army Research Laboratory (2017)
21. Lackey, S. J., Barber, D. J., Reinerman-Jones, L., Badler, N., Hudson, I.: Defining next-generation multi-modal communication in human-robot interaction. In: Human Factors and ERgonomics Society Conference. Las Vegas: HFES (2011)
22. Nigay, L., Coutaz, J. A Design Space for Multimodal Systems: Concurrent Processing and Data Fusion. In: INTERACT 1993 and CHI 1993 Conference on Human Factors in Computing Systems, pp. 172–178 (1993)
23. Oh, J., et al.: Integrated intelligence for human-robot teams. In: Kulić, D., Nakamura, Y., Khatib, O., Venture, G. (eds.) ISER 2016. SPAR, vol. 1, pp. 309–322. Springer, Cham (2017). https://doi.org/10.1007/978-3-319-50115-4_28
24. Open Source Robotics Foundation, 25 January 2018. Gazebo. http://gazebosim.org/. Accessed 14 Feb 2018
25. Parr, L.: Perceptual biases for multimodal cues in chimpanzee (Pan troglodytes) affect recognition. Anim. Cogn. **7**, 171–178 (2004)
26. Partan, S., Marler, P.: Communication goes multimodal. Science **283**(5406), 1272–1273 (1999)
27. Raisamo, R.: Multimodal Human-Computer Interaction: A Constructive and Empirical Study. University of Tampere, Tampere (1999)
28. Reinerman-Jones, L., Taylor, G., Sprouse, K., Barber, D., Hudson, I.: Adaptive automation as a task switching and task congruence challenge. In: Proceedings of the Human Factors and Ergonomics Society Annual Meeting. vol. 55, pp. 197–201. Sage Publications (2011)
29. Sutherland, J., Baillergeon, R., McKane, T.: Cordon and search operations: a deadly game of hide and seek. Air Land Sea Bull. Cordon Search, pp. 4–10 (2010)
30. U.S. Air Force: EOD craftsment balances family, mission, 24 May 2016. from http://www.af.mil/News/Article-Display/Article/779650/eod-craftsman-balances-family-mission/. Accessed 7 Feb 2018
31. U.S. Army Research Laboratory, 17 March 2017. Robotics. U.S. Army Research Laboratory: http://www.arl.army.mil/www/default.cfm?page=392. Accessed 7 Feb 2018
32. U.S. Congress: National Defense Authorization Act for Fiscal Year 2001, Washington, D.C (2001)
33. University of Central Florida.: Mixed Initiative Experimental (MIX) Testbed, 23 July 2013. http://active-ist.sourceforge.net/mix.php?menu=mix. Accessed 02 Sept 2018
34. US Army Research Laboratory Aberdeen Proving Ground United States.: Agent Reasoning Transparency: The Influence of Information Level on Automation Induced Complacency. US Army Research Laboratory Aberdeen Proving Ground United States (2017)
35. Wikipedia: JAUS, 06 July 2017. https://en.wikipedia.org/wiki/JAUS. Accessed 02 Sept 2018

Animacy Perception
Based on One-Dimensional Movement
of a Single Dot

Hidekazu Fukai, Kazunori Terada$^{(\boxtimes)}$, and Manabu Hamaguchi

Gifu University, 1-1 Yanagido, Gifu 501-1193, Japan
{fukai,terada}@gifu-u.ac.jp

Abstract. How humans perceive animacy based on movement is not well understood. In the present study, we conducted an experiment to investigate how humans perceive animacy based on the one-dimensional movement of a single dot. Ten participants were asked to generate 60 s of one-dimensional movement with three assumptions: randomness, inanimacy and animacy. The time-series analysis revealed that the movements generated with the assumption of randomness were similar to white noise, the movements generated with the assumption of inanimacy were periodic, and the power spectra of the movements generated with the assumption of animacy were located between pink and brown noise with trajectories with autocorrelations but no clear periodicity.

Keywords: Animacy perception · Movement of a single dot
Time series analysis

1 Introduction

Humans can distinguish between animate and inanimate motion. Many studies have revealed which characteristics contribute to the perception of animacy. According to previous studies, self-propelled motion, in which a stationary object starts to move without an external force [1–3], movement that violates a physical law [4], goal-directed movement [5], contingent movement [6], the environment and context [4], temporal features and the complexity of interactions [7], and intra-system movement of 2-link mechanisms [8] are reported as types or features of movement that result in a perception of animacy.

Humans do not perceive animacy in unpredictable physical behaviors, such as leaves falling, or in easily predictable mechanical behaviors [9]. We hypothesize that one of the factor for animacy perception is a balance of predictability and unpredictability.

Tremoulet and Feldman [10] showed the motion of a single object to participants on a computer screen and asked them to rate its animacy. The results showed that subjects' animacy ratings were significantly influenced by the magnitude of the change in speed and the angular magnitude of the change in direction, which indicates that both predictability, as perceived based on uniform

© Springer International Publishing AG, part of Springer Nature 2018
S. Yamamoto and H. Mori (Eds.): HIMI 2018, LNCS 10904, pp. 185–193, 2018.
https://doi.org/10.1007/978-3-319-92043-6_16

linear motion, and unpredictability, as perceived based on the sudden changes in speed and angle, contributed to the perception of animacy.

Fukai and Terada [11] conducted an experiment with human participants to clarify the difference between animacy and intentionality in terms of the predictability of behavior. They modeled the behavior of goldfish using an autoregressive model and created movies of a white dot moving on a black background by changing the magnitude of the change in velocity and the rotation angle. The movies were presented to participants, who were asked to chase the white dot with a pen on a tablet to quantify the predictability of its behavior. The participants were also asked to rate the moving dot's animacy and intentionality. The results showed that the subjective impression of animacy increased, and the subjective impression of intentionality decreased as predicting the dot's behavior became more difficult.

Fukai et al. [12] revealed the cause of animacy perception through psychological experiments based on the motion of a simulated double pendulum. The double-pendulum equations lead to a variety of motions, from simple swinging to chaos, depending on their parameters. The participants were instructed to adjust the parameters to the values for which they perceived the most and least animacy. The results indicated that humans do not perceive animacy in periodic motions with little randomness. Furthermore, they reported that the human perception of animacy was comparatively small for most random motions. In contrast, humans perceived animacy strongly in non-stationary random motion with temporal structures, such as motion in which small circular movements occur irregularly.

As mentioned above, although studies suggest that motions with both predictability and unpredictability contribute to the perception of animacy, the property and the extent to which predictability and unpredictability contribute to this perception is not clear. Therefore, in the present study, we use a simple environment in which motion of a single dot is constrained to one dimension to facilitate a time series analysis. In addition, we use a method in which participants generate motion themselves to explore the various motion features that contribute to animacy perception.

2   Method

2.1   Participants

Ten healthy graduate and undergraduate students (10 male, $M_{age} = 23.2$ years, $SD_{age} = 1.4$ years, age range: 21–25 years) participated in the experiment.

2.2   Apparatus

Figure 1 shows the apparatus used in our experiment. A wireless mouse was fixed on a block that slid freely along a rail with a length of 270 mm. The apparatus limited the motion of the participants' hands to the lateral direction.

Fig. 1. Apparatus

Fig. 2. Graphical user interface

The interface used in this experiment is shown in Fig. 2. On the screen, a white dot and the remaining time were shown. The position of the white dot was controlled by the mouse. Therefore, the white dot moved horizontally across the screen following the mouse's movement.

2.3 Procedure

The participants were asked to read a document in which the purpose of the experiment and the method for controlling the apparatus were written. When the experiment started, the screen, as shown in Fig. 2, was shown to the participants. The participants were asked to move the dot on the screen. A one-minute training period was given to the participants. Then, they were asked to move the dot in such a way that they felt the three characteristics were exhibited in the motions they generated. The three characteristics were randomness, inanimacy, and animacy. Participants were asked to move the dot for one minute for each motion category.

188   H. Fukai et al.

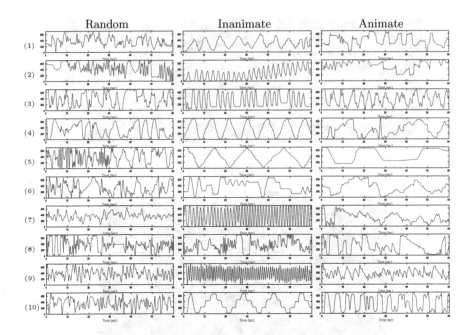

Fig. 3. The trajectories of the x-coordinate of the white dot generated by the ten participants. The left column identifies the participants by number.

2.4   Measurement

The trajectory of the mouse was recorded, and an autocorrelation analysis was performed. The following questionnaire was given to the participants after they finished to movement generation session.

- What type of movement did you imagine when you created the random movement?
- What type of movement did you imagine when you created the inanimate movement?
- What type of movement did you imagine when you created the animate movement?

3   Results

3.1   Trajectory

The trajectories of the x-coordinate of the white dot generated by all ten participants are shown as a function of time in Fig. 3.

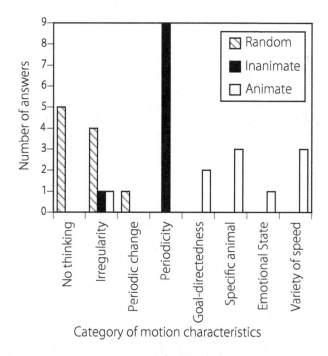

Fig. 4. The motion characteristics assumed by participants.

3.2 Questionnaire

Figure 4 shows a histogram of the answers to the post questionnaire. We defined eight motion characteristics that were imagined by participants when they moved the mouse. Most of the participants assumed nothing and irregularity when they generated the random movement. Most of the participants assumed periodicity when they generated the inanimate movement. Most of the participants assumed goal-directedness, a specific animal, such as a fly or a spider, an emotional state, and a variety of speeds when they generated animate movement.

3.3 Time Series Analysis

A time series analysis was performed on the trajectories generated by the participants to investigate the temporal features of the trajectories. First, we calculated the autocorrelation function of the trajectory of the dot to determine the extent to which its motion was periodic.

Figure 5a, b, and c show the autocorrelation functions for the random, inanimate, and animate motions of the dot, respectively. Note that the data for all ten participants are shown on one graph. The figures show that the autocorrelation is weak in the random condition, both autocorrelation and periodicity are observed in the inanimate condition, and although periodicity is not observed, some trajectories exhibit autocorrelation.

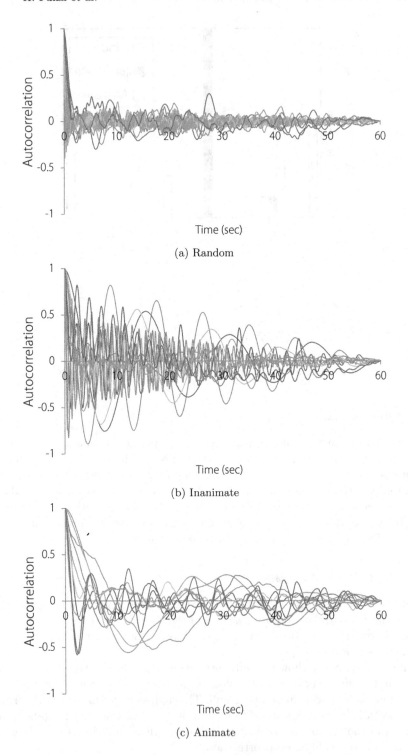

(a) Random

(b) Inanimate

(c) Animate

Fig. 5. Autocorrelation functions of the movements generated by the participants.

Table 1. Means and standard deviations of the autocorrelations.

Condition	Mean	SD
Random	.06	.02
Inanimate	.16	.07
Animate	.13	.08

Table 1 shows the mean and standard deviation of the mean absolute values of the autocorrelation calculated for 60 s for each participant. A one-way ANOVA $(F(2, 27) = 8.70, p < .01)$ confirmed a significant difference in the mean absolute value of the autocorrelation. Fisher's LSD post-hoc test revealed that the mean absolute values of the autocorrelation were significantly higher in the inanimate and animate conditions than in the random condition $(p < .01)$.

The peaks with correlations of at least 0.3 in the autocorrelation function were counted. The mean numbers of peaks were 0.1, 4.7, and 0.1 in the random, inanimate, and animate conditions, respectively.

We calculated the inclination of the regression lines of the power spectral density functions graphed on a double-logarithmic scale up to 1 Hz. The mean inclinations were -0.86, -1.33, and -1.63 for the random, inanimate, and animate conditions, respectively. A one-way ANOVA $(F(2, 27) = 2.73, p = .08)$ confirmed a marginally significant difference in the mean inclination of the regression lines. Fisher's LSD post-hoc test revealed that the mean inclination was greater in the random condition than in the inanimate condition $(p < .05)$.

4   Discussion

In the present study, we asked participants to generate three types of one-dimensional movement that they felt exhibited randomness, inanimacy, and animacy. In this section, we discuss the features of each movement.

For the movements to which participants assigned randomness, subjective reports obtained from the post questionnaire revealed that most of the participants assumed nothing and irregularity. In consistent with the subjective reports, the autocorrelation was low, and the movements were similar to white noise. These results indicate that participants could actually generate random movement.

For the movements to which participants assigned inanimacy, subjective reports revealed that almost all the participants assumed periodicity when they generated the movements. The autocorrelation analysis provided consistent results in which the movements were periodic.

The procedures of the experiment are considered one of the reasons why participants assumed periodicity as a feature of inanimate movement. The opposite of animacy is not necessary periodicity. Random movement is sometimes assumed to be the opposite of animate movement. However, the participants in our study were first asked to generate random movement. Therefore, they excluded randomness when they generated inanimate movement but assumed periodicity.

For the movements to which the participants assigned animacy, subjective reports revealed that the characteristics assumed when they generated the movements were more widely distributed than they were for the other two conditions. This means that people attribute many characteristics to animate movement. However, the following can be said based on the results of the time series analysis. The proportion of high frequency components included in the movement generated when participants imagined animacy was less than it was for the other types of movement. Colored noise, such as $1/f$ noise (pink noise) and $1/f^2$ noise (brown noise), are identified by the change in the power spectral densities of their time series, which is inversely proportional to the frequency of the change. It is known that pink noise is seen in biological phenomena, such as heart rate variability and firefly luminescence patterns. It is also known that pink noise induces predatory behavior in fish [13].

With colored noise, stronger autocorrelations are observed when the attenuation rate of the power spectral density increases. According to the approximation using data up to 1 Hz in our experiments, we found that the frequency spectrum of the motion generated in the animacy condition was between those of pink and brown noise. Furthermore, the exponent of the power approximation was greater in the animacy condition than in the other conditions. Given that the periodicity of the movement which participants imagined expressed inanimacy was strong, the movement produced with imagined animacy exhibited a temporal correlation and was neither random nor periodic. This finding is consistent with the reports of Fukai and Terada [11] and Fukai et al. [12].

Three of the participants noted that they had varied the speed of the movement in the animacy condition. In addition, the two participants who tried to express goal-directedness in the animacy condition did not actually generate linear movement, which is the most goal-directed, but rather, generated fluctuating motion. The trade-offs between exploration and exploitation in a reinforcement learning agent can be considered a mechanism to generate such movements. If the exploration rate is high, past experiences can not be used, and there is a high possibility of missing an unexploited profit that has already been found. The higher the exploitation rate is, the more likely an agent is to miss the new higher profit. Therefore, an appropriate balance of exploitation and exploration is important for agents. Exploration is the cause of random movements, and exploitation is cause of linear and goal-directed movements. There is a possibility that such properties of agents may affect the perception of animacy by their combination of randomness and regularity.

5   Conclusion

We hypothesized that one of the factors for animacy perception is a balance between the predictability and unpredictability that are observed in the behavior of a target. In the present study, we analyzed the time series of the motion of a single dot constrained to one dimension generated by participants to identify the properties and predictability or unpredictability that are related to a perception of animacy. The results showed that periodic changes that have clear temporal structures and white noise without autocorrelation are not perceived as animate movements. We also found that humans perceive animacy from temporal change with a frequency spectrum between those of pink and brown noise. Although, in the present study, we assumed a stationarity when we analyzed the movement, the movements of living organisms are essentially non-stationary and reflect internal state changes. We will examine the characteristics of movement that contribute to animacy perception in detail in the future by assuming internal state transitions.

References

1. Baron-Cohen, S.: Mindblindness: An Essay on Autism and Theory of Mind. The MIT Press, Cambridge (1995)
2. Premack, D., Premack, A.J.: Moral belief: form versus content. In: Mapping the Mind: Domain Specificity in Cognition and Culture, pp. 149–168. Cambridge University Press, Cambridge (1994)
3. Heider, F., Simmel, M.: An experimental study of apparent behavior. Am. J. Psychol. **57**(2), 243–259 (1944)
4. Gelman, R., Durgin, F., Kaufman, L.: Distinguishing between animates and inanimates: not by motion alone. In: Sperber, D., Premack, D., Premack, A.J. (eds.) Causal Cognition: A Multidisciplinary Debate, pp. 150–184. Oxford University Press, Oxford (1995)
5. Dittrich, W.H., Lea, S.E.G.: Visual perception of intentional motion. Perception **23**(3), 253–268 (1994)
6. Bassili, J.N.: Temporal and spatial contingencies in the perception of social events. J. Pers. Soc. Psychol. **33**(6), 680–685 (1976)
7. Santos, N.S., David, N., Bente, G., Vogeley, K.: Parametric induction of animacy experience. Conscious. Cogn. **17**(2), 425–437 (2008)
8. Aono, N., Morita, T., Ueda, K.: Analysis of animacy perceived from movement of a two-link rigid arm. IEICE Trans. Inf. Syst. (Japan. Ed.) **J95-D**(5), 1268–1275 (2012)
9. Terada, K., Iwase, Y., Ito, A.: Verification of three stances discussed by Dennett. Trans. Inst. Electron. Inf. Commun. Eng. A **J95-A**(1), 117–127 (2012)
10. Tremoulet, P.D., Feldman, J.: Perception of animacy from the motion of a single object. Perception **29**(8), 943–951 (2000)
11. Fukai, H., Terada, K.: Modeling of animal movement by AR process and effect of predictability of the behavior on perception of animacy and intentionality. In: The 35th Annual International Conference of the IEEE Engineering in Medicine and Biology Society (EMBC 2013), pp. 4125–4128 (2013)
12. Fukai, H., Terada, K., Takeuchi, Y., Ito, A.: Perceiving animacy from the double-pendulum movement. Technical report of IEICE, vol. 114, pp. 49–53 (2014)
13. Matsunaga, W., Watanabe, E.: Visual motion with pink noise induces predation behaviour. Sci. Rep. **2**(219), 1–7 (2012)

Experimental Observation of Nodding Motion in Remote Communication Using ARM-COMS

Teruaki Ito[1(✉)], Hiroki Kimachi[2], and Tomio Watanabe[3]

[1] Graduate School of Technology, Industrial and Social Sciences,
Tokushima University, 2-1 Minami-Josanjima, Tokushima 770-8506, Japan
tito@tokushima-u.ac.jp
[2] Graduate School of Advanced Technology and Science,
Tokushima University, 2-1 Minami-Josanjima, Tokushima 770-8506, Japan
c501732024@tokushima-u.ac.jp
[3] Faculty of Computer Science and System Engineering,
Okayama Prefectural University, 111 Tsuboki, Souja, Okayama 719-1197, Japan
watanabe@cse.oka-pu.ac.jp

Abstract. Considering the critical issues of remote communication, this study proposes an idea of remote individuals' virtual connection through augmented tele-presence systems called ARM-COMS (ARm-supported eMbodied COmmunication Monitor System). Several ideas of robot-based remote communication systems have been proposed to challenge the telepresence issue of remote participants. However, it does not cover the issue of relationship. An idea of robotic arm-typed system and/or an idea of anthropomorphization draw researchers' attentions to challenge the lack of relationship with remote participants. However, usage of the human body movement of a remote person as a non-verbal message, or cyber-physical media in remote communication is still an open issue. Under these circumstances, this paper describes the system configuration of ARM-COMS based on the proposed idea and discusses the feasibility of the idea using the experimental observations.

Keywords: Cyber-physical communication media · Embodied communication
Augmented tele-presence robotic arm manipulation · Face detection

1 Introduction

TV phone was regarded as a dream for communication tools in SF movies in the old days. However, a smartphone-based video communication tool is now one of the convenient popular tools freely available to mostly everybody [1]. Supporting by ICT (Information and Communication Technology) technologies, further enhancement of better communication is expected. In the meantime, this tool addresses the two types of critical issues, which are the lack of tele-presence feeling and the lack of relationship feeling in remote video communication [5] as opposed to a face-to-face communication.

Several ideas of robot-based remote communication systems have been proposed as one of the solutions to the former issue; these robots include physical telepresence robots [9, 21, 22]. Anthropomorphization [14] is another new idea to show the telepresence of a remote person in communication system. Remote communication can

© Springer International Publishing AG, part of Springer Nature 2018
S. Yamamoto and H. Mori (Eds.): HIMI 2018, LNCS 10904, pp. 194–203, 2018.
https://doi.org/10.1007/978-3-319-92043-6_17

be basically supported by the primitive functions of physical tele-presence robots, such as a face image display of the operator [15], as well as tele-operation function such as remote-drivability to move around [10], or tele-manipulation [10]. However, there are still an open issue to be studied to narrow the gap between robot-based video communication and face-to-face one.

The second issue in the lack of relationship-type feeling in remote video communication is another big challenge. Recently, an idea of robotic arm-type systems draws researchers' attention [25]. For example, Kubi [13], which is a non-mobile arm type robot, allows the remote user to "look around" during video communication by way of commanding Kubi where to aim the tablet with an intuitive remote control over the net. Furthermore, an idea of enhanced motion display has also been reported [16] to show its feasibility over the conventional display. However, the usage of the human body movement of a remote person as a non-verbal message is still an open issue.

This study proposes an idea of human-computer interaction through remote individuals' connection with augmented tele-presence systems called ARM-COMS (ARm-supported eMbodied COmmunication Monitor System) [6, 7, 14]. The challenge of this idea is to use the human body movement of a remote person as a non-verbal message for sharing the connected communication, and to implement a cyber-physical media us-ing ACM-COMS for connected remote communication [8].

2   Overview of ARM-COMS (ARm-supported eMbodied COmmunication Monitor System)

2.1   System Overview of ARM-COMS

Considering the physical entrainment motion in human communication [24], this research challenges these two issues mentioned in the Sect. 1 by the idea of ARM-COMS (ARm-supported eMbodied COmmunication Monitor System) [6]. This paper focuses on the nodding motion as a non-verbal message contents in remote communication using ARM-COMS. Figure 1 shows the system overview of ARM-COM for the experiment in this study. Face detection procedure of a prototype of ARM-COMS is based on the algorithm of FaceNet [20], which includes image processing library OpenCV 3.1.0 [17], machine learning library dlib 18.18 [3], and face detection tool OpenFace [18] which were installed on a control PC with Ubuntu 14.04 [23] as shown in Fig. 2. Using the input image data from USB camera, landmark detection is processed.

ARM-COMS is composed of a tablet PC and a desktop robotic arm. The table PC in ARM-COMS is a typical ICT (Information and Communication Technology) device and the desktop robotic arm works as a manipulator of the tablet, of which position and movements are autonomously manipulated based on the behavior of a human user who communicates with remote person through ACM-COMS. This autonomous manipulation of ARM-COMS is controlled by the head movement, which can be recognized by one of the typical portable sensors, such as a magnetic sensor, gyro-sensor, motion capturing sensor, or a typical cameras, such as Kinect [11] sensor, or a general USB camera.

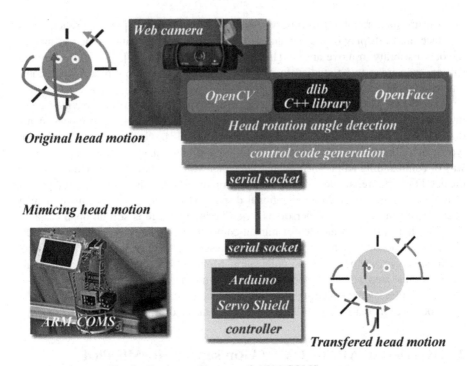

Fig. 1. System architecture of ARM-COMS prototype

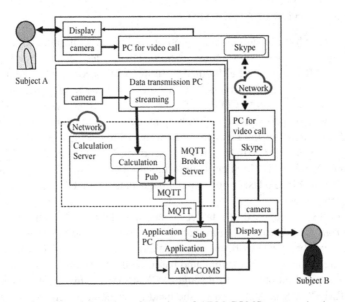

Fig. 2. Network-based configuration of ARM-COMS communication

2.2   System Configuration of ARM-COMS for Network Usage

ARM-COMS is configured to implement network communication as shown in Fig. 2. Head motion of Subject A is used as a non-verbal communication to ARM-COMS which interact with Subject B. Video communication itself was performed by a typical software (Skype). However, the head motion image data is processed by the face detection algorithms mentioned in the Sect. 2.1, which was used to trigger the motion of ARM-COMS installed at the site of subject B.

3   Experimental Comparison

3.1   Experimental Configuration for Nodding Observation

Based on the system configuration shown in Fig. 2, three types of experimental setups were configured, which include (a) face-to-face communication, (b) video communication, and (c) ARM-COMS communication as shown in Fig. 3. The detailed experimental setups are shown in Figs. 4, 5 and 6.

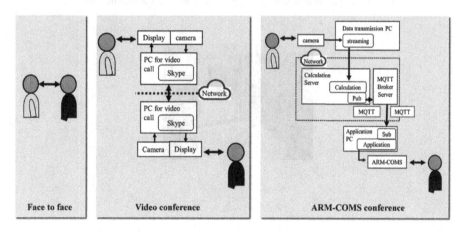

Fig. 3. Three types of experimental setup

Communication experiments were conducted by 8 subjects, which were composed of a pair to make communicate in three types of setups for a short conversation with maximum 2 min conversation in the procedure below.

Experimental procedure:

Step 1: Subject A and B are positioned to see each other
Step 2: Subject A and B start nodding in the beginning of conversation.
Step 3: Subject A and B start short conversation on a topic of breakfast menu.
Step 4: Subject A and B end conversation by nodding greeting.

3.2   Experiments for Face-to-Face Communication

Figure 4 is experimental setup for face-to-face communication. Head-motion of a human subject is detected and traced according to the short conversation. One magnetic receiver (Fastrak RX-2 [4]) is attached to the head of human subject A and another magnetic receiver B is attached to the other subject.

Figure 5 shows a result of this experiment, which shows the clear correspondence to the nodding interaction between the two subjects.

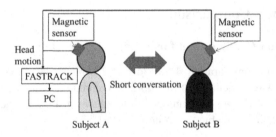

Fig. 4. Experimental configuration of face-to-face communication

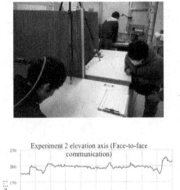

Fig. 5. An experimental result of face-to-face communication

3.3   Experiments for Video Communication

Figure 6 shows the experimental setup for video communication. Head-motion of Subject A and B was detected and traced during the short conversation using magnetic sensor and video imaging as well as gaze point tracking sensor. A general USB camera (Buffalo) captures the image of human subjects during the experiments. A desktop PC

(Windows 7/64) was used for the data collection, whereas a laptop pc (Ubuntu 14.04) was used for ARM-COMS control.

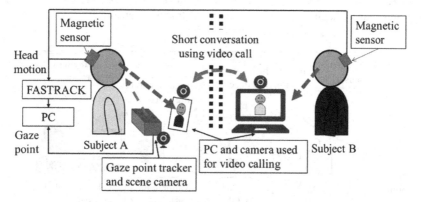

Fig. 6. Experimental setup for video communication

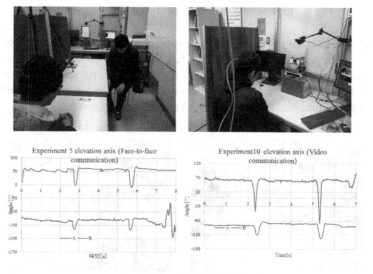

Fig. 7. Experiment 5 & 10

Figure 7 shows some results of this experiment, which shows the clear correspondence to the nodding interaction between the two subjects, both in face-to-face conversation and video conversation.

3.4   Experiments for ARM-COMS Communication

Figure 8 is experimental setup for ARM-COMS communication. Head-motion of Subject A and B was detected and traced during the short conversation using magnetic sensor and video imaging as well as gaze point tracking sensor.

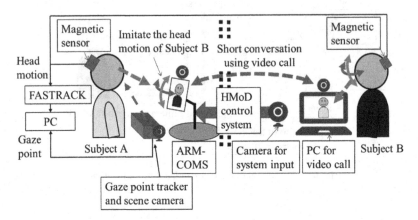

Fig. 8. Experimental setup for ARM-COMS communication

Figure 9 shows some results of this experiment, which shows the clear correspondence to the nodding interaction between the two subjects, both in face-to-face conversation, video communication and ARM-COMS conversation. However, the result did not show any significant difference between video communication and ARM-COMS communication. Further experiments were required.

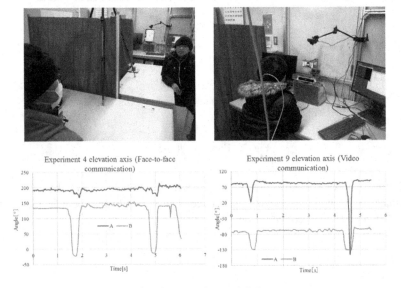

Fig. 9. Experiment 4 & 9

3.5 Results and Discussion

Three types of experiments were conducted to study the feasibility of ARM-COMS communication, namely, face-to-face communication, video communication, and ARM-COMS communication. As shown in the experimental results in the Sects. 3.1, 3.2, 3.3 and 3.4, the experimental setups worked well and three types of experiments were all conducted. Therefore, it could be mentioned that the systems were well configured to implement the idea of this research.

Nodding motion in conversation is very common for Japanese culture, whereas it is not so common for other cultures. Therefore, the subjects including Malaysian and Chinese as well as Japanese attended to the experiments in order to see the difference. The basic instruction was given to the subjects as mentioned in the Sect. 3.1 before the experiment. However, the naturalness of nodding was not similar in subjects, of which difference was clearly recognized by the authors, but was not well analyzed by the experimental data.

According to the head motion data, there was no significant difference between face-to-face communication and video communication for all subjects. Natural nodding style of Japanese subjects was observed both in face-to-face and video communication, which was recognized by head tracking data analysis. Unnatural nodding gesture of non-Japanese subjects was observed both in face-to-face or video communication. However, the unnaturalness of nodding gesture was not recognized by the collected data of the experiments. Since the nodding style is another issue to be studied to show the feasibility of ARM-COMS idea, further design of experimental setup should be considered.

In addition to head motion tracking by magnetic sensor, eye-tracking measurement was also conducted to trace the eye movement [12] during the conversation to see the difference between a typical video communication and ARM-COMS communication. However, eye tracking of subjects could not be traced during nodding motion because the gaze point of subject disappeared out of sight from eye tracker. Therefore, gaze tracking data were not well utilized in the experiments. Further design of experimental setup should be considered.

4 Concluding Remarks

This study proposed an idea of human-computer interaction through remote individuals' connection with augmented tele-presence systems called ARM-COMS (ARm-supported eMbodied COmmunication Monitor System). The challenge of this idea is to use the human body movement of a remote person as a non-verbal message for sharing the connected communication, and to implement a cyber-physical media using ACM-COMS for connected remote communication. Based on the implemented communication platform prototype presented in this paper, three types of experiments were conducted to study the feasibility of the proposed idea. The configuration of the prototype worked well for the experiment. However, Further consideration for design experiments is required to collect measurement data, which will be used for feasibility analysis of the proposed idea.

Acknowledgement. This work was supported by JSPS KAKENHI Grant Numbers JP16K00274. The author would like to acknowledge all members of Collaborative Engineering Labs at Tokushima University, and Center for Technical Support of Tokushima University, for their cooperation to conduct the experiments.

References

1. Abowdm, D.G., Mynatt, D.E.: Charting past, present, and future research in ubiquitous computing. ACM Trans. Comput.-Hum. Interact. (TOCHI) **7**(1), 29–58 (2000)
2. Bertrand, C., Bourdeau, L.: Research interviews by Skype: a new data collection method. In: Esteves, J. (ed.) Proceedings from the 9th European Conference on Research Methods, pp. 70–79. IE Business School, Spain (2010)
3. Dlib C++ libraty. http://dlib.net/
4. FASTRK. http://polhemus.com/motion-tracking/all-trackers/fastrak
5. Greenberg, S.: Peepholes: low cost awareness of one's community. In: 1996 Conference Companion on Human Factors in Computing Systems: Common Ground, Vancouver, British Columbia, Canada, pp. 206–207 (1996)
6. Ito, T., Watanabe, T.: Three key challenges in ARM-COMS for entrainment effect acceleration in remote communication. In: Yamamoto, S. (ed.) HCI 2014. LNCS, vol. 8521, pp. 177–186. Springer, Cham (2014). https://doi.org/10.1007/978-3-319-07731-4_18
7. Ito, T., Watanabe, T.: ARM-COMS for entrainment effect enhancement in remote communication. In: Proceedings of the ASME 2015 International Design Engineering Technical Conferences & Computers and Information Engineering Conference (IDETC/CIE2015), August, Boston, USA, no. DETC2015-47960 (2015)
8. Ito, T., Watanabe, T.: Motion control algorithm of ARM-COMS for entrainment enhancement. In: Yamamoto, S. (ed.) HIMI 2016. LNCS, vol. 9734, pp. 339–346. Springer, Cham (2016). https://doi.org/10.1007/978-3-319-40349-6_32
9. Kashiwabara, T., Osawa, H., Shinozawa, K., Imai, M.: TEROOS: a wearable avatar to enhance joint activities. In: Annual Conference on Human Factors in Computing Systems, pp. 2001–2004 (2012)
10. Kim, K., Bolton, J., Girouard, A., Cooperstock, J., Vertegaal, R.: TeleHuman: effects of 3D perspective on gaze and pose estimation with a life-size cylindrical telepresence pod. In: Proceedings of CHI2012, pp. 2531–2540 (2012)
11. Kinect. https://dev.windows.com/en-us/kinect
12. Krafka, K., Khosla, A., Kellnhofer, P., Kannan, H., Bhandarkar, S., Matusik, W., Torralba, A.: Eye tracking for everyone. In: IEEE Conference on Computer Vision and Pattern Recognition (CVPR) (2016)
13. Kubi. https://www.revolverobotics.com
14. Osawa, T., Matsuda, Y., Ohmura, R., Imai, M.: Embodiment of an agent by anthropomorphization of a common object. Web Intell. Agent Syst.: Int. J. **10**, 345–358 (2012)
15. Otsuka, T., Araki, S., Ishizuka, K., Fujimoto, M., Heinrich, M., Yamato, J.: A realtime multimodal system for analyzing group meetings by combining face pose tracking and speaker diarization. In: Proceedings of the 10th International Conference on Multimodal Interfaces (ICMI 2008), Chania, Crete, Greece, pp. 257–264 (2008)
16. Ohtsuka, S., Oka, S., Kihara, K., Tsuruda, T., Seki, M.: Human-body swing affects visibility of scrolled characters with direction dependency. In: Society for Information Display (SID) 2011 Symposium Digest of Technical Papers, pp. 309–312 (2011)
17. OpenCV. http://opencv.org/

18. OpenFace API documentation. http://cmusatyalab.github.io/openface/
19. Padmavathi, G., Shanmugapriya, D., Kalaivan, M.: A study on vehicle detection and tracking using. Wirel. Sens. Netw. **2**, 173–185 (2010)
20. Schoff, F., Kalenichenko, D., Philbin, J.: FaceNet: a unified embedding for face recognition and clustering. In: IEEE Conference on CVPR 2015, pp. 815–823 (2015)
21. Sirkin, D., Ju, W.: Consistency in physical and on-screen action improves perceptions of telepresence robots. In: HRI 2012 Proceedings of the Seventh Annual ACM/IEEE International Conference on Human-Robot Interaction, pp. 57–64 (2012)
22. Tariq, A.M., Ito, T.: Master-slave robotic arm manipulation for communication robot. In: Japan Society of Mechanical Engineer, Proceedings of 2011 Annual Meeting, vol. 11, no. 1, p. S12013, September 2011
23. Ubuntu. https://www.ubuntu.com/
24. Watanabe, T.: Human-entrained embodied interaction and communication technology. In: Fukuda, S. (ed.) Emotional Engineering, pp. 161–177. Springer, London (2011). https://doi.org/10.1007/978-1-84996-423-4_9
25. Wongphati, M., Matsuda, Y., Osawa, H., Imai, M.: Where do you want to use a robotic arm ? And what do you want from the robot ? In: International Symposium on Robot and Human Interactive Communication, pp. 322–327 (2012)

Hands-Free Interface Using Breath Residual Heat

Kanghoon Lee[1], Sang Hwa Lee[2], and Jong-Il Park[1(✉)]

[1] Hanyang University, Seoul 04763, Korea
aeternalis999@gmail.com, jipark@hanyang.ac.kr
[2] INMC, Seoul National University, Seoul 08826, Korea
lsh529@snu.ac.kr

Abstract. Most user interfaces have been studied based on hand gestures or finger touches, but the interface using the user's hands does not reflect the user's various situations. In this paper, we propose a hands-free user interaction system using a thermal camera. The hands-free interface proposed in this paper exploits user's breath heat and thermal camera, thus it is very useful for users who have difficulty in using their hands. In addition, the thermal camera is not affected by background color and lighting environment, so it can be used in various complex situations. For hands-free interaction, the user creates a residual heat on the surface of the object to interact, and the thermal camera senses the residual heat. This paper has observed that the residual heat from breath is most suitable for the interaction design. For this observation, several different methods were tested for how to generate strong residual heat on the various materials. According to the tests, it was verified that the residual heat generated from breath with hollow rod (straw) is most stable for sensing and interaction. This paper demonstrates its usefulness by implementing an interaction system using camera projection system as an application example.

Keywords: Hands-free · Interface · Residual heat · Breath heat
Thermal camera

1 Introduction

Interfaces can be said to be intermediaries for communication between objects and humans. Especially, the interface called Human-Compute Interaction (HCI) for the interaction between computer and user has been applied to various fields as the computer technology has been developed. As a method for interacting with a computer, a method of recognizing a user's hand or finger contact is widely used. Hand gestures and finger touch recognition can also be done using a touch screen or sensors attached to the hands or arms separately. However, the method using camera has been studied in HCI field. Among them, interfaces using RGB cameras have been studied for a long time, and many methods of using depth camera and infrared camera have been studied. Such cameras may not be suitable depending on the user environment or the recognition performance. Thermal cameras are often used in special applications because of their high cost and size. Recently, as the price has decreased and the size has become

© Springer International Publishing AG, part of Springer Nature 2018
S. Yamamoto and H. Mori (Eds.): HIMI 2018, LNCS 10904, pp. 204–217, 2018.
https://doi.org/10.1007/978-3-319-92043-6_18

smaller, the thermal cameras are utilized in various fields. The thermal camera can detect the temperature by sensing thermal energy radiation. With this feature, the thermal camera is used to detect the motion of user's hands and to sense the residual heat generated by fingertip touches on the target surface. A touch interface using a thermal camera is a method for detecting a residual heat generated by a user. When a user touches a surface with a hand or a fingertip, the user's body temperature is conducted to the contact surface, which generates residual heat on the surface. The sensed residual heat provides some information of position on the surface, movement and gestures of hands, and so on. The residual heat can be generated by breath instead of the fingertip contact. In this case, the users do not need to use their hands for interaction.

This paper deals with the study of interfaces that do not require hands using thermal cameras and breath residual heat. Most HCI interfaces for computer-user interaction are based on fingertip touch or hand movements. However, for users who cannot use their hands temporarily or permanently, there is restriction on the use of the HCI interface. As an alternative method, the interface without hands would be useful for users who have difficulty in using their hands. There is a disadvantage in that it is necessary to be close to the target surface in order to generate residual heat by breath. Also, since the temperature of residual heat by breath is usually lower than that of fingertip contact, it is difficult to distinguish the residual heat by breath from fingertip heat. A hollow rod (straw) was used to increase the residual heat by breath at a distance from the interaction surface. If a user blows the hollow rod with user's mouth, breath will be concentrated and the residual heat temperature will increase on the surface.

In this paper, various experimental evaluations were conducted to verify the feasibility and usefulness of interface method using residual heat by breath. Experimental evaluation was carried out by comparison of residual heat generation methods and various materials surfaces. According to the experimental results, the residual heat generation method using the hollow rod (straw) is most stable in the aspect of high temperature and detection performance.

The remainder of paper is organized as follows. In Sect. 2, some related works are briefly summarized. The characteristics of residual heat by breath is described in Sect. 3, and various comparative observation on residual heat generation methods and surface materials is explained in Sect. 4. Section 5 deals with the implemented interaction system for a hand-free interaction application. Finally, some problems and future researches of proposed system are discussed in Sect. 6.

2 Related Works

Researches on interfaces should take into account the means of interaction and recognition. There is fingertip contact as a typical interaction means. Fingertip contact is a system that touches a wall or table surface with a touch screen and sensor device [1, 14]. In this paper, we have studied camera based surface touch interface.

Types of cameras include a visual camera that captures visible light, an infrared camera that captures near infrared rays, and a thermal camera that captures the energy emitted from far infrared bands. Visual cameras are dependent on the sun or illumination because they can only represent the wavelengths of visible light. Therefore, visual camera based user interaction system has been studied extensively. However, when the intensity of illumination is high or low, comparison of skin color and detection of contour of hand become difficult [2, 8]. The infrared camera is a camera that captures the near infrared rays of 0.7 μm to 1.3 μm wavelength near infrared rays and reflects the object. Compared to a visual camera, it has a low illumination and has the advantage of viewing objects in dark environments. In the case of fingertip or hand shape, it is widely used because it can be seen more clearly than visual camera due to the retro-reflective characteristics of the object. Especially, when using a projector or a screen in a tabletop interaction study, it is easy to use a projector or an LED screen because it is less influenced by illumination [6, 13, 17]. However, infrared cameras require an infrared light source. And when the infrared camera is used outdoors in the daytime, there is a disadvantage that the infrared light source is affected by the sun light.

Usually the infrared camera is a near-infrared camera and the far-infrared camera is called a thermal camera. The thermal camera senses the far-infrared energy of 3 μm–14 μm wavelength band emitted by the object based on absolute temperature 0. Thermal cameras do not need a separate light source because they capture the heat energy emitted by the object. Therefore, there is no problem in using at night and daytime and outdoors. Figure 1 shows an image captured with an infrared camera at night in an outdoor space. The initial thermal imaging camera was developed for use in military sector applications [4]. In recent years, there have been many studies on prevention of overheating in industrial sites, confirmation of cracks, detection of nighttime pedestrians [3], face recognition [9], study of human skin surface change [12], medical field [7, 15], food quality and safety profiling [5], has been utilized.

Fig. 1. Thermal image

In general, the temperature of the hands or fingers is higher than the room temperature, and researches using thermal cameras have been actively conducted [10, 11, 16]. Because heat has the property of being conducted, the user can contact the surface to transmit heat. The heat transferred to the surface causes the contact surface to show a higher temperature than the ambient. We call this a residual heat (see Fig. 2). When the fingertip is brought into contact with the surface to generate residual heat, the contact position can be extracted.

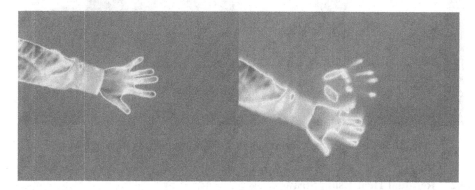

Fig. 2. Thermal image of residual heat

3 Hands-Free Interface Using Breath Heat

The touch interface using the thermal camera is similar to the existing touch interface [10, 11]. The difference is how to find the tip of the finger that touches surface. The touched trace is called the residual heat, which can be found using the thermal camera. The proposed hands-free interface interacts by sensing the residual heat without using a hand. This paper exploits breath to generate sufficient residual heat.

3.1 Residual Heat by Breath

The residual heat generated by fingertip touches can be stably interacted with the thermal camera without being affected by the background or illumination. However, the users who cannot use their hands cannot do this interface.

This paper studies how to interact through the breath heat. The temperature of breath is usually higher than room temperature (on average between 20 and 24°C). Therefore, it is possible to detect residual heat on the surface by blowing breath. The use of breath can generate residual heat on the interaction surface without the use of hands. However, when we look at the image of the residual heat due to the breath in Fig. 3, the temperature is low and spread widely. This is because of spreading of breath air. There are two problems when we breath out. One is that the breathing air has low temperature. The other is that the warm breath air is spread out quickly. The temperature deviation of the breath heat is large due to the problems and users difference. The more the distance between the user and the surface, the more the breath spreads. And as the breath passes through the air, the temperature of breath also decreases. In order to use the residual heat by breath, it is necessary to collecting the breath air and to be close to the surface. However, when the user is very close to the surface, it may be difficult to detect residual heat at the target surface position by optical occlusion.

Fig. 3. Residual heat by usual breath (left: RGB image, right: thermal image)

3.2 Residual Heat Using Straw

In this paper, we use hollow rod (straw) to collect breath heat and to generate sufficient residual heat at the position intended by the user. If you use a straw to breath, the temperature of the residual heat increases because of the concentration of warm air. And since it is 25 cm to 30 cm away from the surface, it is easy to generate residual heat at the target position. Figure 4 shows the generating residual heat using a straw. The users who have difficulty in using their hands can generate residual heat on the target surface exactly.

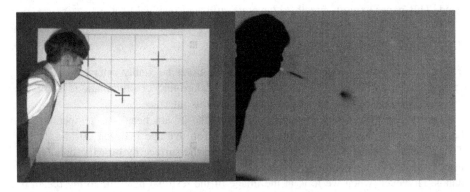

Fig. 4. Generating residual heat by using straw (left: RGB image, right: thermal image)

4 Experimental Analysis of Residual Heat

In order to effectively generate the intended residual heat, basic research on the generation of residual heat is required. The experiments were conducted to generate and detect residual heat by various methods in order to apply it to thermal camera and application design. In the experiments, the temperature of the residual heat was checked and compared right after generation of residual heat. The experiments were conducted indoors, where the temperature was between 20 and 24°C. The participants in the experiments were eight in total, four men and four women. The thermal camera used for the experiments is the VarioCAM hr head 420 (see Fig. 5). The image resolution is 384 × 288 pixels.

Fig. 5. Thermal camera

4.1 Residual Heat Generation Methods

We conducted three ways to generate residual heat on the surface. The three methods are fingertip touch, blowing breath, and blowing breath with a hollow rod (straw). The experiments generated residual heat on the paper attached to the wall. Figure 6 shows three methods to generate residual heat on the surface.

Figure 7 shows a graphical representation of the three types of residual heat peak temperatures (Participants 1). The temperatures of residual heat by fingertip contact are indicated by the blue point and are distributed between 28.17° and 29.93°C. The residual heat temperatures of blowing breath are indicated by the red points and are distributed between 27.69° and 31.64°C. The residual heat temperatures of blowing breath through the hollow rod (straw) are indicated by the green points and are distributed between 31.57° and 34.09°C. As we can see in Fig. 7, the temperatures of the residual heat of blowing breath using the hollow stick are highest among the three ways. In addition, the residual heat exists in the small area. Therefore, the blowing breath with straw is most suitable for detecting with thermal cameras and localizing its position. There are no occlusion problems since the users blow their breath using the straw distant from the surface.

(a) Fingertip touch

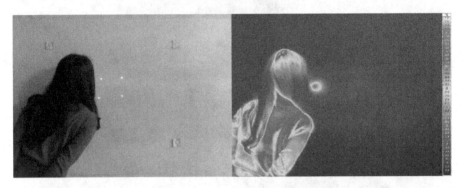

(b) Breath

(c) Breath using the hollow rod(straw)

Fig. 6. Generation residual heat left: RGB image, right: thermal image

4.2   Heat Response Characteristics of Materials

The temperature of residual heat varies differently depending on the materials. There is also a temperature difference in the residual heat because the thermal conductivity is

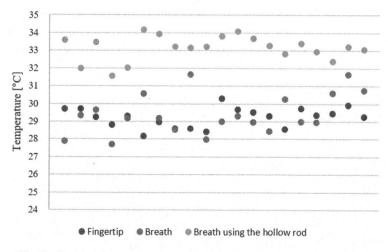

● Fingertip   ● Breath   ● Breath using the hollow rod

Fig. 7. Residual heat temperature comparison graph (generation methods)

different. It reflects the radiated thermal infrared radiation according to the character-istics of the surface [16]. Considering this point, we selected six materials that are expected to be widely used as surfaces. The types of surface materials are paper, acrylic, canvas paper, foam board, Iron plate, and MDF. Each surface material was separated into fingertip contact and breath and breath through the hollow rod to gen-erate residual heat. Figure 8 shows the residual heat generated from each surface material.

The highest temperature values of residual heat generated on the each surface material were classified as fingertip contact and breath through the hollow rod (see Fig. 9). Figure 9-(a) is the fingertip contact and Fig. 9-(b) is the breath with hollow rod. Figure 9 shows that the temperatures of paper and canvas are relatively high. In the case of paper, the most of temperatures is above 30°. In case of iron plate, it shows very low temperature. In the case of iron plat, since the conductivity is high, the fingertip or the high residual heat is immediately transferred to the periphery of the iron plat and the surrounding air. In order to maintain sufficiently high residual heat, long time touch or breath with hollow rod are appropriate. In the experiment, the residual heat generation time is 1.5 s to 2 s.

4.3   User Characteristics of Heat Generation

According to the previous experiments, we found that the temperature distribution of residual heat by fingertip contact was different from each user. This difference is due to the characteristics of the user's physical conditions. People with cold hands have a low temperature of contact residual heat, and vice versa. As a result, some users may fail to generate sufficient residual heat by fingertip contact. However, the temperature inside the human body is almost the same for all users and remains constant. Therefore, the heat generated by internal body heat can generate high residual heat. Figure 10-(a) shows the residual heat temperature difference for two classes of users.

The two types of user temperature distribution according to the residual heat generation method can clearly distinguish the temperature distribution difference in fingertip contact (see Fig. 10-(b)). It can be seen that the difference in temperature

Fig. 8. Thermal image of residual heat (surface material in order from above: acrylic, canvas paper, form board, iron plate, MDF) left: fingertip, right: breath using the hollow rod

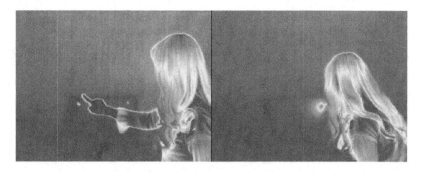

Fig. 8. (*continued*)

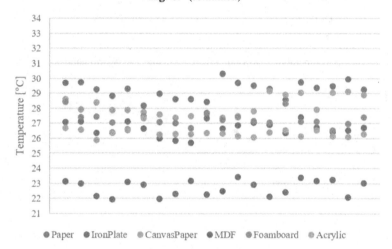

● Paper   ● IronPlate   ● CanvasPaper   ● MDF   ● Foamboard   ● Acrylic

(a) Fingertip Touch

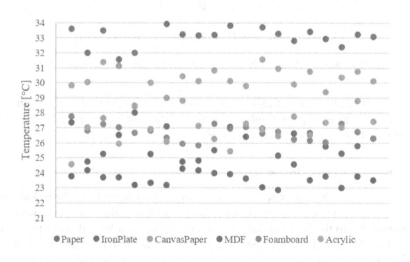

● Paper   ● IronPlate   ● CanvasPaper   ● MDF   ● Foamboard   ● Acrylic

(b) Breath using the hollow rod

Fig. 9. Residual heat temperature comparison graph (surface materials)

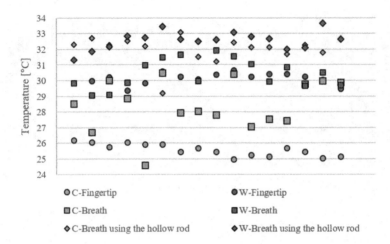

(a) Temperature distribution (C: Cold hands, W: Warm hands)

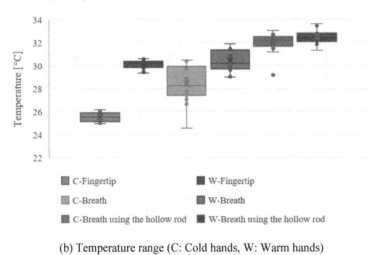

(b) Temperature range (C: Cold hands, W: Warm hands)

Fig. 10. Residual heat temperature comparison graph (user characteristics)

distribution is small in case of using hollow rod (straw). Some temperature values are high or low because the participant unconsciously blows strongly or the finger contact time and intensity are not exactly the same.

4.4   Conclusions and Discussion

We conclude the following conclusions from three experiments. First, the residual heat temperature of the breath using the hollow rod (straw) was higher than the residual heat temperature due to the fingertip contact. Second, the residual heat of fingertip may not be detectable because the temperature is not consistent depending on the characteristics of the users. However, the residual heat of the breath using the hollow rod (straw) is consistent regardless of users. Third, there is the difference of residual heat temperatures with respect to the surface materials. Since the user generates residual heat for a short time, it is advantageous for surface interaction to use paper and canvas paper as surface materials.

Participants gave some comments on finger contact. Especially, in the case of steel plate, the fingertips become colder as the fingertip contact is repeated. The canvas paper and the foam board were said to have a low sense of heterogeneity with soft touch.

5   UI Application Using Residual Heat

We proposed a method that interactive interface using residual heat through a hollow rod for users who are not able to use their hands. In the previous section, residual heat experiments were performed on various surface materials. Experiments have shown that the residual heat by using the hollow rod is suitable for surface interaction.

5.1   Surface Interaction System

In this paper, we have implemented a prototype application to test the usefulness of the proposed interaction technique. The surface-based interaction used a projection-camera system. When the user creates residual heat on the surface of the paper material attached to the wall, the projector projects another image to the residual heat generating position. The conversion from the camera coordinates and the projected coordinates is accomplished by using a four-point homogeneous transformation. The application displays an historical old map at the selected location when the user views the satellite map projected onto the projector and generates residual heat at the desired location. Figure 11 shows a demonstration of a surface based interaction application. Recognizing the residual heat of the breath, it was confirmed that the user indicating location is exactly specified and interacted. Since it is possible to indicate a precise narrow position through a hollow rod to a high residual heat, it has been confirmed that a natural interface is possible without the need of hand movement.

Fig. 11. Demonstration of application

6 Conclusion and Future Work

In this paper, we have proposed a method for user interaction without the use of a hand by using a thermal camera. This method can generate the residual heat with the use of a hollow rod, and detect it with a thermal camera. Thus it can interact without using a hand. To verify that the breath from hollow rod is efficient for generating residual heat, surface materials and residual heat generation methods were examined. Experimental results have shown that the proposed method enables the user who has difficulty in using their hands to interact with the same accuracy and naturalness as by touching them with their fingers. And since thermal cameras are not affected by background and illumination changes, they can be used in various environments. We have developed an application program to demonstrate the usefulness of the proposed interaction system.

The prototype system, which is made as an application example, includes a thermal imaging camera and a projector placed behind the user, so the projector image is hidden when the user generates residual heat. If the user generates residual heat and does not moved, the thermal camera cannot detect the residual heat generated on the surface. In the future work, we will improve the table top based interaction system using the rear screen. In addition, we plan to conduct research on user interfaces to enable various manipulations and interactions.

Acknowlegements. This work was supported by Institute for Information & communications Technology Promotion (IITP) grant funded by the Korea government (MSIT) (No. 2017-0-01849, Development of Core Technology for Real-Time Image Composition in Unstructured In-outdoor Environment)

References

1. Dietz, P., Leigh, D.: DiamondTouch: a multi-user touch technology. In: Proceedings of the 14th Annual ACM Symposium on User Interface Software and Technology, pp. 219–226. ACM (2001)
2. Erol, A., Bebis, G., Nicolescu, M., Boyle, R.D., Twombly, X.: Vision-based hand pose estimation: a review. Comput. Vis. Image Underst. **108**(1–2), 52–73 (2007)
3. Fang, Y., Yamada, K., Ninomiya, Y., Horn, B.K., Masaki, I.: A shape-independent method for pedestrian detection with far-infrared images. IEEE Trans. Veh. Technol. **53**(6), 1679–1697 (2004)
4. Gade, R., Moeslund, T.B.: Thermal cameras and applications: a survey. Mach. Vis. Appl. **25**(1), 245–262 (2014)
5. Gowen, A.A., Tiwari, B.K., Cullen, P.J., McDonnell, K., O'Donnell, C.P.: Applications of thermal imaging in food quality and safety assessment. Trends Food Sci. Technol. **21**(4), 190–200 (2010)
6. Hilliges, O., Izadi, S., Wilson, A.D., Hodges, S., Garcia-Mendoza, A., Butz, A.: Interactions in the air: adding further depth to interactive tabletops. In: Proceedings of UIST 2009, pp. 139–148. ACM Press, New York (2009)
7. Jones, B.F., Plassmann, P.: Digital infrared thermal imaging of human skin. IEEE Eng. Med. Biol. Mag. **21**(6), 41–48 (2002)

8. Kane, S.K., Avrahami, D., Wobbrock, J.O., Harrison, B., Rea, A.D., Philipose, M., LaMarca, A.: Bonfire: a nomadic system for hybrid laptop-tabletop interaction. In: Proceedings of the 22nd Annual ACM Symposium on User Interface Software and Technology, pp. 129–138. ACM (2009)
9. Kong, S.G., Heo, J., Boughorbel, F., Zheng, Y., Abidi, B.R., Koschan, A., Abidi, M.A.: Multiscale fusion of visible and thermal IR images for illumination-invariant face recognition. Int. J. Comput. Vis. **71**(2), 215–233 (2007)
10. Kurz, D.: Thermal touch: thermography-enabled everywhere touch interfaces for mobile augmented reality applications. In: Mixed and Augmented Reality (ISMAR) 2014, pp. 9–16. IEEE (2014)
11. Larson, E., Cohn, G., Gupta, S., Ren, X., Harrison, B., Fox, D., Patel, S.: HeatWave: thermal imaging for surface user interaction. In: Proceedings of the 2011 SIGCHI Conference on Human Factors in Computing Systems, pp. 2565–2574. ACM (2011)
12. Lewis, G.F., Gatto, R.G., Porges, S.W.: A novel method for extracting respiration rate and relative tidal volume from infrared thermography. Psychophysiology **48**(7), 877–887 (2011)
13. Oka, K., Sato, Y., Koike, H.: Real-time tracking of multiple fingertips and gesture recognition for augmented desk interface systems. In: Proceedings of 2002 Fifth IEEE International Conference on Automatic Face and Gesture Recognition, pp. 429–434. IEEE (2002)
14. Rekimoto, J.: SmartSkin: an infrastructure for freehand manipulation on interactive surfaces. In: Proceedings of the 2002 SIGCHI Conference on Human Factors in Computing Systems, pp. 113–120. ACM (2002)
15. Ring, E.F.J., Ammer, K.: Infrared thermal imaging in medicine. Physiol. Meas. **33**(3), R33 (2012)
16. Sahami Shirazi, A., Abdelrahman, Y., Henze, N., Schneegass, S., Khalilbeigi, M., Schmidt, A.: Exploiting thermal reflection for interactive systems. In: Proceedings of the 2017 SIGCHI Conference on Human Factors in Computing Systems, pp. 3483–3492. ACM (2017)
17. Wilson, A.D.: PlayAnywhere: a compact interactive tabletop projection-vision system. In: Proceedings of UIST 2005, pp. 83–92. ACM Press, New York (2005)

A Study of Perception Using Mobile Device for Multi-haptic Feedback

Shuo-Fang Liu, Hsiang-Sheng Cheng$^{(\boxtimes)}$, Ching-Fen Chang,
and Po-Yen Lin

Department of Industrial Design, National Cheng-Kung University, No. 1,
University Road, Tainan City 701, Taiwan (R.O.C.)
chengjohnsonhs@gmail.com

Abstract. As developments are made to mobile devices, advances are also made to vibration feedback technology to help visually impaired and elderly users. At present, mobile devices still use motor technology to provide vibration feedback. Therefore, in order to explore the possible applications of motor vibration feedback, two experiments were carried out in this study. In both of the experiments, four motors were used. In the first experiment, four motors were installed in each corner of two prototype devices of different sizes (5.5 inches and 9.7 inches). These devices were placed on top of a desk and the motors were randomly activated. The subjects then touched the center of the prototypes with their index finger, and had to identify which motor was vibrating. The results showed that age difference had a significant difference in the perception of the vibration position, but the difference between the two sizes was not significant. The second experiment compared the perception of the vibration position in hand-held devices by using a 5.5-inch prototype. The results showed that the different age groups showed a minor difference in how the prototype was used. However, the different ways of using the prototype had a significant difference in the identification of the vibration position.

Keywords: Elderly · Visual · Tactile · Vibrotactile · Haptic · Motor
Stimuli · Mobile device

1 Introduction

With the development of mobile devices, many researchers and developers have committed to providing user more rich and delicate experience. According to recent technology trends and related literature, it is possible to reduce the rate of error in the use of mobile devices by providing voice and tactile feedback [1]. Haptic technology in mobile devices can provide a larger variety of feedback than visual or auditory feedback [2]. Another study also showed that tactile feedback not only improves the use of virtual button, but also provides a higher level of satisfaction when using the touch screen [3].

In recent years, haptic feedback has greatly improved in mobile devices. The vibration technology in the iPhone, named the Taptic Engine, can control haptic feedback through computer programs to stimulate press force. This is also known as 3D

© Springer International Publishing AG, part of Springer Nature 2018
S. Yamamoto and H. Mori (Eds.): HIMI 2018, LNCS 10904, pp. 218–226, 2018.
https://doi.org/10.1007/978-3-319-92043-6_19

touch [4]. The Taptic Engine is a type of linear resonant actuator, which also has the ability of displaying direction. By increasing the different types of vibration patterns and using spoken feedback, the operation of mobile devices for elderly or visually impaired users will be more accurate [5]. The use of haptic feedback techniques as a supplement to mobile devices will not only improve the quality of life for these types of users, but it will also allow them to live with a degree of independence. In the medical field, by integrating tactile feed-back with a robotic arm system helps operators to control the grasping force appropriately when conducting minimally invasive surgery. This can significantly reduce the damage done to body tissue [6].

Vibrotactile is used in different devices to convey information such as navigation or warnings without distracting the user [11]. Some of the uses of vibrotactile include vibrating vests [10], vibrotactile sleeve-armbands [9], or using the vibration mode of the phone to let the user know the right direction [12]. Although many techniques of tactile feedback are not widely available yet, studies show that tactile feedback may be applied to various parts of the body and can be developed in other different functions, providing different information especially in noisy environments.

Therefore, this study will investigate the use of haptic feedback with motor technology to discover the feasibility of haptic feedback in multi-vibration motors. The study will also aim to understand the tactile perception in different types of mobile devices, and whether handheld devices can convey tactile feedback more effectively. The aim is to increase the haptic feedback variety and explore the future applicability of haptic feedback technology.

2 Method

This study is divided into two parts: experiment 1 and experiment 2. The purpose of experiment 1 was to find out whether there were any differences between the perceptions of multi-point haptic feedback between two sizes of mobile devices. Acrylic panels were used to simulate a mobile phone and a tablet while vibration motors were set in the four corners of the panels (Fig. 1). Participants placed their index finger in the center of panel to complete the perception test (explained below). The aim of experiment 2 was to test whether handheld devices can help users to sense the vibration position more accurately. A semi-structured questionnaire was used as a study reference.

2.1 Apparatus

According to recent literature, middle-aged people often choose to use larger-sized mobile devices because, one, the larger screen allows the interface to contain more information; and, two, if the screen is too small, it will lead to only displaying incomplete information, and thus affect its usability [7]. Thus, in this study larger mobile devices were selected. The smaller device had a 5.5-inch smart phone panel and the larger device had a 9.7-inch tablet panel, a 10 mm × 3.4 mm shaftless vibration motor (Fig. 1) pasted onto the back of the prototypes in each corner. Bluetooth was

Specification	Value
Voltage [V]	3
Voltage Range [V]	2.5~3.8
Rated Speed [rpm]	12000
Rated Current [mA]	75
Start Voltage [V]	2.3
Start Current [mA]	85
Terminal Resistance [Ohm]	75
Vibration Amplitude [G]	0.8

Fig. 1. Motor and specification

used to connect the Arduino module and remote control by computer. To reduce the sound of vibration, a sponge was placed under the prototype (Fig. 2).

Fig. 2. Apparatus of panel and haptic system

2.2   Subject

There were 36 participants in experiment 1. 16 participants were aged 20–30 years and 20 subjects were over the age of 50. There were 22 participants in experiment 2. 11 of these participants were aged 20–30 years and 11 were over 50. All participants had experience of using smart phones and tablets and have normal behavior ability.

2.3   Experiment 1: Vibration Perception in Two Sizes of Prototype

The purpose of experiment 1 was to explore whether the participants could perceive multi-point vibration feedback in mobile devices of different sizes. This emulated the usual behavior of most mobile device users in two main ways. First, the tactile receptors in the fingertip are more sensitive than any other part of the hand [9], which means people choose to use their fingertips because they provide the most feedback and allow for more careful interactions. Second, because many virtual buttons on touch screens are small, users tend to use their index finger to touch screen rather than the thumb [1]. It should also be noted that the index figure allows for the most control, so participants were asked to put their index finger on the center of prototypes (Fig. 3).

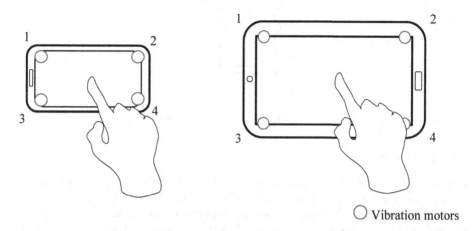

Fig. 3. 5.5 inch phone panel and 9.7 inch tab panel

Procedure. Before the formal test, participants were asked to place their index finger in the center of the prototype while the four motors were activated in vibrate four motors in sequence. The motors are programmed to vibrate for 100 ms at 400 ms intervals at 168 Hz for a total of 2.5 s. This informal test allowed the participants to practice feeling the difference between the different positions of the vibrations. In the formal test, the four motors were randomly activated, and the participants had to identify the position of the vibration based on their feeling for a total of six times. Then the participants' answers were compared with the correct answers.

2.4   Experiment 2: Hand-Held Way to Feel the Vibration Position

The purpose of Experiment 2 was to explore whether holding the 5.5-inch prototype (Fig. 4) in the hand could produce more accurate vibration position identification than when the device is placed on a desk. (Experiment 1).

Procedure. Before the formal test, the participants took part in an informal test. As in experiment 1, the participants practiced experiencing and identifying the vibrations before taking the formal test. In the formal test, the four motors were randomly activated using the same vibration specifications as experiment 1. The participants were asked to identify the position of each vibration based on their tactile sense six times. Each participant's answers were then with the correct answers.

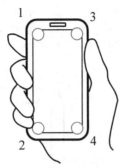

Fig. 4. Hand-held 5.5-inch prototype

2.5   Limitation

One of the limitations of the study was the authenticity of the prototype. Because there are no mobile devices with four motors on the market, there had to be prototypes created for this study. Although care was taken to provide an authentic prototype, the material used in the prototype may have affected the vibration. Another limitation of the study was the degree of resonance. Unfortunately, the degree of resonance in this study was different from the degree of resonance in actual mobile devices. A final limitation is the type of motor used. In this study ERM motors were used rather than LRA motors. This could be a limitation as the perception of vibration could be related to the type of motor.

3   Result

3.1   Experiment 1

The results show that if the location of the four motors are used for analysis, the correct proportion of two sizes were only 0.28 and 0.4, for different ages are 0.42 and 0.29. Because the proportion weren't high enough, the study changed to use left and right side, and up and down side for analysis (Fig. 5). The results show that the accuracy rates of different ages were 0.67 and 0.53 for the horizontal analysis, while the correct rates for different sizes were 0.59 and 0.58 respectively. In the vertical analysis, the correct rates for different age groups were 0.61 and 0.51 respectively, while the accuracy of different sizes was 0.5 and 0.61 (Figs. 6 and 7). Because the highest average correct proportion rate was obtained from the horizontal analysis, the two-way ANOVA (Table 1) used the data from the horizontal analysis. The result is that the age difference has a significant effect on the perception of the vibration position ($P < 0.05$). The perception of different sizes is not significant ($P > 0.05$).

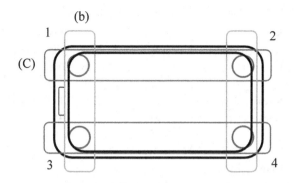

Fig. 5. (b) left and right side (c) up and down side

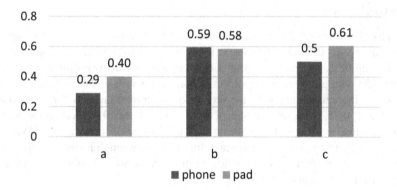

Fig. 6. (a) 4 corner analysis (b) left and right analysis (c) up and down analysis

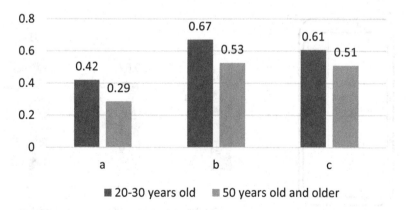

Fig. 7. (a) 4 corner analysis (b) left and right analysis (c) up and down analysis

Table 1. Two-way ANOVA analysis

Source	df	Mean square	F	P value
Adjusted model	3	.136	2.033	.117
Intercept	1	25.334	379.160	.000
Ages	1	.367	5.498	.022
Sizes	1	.005	.070	.792
Ages * sizes	1	.038	.569	.453
Error	68	.067		
Total	72			
Corrected total	71			

R square = .082 (adjusted R square = .042)

3.2   Experiment 2

The results of experiment 2 showed that the correct proportion of the younger users and the older users was 0.61 and 0.44 respectively with the average proportion of correct answers being 0.52. The proportion of correct answers for the two different ages were 0.67 and 0.55 with an average proportion of correct answers of 0.61 for the horizontal analysis. For the vertical analysis, the proportion of correct answers for the two ages was 0.91 and 0.85, with an average of 0.88 (Figs. 8 and 9). Because the vertical analysis provided the highest average, it was used to run a one-way ANOVA. The findings show that there was no significant difference between different ages (p > 0.05) (Table 2). However, there was a significant difference between two ways of using the prototype (in the hand and on the desk) (Table 3).

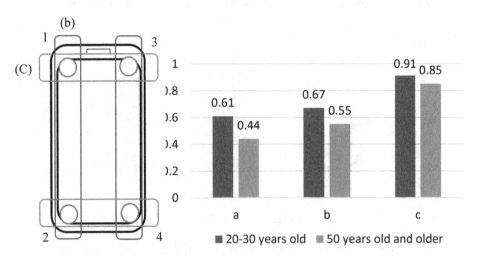

Fig. 8. (b) left and right side (c) up and down side   **Fig. 9.** (a) 4 corner analysis (b) left and right analysis (c) up and down analysis

Table 2. Different ages one-way ANOVA

	Sum square	df	Mean square	F	P value
Between groups	.020	1	.020	.930	.346
Within groups	.434	20	.022		
Total	.455	21			

Table 3. One-way ANOVA with experiment 1 data

	Sum square	df	Mean square	F	P value
Between groups	1.104	1	1.104	21.975	.000
Within groups	2.813	56	.050		
Total	3.917	57			

4 Conclusion and Future Work

According to recent literature, it is known that the user operation satisfaction can be increased by providing different kinds of haptic feedback. Elderly users inevitably suffer from visual and auditory issues and the decline of cognitive and motor function use. These issues may cause the elderly to feel anxious when using mobile devices. However, current mobile devices contain a variety of accessibility features, such as feedback technology, which can help elderly users to cope with those physical changes. For example, in the past, a noisy environment would make it difficult for an elderly user to receive auditory feedback, but now, haptic feedback can provide the feed-back needed to successfully use the device [7]. This feedback can be increased by providing elderly users with a more intensive vibration to increase finger touch stimulation [8].

Although there have been many motor-related studies, there have not been much further investigate into the size of the mobile device. While this study did not find a significant difference between the different in size, in the future, the size of the mobile devices of different brands like Apple or Samsung could be further studied. This type of study may make it possible to find out the size of mobile devices that can deliver accurate vibrations to correctly stimulate the tactile senses.

In experiment 2, the position of the mobile device was compared. It was found that users were able to identify the location of the vibrations with more accuracy when the device was held in the hand rather than on a flat surface. This could be due to having more of a physical connection with the device. If different vibration modes can be added according to the result, different experiences and information can be added to the tactile feedback. If this is applied in mobile devices, more diverse changes can be made with different gestures.

It was found that it was easier to differentiate between the longer sides of the devices rather than the shorter sides. If further research is conducted in the future, it is possible to reduce the time cost of the research simply by arranging a motor on each long side of the mobile device. The results also show that age differences have significant differences in perception of the vibration position. Due to time constraints, the study was not able to investigate if the frequency has a correlation with the perception of the position or not. Further studies into this area could investigate different frequencies as the research direction in order to explore what kind of frequency can increase the usability of tactile feedback.

Vibration motor technology continues to evolve. The delicate vibration of the linear motor can show the change of the pressing force. And the mobile device generally has only one vibration motor in the market. Although the linear motor can show a little direction, it is still not clear enough. Therefore, if multiple linear motors can be used in this study, the correct proportion of this study may be improved. Another suggestion is that there may be a way to place motors in other patterns, so that the vibration shows the direction and linear changes more obvious.

At present, the technology of tactile feedback is constantly improving. As the area is developing, more research can only aid developments and ensure that they are heading down a production and meaningful path.

Acknowledgements. We are thankful for the financial support from The Ministry of Science and Technology (MOST), Taiwan. The grant MOST 106-2221-E-006-156.

References

1. Lee, S., Zhai, S.: The performance of touch screen soft buttons (2009)
2. Banter, B.: Touch screens and touch surfaces are enriched by haptic force-feedback. Inform. Disp. **26**(3), 26–30 (2010)
3. Koskinen, E., Kaaresoja, T., Laitinen, P.: Feel-good touch: finding the most pleasant tactile feedback for a mobile touch screen button (2008)
4. Dempsey, P.: The teardown Apple iPhone 7 smartphone. E&T Mag. **11**(11) (2016). http://ieeexplore.ieee.org/stamp/stamp.jsp?arnumber=7791546&tag=1
5. Yatani, K., Truong, K.N.: SemFeel: a user interface with semantic tactile feedback for mobile touch-screen devices (2009)
6. Wottawa, C.R., Genovese, B., Nowroozil, B.N., Hart, S.D., Bisley, J.W., Grundfest, W.S., Dutson, E.P.: Evaluating tactile feedback in robotic surgery for potential clinical application using an animal model. Surg. Endosc. **30**, 3198–3209 (2016)
7. Caprani, N., O'Connor, N.E., Gurrin, C.: Touch screens for the older user (2012)
8. Liu, S.-F., Chang, C.-F., Wang, M.-H., Lai, H.-H.: A study of the factors affecting the usability of smart phone screen protectors for the elderly. In: Zhou, J., Salvendy, G. (eds.) ITAP 2016. LNCS, vol. 9754, pp. 457–465. Springer, Cham (2016). https://doi.org/10.1007/978-3-319-39943-0_44
9. Kandel, E.R., Schwartz, J.H., Jessell, T.M.: Principles of Neural Science (1991)
10. Hung, C.-T., Croft, E.A., Van der Loos, H.F.M.: A wearable vibrotactile device for upper-limb bilateral motion training in stroke rehabilitation: a case study (2015)
11. Elliott, L.R., van Erp, J.B.F., Redden, E.S., Duistermaat, M.: Field-based validation of a tactile navigation device (2010)
12. Pielot, M., Poppinga, B., Heuten, W., Boll, S.: PocketNavigator: studying tactile navigation systems in-situ (2012)
13. Williamson, J., Robinson, S., Stewart, C., Murray-Smith, R., Jones, M., Brewster, S.: Social gravity: a virtual elastic tether for casual, privacy-preserving pedestrian rendezvous (2010)

Realizing Multi-Touch-Like Gestures in 3D Space

Chunmeng Lu$^{(\boxtimes)}$, Li Zhou, and Jiro Tanaka

Graduate School of Information, Production and System,
Waseda University, Tokyo, Japan
luchunmeng@fuji.waseda.jp

Abstract. In this paper, our purpose is extending 2D multi-touch inter-action to 3D space and presenting a universal multi-touch gestures for 3D space. We described a system that allows people to use their familiar multi-touch gestures in 3D space without touching surface. We called these midair gestures in 3D as *3D multi-touch-like gestures*. There is no object or surface for user to touch in 3D space, so we use depth camera to detect fingers' state and estimate whether finger in the "click down" or "click up", which show user's intention to interact with system. We use machine learning to recognize hand shapes. While we do not need to precessing the recognition all the time, we only recognize hand shape between "click down" or "click up".

Keywords: Gesture · Human-computer interaction · Machine learning

1 Introduction

Multi-touch extends the vocabulary of interaction and has been used in many devices, such as smart phone and laptop. People use 2D multi-touch gestures to accomplish interaction with these devices. After learning to use a set of 2D multi-touch gestures, people could perform the same gestures in different kinds of machines. The universality and ease of use make multi-touch interaction become a revolutionary technology.

Future human-computer interfaces will enable more nature, intuitive com-munication between people and all kinds of sensor-based devices. Gesture-based interaction, as a kind of nature interaction, will be popular in the future [11]. Obviously the present multi-touch gestures have some limitations, because many devices will not provide object surfaces for users to touch, like some VR and AR devices. Thus non-touch interaction system cannot apply the 2D multi-touch ges-tures directly. In this case, a gesture-based interaction system needs to realize the hand gestures in 3D space.

In our study, inspired by the 2D multi-touch interaction, we intend to design a new system to realize 3D multi-touch-like gestures for 3D space. "Multi-touch-like gestures" here means that the hand shapes and functions of 3D gestures are similar to 2D multi-touch gestures. Our system will allow users to perform

© Springer International Publishing AG, part of Springer Nature 2018
S. Yamamoto and H. Mori (Eds.): HIMI 2018, LNCS 10904, pp. 227–239, 2018.
https://doi.org/10.1007/978-3-319-92043-6_20

their familiar multi-touch gestures in 3D space. In the system, we extend five typical 2D multi-touch gestures to 3D space to become multi-touch-like gestures: zoom in/out, rotate, scroll, swipe and drag, as showed in Fig. 1. People are often comfortable with traditional interaction methods, and our method that extending 2D multi-touch gestures to 3D space will decrease the difficulty and time to learn a new interaction style for people.

In order to recognize these 3D multi-touch-like gestures, we present a method using machine learning. However, only using machine learning is not good enough. It is because machine learning can recognize the hand shape, however, it does not tell when the gestures start and when they end. If we cannot recognize the exact timing of gestures, it makes difficult to return the response to the given gesture in the right timing. So we also need to realize the state of fingers. We use depth camera to detect the state of fingers to realize the starting and ending of gestures.

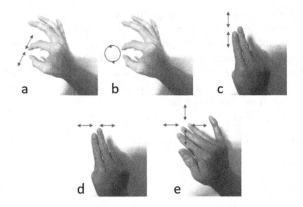

Fig. 1. Five 3D multi-touch-like gestures: (a) zoom in/out, (b) rotate, (c) scroll, (d) swipe and (e) drag

2   Related Works

Gesture-based interaction provides a nature way for user to interact with machines and has been researched for over three decades. There are two kinds of approaches to capture gesture data and recognize gestures: "data-glove based" and "vision-based". When applying data-glove based method, users need to wearing a glove-like device which is equipped sensor to collect data of hand and finger motions [5]. However, the extra devices are not cheap and not convenient. Thus, applying vision-based methods can be a good choose.

Many researches show that using video camera and depth sensor to recognize hand gesture is a effective way [5,9,10]. Wilson [12] presented a novel touch screen technology. They use two video cameras to capture hand and process the images. In their system, the a pair of video cameras could get depth information of hand gesture and give feedback for interface. Boussemart et al. [3] present a framework for 3D visualization and manipulation in an immersive space. Their work can be used in AR and VR system. Wachs et al. [11] summarizes the requirements of hand-gesture interfaces and the challenges when applying hand gestures in different application. They divide these applications into four classes: medical systems and assistive technologies; crisis management and disaster relief; entertainment and human-robot interaction. These applications reveal the rich vocabulary of hand gesture.

In our system, the important part is realizing the state of fingers. This part will tell system when to start analyze hand gestures and when to end. There are also previous works to research this issue. Karam et al. [6–8] carry out a series of studies about finger click detection in 3D space. They present two-hand interactive menu and introduce how to use depth based selection techniques [6]. They design a selection mechanism approach, "three fingers clicking gesture", to detect intention from user [7]. Their approach is using depth camera to capture hand shapes and analyze the relative positions of the fingers. They present "Xpli Click" to improved midair finger click detection method [8]. The "Xpli Click" allow user to perform more nature click action in their system.

3   System Design

Our system combines theses three parts to realizing multi-touch-like gestures in 3D space:

(1) Mechanical click detection.
(2) Hand shape recognition.
(3) Motion recognition.

In the first part, mechanical click detection, system will detect whether user want to interact with machines. In the second part, hand shape recognition, system will realize which hand shape that user performs. And in the third part, motion recognition, system will recognize the movement of fingers to find which gesture that user want to use.

3.1   "Click Down" and "Click Up" in 3D Space

In the case of 2D gesture, we start gesture by touching the screen and end gesture by leaving hands. Similarly, there is a click action in mouse gesture.

Therefore, we need to catch the timing of when the gesture starts and ends in our 3D gestures. We introduce click/touch action in our 3D gestures for such purpose. In 2D multi-touch, system could recognize gestures by detecting click action and movements of fingers on surface. While in 3D space, there is no midair physical surface for user to click.

In order to detect the timing of starting and ending of 3D gestures in 3D space, "click state" and "neutral state" are defined to show whether user want to perform gestures [8]. In 3D space, user's hand moves freely in "neutral state" when having no intention to interact with computer. As for "click state", we define "click down" and "click up" to distinguish the states. "Click down" means that user has the intention to interact with computer and fingers are getting into "click state". "Click up" means fingers are going back to "neutral state".

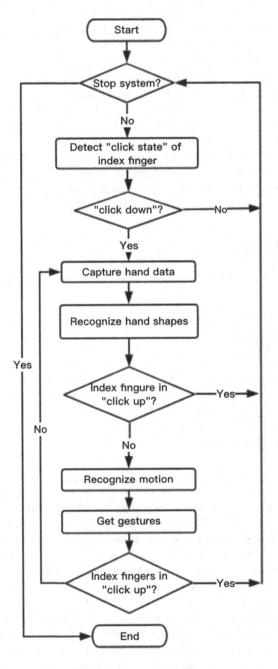

Fig. 2. System overview

3.2 System Overview

From Fig. 1, we could find that index finger is used in these five gestures that we need to realize. It means that we could only detect the state of index finger to know whether user is performing a gesture.

The system is designed to follow these steps, as showed in Fig. 2:

(1) System detects the "click state" of index finger all the time.
(2) Once detecting the "click down", system will capture hand data and start machine learning to recognize hand shape.
(3) If getting a hand shape of one of the five gestures, system will capture the motion of the fingertips and get gestures until detecting index fingers in "click up" or stopping system. If not, system will continue recognize hand shape until getting a hand shape of the five gestures or detecting index finger in "click up" or stopping system.

4 Mechanical Click Detection

In our system, we detect "click down" and "click up" of index finger by mechanical calculation. We could calculate the angle θ related with palm center and index finger, as showed in Fig. 3. There is a range of the angle θ when index finger in "click down" [8]. So, we could calculate angle θ to confirm whether index finger is in "click down" or "click up".

The mechanical calculation follows these steps:

(1) Use Leap Motion as depth sensor to capture the coordinates of fingertip of index finger and palm center.
(2) Create the vector from palm center to fingertip of index finger basing on the coordinates.
(3) Think of the palm as a plane and calculate the angle θ between palm and the vector in step (2).
(4) If the angle in the range, index finger is in "click down". Otherwise, index finger is in "click up".

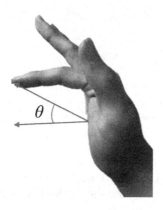

Fig. 3. Angle θ

5   Hand Shape Recognition

Multi-touch-like gestures start from "click down" and end at "click up". Between them, there are hand shape performing and movement. Hand shape and fingers movements actually decide gestures.

System needs to classify the hand shapes. We use SVM (Support Vector Machine) to achieve the purpose.

Adapting SVM method to our system, we design four steps for hand shape recognition:

(1) data collection
(2) normalization and scale
(3) model training
(4) predicting.

5.1   Data Collection

In the first step, data collection, we need to decide what kind of data of hand that we need to capture.

In a hand of human, there are nineteen bones related to five fingers. We could use the endpoints of these bones and palm center as "key points" to describe these hand shapes, as showed in Fig. 4. There are total 26 key points. Then we use depth camera to capture the coordinates of key points of user's hand.

Fig. 4. 26 Key Points: two blue points represent palm center and wrist joint; red points represent the endpoints of bones in hand (Color figure online)

In one frame of Leap Motion, all coordinates of key points become a group of original data. As we use Leap Motion, the center of Leap Motion is the origin of coordinate system, and the data shows the key points' positions relative to the center of Leap Motion. Using the data, we could draw the skeleton of hand.

5.2 Normalization and Scale

In 3D space over Leap Motion, hand moves around freely. Thus the data of hand shape can be captured in any position in that 3D space. So we can make the palm center as the origin coordinate. Then we calculate other key points positions relative to palm center.

Data normalization follows these steps:

(1) Translate all the points until the palm center is on the origin coordinate.
(2) Rotate the points around the palm center until the palm parallel to the x-z axis plane.
(3) Rotate again the points around the y coordinate axis until the palm points the - z axis.

There is still a problem that the sizes of hand model are different because of the different distances between hand and Leap Motion. So we scale all the data to $[-1, 1]$. After that, we also could use the new data to draw the skeleton of hand and exclude the effect of Leap Motion's position. This step will eliminate noise and improve accuracy.

5.3 Model Training

The third step of hand shape recognition is model training.

Our work is to recognize 5 kinds of gestures: zoom, rotate, scroll, swipe and drag. While the zoom and rotate gestures have the same hand shape when moving, we mark this hand shape as hand shape 1. And there is the same situation with scroll and swipe, and we mark this hand shape as hand shape 2. We also mark the hand shape of drag as hand shape 3. So we only need to classify 3 kinds of hand shapes.

There are many SVM models and we use Classification SVM Type 1 (also known as C-SVM classification) in our system. As there are three kinds hand shapes, we use one-vs.-one (OvO) method to classify them. In OvO method, we need to get three classifiers: classifier 1 for hand shapes 1 and 2, classifier 2 for hand shapes 1 and 3, and classifier 3 for hand shapes 2 and 3. Then combining the classifiers, we will get a 3-label classifier model.

There are three steps when applying SMV method: (1) making training set; (2) getting 3-label classifier model through training; (3) predicting hand shapes with trained model in our system. In this three parts, making training set and getting 3-label classifier model through training have to be finished before building our system.

Here are the steps to get the 3-label classifier model:

(1) Capture 50 groups coordinates of the key points for every hand shape.
(2) Normalize and scale the data and get 3 groups training sets.
(3) Through training sets, get classifier 1, classifier 2 and classifier 3.
(4) Combine this three classifiers to get a 3-label classifier model.

The model training part is preparation work for system. The 3-label classifier model will be used to realize hand shapes in time in our system.

5.4 Predicting

The final part of hand shape recognition is predicting.

We use 3-label classifier model to predict hand shapes. The 3-label classifier model works following these steps:

(1) Input a hand shape.
(2) Start classifier 1, 2 and 3 in order. Then we will get three result numbers of labels of hand shapes.
(3) Count the times of occurrences of the labels.
(4) Put out the highest times of occurrences of label, and label shows which gesture is performed.

In this part, we collect 20 data groups of every hand shapes to test the accuracy of the 3-label classier model. The total 60 data groups become the testing set.

The predicting part plays an important role in our system. In the system, once detecting "click down", system collects a group of hand data immediately. Then using 3-label classifier model that we get in model training part, predict part will tell system what hand shape that this captured group of data represents.

6 Motion Recognition

After knowing the hand shape, the system needs motion recognition because gestures are defined by both hand shape and hand motion together.

Once getting the hand shape, system starts motion recognition step. After the frame when hand shape recognized, system will continuously calculate and record the positions of fingers in every frame captured by depth camera. For different hand shapes, system detects motion of different fingertips to realize gestures, as showed in Fig. 5.

In case of knowing the hand shape 1, system will detects movements of thumb and index finger. We create a vector v from thumb fingertip to index fingertip. Firstly, we calculate and record the v in the frame where system gets hand shape 1. We mark this original v as v_0. After that, in every frame, We compares v with v_0. If v is longer/shorter than v_0, gesture will be zoom in/out and system calculates the distance d between the two fingers. If System also calculate the angle between v and v_0 in every frame. If fingertips of thumb and index finger rotate around a center, the angle between v and v_0 will change continuously and system will realizing gesture Rotate.

In case of knowing the hand shape 2, system will detects movements of index finger and middle finger. In every frame, system calculates the coordinate of the midpoint of index finger and middle finger. We mark this midpoint as p_0. System calculates the relative distance d_0 between the p_0 in present frame and the p_0 in the frame when system gets hand shape 2. The d_0 will be used for interaction. Besides, system could realize movement directions through comparing the coordinates the p_0 in present frame and the p_0 in previous frame. If p_0 moves up or down, the gesture can be scroll; if p_0 moves left or right, the gesture can be swipe; if p_0 moves to other directions, system will not give response.

In case of knowing the hand shape 3, system will detects movements of index finger, middle finger and ring finger. In every frame, system calculates the coordinate of the midpoint of index finger, middle finger and ring finger. We mark this midpoint as p_1. System calculates the relative distance d_1 and realize the movement direction p_1 through comparing the p_1 in present frame and the p_1 in the frame where system gets hand shape 3. d_1 and p_1 will be used for interaction.

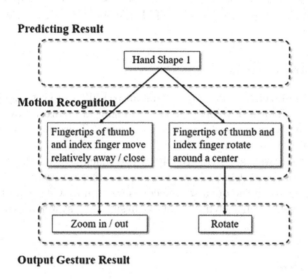

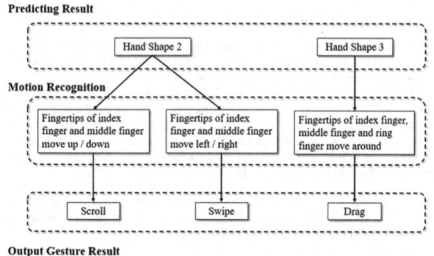

Fig. 5. Motion recognition and getting output

7 Implementation

7.1 Hardware Overview

In our system, we use Leap Motion as the depth camera [1]. Leap Motion could recognize and track hands and fingers in a 3D space. The 3D space is over the sensors and the best scope of sensor is 25 mm to 60 mm. The average frame rate of sensor is about 90 fps. We use APIs from Leap Motion to recognize the joints of hands, which allow us to collect data of key points in system. Leap Motion works with a Thinkpad laptop (Intel(R) Core(TM) i7-6500 CPU @2.5 GHz, RAM 8 GB, 64 bit windows 10).

7.2 Software Overview

We use Qt [2] as the development environment to build our system. Qt provides a cross-platform software development environment to create innovative devices, modern UIs and applications for multiple screens. This character will help us apply our system in other devices.

In our system, we use LIBSVM [4] as the machine learning tools in hand shape recognition part. LIBSVM is an open source library for support vector machines. We combine the open source codes written by C++ and Leap Motion APIs in Qt development environment to design the system, experiences and application.

8 Evaluation

8.1 Accuracy Rate of Click Detection

In system, hand shape recognition starts when detecting "click down" of index finger and ends when detecting "click up" of index finger. Therefor accuracy of click detection decides the usability of our system.

When interacting with system, hands will perform click action in state of stopping and in state of motion. Besides, different users are habituated to different speeds. Considering these two factors, we design a experience to evaluation the accuracy of click detection. In the experience, we choose three click speeds: low speed is one click per two second, normal speed is one click per second and fast speed is two clicks per second. Here are the experience steps:

(1) User move hand around in a 3D space over depth sensor and make click actions with index finger randamly for 100 times in fast speed, 100 times in normal speed and 100 times in low speed.
(2) User stops his hand over Leap Motion and make click actions with index finger randomly for 100 times in fast speed, 100 times in normal speed and 100 times in low speed.

The Accuracy rate are showed in Table 1.

Table 1. Test accuracy rate of click detection

State	Moving			Stopping		
Click speed	Fast	Normal	Low	Fast	Normal	Low
Accuracy rate	96%	98%	98%	99%	100%	100%

8.2 Accuracy Rate of Hand Shape Recognition

The system needs to start SVM method recognize the hand shapes after detecting "click down" of index finger. So there are two situations that recognizing hand shape of a hand in state of motion and in state of stopping.

Here are the experience steps:

(1) A user moves his hand around in a 3D space over the depth sensor and perform the three hand shapes. He needs to perform every hand shape for 100 times.
(2) User stops his hand over depth sensor and perform every hand shape for 100 times.

The accuracy rates of hand shape recognition are showed in Table 2.

Table 2. Test accuracy rate of hand shape recognition

State	Moving			Stopping		
Accuracy rate	93%	96%	95%	95%	98%	96%

8.3 Accuracy Rate of Realizing Multi-Touch-Like Gestures in 3D Space

As explained in motion recognition part, we need to realizing the five gestures basing on movement trajectories. We designed a experience to test the accuracy rate of realizing the five gestures.

In this experience, user put his hand with neutral stat over the sensor. Then he performs the five gestures: zoom in/out, rotate, scroll, swipe and drag. User needs to perform every gestures for 100 times one by one. As for zoom in/out, user performs zoom in and zoom out for 50 times respectively. In every times, the system outputs which gesture it captures. Table 3 shows the accuracy rate of realizing these five multi-touch gestures.

8.4 Simple Application

We design a picture viewer to test the usability of our system. As showed in Fig. 6, we place Leap Motion on the touchpad of a laptop to replace touchpad and mouse. User performs 3D gestures over Leap Motion without touch anything.

In this simple application, the five typical gestures: zoom in/out, rotate, scroll, swipe and drag are applied in the application to control the picture:

Table 3. Accuracy rate of realizing five gestures

Gestures	Zoom in	Zoom out	Rotate	Scroll	Swipe	drag
Accuracy rate	96%	95%	95%	98%	98%	98%

(1) When user performs gesture zoom in/out, system will detect the distance d between the thumb fingertip and index fingertip. In every frame, system will calculate the difference value between d in present frame and d in the frame when system gets hand shape. As the window size of picture viewer is not big, in this application, if the difference value is extended or reduced by $x$cm, the length of diagonal of picture in viewer will be extended or reduced $2x$cm.

(2) When user performs gesture rotate, picture will rotate following the angle between v and v_0 in every frame, which is described in Motion Recognition part.

(3) When user performs gesture scroll, swipe or drag, picture will move following the d_0 or d_1, which is described in Motion Recognition part. If d_0 or d_1 is extended or reduced by $x$cm, the picture in viewer will move $x$cm to corresponding direction.

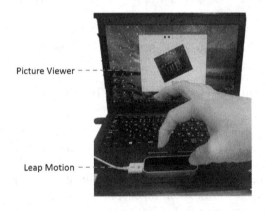

Fig. 6. Picture viewer: user is performing rotate gesture

9   Conclusion

With the rapid development of 3D technology, many researchers design new 3D gestures for their own system. Thus, people need to learn different gestures in different systems. In this situation, a universal 3D gesture set is need for 3D interaction system. In this paper, the experiences and simple application reveal the availability of our idea that expands the vocabulary of 2D multi-touch gestures to 3D space to improve non-touch interactions. In recent years, augmented reality and virtual reality become more and more popular. In future work, we will apply the multi-touch-like gestures in these fields to enhance the interaction with virtual environment.

References

1. Leap Motion. https://www.leapmotion.com. Accessed 5 Feb 2018
2. Qt. https://www.qt.io. Accessed 5 Feb 2018
3. Boussemart, Y., Rioux, F., Rudzicz, F., Wozniewski, M., Cooperstock, J.R.: A framework for 3D visualisation and manipulation in an immersive space using an untethered bimanual gestural interface. In: Proceedings of the ACM Symposium on Virtual Reality Software and Technology, VRST 2004, pp. 162–165. ACM, New York (2004)
4. Chang, C.C., Lin, C.J.: LIBSVM: a library for support vector machines. ACM Trans. Intell. Syst. Technol. **2**(3), 27:1–27:27 (2011)
5. Garg, P., Aggarwal, N., Sofat, S.: Vision based hand gesture recognition. World Acad. Sci. Eng. Technol. **49**(1), 972–977 (2009)
6. Karam, H., Tanaka, J.: Two-handed interactive menu: an application of asymmetric bimanual gestures and depth based selection techniques. In: Yamamoto, S. (ed.) HCI 2014. LNCS, vol. 8521, pp. 187–198. Springer, Cham (2014). https://doi.org/10.1007/978-3-319-07731-4_19
7. Karam, H., Tanaka, J.: Finger click detection using a depth camera. Procedia Manuf. **3**, 5381–5388 (2015)
8. Karam, H., Tanaka, J.: An algorithm to detect midair multi-clicks gestures. Inf. Media Technol. **12**, 340–351 (2017)
9. LaViola, Jr., J.J.: An introduction to 3D gestural interfaces. In: ACM SIGGRAPH 2014 Courses, SIGGRAPH 2014, 42 p. ACM, New York (2014)
10. Lee, U., Tanaka, J.: Finger identification and hand gesture recognition techniques for natural user interface. In: Proceedings of the 11th Asia Pacific Conference on Computer Human Interaction, pp. 274–279. ACM (2013)
11. Wachs, J.P., Kölsch, M., Stern, H., Edan, Y.: Vision-based hand-gesture applications. Commun. ACM **54**(2), 60–71 (2011)
12. Wilson, A.D.: TouchLight: an imaging touch screen and display for gesture-based interaction. In: Proceedings of the 6th International Conference on Multimodal Interfaces, ICMI 2004, pp. 69–76. ACM, New York (2004)

Effects of Background Noise and Visual Training on 3D Audio

Christian A. Niermann$^{(\boxtimes)}$

German Aerospace Center (DLR), Institute of Flight Guidance,
Lilienthalplatz 7, 38108 Braunschweig, Germany
christian.niermann@dlr.de

Abstract. Spatial audio or 3D audio as an information channel is increasingly used in various domains. Compared to the multitude of synthetic visual systems and 3D representations, audio interfaces are underrepresented in modern aircraft cockpits. Civil commercial aircraft rarely use spatial audio as a supplementary directional information source. Although, different research approaches deal with the benefits of spatial audio. In 3D audio simulator trials, pilots express concern over distractions from background noise and possibly mandatory training requirements. To resolve this, the author developed and tested a 3D audio system to support pilots in future cockpits, called *Spatial Pilot Audio Assistance (SPAACE)*.

The experiment took place at the German Aerospace Center's Apron and Tower Simulator. The developed system creates a three-dimensional audio environment based on normal non-spatial audio. The 27 participants heard the sound through an off-the-shelf aviation-like stereo headset. The main subject of investigation was to evaluate if air traffic control background noise affects spatial perception. The non-normally distributed location error with background noise ($Mdn = 6.70°$) happened to be lower than the location error without air traffic control background noise ($Mdn = 7.48°$). The evaluation the effect of visual feedback-based training was the second part of the experiment. In comparing the training session with the no-training session, the location error with training ($Mdn = 6.51°$) is only moderately lower than the location error without training ($Mdn = 7.96°$).

The results show that humans can perceive the SPAACE audio with high precision, even with distracting background noise as in a busy cockpit environment. The effect of training was not as high as expected, primarily due to the already precise localization baseline without training.

Keywords: SPAACE · Disturbance · Aircraft · Spatial audio
Cockpit · Human machine interface

1 Introduction

A modern flight deck with large visual-display units provides pilots with a massive amount of information. Although humans perceive with various senses, most

© Springer International Publishing AG, part of Springer Nature 2018
S. Yamamoto and H. Mori (Eds.): HIMI 2018, LNCS 10904, pp. 240–253, 2018.
https://doi.org/10.1007/978-3-319-92043-6_21

information is gathered visually [1]. An increasing number of pilot assistance systems in current and prospective cockpits will raise further challenges in human-machine interactions [2,3]. Perception issues will worsen with new and advancing complex missions, which will require pilots to manage greater amounts of data, fly according to higher precision standards, and adopt new responsibilities [4]. Currently, new information systems like head-up or head-worn displays and the increasing number of high-resolution displays in the cockpit mainly target pilots' visual perception [1,5]. With the expanding number of systems that the cockpit crew must use, manage, or monitor, it creates new operational burdens and types of failure modes in the overall human-machine system [6,7]. Besides the well-known ways to present information visually, audio seems to be a reasonable way to support flight crews. In present commercial aircraft, audio only provides simple warnings or informational sounds, mainly drawing the pilot's attention to a designated display. In comparison to an increasing number of synthetic vision systems and visual 3D presentation, avionics systems rarely use spatial or 3D audio cues the moment [8]. In civil commercial airliners, it is not present. With the increasing complexity of operations, pilot assistance systems must be designed to relieve the already overloaded visual channel in order to improve safety [9]. Assuming this, research in the domain of audio is necessary. Nevertheless, audio research has been sparse in aviation and mostly covers spatial audio with a set of loudspeakers around participants' heads or simple left-right volume differences in the headset [10]. However, several studies have suggested a multitude of applications for the use of 3D audio in the cockpit [1,9,11,12]. The author has developed and tested a 3D audio system called *Spatial Pilot Audio Assistance(SPAACE)* to support pilots in future cockpits. The localization precision was tested in various setups, and trials show that participants can locate 3D audio with high precision [13,14]. Thus, the audio system is worth considering as an additional or supplementary information channel in future aircrafts. However, all previous experiments were conducted in a clean experimental room environment without distraction or specific training. Both points must be considered when thinking about a pilot assistance system. This paper aims to fill this gap and introduce the design and results of a psychoacoustic 3D audio experiment with a focus on the effects of background noise and visual-based training on 3D audio. This paper is organized into five parts. The first chapter gives a short introduction to the existing research. The following chapter describes the experimental setup and the developed software. The findings of the experiment are presented in the results section and are then discussed. Conclusions and outlines of future work are given in the final chapter.

2 Research Question

As mentioned, most information on flight decks is still presented visually [8]. Human visual processing can become overloaded in high-workload flight phases. This happens especially when flying a helicopter at low altitudes in a degraded visual environment such as a brown-out, white-out, or during night operations. In the domain of fixed-wing aircrafts, high-workload flight phases can be found during takeoff, approach, or in the case of a system malfunction. Several advanced technology concepts that support pilots during flight are integrated into the cockpits side by side, and each is proven to benefit safety and performance. However, those systems convey no, or only limited, audio information.

It is apparent, that a cross-modal time-sharing technology improves situational awareness, dividing attention between the eyes and ears instead of two visual sources or two auditory sources. Previous research has shown that during high visual attention tasks like flying, auditory information is better and more quickly recognized than additional visual cues [12, 15–18]. It becomes conceivable that audio has a positive effect in high-workload situations in the cockpit.

When introducing spatial audio to the cockpit, several boundaries are concerned. This work addresses two of them. One is the impact of distraction from background noise on localization performance. Obviously, modern airplane and helicopter cockpits are quieter compared to those in earlier days. Cabin noise level is around 85 dB(A) for transition flight in a common helicopter and average 70 dB(A) in an airliner [19, 20]. Additionally new hearing protection with active noise reduction helps to lower external noise. However, since long verbal communication with air traffic control (ATC) or inter-crew communication still plays a major role in cockpit work, this dynamic distraction must be considered.

The second question is the influence of visual feedback as training. Commercial and military pilots are highly trained professionals. They become familiar with all components and systems of their future aircraft during training. Following the literature [21, 22], it is worth testing the influence of spatial audio as a part of this training.

3   Method

The experiment took place at the Institute of Flight Guidance at German Aerospace Center (DLR), Braunschweig, Germany. Participants were randomly selected from the employees and students at the research facility without focusing on an aviation background. In total, 27 participants, 4 female and 23 male, with ages ranging from 22 to 54 years ($m = 34.26, SD = 8.89$), volunteered for the experiment. All participants declared unobstructed hearing abilities. As the results from previous experiments showed no significant impact on localization performance from different hearing performance [14], audiometry testing was not conducted this time. A 360-degree round room, normally used as Apron and Tower Simulator (ATS), was used for the experiment. The inside simulator wall was in a monotone light blue color with a reference mark for 0 degrees for orientation and the initial calibration of the system. As shown in Fig. 1, the participants sat on a swivel chair in the center of the room. The shape of the room gave participants the ability of a free 360-degree movement around the vertical axis with a 360-degree field of vision. The combination with a head tracker gives participants the possibility of natural head movement, which is essential for localizing virtual sounds sources [14, 23]. The experiment operator sat in the same room at approximately 100-degrees, 2 m away from the participant. Due to the running projectors in the ATS, a constant noise of 48 dB(A) (equal to quiet suburb at daytime) was present during the experiment.

Fig. 1. The 360-degree tower simulator during the experiment. The participant wears a headphone with an attached head tracker to control the sound position and the digital red ball on the wall. (Color figure online)

Two audio objects with a frequency of 440 Hz, which is the pitch *tone A*, form the target sound. An introductory sound, rising in volume, emanates in a second sound, the actual signal tone. This signal tone was generated based on a sine tone. The characteristics are the short, hard transient and the clearly audible release time. Harmonics were added by distortion for better acoustic performance. To create the introductory sound, the actual signal tone was copied and inverted to play backwards. This sound was placed before the signal tone, whereby a direct transition was created. The length of the introductory sound has been manipulated by fading. The created target sound can be described as a warm, friendly *Bing*, motivating participants to follow it, without becoming annoying, with a clear, synthetic character. The sound remained unchanged throughout the experiment and was played repeatedly for 1 s with a 1 s pause until the participant pressed a button and determined the sound's location. The randomly distributed target sound was located clockwise from 0 degrees to 350 degrees in 10-degree intervals. All tones throughout the experiment were located at the participant's eye level, with a fixed elevation angle of 0 degrees for the horizontal plane. Participants were informed of the characteristics of the target sound.

All sounds in the experiment were played using off-the-shelf, over-ear Beyerdynamic DT 880 stereo headphones. This semi-open headphone has a frequency range from 5 Hz to 35 000 Hz and no built-in 3D audio features. A Carl Zeiss

Cinemizer head tracker was attached to the headphone to transmit the participants' head movements. This head tracker sent information to the DLR experiment software SPAACE. Figure 2 illustrates the structure with the head tracker
linked to the headphones and the ATS.

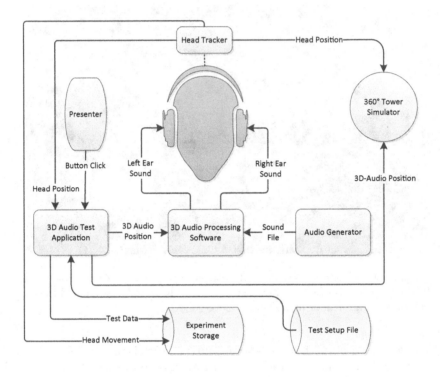

Fig. 2. Structure of the experiment test system.

By continuously sending head tracking information to the 3D audio test
application, recalculated sounds were played inside the participants' headphones,
giving the impression of spatial sounds at a steady position in relation to the
screen's 0-degree line. During the experiment, the participants were instructed
to point to the perceived sound position. By rotating their whole body and
head on the swivel chair, participants moved a virtual red ball on the wall of
the simulator. At the selected position, participants confirmed their input by
clicking the button on the wireless presenter. For the experiment, 20 sound
positions were defined. The first six angles always started in the order at 90, 30,
270, 330, 150 and 210 degrees. The next 14 angles were defined randomly. The
presented sound had an offset of at least 40 degrees from the preceding sound.
For the test sessions, two distinct and independent angle sets were defined.

The experiment was split into two sequences, and each sequence was separated into three sessions. As Table 1 shows, every participant started with the
sequence-*No Training* and went through three sessions. After each session, a

questionnaire was presented followed by the sequence- *With Training*, which was also finalized by questionnaires. The experiment finished in a final questionnaire and an open interview. The sequences and sessions are explained in detail.

Table 1. Procedure of the experiment.

	Sequence - No training	Sequence - With training
0	Introduction	Introduction
A	No background	No background
B	ATC background	ATC background

Sequence - No Training: During the first sequence, participants received no feedback about their localization performance. They completed all three sessions without visual feedback. This sequence was used as a baseline for how accurately participants could locate the spatial sound.

Sequence - With Training: The second sequence was set up to evaluate the impact of visual feedback as training on location error. Visual feedback was given only during the second sequence. After participants affirmed the perceived sound location, a yellow ball appeared on the ATS screen at the real target sound position. By comparing the red ball (the participant's perceived sound location) with the yellow ball (the real target sound location), participants could see their localization offset. Participants were instructed to correct further sound localizations accordingly. The offset information was given directly after each of the 20 sound positions.

Both sequences were split into three sessions: Introduction, No Background, and ATC Background.

0 - Introduction: Prior the two main sessions, a brief introduction session was conducted. Participants became familiar to the 3D audio and the research procedure. Each introduction session comprised five spatial sound positions, which had to be located. Further, this session provided the opportunity for questions.

A - No Background: In this session, only the target sound was played without any background distraction or further auditory information. Each participant had to locate 20 predefined audio positions. The session provided basic information on location error rates without external disturbance and was used as a reference for the last session.

B - ATC Background: As stated in the introduction, aircrafts have become quieter. However, ATC and crew communication still takes place. Thus, this session evaluated if background voices at low volume (half the volume of the target sounds) distracted participants or affected localization performance. During this session, a non-spatial ATC recording was played continuously, imitating a realistic aviation environment, and participants had to locate 20 target sounds. Concentrating solely on the accuracy of target sound localization, participants were advised not to react to the ATC instructions. Further, it was evaluated if background voices, with a range between 500 Hz and 3000 Hz [24], can mask a target sound in the same frequency band.

During the experiment, SPAACE logged all audio positions, related head tracker information, and perceived and target sound location. The applicable data was later imported into SPSS for analysis and visualization.

4   Results

For this experiment, homogeneity of variance was assumed, and the main sessions, A and B, were evaluated. The introduction sessions 0, were used only as an opening for participants and were excluded from statistical evaluation. Target sound positions are written as x *degree*, calculated values in $x°$.

During the whole experiment, a total of 2430 sound positions were evaluated, and eight were corrected due to a system malfunction. Five participants experienced front-back confusion at sound angle 160 degrees and 180 degrees. These were treated as spikes and have been manually corrected by calculating the mean location error for each affected participant. Certainly, the impact of front-back confusion is critical for understanding the possible risks of 3D audio and was therefore analyzed separately.

For better understanding of the impact of ATC background noise and visual-based training on 3D audio, the overall localization capability in this experiment is presented. Figure 3 presents the mean localization error for all sessions. Although the results show a relatively constant allocation, participants improved over time, leading to a decreasing mean in location error from $m = 10.09°$ $(SD = 11.90°)$ to $m = 6.56°$ $(SD = 5.26°)$. The mean absolute location error ranged from $m = 1.91°$ $(SD = 3.17°$ at 0-degree) to $m = 10.66°$ $(SD = 11.79°$ at 320-degree), overall absolute location error of all target sound angles scattered around $m = 8.16°$ $(SD = 8.43°)$. The results show that participants located target sounds at 180 degrees with high accuracy $(m = 4.87°, SD = 4.72°)$. Participants had the highest deviation between the target and perceived sound angles at the front-left and front-right side at 30 degrees $(m = 9.80°, SD = 13.60°)$ and 320 degrees $(m = 10.66°, SD = 11.79°)$.

Figure 4 depicts the high performance of participants for target angles 0 degrees $(m = 1.19°, SD = 3.15°)$ and 180 degrees $(m = 4.87°, SD = 4.72°)$. The location errors at 90 degrees $(m = 8.82°, SD = 6.37°)$ and 270 degrees $(m = 9.66°, SD = 7.07°)$ scattered recognizably more.

4.1   Background Noise

To evaluate if ATC background noise affects spatial perception, the results of session A (no background) are compared to those of session B (ATC background). Since the results are not normally distributed (Shapiro-Wilk test: $p < 0.05$), the Wilcoxon signed-rank test is used.

Figure 5 presents the localization error for session A and session B. The results suggest that background noise has a positive influence on localization error. The error reduces with ATC background to $Mdn = 6.70°, z =$

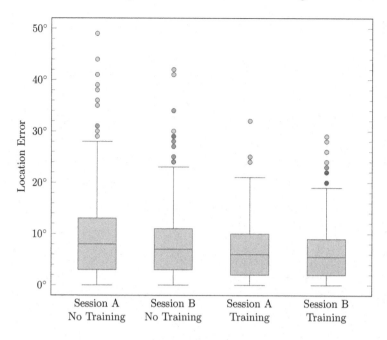

Fig. 3. Comparison of absolute location errors by session and sequence.

$-2.27, p < 0.05$; compared to the slightly higher error without background noise ($Mdn = 7.48°$). Although the results are not normally distributed, by comprising the average localization error it becomes visible that the error reduces from $m = 8.40°, SD = 9.34°$ (no background) to $m = 7.37°, SD = 6.34°$ with ATC background. These findings are reflected in the questionnaire results, 17 participants (61%) out of 28 participants reported that the ATC background did not interfere with the 3D audio presented with SPAACE. Only one participant reported problems with localization due to the presence of background voices, but this was not detectable from the allocated localization results.

One further effect can be identified from the time duration participants needed to locate target sounds. The time measurement started with the first 3D target sound and stopped when participants pressed the button to confirm the perceived sound location. Although there was no time limit in any session and participants were not instructed about the time tracking, the mean time to locate a target sound was fast with low variance among all participants ($m = 12.54$, $SD = 6.84$ s). Splitting the needed, not normally distributed (Shapiro-Wilk test: $p < 0.05$), time between session A (no background) compared to session B (ATC background), it is apparent that the time to locate the target sound with ATC background ($Mdn = 11.51$ s) was almost identical to the session without background distractions ($Mdn = 11.50$ s, $z = -2.28, p < 0.05$).

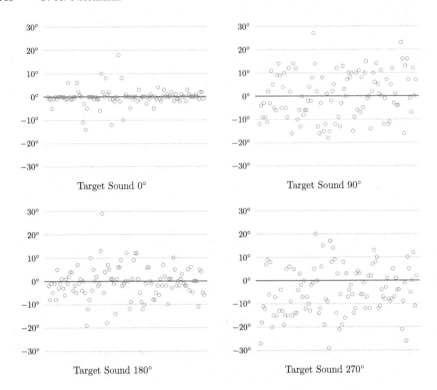

Fig. 4. Scattering of location error around 0, 90, 180, and 270 degree target sound position. The orange line represents the target sound position. The blue circles mark participants' determined sound locations. (Color figure online)

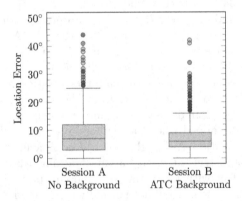

Fig. 5. Comparison of the mean absolute location error between session A and session B. The blue circles represent outliers ($> 1.5\ IQR$). (Color figure online)

4.2 Visual Training

The second objective of the investigation is the influence of visual feedback as training on the tracking performance. Comparing the no-training sequence with the training sequence, according the Wilcoxon signed-rank test, the location error without training ($Mdn = 7.96°$) is significantly higher. As expected, training has a positive impact on location error ($Mdn = 6.51°, z = −4.33, p <$ 0.05). Figure 6 shows how participants improved during sessions with training ($m = 6.86°, SD = 1.97°$) compared to sessions without training ($m = 9.47°$, $SD = 4.86°$).

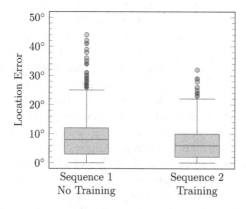

Fig. 6. Comparison of absolute location error by separating sequence 1 and sequence 2. The blue circles represent outliers ($> 1.5\ IQR$). (Color figure online)

To understand the effects of feedback as training, the localization error is analyzed for each participant. Comparing the mean location error of the first half with the corresponding mean location error of the second half, a positive effect on the location error is detectable. In numbers, 19 participants (70%) reduced their location error in the second half of the no background session, whereas 15 participants (56%) reduced their location error during the second half of the ATC background session. The questionnaires show that more than 40% of all participants reported rarely or never being able to correct subsequent 3D localizations with the help of visual feedback. However, 50% of participants judged the visual feedback as helpful in understanding their localization error. Further, less than 20% of all participants were confused sometimes by the visual feedback. Although the participants did not adequately value the feedback in subsequent rounds, there was a statistically significant improvement.

Again, participants' times to locate target sounds were measured. According to the Wilcoxon signed-rank test (Shapiro-Wilk test: $p < 0.05$), the time needed to locate the target sound with training was significantly higher ($Mdn =$ 12.02 s) than during the sequence without visual training ($Mdn = 10.18$ s),

$z = -3.7, p < 0.05$. By comparing sequence 1 (no training) with sequence 2 (with training), a worsening from $m = 11.23$ s $(SD = 3.94$ s) to $m = 13.76$ s $(SD = 6.33$ s) can be observed.

5 Discussion

Based on the findings of this research, the effects of background noise and visual training on 3D audio are evaluated.

As the participants' absolute mean location error during this study was $m = 8.16°$ $(SD = 8.43°)$, it can be assumed that humans can locate spatial sounds with the developed SPAACE system within a range of $\pm10°$. This accuracy is similar to free-field findings and listening tests with an array of loudspeakers. Concerning background noise generated by voices, this experiment shows no increase in location error during sessions with ATC background noise $(m = 7.37°, SD = 6.34°)$ compared to sessions without background noise $(m = 8.40°, SD = 9.34°)$. In addition, 61% of participants reported that ATC background noise rarely or never interfered with spatial target sounds. Besides, background noise neither affected spatial perception nor masked the target sound. Participants performed better during sessions with background noise compared to sessions without background noise. These findings match other research, e. g. stated by Godfroy-Cooper [25], and all participants agreed that it was easy to discriminate the sonifications from the background noise. Further, Ericson shows that the advantages of spatial separation are greatest when listeners are subjected to an ambient noise field [26]. With these results, multiple spatial applications in the cockpit are feasible. The pure presence of background communication seems not to have a negative influence. Future trials may consider active rather than passive ATC background, demanding that participants react to the background noise.

Training improves spatial perception. This becomes evident when comparing the no training sequences with the training sequences. The experiment shows an average improvement in location error of 2.61° (from $m = 9.47°, SD = 4.86°$ down to $m = 6.86°, SD = 1.97°$). Further, participants improved during sessions without background noise by 70.37% and during sessions with ATC background by 55.56%. However, overall improvements within the training sequence were only marginal (session A $m = 6.71°, SD = 5.25°$ and session B $m = 6.56°, SD = 5.26°$). The low improvement between session A and session B in sequence 2 seems to be a result of the overall good performance of participants close to natural human limits. As Majdak et al. [22] state, there is always a remaining error in localization, even when subjects receive extensive training. The improvement of participants' localization was significant with visual feedback as training, which is consistent with other studies [1,22,27].

Information about the time taken to locate spatial audio was also collected during the study. As participants were not informed about time limits, they could locate spatial sounds without time-pressure. On the other hand, participants were not told to speed up, so the time duration of task should not be overrated.

The overall mean time needed to locate spatial target sounds was $m = 12.54\,s$ ($SD = 6.84\,s$), with participants needing more time during sessions with training. It can be assumed that the increase in time during visual feedback sessions is a result of a more cognizant execution of tasks when receiving training in the form of visual feedback. Overall, it can be assumed that without any time pressure, humans can locate spatial target sounds in this setup and for the given task within 15 s. Further analysis of the experiment should investigate how time and accuracy correlate.

6 Conclusion

This experiment has been conducted without any audiometry testing for the participants, but major hearing impairments would have been visible in the results. Minor hearing impairments cannot be excluded. However, since this experiment targets later use in aviation, audiometry testing for aviation personnel, according to the 'Commission Regulation (EU) No 1178/2011' 2011, is only done every five years (or every two years for personnel older than 40), so minor hearing impairment in aviation personnel can never be excluded completely. Also, the results from previous experiments show no significant impact of different hearing performance on localization performance [14]. The results build a base for future 3D audio applications in aviation.

Today, cockpits rely mainly on visual warnings as a primary information source. However, by looking at the underlying theories of human perception, e. g. the multiple resource theory, humans are better at dividing their attention across separate pools of information-processing resources (cross-modal), rather than using a single sensory channel (intra-modal) [18]. Presently, mono-aural message systems are already incorporated in modern cockpits to distribute information across different sensory channels. By using spatial audio, systems are able to attach relevant information to the aural alert, e. g. directional information during Traffic Alert and Collision Avoidance System (TCAS) alerts or Enhanced Ground Proximity Warning System (EGPWS) warnings. Especially under high workloads, spatial audio can complement visual instruments to relieve the visual channel.

The experiment shows humans' ability to locate spatial sounds and provides a basis for developing aviation-related 3D audio applications. The main findings are:

- the overall mean location error of participants is $8.16°$ ($SD = 8.43°$)
- ATC background noise has no negative influence on sound localization
- visual feedback as training improves participants' localization ability
- the overall mean time to locate the target sounds is 12.54 s

According to the research results, it can be assumed that 3D audio can complement visual instruments to improve information distribution and enhance situational awareness. By designing applications according to human limitations, 3D audio is a technology capable of supporting flight crews.

Further studies must be conducted to investigate the impact of visual feedback. Also, research must be done in a demanding environment, e.g. during simulated flights, to examine the effects of stress on location error. Additionally, further research should be conducted to examine people's ability to react to different auditory information simultaneously and respond to ATC or inter-crew communication while locating target sounds.

References

1. Veltman, J.A., Oving, A.B., Bronkhorst, A.W.: 3-D audio in the fighter cockpit improves task performance. Int. J. Aviat. Psychol. **14**(3), 239–256 (2004). ISSN 1050-8414
2. Jovanovic, M., Starcevic, D., Obrenovic, Z.: Designing aircraft cockpit displays: borrowing from multimodal user interfaces. In: Gavrilova, M.L., Tan, C.J.K. (eds.) Transactions on Computational Science III. LNCS, vol. 5300, pp. 55–65. Springer, Heidelberg (2009). https://doi.org/10.1007/978-3-642-00212-0_3. ISBN 978-3-642-00211-3
3. EASA: European Plan for Aviation Safety (EPAS). 2017–2021. 2017. 107 pp. https://www.easa.europa.eu/system/files/dfu/EPAS_2017-2021.pdf. Accessed 09 Sept 2017
4. Riggs, S.L., et al.: Multimodal information presentation in support of NextGen operations. Int. J. Aerosp. Psychol. **27**(1–2), 29–43 (2017). https://doi.org/10.1080/10508414.2017.1365608. ISSN 2472-1840
5. Parker, S.P.A., et al.: Effects of supplementing head-down displays with 3-D audio during visual target acquisition. Int. J. Aviat. Psychol. **14**, 277–295 (2004). ISSN 1050-8414
6. Oving, A.B., Veltman, J.A., Bronkhorst, A.W.: Effectiveness of 3-D audio for warnings in the cockpit. Int. J. Aviat. Psychol. **14**(3), 257–276 (2004). ISSN 1050–8414
7. Spence, C., Ho, C.: Tactile and multisensory spatial warning signals for drivers. IEEE Trans. Haptics **1**, 121–129 (2008)
8. Koteskey, R.W., Wu, S.-C., Battiste, V.: Enhanced audio for NextGen flight decks. In: Landry, S.J. (ed.) Advances in Human Aspects of Aviation. Advances in Human Factors and Ergonomics Series. CRC Press, Boca Raton (2012). ISBN 978-1-439-87117-1
9. Begault, D.R., Pittman, M.T.: Three-dimensional audio versus head-down traffic alert and collision avoidance system displays. Int. J. Aviat. Psychol. **6**(1), 79–93 (1996). ISSN 1050-8414
10. Simpson, B.D., et al.: In-flight navigation using head-coupled and aircraft-coupled spatial audio cues. In: Proceedings of the Human Factors and Ergonomics Society Annual Meeting, vol. 51, no. 19, pp. 1341–1344 (2007). https://doi.org/10.1177/154193120705101914. ISSN 1071–1813
11. Haas, E.C.: Can 3-D auditory warnings enhance helicopter cockpit safety? In: Proceedings of the Human Factors and Ergonomics Society Annual Meeting, vol. 42, no. 15, pp. 1117–1121 (1998). https://doi.org/10.1177/154193129804201513. ISSN 1071–1813
12. Niermann, C.A.: Can spatial audio support pilots? 3D-audio for future pilot-assistance systems. In: 2015 IEEE/AIAA 34th Digital Avionics Systems Conference (DASC) (2015). https://doi.org/10.1109/DASC.2015.7311401. http://ieeexplore.ieee.org/iel7/7301991/7311321/07311401.pdf?arnumber=7311401

13. Niermann, C.A.: Potential of 3D audio as human-computer interface in future aircraft. In: Harris, D. (ed.) EPCE 2016. LNCS (LNAI), vol. 9736, pp. 429–438. Springer, Cham (2016). https://doi.org/10.1007/978-3-319-40030-3_42. ISBN 978-3-319-40029-7
14. Niermann, C.A.: Can 3D-audio improve state-of-the-art pilot-assistance systems? In: International Council of the Aeronautical Sciences (2016)
15. Perrott, D.R., et al.: Aurally aided visual search under virtual and free field listening conditions. Hum. Factors **38**(4), 702–716 (1996). ISSN 0018-7208
16. Flanagan, P., et al.: Aurally and visually guided visual search in a virtual environment. Hum. Factors **40**(3), 461–468 (1998). ISSN 0018-7208
17. Nelson, W.T., et al.: Monitoring the simultaneous presentation of spatialized speech signals in a virtual environment. In: Proceedings of the 1998 IMAGE Conference, Scottsdale, AZ, pp. 159–166 (1998)
18. Wickens, C.D.: Multiple resources and performance prediction. Theor. Issues Ergon. Sci. **3**(2), 159–177 (2002). https://doi.org/10.1080/14639220210123806. ISSN 1463-922X
19. Lower, M.C., Bagshaw, M.: Noise levels and communications on the flight decks of civil aircraft. In: The 1996 International Congress on Noise Control Engineering. Proceedings of Internoise 1996, Liverpool, pp. 349–352 (1996)
20. Simon, F., et al.: Activities of European research laboratories regarding helicopter internal noise. AerospaceLab (2014). https://doi.org/10.12762/2014.AL07-04
21. Kenneth, W., et al.: Efficient and effective use of low-cost 3D audio systems. In: International Conference on Auditory Display, Kyoto, Japan (2002)
22. Majdak, P., Goupell, M.J., Laback, B.: 3-D localization of virtual sound sources: effects of visual environment, pointing method, and training. Atten. Percept. Psychophys. **72**(2), 454–469 (2010). https://doi.org/10.3758/APP.72.2.45. ISSN 1943-393X
23. Bronkhorst, A.W.: Localization of real and virtual sound sources. J. Acoust. Soc. Am. **98**(5), 2542–2553 (1995)
24. Antunano, M.J., Spanyers, J.P.: Hearing and noise in aviation. Medical facts for pilots (AM-400-98/3 2002)
25. Godfroy-Cooper, M., et al.: 3D-sonification for obstacle avoidance in brownout conditions. In: AHS International Forum 73rd Annual Forum & Technology Display (2017)
26. Ericson, M.A., Brungart, D.S., Simpson, B.D.: Factors that inuence intelligibility in multitalker speech displays. Int. J. Aviat. Psychol. **14**(3), 313–334 (2004). https://doi.org/10.1207/s15327108ijap1403_6. ISSN 1050-8414
27. Zahorik, P.: Localization accuracy in 3D sound displays: the role of visual feedback training. In: ARL's 5th Federated Laboratory Annual Symposium 2001, pp. 17–22 (2001)
28. Commission Regulation (EU) No 1178/2011 laying down technical requirements and administrative procedures related to civil aviation aircrew pursuant to Regulation (EC) No 216/2008 of the European Parliament and of the Council. Official Journal. L 311. 2011. 193 pp. Accessed 15 Apr 2017

Development of an End Effector Capable of Intuitive Grasp Operation for SPIDAR-W

Kanata Nozawa[1](✉), Ryuki Tsukikawa[1], Takehiko Yamaguchi[2], Makoto Sato[3], and Tetsuya Harada[1]

[1] Tokyo University of Science, 6-3-1 Niijuku, Katsushika, Tokyo, Japan
8114085@ed.tus.ac.jp
[2] Tokyo University of Science, Suwa, 5000-1 Toyohira, Chino, Nagano, Japan
[3] Tokyo Metropolitan University, 6-6 Asahigaoka, Hino, Tokyo, Japan

Abstract. This paper proposes a new grasp operation end effector for the wearable 6 DoF haptic device SPIDAR-W. With the new end effector, users can intuitively perform grasp operations in a virtual environment. Traditional end effectors with a button type interface, are only able to be held by hand and a button must be pushed to lift a virtual object. With the new end effector, the hand can be opened and closed naturally as well as lifting a virtual object, grasping it with significant force. The experiment was made with a pressure sensor monitoring the gripping force and a Velcro belt fixing the end effector to the hand. Performance measurements were made using the new end effector. As a result, users were able to perform grasp operations to some extent arbitrarily. However, there were some unintended operational errors as well as points of improvement that were noted.

Keywords: Haptic and tactile interaction · Virtual reality

1 Introduction

1.1 Research Background

In recent years, virtual reality (VR) technology has attracted significant attention. In 2016, head mounted displays (HMD) such as Oculus Rift [1], HTC Vive [2] and PlayStation VR [3] were released allowing the use of VR devices on a daily basis.

The term "virtual reality" originated in 1989, and since the 1980s, the emergence of VR research has emerged in various disciplines such as computer interfaces, simulation and robotics [4].

VR technology requires the satisfaction of three elements: (1) *"three-dimensional spatiality"*, (2) *"real-time interactivity"* and (3) *"self-projectability."* Self-projectability, allows the user to obtain a "state without inconsistency between sensory modalities [4]."

Sensory modality is an integrated perception, consisting of visual and auditory sensations, such as somatosensory and vestibular. In other words, it is essential that both the user's perception of arm position and acceleration be recognized by a sense of self-acceptance, and that both are consistent with accompanying audiovisual information.

S. Yamamoto and H. Mori (Eds.): HIMI 2018, LNCS 10904, pp. 254–266, 2018.
https://doi.org/10.1007/978-3-319-92043-6_22

HMDs such as the Oculus Rift and HTC Vive are wearable, visual displays, both worn and used on the head. Images are displayed in such a manner to cover the user's field of vision, presenting, to the user, an immersive experience with high three-dimensional spatiality. Recent HMDs also have self-position estimation functions, allowing the possibility to increase self-projectability via the combination of movement in real space with movement in the projection.

HMD, a technology for visual perception, is currently on the market, but technologies that let the user experience sensations such as smell, taste, acceleration and force sense are currently in production.

Among these sensations, the sensation of force and tactility is thought to be an important ability in human interaction. Force/tactile sensations will play an important role in many areas of VR exploitation – e.g., medical, manufacturing, art, education and entertainment. For example, force/tactile sensations have a number of applications in medical fields, especially in surgical simulations, with the aim of increasing surgical success rates of surgery.

Devices such as the Phantom [5] and Falcon [6] simulate this sense of force. Although these devices are able to obtain feedback on force, the user is not able to use the device while mobile. A device that solves this problem is the *SPIDAR-W* [7].

SPIDAR - W is a device worn on the body, allowing users to obtain force feedback while walking. With SPIDAR-W, force measurements are sent directly to the end effector, held in hand by the user. This mechanism is capable of simulating the forces involved with hitting an object, but is incapable of simulating the *grasp force*, i.e., while the user grasps an object.

If, in the future, SPIDAR – W technology is able to simulate this grasp force, there are numerous applications to various fields. For example, the self-projectability of VR sports training can be further enhanced. In the immersive social network service or the teleexistance technology which can be interacted as if it is on the spot with the acquaintance of a remote place, user will be able to get close communication.

1.2   Research Objectives

This paper documents the attempt to add a new grasp force function to the SPIDAR-W, a wearable, force, display device. In order to increase the "self-projectability of grasp operation", a device has been developed, which provides a correct calculation result to the user's grasp operation.

1.3   Research Outline

First, a new end effector for grasp operation was developed for the SPIDAR-W. Next, the performance and points of improvement of the manufactured end effector were evaluated by a number of experiments. In the experiment, the self-projectability of the grasp operation was evaluated by measurement. Finally, experimental results were analyzed and the performance of the device as well as future issues were examined.

2 End Effector for the SPIDAR-W

2.1 SPIDAR-W

SPIDAR-W is a wire drive type 6 DoF force display device that is controlled by two end effectors, each with eight motor units, enabling a dual-hand force feedback. A thread is used for the force feedback, and a force of 6 DoF is provided by the resultant force of the tension generated by the motor. By moving the end effector in the SPIDAR-W, both hands can be manipulated in VR space with the corresponding force in VR space transferred to both hands of the user. SPIDAR-W consists of an aluminum frame, motor unit and SPIDAR circuit board. Figure 1 displays the SPIDAR-W installation scheme and Fig. 2 displays the outline of SPIDAR-W.

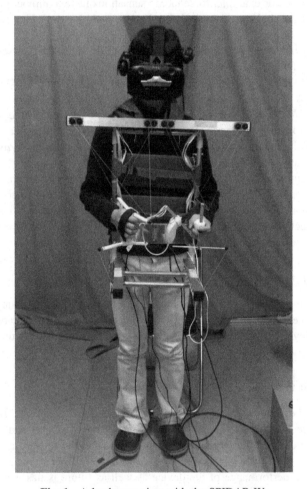

Fig. 1. A landscape view with the SPIDAR-W.

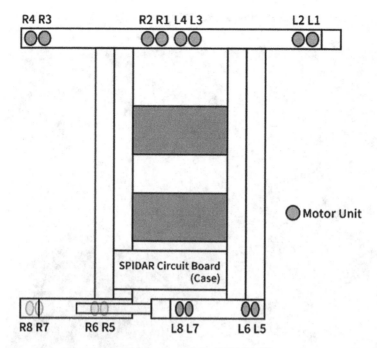

Fig. 2. Outline of the SPIDAR-W.

The motor unit is composed of a DC motor (MABUCHI), a pulley with slit and an encoder. The SPIDAR circuit board is connected to the PC via USB cable.

The HMD (HTC Vive) is attached to the user's head and the VR space is projected. The HMD, in Fig. 1, has a stereo camera but was not used for experiments performed here.

The HMD can perform 6 DoF tracking routines in the space surrounded by the two sensors. However, the coordinate system of the HMD and the coordinate system of the SPIDAR-W are independent.

2.2 Current End Effector

The end effector is a device that links the hand of the user to the hand simulated in VR space. With the current end effector, it is possible to grab or release objects in VR space using buttons. The current end effector is composed of four end points for the fixation of the thread, a grip and two buttons. Dimensions of the end effector are 15.5 cm in height and 19.5 cm in width. The four end points are placed upon the apex of the regular tetrahedron and are connected by an aluminum pipe. Due to the sensitivity of the grip, it is based on the expansion controller for the Nintendo Wii video game system. The grip has the two buttons, each used for calibration and grasp operation. The current end effector is shown in Fig. 3.

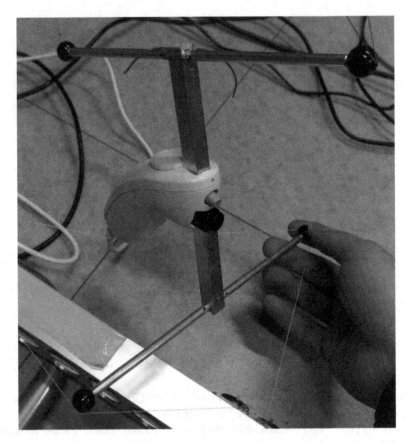

Fig. 3. The current end effector.

2.3 Grasp Operation End Effector

Since the current end effector always needs to be held, the user perceives that (s)he is continuously gripping the object. To solve this problem, a new end effector was designed, capable of achieving both a "grasp" state while grasping the object and an "open" state while the hand is open. An "open" state was achieved using velcro tape.

The use of a pressure sensor, instead of a button, permitted a more intuitive grasping operation. By using the pressure sensor, not only the ON/OFF of the grasp operation but also the grasping strength can be reflected in VR space.

Similar to the current end effector, the grasp operation end effector is composed of four end points to fix the thread. Figure 4 displays the new end effector for grasp operation.

The semi-cylindrical end effector body was constructed using a Value 3D MagiX MF-1100 (MUTOH) 3D printer, with an ABS HG 1.75 mm filament. The size and arrangement of the endpoints were identical to that of the current end effector. The end effector was fixed with Velcro tape for easy attachment to the palm of the hand, with

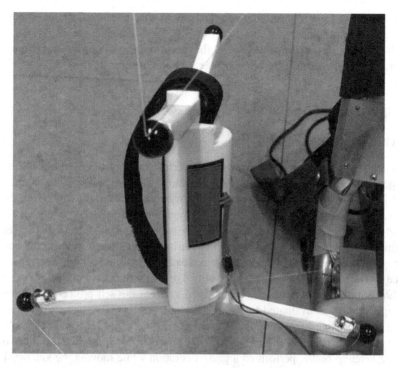

Fig. 4. The new end effector for grasp operation.

the pressure sensor designed to touch the middle phalanx of the finger. Figure 5 displays both the mounted configuration and performance of the grasp operation for the end effector.

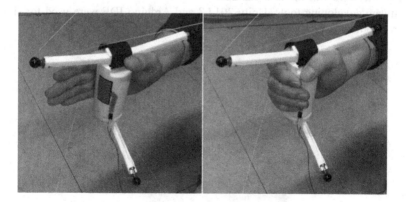

Fig. 5. Performing a grasp operation.

Pressure sensor FSR - 406 (INTERLINK ELECTRONICS) was used as the input sensor. In this pressure sensor, resistance values change in accordance with the contact area, allowing the detection of continuous pressure changes. The pressure is digitally converted to 1024 steps through Arduino Uno so that it can reflect the strength of grasping in the simulation.

Subsequently, in order to evaluate the performance of the manufactured end effector, the experiment of the next chapter was conducted.

3   Experimental Methods

3.1   Outline of Experiment

Experiments were conducted to evaluate the performance of the new end effector. In the experiment, the subjects tried to grasp a bottle of ketchup and draw the figure of an omelet in the VR environment. In accordance with the input of pressure from the end effector, the experimenter distinguished between three states: (1) not gripping the ketchup, (2) only lifting the ketchup and (3) squeezing the ketchup bottle. In this experiment, the degree of coincidence between operations performed by the subjects and the operations simulated in VR was evaluated. We found a high degree of coincidence between grasp operations performed by the subject and grasp operations reproduced in the VR environment. That is to say that the self-projectability between grasp operations was evaluated to be higher.

In a second experiment (Fig. 6), a "rectangle", "star", and "character (TUS)" were used as guidelines (Fig. 6), and each trial was performed twice. In each trial, the subject lifted the ketchup bottle, performed a grasp operation while moving the ketchup bottle in order to trace the guideline. For the simplicity of the simulation, spilt ketchup was represented by individual drops falling according to the grasping strength. At this time, the subject was instructed to "operate as quickly and accurately as possible, without making gaps, do not turn back, without letting go". A sufficient amount of time to practice the task was given to each user in order to get used to the grasping operation, and only afterward was the task started.

Software developing tools, Unity 2017.3 and Arduino IDE were used for simulation development.

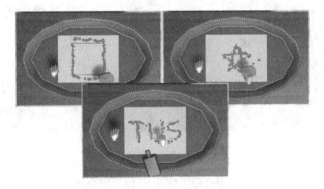

Fig. 6. Example of results.

3.2   System Configuration

Subjects were asked to perform experiments with the HMD (HTC Vive) and SPIDAR-W. The HMD and the end effector each track 6 DoF, but since their respective coordinate systems are independent, we allowed subjects to work facing forward.

The control board of the SPIDAR-W, HMD, and Arduino Uno, with the pressure sensor, were connected to the PC, respectively, and each device was controlled with an application developed by Unity. Serial communication was used for communication between Arduino Uno and Unity. Figure 7 shows a simulation screen of the experiment projected onto the subject's HMD.

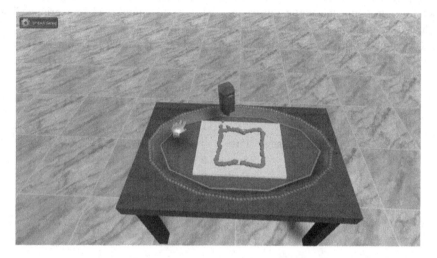

Fig. 7. Work space presented in the HMD.

A conventional end effector was used for the left hand of the SPIDAR-W while the new end effector was used for the right hand. Calibration of end effector position and bottle position in VR space as well as the end of the measurement working time were performed by button input for the end effector on the left hand. The input value of the pressure sensor was handled by using 1024 A/D converted values. This was converted to a scale of 0 to 1: (1) a state of releasing from 0 to 0.05, (2) a state of grasping from 0.05 to 0.5 and (3) a state of squeezing from 0.5 to 1.

3.3   Measurement Item

In each experimental trial, a guideline was displayed first, the end effector was brought to the home position and a calibration was performed. After the bottle was lifted, the operation started, the subjects traced the guidelines as instructed and the distance of each drop was measured.

Items measured were the distance between each droplet and the guideline and the number of times the bottle was dropped.

The distance between the drop and the guideline was measured with an accuracy of half of the drop radius, and the distance was recorded for each trial.

From each data measurement, voluntary operation was confirmed with bottle error as bottle error and drop error of an unintentional drop as drop error.

4   Results and Discussion

4.1   Bottle Error

When performing the task, the user was instructed to do "without releasing the bottle." Therefore, dropping the bottle during the task, i.e., the input was "not gripped," is evaluated as an unintended input by the user.

The number of times the subject dropped the bottle is shown in Table 1 below.

Table 1. Bottle error count for each trial

Guideline		Examinee no.				
		1	2	3	4	5
Rectangle	First	0	0	1	1	0
	Second	0	1	0	0	0
Star	First	0	0	0	0	0
	Second	1	0	0	0	0
Character	First	0	0	1	1	1
	Second	0	0	1	0	0

Table 1 shows that bottle removal is occurring throughout the experiment. Since the maximum number of dropouts is one per trial, the user may be accustomed to the sensation of dropping, or they may be cautious about sensor handling. Unintended input could be made with regard to bottle error, since a maximum of 3 dropouts per person occurred in 6 trials.

Bottle error occurs between the states of "non-grasping" and "grasping" at the input of the pressure sensor and it is therefore necessary to review the threshold value. By establishing a threshold value, it is possible to determine a method to measure the actual gripping strength.

4.2   Drop Error

When judging the accuracy of the traced guidelines, if the path drawn by the user deviates from the guideline task has not been completed. However, it is not possible to judge whether or not the subject failed due to an unintended input or some other reason. Contrarily, when a droplet is significantly deviated from the guideline, we can determined that grip input corresponds to an action or timing not intended by the user. It is due to the fact that the subject is required to trace only the guideline composed of lines.

For each trial, the drop distance furthest from the subject's trajectory is summarized in Table 2 in units of measurement accuracy.

Table 2. Drop error distance for each trial.

Guideline		Examinee no.				
		1	2	3	4	5
Rectangle	First	0	15	0	1	4
	Second	6	0	1	1	4
Star	First	12	0	2	1	1
	Second	10	3	1	3	1
Character	First	0	0	3	2	0
	Second	0	1	0	1	0

The larger the value of the drop error, the more drops that are placed either singly or multiply, indicating that they were dropped at a position distant from the trajectory drawn by the subject.

An example of when a large drop error is occurring is shown in Fig. 8.

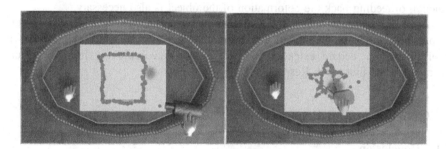

Fig. 8. Drop error example

In Table 2, drop errors as large as a drop diameter of 2 or 3 are recorded, enough to conclude that drop generation is unintended by the subject.

These drop errors occur between "grasping" and "squeezing" at the input of the pressure sensor, and it is therefore important to review the threshold value for bottle error.

4.3 Questionnaire About Device

In a questionnaire about the device, several suggestions were made on the simulation of the sense of the grasping force and the responsiveness of the pressure sensor.

In response to the question "Which figure was the most difficult and why?", 4 out of 5 answered the star and 1 person said the character. The users who chose the star mentioned that the "figures were too complex such as many sharp corners" and "the figure was relatively small." The user who chose character stated that "It was easy because there are many straight lines in other figures". The character is thought to be the most difficult figure because the intermediate grasp operation, "lifting but not squeezing," is required unlike the other two guidelines. However, from the question-naire results, it can be inferred that this operation was relatively easy. From these results, we consider responsiveness of the pressure sensor was calibrated relatively accurately.

For the next four items, we conducted a numerical evaluation in 7 stages (1–7). The response, "Whether it felt like the actual feeling of gripping ketchup," was evaluated at 3.2, with the actual gripping feeling not yet reproduced. The response, "Whether the gripping motion could be done naturally", was evaluated at 5.6, which corresponds to the idea that it was relatively intuitive to handle the pressure sensor. The responses, "I was able to put ketchup in the position I expected" even though "I was not able to put out the amount of ketchup as expected", was evaluated at 4.4, it was neither good nor bad.

Some impressions were also seen in impressions on tasks and devices. Regarding the opinion that many ketchup bottles obtained were "deformed when the actual ketchup was depressed, but not punctured in the experiment," the visually ketchup bottle is deformed, while the force surface does not deform the shape of the device surface Therefore, it seems that there was a difference between force and visual sense modality. Since this leads to a decrease in self-projectability, it can be said that a function of feeding back the deformation of the object is also necessary.

There are multiple opinions that "Learning by experience and learning in the second half was easy" Although this suggests that it can be handled intuitively if accustomed to a certain extent, it can not be handled intuitively from the actual operation at the first touch Respectively.

5   Conclusion and Prospects

5.1   Conclusion

The purpose of this study, in addition to the introduction of the SPIDAR - W wearable force environment, was to increase the immersional degree of grasp operations. In the newly developed end effector, it is now possible to simulate changes in the grasp operation with natural actions, such as fixing the hand using the belt or inputting arbitrary grip conditions via the pressure sensor.

Both bottle and drop errors occurred in the experiments, thus it is necessary to make adjustments to the pressure sensor and equipment used so that more intuitive operations can be performed using threshold settings of "non-grasping", "grasping." Despite numerous errors, the newly, developed end effector is capable of replicating intuitive grasping operations, which raised self-projectability related to the sense of grasp. Fur-thermore, despite various improvements, we found the existence of problems during the precise replication of grasp operations and the representation of grasp forces.

5.2   Prospect

Many improvements were observed in the grasp operation end effector developed in this study.

First, excluded for the sake of simplicity, the addition of a force parameter representing the shape, size, softness and deformation of an object can be used. Although there are limitations on the methods that can be used due to weight and volume relationships, in terms of precisely reproducing the grasp force, it is a task that needs to be evaluated eventually.

In this study, we conducted experiments under the assumption that no error occurs in the actual work. However, even if the same task is done in reality, there is a possibility of the occurrence of a poor degree of agreement with the guideline as well as bottle and drop error. In terms of strictly reproducing real grasp operations, we believe that it is necessary to perform the same task in real life, to be able to compare to simulated scenarios and evaluate the similarities and differences.

The threshold used for input discrimination of "non-grasping", "grasping" and "squeezing" in the pressure sensor was arbitrarily applied, and thus, could be a source of bottle and drop error, indicated by the results. Accurate grasp operation simulation is accomplished by measuring the grip force when a human actually holds a ketchup bottle as well as the grip force when pushing ketchup out of the bottle and setting the same pressure as the threshold. However, in the current end effector since the pressure sensor does not deform by grasping, the relationship between deformation of the actual bottle and pressure is not expressible. It is therefore necessary to examine threshold values that can be operated in the most intuitive manner possible.

The present device only senses pressure after it is touched by a gripped object, "coincidence between sensory modalities of hand movement" for self-projection properties are not satisfied. In order to satisfy this condition, visual information, for example the degree of bending of a finger, acquired by the use of proximity sensor is needed. The "coincidence between sensory modalities of hand position/posture", on a larger scale, corresponds to the need of matching the coordinate system of HMD and SPIDAR - W as well as to always draw the hand object at the actual hand position. A conceivable solution, such as incorporating SPIDAR-W into a general-purpose tracking device, can give the SPIDAR-W the ability to estimate its own position, which can be used in the HMD.

In order to simplify the simulation, we did not visually correct the bottle grasping. In the current SPIDAR - W system, you can grab things when the drawn hand overlaps with the object. This is different from the actual movement. Visual information indicating that the hand is buried in a bottle and does not actually penetrate is one factor that lowers the self-projection property. In order to improve this, a contact determination needs to be added to the hand object itself after reflecting the actual position, the posture of the hand and the degree of curvature of the finger in the drawing.

References

1. Oculus Rift | Oculus. https://www.oculus.com/rift/. Accessed 13 Feb 2018
2. VIVE. https://www.vive.com/. Accessed 13 Feb 2018
3. Playstation VR. https://www.playstation.com/en-us/explore/playstation-vr/. Accessed 13 Feb 2018
4. The Virtual Reality Society of Japan: Virtual Reality Science, 6th edn. Corona publishing, Japan (2016)
5. Dimension Force Device. http://www.nihonbinary.co.jp/Products/VR/Haptic/Phantom/. Accessed 13 Feb 2018
6. Ikeda, K.: Haptic device with parallel mechanism. J. Robot. Soc. Jpn. **30**(2), 52–53 (2012)
7. Nagai, K., Qian, Y., Akahane, K., Sato, M.: Wire Driven Wearable 6DOF Haptic Device "SPIDAR-W", IPSJ Interaction 2016, pp. 315–320 (2016)

Proposal of Interaction Using Breath on Tablet Device

Makoto Oka[✉] and Hirohiko Mori

Tokyo City University, Tokyo, Japan
{moka,hmori}@tcu.ac.jp

Abstract. We would like to propose an interaction which is operated by blowing a breath on a screen of information terminal. Therefore, we propose and evaluate a device for detecting breath and an algorithm for identifying the breath. While it has been studied conventionally about expiration input device operated by a breath, users are not supposed to blow a breath on a touch panel like an ordinary manual operation on the touch panel but required to blow a breath toward a dedicated input sensor. In our proposed system, it has become possible for a user to perform operations such as selection and determination of objects displayed on a screen by detecting a breath blown out toward a screen of information terminal. In this study, a breath interaction is proposed by allocating various breaths to various operations of a tablet terminal.

Keywords: Tablet device · Breath · Interaction · User interface

1 Introduction

In recent years, it has become possible to use information terminal at many locations under various conditions due to spread of smartphone. The contributing factors include improvement in portability by miniaturization and operability in touch panel. Along with spread of large-size touch panel for liquid crystal display, not only small-size information terminals such as smartphone but also those with large size such as large-screen tablet terminals have achieved widespread use. Tablet terminal is excellent in its large screen capable of handling a lot of information at a time. However, it is difficult to operate a large and heavy tablet terminal just in one hand. It is also difficult to use a tablet terminal "in a state that one hand is detained" by other purposes such as to hang on a strap and hold a baggage. Conventional operating method has a problem that use of large size information terminal is restricted. In addition, it is impossible to operate such tablets either by one hand or both hands when the hands get dirty by cooking or they are gloved due to cold weather.

In this study, we propose an interaction for operating objects displayed on a screen by blowing a breath on a screen of information terminal. It is aimed to make it possible to operate even "in a state that hands are detained by other purposes" by allocating "various breaths" to each operation of a tablet terminal with a large screen. Moreover, availability of interaction based on a breath is evaluated by implementing applications compliant with the interaction in a tablet terminal.

S. Yamamoto and H. Mori (Eds.): HIMI 2018, LNCS 10904, pp. 267–278, 2018.
https://doi.org/10.1007/978-3-319-92043-6_23

2 Related Studies

Several studies have been made for operation of information terminal using something other than fingers. For example, Zarek et al. have proposed SNOUT for operating smartphone with the nose [1]. Since it is difficult to expect all application developers to design a screen exclusively for the nose, they have resolved the problem by virtually superposing layers with selection determination area for nose interaction on the screen in order for operation by the nose to become easier. They have described as a result of accuracy verification of target selection by the nose that it is possible to select objects by the nose as long as the determination area is about one-and-a-half times as large as object size displayed on the screen. However, they have also explained that eyestrain is caused due to the close distance between the eyes and the screen.

There is also a study on operation using tilt and movement of information terminal. Some system operates a selection pointer (like a mouse cursor) in such a way to roll a virtual ball on the screen by detecting a tilt of information terminal with an acceleration or gyroscope sensor [2]. Such an interaction has been mounted that performs cancelling operation by a gesture to shake the terminal (shake) [3–5]. There are also systems other than that to perform various operations by gestures by a hand holding an information terminal.

Even though there are some previous studies on a device to operate using a breath, the purpose and requirements of the device are different. It has been conventionally studied on expiration (inspiration) input device to operate with a breath. Many of their purposes aim at providing input operation support for severely disabled people with physical inability and entertainment system under VR and MR environment. A device called as expiration (inspiration) switch has been put into practice in a study aiming at facilitating input operation by severely disabled people. Expiration switch has made it possible for users to perform input operation with the switch by blowing (inhaling) a breath from a tube in their mouth. Information to be input by this switch is limited to ON-OFF binary information, however, it is not able to make operation of PC easy.

Kitayama and Nakagawa have proposed an expiration mouse by further developing the expiration switch [6]. Different from conventional expiration switch, the expiration mouse has realized a way of operation by a mouse cursor by utilizing strong/weak breath. Because strong/weak expiration and inspiration, i.e. four elements in total, are allocated to transfer in upper/lower and left/right directions, however, it is impossible to directly select objects to be operated.

In relation to a contact type device with a tube to be held in the user's mouth, Kume and Dong have pointed out problems on restriction of the mouth as well as a sanitary risk [7]. Therefore, they have proposed a non-contact type device equipped with a load cell and detected a breath from wind pressure using the load cell. In case of non-contact type, however, it is required for a user to blow a breath at a sensor installed on the device.

In a study conducted by Iga et al., expiration and inspiration was detected in a state of non-contact using audio signals of a microphone mounted on a headset [8]. In addition, performing a pointing operation depending on a direction of the head part using a magnetic sensor attached on the head set, the operation using a breath was reflected on an object targeted by the user. Since the device is a wearable type, it has become unnecessary to target at the sensor part every time when blowing a breath but users have to wear it for use. Further, a direction to which a breath is actually blown may shift from a position of the targeted object to be operated because the pointing operation is performed by changing a direction of the head.

3 Our Goal

In this study, we propose a system in which a user blow a breath on the actual object to be operated without wearing and contacting the device. While interaction based on breath has been proposed by conventional studies, breath is used just for generating on-off binary elements and has no direct relationship with objects to be operated. We would like to propose a framework to operate by directly blowing a breath on illustrations and letters displayed on a touch panel just like a way we directly touch them with a finger. Therefore, we propose a system to identify position of breath by multiple sensors mounted on a tablet terminal and implement and assess the system. Then, we create an application for a tablet terminal which is operated with a breath and discuss the practical utility based on the assessment.

4 Breath Interaction

4.1 Properties of Breath

As breath is a common interaction used also for daily living, people take actions using a breath such as to "blow dusts away" and "warm hands". It is believed to be possible to make users imagine operating method intended by a designer without an instruction manual by showing them metaphors which resemble such interaction based on breath. Breathing is a motion that is required for maintenance of life and can be controlled by human as well. Therefore, it is believed that problems on information terminal operation restricted by situation and environment may be settled by implementing operation by using breath which is always available for human.

As interaction of common information terminals is represented by direct operation, operation and its object correspond in a one-on-one manner in many of them. According to a study by Iga et al. [8], breath has a characteristics that one operation is performed for multiple objects in a one-to-many manner such as an action "to blow out multiple candles".

4.2 Elements of Breath

In order to allocate various types of breaths to multiple operations, how many types of breaths exist is considered at first. In Table 1, actions of human to "blow" a breath are classified. In case of playing wind instruments, sound is expressed differently by blowing a breath strongly/weakly for a long/short period of time with the mouth contacting with the instruments. In case of blowing a candle out, breath is blown strongly for a short period of time without contacting the mouth with the candle. In case of just flickering the candle, breath is blown weakly for rather a long period of time. Position for blowing a breath is important when there are multiple candles. Thus, breath is blown aiming at objects to be operated by selecting number of objects and range, and duration of the effect to be maintained may vary depending on the length of blowing of a breath. Therefore, we have determined it possible to achieve non-contact breath operation by detecting position, strength, and length of a breath.

Table 1. Types of breaths.

State	Operational object	Operation
contact	Balloon	Blow up a balloon
contact	Wind instrument	Play (Blow) a musical instrument
contactless	Candle	Blow out a candle
contactless	Dust	Puff (Blow) away dust

Then, we would like to examine an action to "inhale" a breath. It is a common action to take in something with the mouth contacting with, such as an action to take in beverages using a straw. On the other hand, as there is no example to "take in something targeted" without contacting with the mouth, operation of inhaling has not been mounted in the current non-contact type device. Measuring position, strength, and length by a sensor when a breath is blown, multiple operations with a breath are implemented.

5   Proposed System

5.1   Overview of System

Overview of a system used in this study is shown in a Fig. 1. Capacitor microphones were arranged as a sensor evenly at 12 points of the end part of a tablet terminal screen. Breath is identified by using values of microphones. Based on identification of presence/absence state of blown breath, position, strength and length of a breath is identified if the breath was present.

5.2   System Configuration

The system was implemented in a tablet terminal Acer ICONIA TAB A500. A liquid crystal display of the tablet terminal is 13.6 cm long and 21.7 cm wide and capacitor microphones were arranged around and parallel to the screen. The capacitor microphones are placed facing toward center of the screen. Arduino, a microcomputer board, controls the 12 capacitor microphones. By sampling of microphones at 180 Hz, values obtained from the microphones are forwarded to a tablet terminal connected in series with Arduino. The tablet terminal identifies position, strength, and length from values obtained.

Fig. 1. Overview of a tablet system

5.3   Breath Detection Treatment

How to treat detection is shown below from presence/absence, length and strength to position of breath in order.

Presence/absence of Breath

Presence/absence of a breath is treated by each microphone. Based on values obtained from each microphone, presence/absence of a breath is determined using values measured 30 times (approx. 0.17 s) in the past from the time of determination. It was determined that "breath was present" if number of microphone's values exceeding $\pm$ 0.49 V from the value of 2.5 V at the time of no sound is 10 or larger out of 30. If breath was determined to be present by any one of the 12 microphones, breath is regarded to be present in the whole system.

Length

Length of a breath is obtained by counting them up if "breath was determined to be present" for more than one microphones. Length of a breath is counted out by continuously counting them up until "breath is determined to be absent".

Strength

Strength of a breath in each microphone is obtained by integrating amount of change during the past 30 times of measurements. In other words, absolute values of differences between 2.5 V at the time of no sound and measured value are added. The whole strength of a breath is obtained by adding up strengths detected by all microphones after calculating strength detected by each microphone.

Assuming that strength detected by each microphone is M_i (i = 1 to 12), strength detected by microphone 1 is obtained by formula (1).

$$M_i = \int_{-30}^{0} |Strength\ Value - 2.5V| \tag{1}$$

Strength of the whole breath BS is obtained by formula (2).

$$BS = \sum_{i=1}^{12} M_i \tag{2}$$

Strengths detected by all microphones are treated to be added because the stronger is a breath the larger is the change in value observed in multiple microphones.

Position
In order to calculate coordinate of a breath, a coordinate was allocated virtually to microphones as shown in the Fig. 2. Assuming strength of a breath detected by each microphone M_i as a weight, position for a breath to be blown is identified by measuring gravity of a breath. In such a case, values measured by each microphone 30 times in the past are used and treatment is performed such that the newer is the value the stronger is its impact on the result.

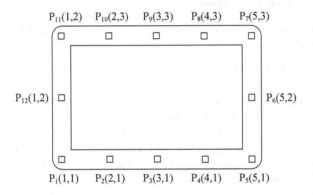

Fig. 2. The position of microphone

Coordinate of a breath B (B_x, B_y) is obtained by formulas (3) and (4). Possible values of B_x and B_y range from 1 to 5 and from 1 to 3, respectively.

$$B_x = \frac{\sum_{i=1}^{12} (P_{ix} \times M_i)}{\sum_{i=1}^{12} M_i} \tag{3}$$

$$B_y = \frac{\sum_{i=1}^{12} (P_{iy} \times M_i)}{\sum_{i=1}^{12} M_i} \tag{4}$$

Here, coordinate of a breath T (T_x, T_y) on the screen is obtained by formulas (5) and (6) using horizontal and vertical resolutions 1280 px and 752 px of the screen of tablet terminal to be implemented.

$$T_x = \frac{B_x}{5} \times 1280 \tag{5}$$

$$T_y = \frac{B_y}{3} \times 752 \qquad (6)$$

6 Assessment Experiment of Breath Interaction

6.1 Objective

Our objective is to detect coordinate, strength, and length of a breath blown by human. Detection accuracy is believed to differ depending on several elements such as deviation from the target caused by inaccurate breathing by human and the way of blowing a breath. Using a device and algorism with a setting that human blows a breath, detection accuracy is measured for position, strength, and length of a breath.

6.2 Objects to Be Measured and Experiment Method

Subjects were 7 male and female university students. They were supposed to hold a tablet terminal with both hands and blow a breath on the tablet terminal. They were instructed how to hold and tilt the tablet terminal as well as to blow their breath with a reasonable method. Distance between the screen of tablet terminal and their face was measured during the experiment. As experiment is performed separately for maximum length, blowing strength, and coordinate accuracy of a breath, experimental method is explained below one by one.

Assessment Experiment of Detection of Breath Position
Subjects blow their breath aiming at the center of target displayed on the screen. Upon detection of a breath by the system, position of the target is changed in a random manner. The procedure was determined to be tried 200 times.

Assessment Experiment of Detection of Breath Strength
Subjects were asked to blow their breath by each of five grades (weakest-weak-medium-strong-strongest), respectively. They repeat to blow their breath by a strength as indicated on the screen. The system does not feedback the strength of a breath blown by the subjects. The procedure was determined to be tried 200 times.

Assessment Experiment of Detection of Breath Length
Instructing subjects to blow a breath as long as possible, how long they are able to blow a breath is measured. Since subjects are required to continuously blow their breath strongly enough for the system to response, whether the system has continuously detected the breath was displayed on the screen. The procedure was determined to be tried 10 times.

6.3 Results

Distance between the tablet terminal screen and their face during blowing a breath was 15 to 20 cm which was closer than that of ordinary use. Any of the subjects blew a breath by bringing the tablet terminal closer to their face. No large difference was observed in the distance between the tablet terminal screen and face of subjects.

Coordinate

As a result of tallying breath detection errors of seven subjects, average value and standard deviation was 2.52 cm and 1.56, respectively. Since the values of detection errors are normally distributed according to a normal probability paper, detection error to obtain 95% of confidence interval is required to tolerate 5.26 cm. Therefore, we design a system in this study based on an assumption that detection error of 5.26 cm by radius from center of target is generated by operation of a breath to be implemented.

Strength

Subjects were asked to blow their breath by each strength of five grades. The Fig. 3 shows a result of a subject who were able to most successfully blow a breath by selected strength. It has been determined to be difficult for the system to achieve confidence level of 95% for selective blowing at five grades. It was possible for all subjects to selectively blow with a confidence level of 95% by two grades (weakest and strongest). Therefore, interaction is implemented with two grades of breath strength in the system.

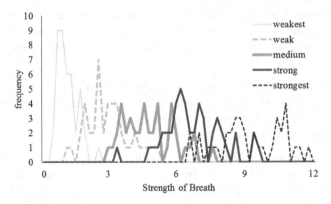

Fig. 3. Result of blow a breath by each strength of five grades.

Length

Average length of longest breath is shown by subject in the Table 2. As it is preferable that many users operate breathing without difficulty in the interaction to be proposed in the study, the shortest result of 2.6 s is set as a condition to determine "long breath".

Table 2. Average length of longest breath.

Subject	1	2	3	4	5	6	7
Average length of longest breath [ms]	5423	4001	3381	2665	5204	4327	2863

6.4   Types of Breath Used for Interaction

Position (coordinate), strength, and length of a breath are implemented by the conditions below which were ever obtained:

1. Tolerate detection error of 5.26 cm by radius from a center of icon and button to be operated.
2. Breath strength: Selectively blow by two grades of weak and strong.
3. Breath length: Determined to be "short breath" if it is 2.6 s or less and "long breath" if it exceeds 2.6 s.

7   Demonstration Experiment Using Application

In order to mount an interaction using a breath in a tablet terminal, we have developed three applications dedicated for breath operation. Qualitative assessment is performed for operability using each application.

7.1   Home Screen (Launcher Application)

A launcher application was implemented by using metaphors of a windmill to be rotated by wind and a candle flame to be put out by strong wind (see Fig. 4). A screen layout with icons arranged annularly was adopted. The icons arranged annularly rotate clockwise or counterclockwise by selectively blow a breath toward left or right. It is possible to select (activate) an icon in the direction of 6 o'clock (lower side of screen) by blowing a breath strongly.

Fig. 4. Home screen by using metaphors of a windmill to be rotated by wind.

7.2   Browsing of Electronic Books

An electronic book browsing application was implemented by using a metaphor of fallen leaves blown away by wind (see Fig. 5). For screen layout to browse books, two modes, i.e. reading mode and thumb nail mode, were prepared. In the reading mode, pages of a book are displayed on the left and right side of the screen in a double-spread state. Thumb nail mode is capable of displaying many pages of a book. It is possible to turn pages just like blowing them away by blowing a breath in the left or right direction. The two modes are switched by blowing a breath strongly.

Fig. 5. Reading mode in electronic books.

7.3   Map Application and Web Browser

Map and web browser applications were implemented by using a metaphor of fallen leaves blown away by wind (see Fig. 6). Scrolling is performed in such a way that coordinate is transferred at the center of screen by blowing a breath on it. In addition, it is possible to zoom out just like blowing away by blowing a breath strongly as well as to zoom in by blowing a breath longer.

Fig. 6. Map application.

7.4   Experiments

Experiments were performed without notifying subjects what types of interaction was available. We have observed how and by what types of breathing the subjects tried to obtain operation results. In order to assess whether metaphors used for the applications are available, we asked subjects to imagine and explain how the applications worked by blowing a breath just by showing them the metaphors of windmill, candle and fallen leaves. Then, we asked them to imagine and explain how operations are performed with breathing by showing them screens of home selection, electronic books, maps and browser. Finally, asking them to operate with breathing, we asked them to make remarks about any part with something different they felt from operations they had imagined.

7.5   Results

At first, subjects tried to express "tap", "double tap" and "flick" by a breath while being unable to notice the new interaction based on the assumption of breath. In the process, however, they have learned the various metaphors based on a breath immediately. It has been revealed that they are able to use the tablet terminal freely by the interaction using a breath once they learned the way of operation.

Subjects often used position and strength of a breath but did not use a breath length for the operation. They described they thought long breath as a measure just to blow on an object at a distance or to continue such actions that were activated by a short breath and did not think that another interaction was assigned to a long breath.

Subjects felt something strange for the fact that there was no operation of "inhaling a breath". They recognize breath as a set of actions to "blow" and "inhale" regarding "inhaling" as a reverse operation of "blowing". Therefore, it is required to review a set of operations to "blow" and "inhale" in constructing an interaction based on a breath in the future.

We have received as one of characteristic opinions from the subjects that they feel existence of themselves when operating breath interaction compared with a case of touch interaction. There was also an opinion about a metaphor of candle that they had a strong awareness of relationship of subjects themselves blowing a breath with a candle as an object to blow a breath. They had another opinion about a screen of map that they had a sense to have looked down a geography from above the sky as if they themselves were a satellite. Therefore, it is believed that breath interaction may provide users with physicality in addition to a sense of unity as if they stay in the same space as the operation object.

8   Conclusion

In this study, we have proposed and assessed a device to operate by blowing a breath and algorism. It has been proved it possible to detect position, strength, and length of a breath in a non-contact and non-wearing state by a device using microphones. It has been also suggested to be possible to realize operations with a breath by blowing a breath on the object of operation itself displayed on a screen.

In consideration of the fact that subjects imagined operation of both "blowing" and "inhaling" first, it is required for us to consider to mount "inhaling" operation which is imagined by subjects as a reverse operation of "blowing" while detecting "inhaling".

Since it was implemented in a tablet terminal, we received many opinions from subjects compared with touch interaction. Therefore, it is also required to compare with touch interaction in difference of operational feeling and possibility to operate without being limited by conditions and environment.

References

1. Zarek, A., Wigdor, D., Singh, K.: SNOUT: one-handed use of capacitive touch devices. In: AVI 2012 Proceedings of the International Working Conference on Advanced Visual Interfaces, pp. 140–147 (2012)
2. Constantin, C.I., MacKenzie, I.S.: Tilt-controlled mobile games: velocity-control vs. position-control. In: Games Media Entertainment (GEM) (2014)
3. Bartlett, J.F.: Rock 'n' scroll is here to stay [user interface]. IEEE Comput. Graph. Appl. Issue **3**, 40–45 (2000)
4. Hudson, S.E., Harrison, C., Harrison, B.L., LaMarca, A.: Whack gestures: inexact and inattentive interaction with mobile devices. In: TEI 2010 Proceedings of the Fourth International Conference on Tangible, Embedded, and Embodied Interaction, pp. 109–112 (2010)
5. Roudaut, A., Baglioni, M., Lecolinet, E.: TimeTilt: using sensor-based gestures to travel through multiple applications on a mobile device. In: Gross, T., Gulliksen, J., Kotzé, P., Oestreicher, L., Palanque, P., Prates, R.O., Winckler, M. (eds.) INTERACT 2009. LNCS, vol. 5726, pp. 830–834. Springer, Heidelberg (2009). https://doi.org/10.1007/978-3-642-03655-2_90
6. Kitayama, I., Nakagawa, H.: Development of the human interface device using flow rate sensor (breath mouse). In: Memoirs of the Faculty of Biology-Oriented Science and Technology of Kinki University, vol. 30, pp. 17–27 (2012). (in Japanese)
7. Kume, Y., Dong, X.: A feasibility study of input interface by expiratory flow pressure. Inst. Image Inf. Telev. Eng. **40**(9), 29–32 (2016). (in Japanese)
8. Iga, S., Itoh, E., Yasumura, M.: Kirifuki: breathe in/out user interface for manipulating GUI. In: SIGCHI in Information Processing Society of Japan, vol. 2000, no. 12, 49–54 (2000). (in Japanese)

Effectiveness of Visual Non-verbal Information on Feeling and Degree of Transmission in Face-to-Face Communication

Masashi Okubo[(⊠)] and Akeo Terada

Doshisha University, 1-3 Miyakodani, Tatara, Kyoto 6100321, Japan
mokubo@mail.doshisha.ac.jp

Abstract. Recently, the importance of non-verbal information is getting attention. Generally, it is believed that the more non-verbal information is exchanged, the better partner's message can be understood. In this field, much research on effectiveness of non-verbal information in communication is performed. However, among these presents doubts about this effect. Prof. Sugiya investigated quality of information's transmission from two points of view; degree of transmission and feeling of transmission, and she suggests that non-verbal information sometimes does not help us to understand partner's message. We try to verify effectiveness of non-verbal information and types of communication on feeling or degree of transmission from these views. For this purpose, two experiments were conducted. The experimental results of the three communication modes—text chat, voice chat, and face-to-face communication—showed that the degree of transmission was lowest in face-to-face communication as evaluated with the listeners' test accuracy rates and consistency of character impressions. Conversely, according to the questionnaire results, feeling of transmission was ranked highest for face-to-face communication, followed by voice chat, and lastly text chat. These results suggested that the communicability of information should be considered using feeling of transmission and degree of transmission as two separate factors.

Keywords: Non-verbal information · Human communication
Communication chanel

1 Introduction

Modes of human communication are classified into two general categories: face-to-face communication and remote communication [1]. In recent years, the frequency of remote communication via email, video calls, and other media has greatly increased due to the proliferation of the internet and smartphones. In face-to-face communication, just as the speaker's verbal information is transmitted, so is non-verbal information such as both the speaker's and listener's facial expression, body language, and tone of voice. At the same time, methods of remote communication, such as text chat, are limited in the amount of non-verbal information they provide, and consequently pictographs and emoticons are used to compensate for this deficiency.

© Springer International Publishing AG, part of Springer Nature 2018
S. Yamamoto and H. Mori (Eds.): HIMI 2018, LNCS 10904, pp. 279–290, 2018.
https://doi.org/10.1007/978-3-319-92043-6_24

The importance of non-verbal information in communicating has been explored in numerous studies, and non-verbal information is believed to be essential to understanding conversational content [2, 3],. Conversely, research that is skeptical of this importance has also been conducted. Kimura and Tsuzuki suggested that text-only CMC (Computer Mediated Communication) is easier to communicate with than face-to-face methods as a result of CMC acting as a sort of filtration system that eliminates non-verbal information such as eye contact and gestures present when face-to-face [4]. In another study, Sugitani examined the communicability of information from two perspectives: the degree of transmission and the feeling of transmission [5]. The results of Sugitani's experiment suggest that the communicability of information should be analyzed using these two perspectives separately.

2 Non-verbal Information and Feeling and Degree of Transmission

2.1 Importance of Non-verbal Information

65% to 93% of a message is delivered by non-verbal information when communicating, as hypothesized by multiple researchers [1]. Sugitani's research indicated that among various forms of communication, most people felt communicating information was easiest face-to-face [5]. Furthermore, approximately 90% of these people gave the fact that they can see their partner's facial expression as the reason. These results indicate that many people believe in the importance of non-verbal information in communication.

2.2 Non-verbal Information in Communication Modes

Table 1 shows the non-verbal information exchanged in each communication mode. The amount and type of non-verbal information exchanged varies based on the mode. However, the specific advantages of each communication mode—whether it be face-to-face, email, chat, video calls, etc.—have not been confirmed [6].

Table 1. Non-verbal information exchanged in each mode of communication

Communication	Expression	Gaze	Posture	Motion	Voice	Symbol
Text chat	–	–	–	–	–	✓
Voice chat	–	–	–	–	✓	–
Face-to-face	✓	✓	✓	✓	✓	✓

2.3 Related Studies

Sugitani examined the communicability of information with two perspectives: the degree of transmission and the feeling of transmission [5]. The degree of transmission indicates the extent that a speaker's dialogue is correctly received by their listener, and the feeling of transmission indicates the extent that the speaker and listener feel that the information was correctly shared. The experiment results showed that text chat and voice chat had lower feelings of transmission than face-to-face communication, but also had higher degrees of transmission. In other words, this suggests that the feeling of transmission increases with an increase in non-verbal information, as shown in Fig. 1, but the degree of transmission may actually decrease. However, the only differing factor between face-to-face communication, voice chat, and text chat is the characteristics of their non-verbal information. Therefore, it is crucial to test this trend by altering the amount of the particular non-verbal information in each mode of communication.

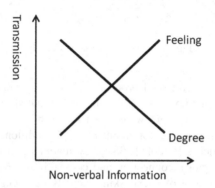

Fig. 1. Relationship between amount of non-verbal information and feeling/degree of transmission

2.4 Study Objectives

This study aimed to more thoroughly test the effects of non-verbal information on feeling and degree of transmission by controlling and altering communication mode and the amount of visual non-verbal information relayed.

3 Effects of Communication Mode on Feeling and Degree of Transmission

3.1 Outline

Based on Sugitani's study, this study tested the effects of non-verbal information on feeling and degree of transmission among three communication modes: text chat, voice chat, and face-to-face.

3.2 Verification Experiment

Experimental Method

Figure 2 shows the conditions of the experiments. The participants were paired into groups of two with one person taking the role of speaker and the other listener. The speaker then relayed to the listener a story they were told. This task was performed once via text chat, once via voice chat, and once face-to-face.

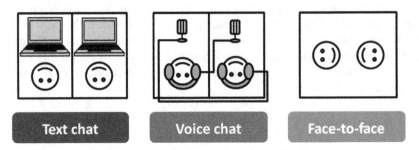

Fig. 2. Experimental conditions I

Figure 3 shows the experimental procedure. In the experiment, i. the participant with the role of speaker memorized a story, ii. relayed that story to listener, and iii. a confirmation test and questionnaire were given for each of the three segments (text chat, voice chat, and face-to-face communication). In addition, as preliminary questionnaires, the two participants took KiSS-18 evaluations and were made to answer questions regarding the pair's connection and compatibility. KiSS-18 is a scale that measures a person's mastery of social skills, and its high degree of reliability and validity has been demonstrated [7, 8].

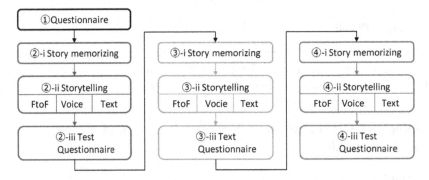

Fig. 3. Experimental procedure

Table 2 shows the questionnaire fields. The questionnaire contained fields regarding the climate of the dialogue and impressions of the story's characters that participants rated on a scale of one to seven.

The confirmation test investigated the listener on the story's keywords given to the speaker at the beginning of the experiment. The test format was multiple choice instead of a free writing format so as to prevent the researcher's subjectivity from affecting the grading of the test.

Table 2. Questionnaire fields

Fields regarding the climate of the dialogue
Q1. I felt the story was understood by my partner/I felt I understood my partner's story
Q2. I felt it was easy to communicate/listen
Q3. I enjoyed the dialogue
Q4. I felt connected with my partner during the dialogue
Q5. I was able to concentrate on the dialogue
Fields regarding impressions of the story's characters
Q1. Rate your impressions of the following three characters
Q2. I felt the story was entertaining
Q3. The story moved me

Evaluation Method

Degree of transmission was evaluated through the accuracy of the transmitted details and sentimental information. The accuracy rates of the listeners' confirmation test results were used to evaluate the degree of transmission of the story's contents. This accuracy rate is the ratio of the number of questions that the listener answered correctly out of the number of questions whose keywords were judged by the researcher to be transmitted correctly by the speaker. In addition, the questionnaire results regarding the impressions of the story characters were used to evaluate the degree of transmission of the sentimental information, and the degree of consistency in impressions was measured. Specifically, the absolute value was taken of the difference between the speaker's and listener's impressions of three story characters, and these values were compared among the three experimental conditions. A higher number indicates a higher discrepancy between the speaker's and listener's impressions. Sense of transmission was evaluated using the questionnaire results regarding the dialogue content.

Experimental Results

Figure 4 shows the accuracy rate results from the listeners' confirmation tests. Little difference was observed between the text chat and voice chat results, and while the face-to-face communication results were slightly lower, no statistically significant

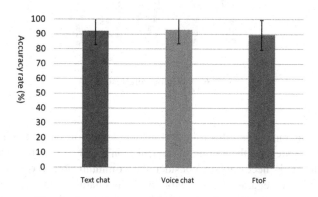

Fig. 4. Accuracy rate of listeners' confirmation tests

difference was observed. Figure 5 shows the absolute value of the difference between the speakers' and listeners' impressions of the story characters. Face-to-face communication demonstrated a greater gap between the speakers' and listeners' impressions than text chat and voice chat.

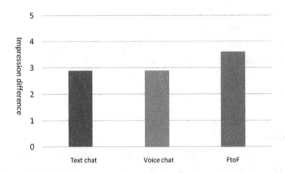

Fig. 5. Absolute value of difference between speakers' and listeners' impressions of characters

Figure 6 shows the combined results of the speakers' and listeners' questionnaires. Feeling of transmission is believed to be heavily influenced by the ease of communicating or listening and the participant's feeling of connection with their partner. Question 1 showed no statistically significant difference, but other questions did. Therefore, it is believed that feeling of transmission was rated highest for face-to-face communication, followed by voice chat, and lastly by text chat.

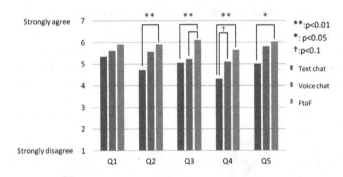

Fig. 6. Questionnaire results

Discussions

Table 3 shows the ranked results of the feeling and degree of transmission of the three communication modes. As the feeling of transmission and degree of transmission show different results, it is necessary to consider the communicability of information by treating these as two separate factors. Moreover, the characteristics of the non-verbal information present in the three modes of communication are fundamentally different.

In order to verify the trends shown in Fig. 1, it is necessary conduct verification experiments that focus on one single mode of communication at a time and quantitatively control that communication mode's non-verbal information.

Table 3. Evaluation of feeling/degree of transmission for each communication mode

Rank	Text chat	Voice chat	Face-to-face
DoT by contents	2	1	3
DoT by impression	1	1	3
FoT	3	2	1

4 Effects of Visual Non-verbal Information on Feeling and Degree of Transmission

4.1 Outline

Based on the analysis in the previous section, this section focuses on the face-to-face mode of communication and outlines the proposed method for the quantitative controlling of visual non-verbal information. The effects of visual non-verbal information on the feeling and degree of transmission in face-to-face communication were tested using this proposed method.

4.2 Control Method for Visual Non-verbal Information

Proposed Method

In the face-to-face communication mode, the sender and receiver of non-verbal information sat across from each other and were separated by fabric. It was hypothesized that the amount of visual non-verbal information passed on to the receiver can be controlled quantitatively by increasing or decreasing the number of sheets of fabric, and the validity of this hypothesis was tested in an evaluation experiment.

Evaluation Experiment

In the experiment, participants were shown through the fabric a total of 12 videos of person displaying facial expressions, body gestures, and hand gestures. Participants were made to answer the question, "Which facial expression (or head/hand gesture) is the person in the video making?" for each video. The videos contained no sound.

Figure 7 shows the experimental conditions. White sheets of tulle fabric measuring 200 cm height by 115 cm width with denier 50 were hung halfway between the participants and the video. By increasing or decreasing the number of sheets of fabric, the amount of visual non-verbal information transmitted was altered. A preliminary experiment was performed to determine the appropriate number of sheets of fabric, leading to the decision of seven patterns. The order of the videos and experimental conditions was randomly changed to create 14 patterns in which the experiment was performed. The experiment was evaluated and the seven experimental conditions were compared using the accuracy rate of answers to the test questions. The test format was multiple choice instead of a free writing format so as to prevent the researcher's subjectivity from affecting the grading of the test.

1: 4 sheets	2: 8 sheets	3: 10 sheets	4: 12 sheets
5: 14 sheets	6: 16 sheets	7: 18 sheets	w/o sheet

Fig. 7. Experimental conditions II

Experimental Results

Figure 8 shows the average accuracy rate of answers to the confirmation test questions. The Bonferroni method was employed to assess the results. The accuracy rate for four sheets of fabric and eight sheets of fabric was almost 100%, and statistically significant differences of 95% and 99% were observed for all other numbers of sheets of fabric. Additionally, the accuracy rate decreased by approximately 20% as the number of sheets of fabric increased by two for each experimental condition beyond eight sheets. Therefore, by using the proposed method, it was evident that the amount of visual non-verbal information relayed can be controlled incrementally. In addition, average accuracy rates were tested by separating facial expressions and gestures (hand and head movements). Results indicated that 12 sheets of fabric was a unique experimental condition in which facial expressions were virtually impossible to read, but gestures could be sufficiently read.

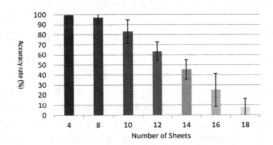

Fig. 8. Accuracy rate of confirmation test on video content

4.3   Verification Experiment

Experimental Method

Using the method explained previously, a verification experiment was performed to investigate the effects of visual non-verbal information on feeling and degree of transmission. Participants were paired into groups of two. The tasks performed and the evaluation criteria were the same as those outlined in the previous section. The number of sheets of fabric were increased or decreased to 8 sheets, 10 sheets, 12 sheets, 14 sheets, and 16 sheets to conduct the telling of the story, done one at a time. The experimental conditions (number of sheets) was determined using the results outlined in the previous section. However, in order to reduce the burden on participants, each was made to perform one pattern containing only three different experimental conditions: either Pattern A (8 sheets, 10 sheets, and 14 sheets) or Pattern B (8 sheets, 12 sheets, and 16 sheets).

Experimental Results and Analysis
While there were two patterns, Pattern A and Pattern B, both patterns shared the eight-sheet segment, and the results of the eight-sheet segment were used as the value of 1 in the figures and proportions explained below and were also used to integrate the results of both patterns.

Figure 9 shows the proportion of the listeners' accuracy rates, and Fig. 10 shows the proportion of the absolute values of the difference in impressions of the story characters. Neither figure shows statistically significant differences, but although the listeners' accuracy was highest for the 12-sheet segment, the same segment shows the greatest difference in impressions of the story characters. Therefore, there is a possibility that the degree of transmission of information regarding the dialogue content and the degree of transmission of sentimental information show opposite trends. These results suggest that the total amount of verbal and non-verbal information that can be received and interpreted may be limited for each individual.

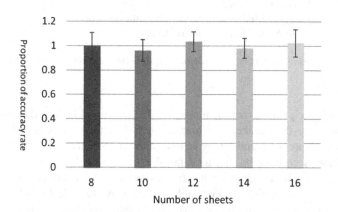

Fig. 9. Proportion of listeners' accuracy rate

Figure 11 shows the proportions of the speakers' and listeners' answers to the questionnaire. The speakers' feeling of transmission in the 12-sheet segment (Question 1) was rated low, and their ease and enjoyment when communicating (Questions 2 and 4) in the 10-sheet segment was rated high. Conversely, the listeners' ratings for the questionnaire fields of the 12-sheet segment were all high. It cannot be concluded for

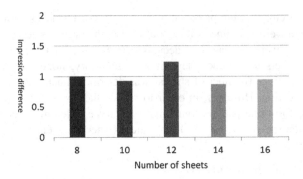

Fig. 10. Proportion of absolute values of difference in impressions of story characters

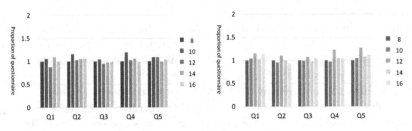

Q1. I felt the story was understood by my partner.
   I felt I understood my partner's story.
Q2. I felt it was easy to communicate/listen.
Q3. I enjoyed the dialogue.
Q4. I felt connected with my partner during the dialogue.
Q5. I was able to concentrate on the dialogue.

Fig. 11. Proportions of speakers' and listeners' questionnaire answers(Left: speakers, Right: listeners)

certain as the participants' experiment patterns (Pattern A and Pattern B) differed, but it may be easiest for the speaker to communicate when they can somewhat see their partner's facial expression, and it may be easiest for the listener to listen when they can see their partner's gestures but not their facial expressions.

Table 4 shows the rankings of the five experimental conditions for feeling and degree of transmission which are FoT and DoT. The speakers' feeling of transmission displayed a strong positive correlation with consistency in character impressions. On the other hand, the listeners' feeling of transmission displayed a strong positive correlation with the listeners' accuracy rates. However, feeling and degree of transmission did not show a proportional correlation with the non-verbal information shown in Fig. 1.

Table 4. Rankings of each experimental condition for feeling/degree of transmission

Rank	8 sheets	10 sheets	12 sheets	14 sheets	16 sheets
DoT by contents	3	5	1	4	2
DoT by impression	4	2	5	1	3
FoT by talker	3	2	5	1	4
FoT by listener	5	3	1	4	2

5 Conclusion

This study investigated the effects of non-verbal information on the feeling and degree of transmission by controlling the visual non-verbal information and communication mode. The experimental results of the three communication modes—text chat, voice chat, and face-to-face communication—showed that the degree of transmission was lowest in face-to-face communication as evaluated with the listeners' test accuracy rates and consistency of character impressions. Conversely, according to the questionnaire results, feeling of transmission was ranked highest for face-to-face communication, followed by voice chat, and lastly text chat. These results suggested that the communicability of information should be considered using feeling of transmission and degree of transmission as two separate factors. Moreover, a negative correlation between the degree of transmission of the story content and consistency of story character impressions was observed in the experiment results performed by controlling visual non-verbal information. These results suggested that the total amount of verbal and non-verbal information that can be received and interpreted may be limited for each individual. It is conceivable that future research may need to be performed using tasks involving conversing freely or communicating to build agreement, or perhaps by controlling non-verbal information of different characteristics to further clarify the effects of non-verbal information on feeling and degree of transmission. Based on the results of such studies, it could be possible to arrange the most suitable forms of communication according to one's aims by including or controlling specific non-verbal information in the future.

References

1. Kurokawa, T: Non-verbal Interface. Ohm press, Monroe (1994). (in Japanese)
2. Birdwhistell, R.L.: Kinesics and Context. University of Pennsylvania Press, Philadelphia (1970)
3. Mehrabian, A.: Communication without words. Psychol. Today **2**(4), 52–55 (1968)
4. Kimura, Y., Tsuzuki, T.: Group decision making and communication mode. Jpn. J. Exp. Soc. Psychol. **38**(2), 183–192 (1998). (in Japanese)
5. Sugitani, Y.: Opinion book about the difference of the information transmission in Internet communication and the face-to-face communication (in Japanese). https://www.kantei.go.jp/jp/singi/it2/kaikaku/dai3/siryou3_2_2.pdf

6. Shibuya, K., Fukuzumi, S., Sakamoto, A.: Comparison subjective self-disclosure by CMC with by FTF– Survey study and laboratory experiment. In: Human Interface Symposium 2016, pp. 601-609 (2016)
7. Goldstein, A.P., Sprafkin, R.P., Gershaw, N.J., Klein, P.: The adolescent: social skill training through structured learning. In: Carledge, G., Milburn, J.F. (eds.), Teaching Social Skills to Children. Pergamon Press, Oxford
8. Kikuchi, A.: Notes on the Researches Using KiSS-18. Bull. Fac. Soc. Welf. Iwate Prefect. Univ. 6(2), 41–51 (2004)

Investigation of Sign Language Recognition Performance by Integration of Multiple Feature Elements and Classifiers

Tatsunori Ozawa[1], Yuna Okayasu[1,2], Maitai Dahlan[3],
Hiromitsu Nishimura[1,2], and Hiroshi Tanaka[1,2(✉)]

[1] Course of Information and Computer Sciences, Graduate School of Kanagawa
Institute of Technology, 1030 Shimo-ogino, Atsugi-shi, Kanagawa, Japan
{s1785014,s1421150}@cce.kanagawa-it.ac.jp,
nisimura@ic.kanagawa-it.ac.jp
[2] Department of Information and Computer Sciences,
Kanagawa Institute of Technology, Atsugi, Japan
h_tanaka@ic.kanagawa-it.ac.jp
[3] Course of Mechanical Engineering, Graduate School of Chulalongkorn
University, Bangkok, Thailand
maitai8town@gmail.com

Abstract. Sign languages are used by healthy individuals when communicating with those who are hearing or speech impaired as well by those with hearing or speech impediments. It is quite difficult to acquire sign language skills since there are vast number of sign language words and some signing motions are very complex. Several attempts at machine translation have been investigated for a limited number of sign language motions by using KINECT and a data glove, which is equipped with a strain gauge to monitor the angles at which fingers are bent, to detect hand motions and hand shapes.

One of the key features of our proposed method is using an optical camera and colored gloves for detection of sign language motion. The optical camera is implemented in a smartphone. This makes it possible to remove the limitation of using area and occasion as a machine translation tool.

The authors propose two new schemes. One is to add the two feature elements, that is, hand direction obtained from the angle between the wrist and fingertips, and hand rotation calculated from the visible size of the palm and wrist incorporating the four conventional elements comprising motion trajectory, motion velocity, hand position and hand shape. The other is integrating the results which is obtained by each classifier to enhance the recognition performance. The six kinds of classifiers have been applied to 35 sign language motions.

A total of 3150 pieces of motion data, that is, 2100 pieces of motion data as training data and 1050 pieces as evaluation data, were used to evaluate the proposed method. The recognition results were examined by integrating the feature elements and classifier. The success rate for 35 words was respectively 76.2% and 94.2%, for the selection of the first ranked answer, and the selection of the first, second or third ranked answers. These values suggest that the proposed method could be used as a review tool for assessing how well learner have mastered sign language motions.

© Springer International Publishing AG, part of Springer Nature 2018
S. Yamamoto and H. Mori (Eds.): HIMI 2018, LNCS 10904, pp. 291–305, 2018.
https://doi.org/10.1007/978-3-319-92043-6_25

Keywords: Sign language · Color gloves · Optical camera · Classifiers
Feature element · Ensemble learning

1 Introduction

Sign language recognition is indispensable in communication with and between hearing impaired people. It is quite difficult to acquire sign language skills since there are vast number of sign language words and some signing motions are very complicated. Furthermore, even if one person has learned sign languages, communication is impossible unless these signs can be recognized. Several gesture recognition systems [1] have been proposed for a limited number of sign language motions by using KINECT [2] and a data glove [3], which is equipped with a strain gauge to monitor the angles at which fingers are bent, to detect hand motions and hand shapes.

One of the key features of our proposed method is the use of an optical camera and colored gloves for detection of sign language motions. The optical camera is implemented in a smartphone. In motion recognition that is not limited to sign language, the widely available Toolkit [4] has also been released, but technology to recognize the movement of fingertips like sign language using only optical camera images has yet to be developed. Studies that did not use colored gloves but used different means to recognize the shape of static hands have been reported [5]. However, we are aiming to develop a sign language recognition technique using optical cameras and colored gloves, giving priority to ease of introduction and the capacity to detect high-speed movements.

We proposed a method for recognizing sign language motion from hand position information acquired by an optical camera using the hidden Markov model (HMM) at HCII'2017 conference [6]. To achieve better recognition performance, we propose two new feature elements to accurately describe the sign language motions. We also propose several different recognition methods and examine a scheme to integrate the recognition results in this paper. It is known that superior recognition performance can be achieved by ensemble learning [7] that combines multiple recognition methods. Our proposal is based on the idea that classifiers work in a complimentary fashion in relation to each other since each classifier is designed based on different classification criteria.

2 Motion Detection and Data Creation

2.1 Colored Gloves

Because identifying each finger is one of the crucial factors for hand shape recognition, colored gloves are proposed for hand shape recognition [8]. The tip of each finger of the glove has a distinct color. This makes it easy to discriminate each finger and results in reliable recognition of hand shapes. If we use wrist bands for both hands, the right and left hand are easily distinguished. The entire hand motion can be detected by the movement of colored wrist bands. In addition, the palms of the hands can be identified by the presence of colored regions. Therefore, the issues outlined in (1)–(4) below can be overcome by using colored gloves.

The colored gloves we designed are shown in Fig. 1. Five colors are used so as to uniquely discriminate each finger, different additional colors distinguish each wrist, and green patches locate the palms of the hands. Thus, a total of eight colors are proposed to facilitate sign language recognition.

(1) Identification of each finger and both hands
(2) Motion detection of wrists/hands and fingers
(3) Hand shape recognition
(4) Discrimination between the palm and the back of the hand.

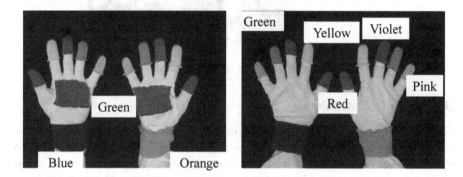

Fig. 1. Colored gloves (Color figure online)

2.2   Current Application Target

While automatic translation remains our final goal, it is too difficult to realize with current technology. Therefore, the authors are now trying to produce a kind of learning tool for sign language. Video data relevant to sign language can be obtained from a web site [9, 10]. Figure 2 shows an image from an instruction video demonstrating sign language motion. A learner memorizes the motions of each sign from this video.

However, it is quite difficult not only to memorize the motion but also to confirm the validity of the motion memorized from the video. The authors are investigating a sign language recognition method that incorporates a tool for checking the learned motion. After memorizing the sign language motion, the learner displays the same motion in front of a web camera connected to a PC. If the PC recognizes his/her motion, the result shows on the display. The learner can evaluate his/her hand and finger movements meaning this system can be used as a review tool for sign language. Although it would be ideal for the recognition success rate to be relatively high for learner's review, this cannot be achieved at this stage. Therefore, our current goal for review tool is to achieve an about 80% success rate.

Fig. 2. Instructional video for sign language motions

2.3 Motion Data Acquisition

It is important to gather accurate motion data. Therefore, the authors asked for the cooperation of the person in charge of making the motion video SmartDeaf [11] to compile the set of motion data used in this investigation. Figure 3 shows an actual motion data capturing scene. To shorten the duration, the motions of two signers were captured simultaneously. The supervisor checked the motions while the signers were performing and confirmed the accuracy of the recorded motion data after the signer had completed the motion.

The conditions under which the motion data were captured are as follows [6].

(1) Camera image resolution of 800 × 600 pixels.
(2) Illumination is set at about 200 lx for both the camera side and signer side.
(3) Frame rate is 30 fps (frames per second). This is the maximum rate for a standard Web camera and smartphone.
(4) The distance between the camera and signer is one meter, as this distance is considered to coincide with a real-life situations.
(5) The color of the signer's clothes and the background wall is black to facilitate easy detection of the colored region of colored gloves.
(6) The height in the field of view of the camera is set at a position such that the wrists of the signer cannot be detected when he/she lowers his/her arm in order to make clear the beginning and the end of a sign language motion.

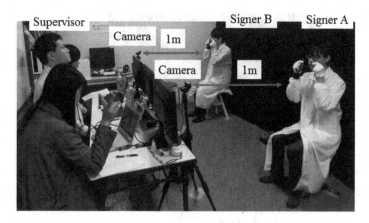

Fig. 3. Scene of data acquisition

2.4   Recognition Data Creation

2.4.1   Color Extraction and Feature Element

Each color region is detected by using colored gloves and an optical camera. An example of the detection of patterns made by color extraction is shown in Fig. 4. Their extraction is based on the hue and saturation values set in the calibration. The center of gravity of each colored region is used to specify the location of each part. The colored region size can be obtained from this image, and this size is also used for creating feature elements. The motion of the blue region assigned to the wrist can be interpreted as the motion of the entire hand.

We have tried to extract many kinds of information from sign language motion as feature elements in order to maintain a high recognition level. The feature data that identify each sign language motion is one of the crucial factors for determining recognition performance. We obtain the following feature data from the position of each colored region and the number of pixels, that is, the region size. The features and their elements are summarized in Table 1.

(1) Hand trajectory, i.e. the shape of the motion
(2) Hand position
(3) Hand velocity
(4) Hand shape, i.e. relative finger location
(5) Hand direction, i.e. hand angle
(6) Hand rotation, i.e. whether from the palm to the back of the hand or vice versa.

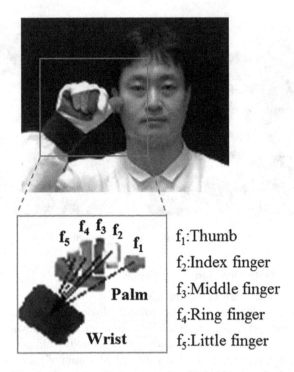

| f_1:Thumb |
| f_2:Index finger |
| f_3:Middle finger |
| f_4:Ring finger |
| f_5:Little finger |

Fig. 4. Example of color extraction (Color figure online)

Table 1. Feature elements

Features	Calculation method	Number of dimensions
Trajectory	$tx_i = (x_i - \bar{x})/A$ $A = \sqrt{\frac{1}{n}\sum_{i-1}^{n}\left((x_i - \bar{x})^2 + (y_i - \bar{y})^2\right)}$ $ty_i = (y_i - \bar{y})/A$	2
Position	$px_i = x_i/800$ $py_i = y_i/600$	2
Velocity	$dx_i = x_i - x_{i-1}$ $dy_i = y_i - y_{i-1}$	2
Shape	$d_{ji} = \sqrt{\left(fx_{ji} - x_i\right)^2 + \left(fy_{ji} - y_i\right)^2}$ $j = 1,2,3,4,5$	5
Direction	$\alpha_i = a \tan 2\left(fy_{ji} - y_i, fx_{ij} - x_i\right)$ j is priority order of $\left\{ \begin{array}{l} \text{Index finger, Middle finger,} \\ \text{Ring finger, Little finger, Thumb} \end{array} \right\}$	1
Rotation	$\beta_i = \left((Area\,of\,wrist)_i, (Area\,of\,palm)_i\right)$	2

(x, y): Wrist position (f_x, f_y): Finger position
i: Frame number, j: Finger number (1–5)

2.4.2   Preprocessing

(1) Interpolation

The center of gravity of the wrist is obtained by detecting the blue color of the wrist band. The position information of the hand is regarded as being equivalent to this wrist position. However, the position information sometimes cannot be obtained due to changes in illumination conditions resulting from movement during sign language motion, occlusion, and so on. We encountered situations where blue could not be extracted.

This problem is not merely a matter of the wrist but also affects the feature elements to be calculated using hand position. Feature elements with the exception of rotation need this information. Linear interpolation is applied when the position of the hand cannot be obtained. An example of linear interpolation is shown in Fig. 5. The hand position in x and y coordinates is shown in this figure. The interpolated values in each frame were used to calculate the feature elements.

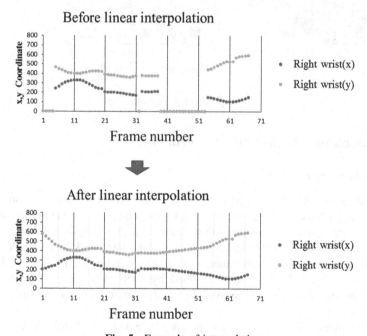

Fig. 5. Example of interpolation

(2) Element length adjustment

Besides classification by HMM, it is necessary to equalize the number of elements of the feature elements vector that is data input into the classifier. The time required for each sign language motion is different. Moreover, even if it is the same word, the same number of vector elements is not realistic from the viewpoint of individual differences and repeatability.

The authors adjusted the number of vector elements according to the method shown in Fig. 6. Some elements in the vector were divided into groups, and their average was calculated to represent this group. This becomes a new data set having the same element length. Based on the results of a preliminary examination to ascertain recognition performance, the number of vector elements was set to ten.

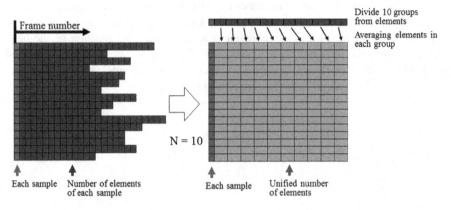

Fig. 6. Adjustment of element length

3   Recognition Method and Experiments

3.1   Recognition Method

One of our new proposals is adding two new feature elements, namely, hand rotation and hand direction, as described in Sect. 2.4.1. The other is to use the plural classifiers to enhance the discrimination performance for each sign language motion. This is based on the idea of ensemble learning [7] that classifiers work in a complementary fashion in relation to each other since each classifier is designed based on different classification criteria. If the results obtained by each classifier are integrated, the classification performance is improved by compensating for deficits in individual classifiers.

Six classifiers were selected in our proposed method. The hidden Markov model toolkit (HTK) [12] which is based on the hidden Markov model (HMM), and support vector machine (SVM) implementing LIBSVM [13], were used as classification tools. In addition, the function provided in MATLAB was used for applying other classifiers. Decision tree (DT), discriminant analysis (DA), linear classification method (LCM) and k-NN (k nearest neighbor) method were applied by using MATLAB function "fitce-coc" [14].

Figure 7 shows the recognition method we propose in this paper. Our proposal is to integrate the classification results by each feature element and each classifier. The likelihood or probability for each sign language motion (hereinafter referred to as a "word") is obtained as the output of each classifier by using each feature element. We use six classifiers and six feature elements, therefore 36 results, that is, likelihood or probability is obtained by each classifier and feature element. The ranking is created based on this value. Integration means that the ranking (order of the candidate answer by classification results) is added up, and the recognition result is the word that obtained the smallest value.

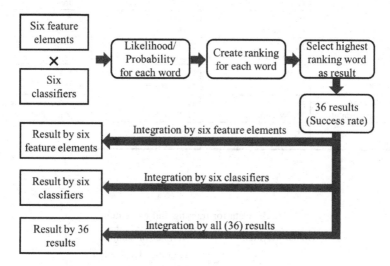

Fig. 7. Recognition method by feature elements and classifiers

3.2   Experimental Results

Thirty-five sign language words were selected for evaluation using the proposed method. Selected words are shown in Table 2. The learning video content in the SmartDeaf is divided into each category based on the usage area and occasion. Nearly 100 words are included in each category. The category of "Health and Diseases" is extremely important to hearing impaired people, therefore we selected this category. In this category, 35 words require right-hand motions only, and all these words were selected.

The data used for training and evaluation are shown in Table 3. The total of 3150 pieces of motion data, that is, 2100 pieces of motion data as training data and 1050 pieces as evaluation data, were used to evaluate the proposed method. There were 11 signers people, with 10 or 20 samples per signer and word. The training was carried out by using 20 samples signed by three signers, that is, 60 samples for each word. Ten samples from three signers were selected for evaluation. The data for training and data for evaluation were obtained from different signers. Here, in the HMM scheme, the appropriate number of the state and the initial parameters of models for each word were clarified [15] before evaluation, and these values are used in this investigation.

Table 2. Target sign language words (35 words)

1. アトピー 1. Atopic	2. おしっこ 2. Urinary	3. ガン 3. Cancer	4. コンタクト 4. Contact lens
5. 喘息 5. Asthma	6. 体調 6. Physical condition	7. ハゲ 7. Bald	8. 発熱 8. Fever
9. 病気 9. Sickness	10. 盲腸 10. Cecum	11. 顔が赤い 11. Blushing	12. カテーテル 12. Catheter
13. 禁煙 13. No smoking	14. 喫煙 14. Smoking	15. 薬を飲む 15. Take medicine	16. 呼吸 16. Breath
17. 耳鼻科 17. Otolaryngology	18. 頭痛 18. Headache	19. 摘出 19. Remove	20. 糖尿病 20. Diabetes
21. 脳卒中 21. Stroke	22. 吐き気 22. Nausea	23. 鼻水 23. Runny nose	24. 昼寝 24. Nap
25. 虫歯 25. Tooth decay	26. 命 26. Life	27. インフルエンザ 27. Influenza	28. 風邪 28. Cold
29. ギリギリ痛む 29. "Giri Giri" Painful	30. 検出 30. Detection	31. 腰 31. Waist	32. ズキンズキン痛む 32. "Zukin Zukin" Painful
33. 毒 33. Poison	34. 捻挫 34. Sprain	35. 冷や汗 35. Cold sweat	

Table 3. Data for training and evaluation

Signer	Words	Number of samples	
		Training	Evaluation
A	#1–#35	20	-
B	#1–#25	20	-
C	#1–#25	20	-
D	#1–#25	-	10
E	#1–#25	-	10
F	#1–#25	-	10
G	#26–#35	20	-
H	#26–#35	20	-
I	#26–#35	-	10
J	#26–#35	-	10
K	#26–#35	-	10
	Total of samples	2,100	1,050

The results are summarized in Table 4. The number indicates the success rate by percentage. Each number in this table was derived by each feature element and classifier. The success rate by one feature element and one classifier ranged from 17.3 to 52.8% for 35 words. The bottom row is the integration results obtained by six features, and the right-hand column shows the results obtained by six classifiers. Integration means that the ranking (the order of candidate answer) is added up, and the recognition result is the word that obtained the smallest value. The integrated result by six feature elements and each classier ranged from 53.6 to 73.1%, the result by six classifiers and each feature element ranged from 33.6 to 51.8%, and the result for total integration, that is, the result obtained by six classifiers and six feature elements was 76.2%. It was verified that integrating each feature element contributes significantly to raising recognition performance. A 3.1% performance enhancement was achieved, which was the difference between the integrated result (76.2%) and the result by SVM (73.1%), which was the best result for a single classifier.

For all classifiers, the integrated result by feature elements improved classification performance markedly, demonstrating that this method can effectively enhance performance. Although an outstanding outcome cannot be obtained by combining each classifier compared with that of integrating feature elements, enhancement can be brought about by integrating classifiers.

Table 4. Recognition result by integration of highest ranking results

Training data A, B, C, J, K & evaluation data D, E, F, L, M, N							
Feature elements	Classifiers						
	HMM	SVM	DT	DA	LCM	KNN	Classifier integrated results
Trajectory	17.3	46.4	34.2	48.5	42.8	43.5	42.7
Position	25.4	50.5	42.5	52.8	44.4	47.1	51.4
Velocity	24.0	46.8	36.6	51.7	46.1	47.0	51.8
Shape	26.8	46.2	38.4	45.5	40.4	42.0	50.6
Direction	18.8	29.4	27.6	29.9	28.1	33.3	33.6
Rotation	14.4	42.3	35.1	42.4	42.5	38.4	46.0
Element integrated results	53.6	73.1	70.1	72.5	70.8	71.8	**76.2**

4 Discussions for Experimental Results

The confusion matrix for the integrated six elements and six classifiers is shown in Table 5. In the case that ranking one is not unique after integration, summing up was not conducted in making this table. Therefore, the summation value of some rows, that is, Words 24, 25, 27 and 29 is not 30. Each word has the highest probability as a recognition result except Words 28 and 29. Good results could not be obtained for these words, the number of correct answers was 8 and 2 for 30 samples, that is, three signers and 10 samples respectively.

Word 28 was mistakenly taken for Words 27 and 33. The representative scenes of these words are shown in Fig. 8. The differences among these three words involve small motions conducted in front of the face. The little finger is bent just in Word 27.

Word 29 was mistakenly taken for Word 32. The scenes where these words were demonstrated are shown in Fig. 9. The difference in how these two words are signed is

Table 5. Confusion matrix for the 35 targeted words

Rank1 (Evaluation data, 35 Words) \ Training data (35 Words)	W1	W2	W3	W4	W5	W6	W7	W8	W9	W10	W11	W12	W13	W14	W15	W16	W17	W18	W19	W20	W21	W22	W23	W24	W25	W26	W27	W28	W29	W30	W31	W32	W33	W34	W35
W1	29	0	0	0	0	0	0	0	1	0	0	0	0	0	0	0	0	0	0	0	0	0	0	0	0	0	0	0	0	0	0	0	0	0	0
W2	0	16	0	0	9	0	0	0	0	0	0	0	0	0	0	0	0	0	0	0	0	0	0	0	0	0	0	0	0	0	0	0	0	0	0
W3	1	0	27	0	0	0	0	0	0	0	0	0	0	0	0	0	0	0	0	0	0	0	0	0	0	0	0	0	0	0	0	0	0	0	0
W4	0	0	0	14	0	0	0	0	0	0	0	0	0	0	0	0	0	2	0	0	0	0	0	0	0	0	0	0	0	0	0	0	0	0	0
W5	0	1	0	0	29	0	0	0	0	0	0	0	0	0	0	0	0	0	0	0	0	0	0	0	0	0	0	0	0	0	0	0	0	0	0
W6	0	0	0	0	0	30	0	0	0	0	0	0	0	0	0	0	0	0	0	0	0	0	0	0	0	0	0	0	0	0	0	0	0	0	0
W7	0	0	0	0	0	0	30	0	0	0	0	0	0	0	0	0	0	0	0	0	0	0	0	0	0	0	0	0	0	0	0	0	0	0	0
W8	0	0	0	0	0	0	0	20	0	0	0	0	0	0	0	0	0	0	0	0	0	0	0	0	0	0	0	0	0	0	0	0	0	0	0
W9	0	0	0	0	0	0	0	0	30	0	0	0	0	0	0	0	0	0	0	2	0	0	0	0	0	0	0	0	0	0	0	0	0	0	0
W10	1	0	0	0	0	0	0	0	0	30	0	0	0	0	0	0	0	0	0	0	0	0	0	0	0	0	0	0	0	0	0	0	0	0	0
W11	0	0	0	0	0	0	0	0	0	0	30	0	0	0	0	0	0	0	0	0	0	0	0	0	0	0	0	0	2	0	0	0	0	0	0
W12	0	0	0	0	0	0	0	0	0	0	0	30	0	0	0	0	0	0	0	0	0	0	0	0	0	0	0	0	0	0	0	0	0	0	0
W13	0	0	0	2	0	0	0	0	0	0	0	0	24	0	0	0	0	0	0	0	0	0	0	0	0	0	0	0	0	0	0	0	0	0	0
W14	0	0	0	6	0	0	0	0	0	0	0	0	0	20	0	0	0	0	0	0	0	0	0	0	0	0	0	0	0	0	0	0	0	0	0
W15	0	0	0	0	0	0	0	1	0	0	0	0	0	0	21	0	0	0	0	7	0	0	0	0	0	0	0	8	0	0	0	0	0	0	0
W16	0	0	0	0	0	0	0	0	0	0	0	0	0	5	5	21	0	0	0	1	0	0	0	0	0	0	0	0	0	0	0	0	0	0	0
W17	0	0	0	0	0	0	0	0	0	0	0	0	0	0	1	6	30	0	0	0	0	0	0	0	0	0	0	0	0	0	0	0	0	0	0
W18	0	0	0	0	0	0	0	0	0	0	0	0	0	0	0	0	0	23	0	0	0	0	0	0	0	0	0	0	0	0	0	0	0	0	3
W19	0	0	0	0	0	0	0	0	0	0	0	0	0	0	0	0	0	0	25	0	0	0	0	0	0	0	0	0	0	0	0	0	0	0	3
W20	0	0	0	0	0	0	0	0	0	0	0	0	0	0	0	0	0	0	0	28	0	0	0	0	0	0	0	0	0	0	0	0	0	0	0
W21	0	0	0	0	0	0	0	0	0	0	0	0	0	0	0	0	0	0	0	0	27	0	0	0	0	0	0	0	0	0	0	0	0	0	0
W22	0	0	0	0	0	0	0	0	0	0	0	0	0	0	0	0	0	0	0	0	0	20	0	0	0	9	0	0	0	0	0	0	0	0	0
W23	0	0	0	0	0	0	0	0	0	0	0	0	0	0	0	0	0	0	0	0	0	4	18	0	0	0	0	0	0	0	0	0	0	0	0
W24	0	0	0	11	0	0	0	0	0	0	0	0	0	0	0	0	0	0	0	0	0	0	0	25	0	0	0	0	0	0	0	0	0	0	0
W25	0	0	0	0	0	0	0	0	0	0	0	0	0	0	0	0	0	0	0	0	0	0	0	0	16	0	0	0	0	0	0	0	0	0	0
W26	0	0	0	0	0	0	0	0	0	0	0	0	0	0	0	0	0	0	0	0	0	0	0	0	0	27	0	0	0	0	0	0	0	0	0
W27	0	0	0	0	0	0	0	0	0	0	0	0	0	0	0	0	0	0	0	0	0	0	0	0	0	0	24	0	0	0	0	0	0	0	0
W28	0	0	0	5	0	0	0	0	0	0	0	0	0	0	0	0	0	0	0	0	0	0	0	0	0	0	13	8	0	0	0	0	0	0	0
W29	0	0	0	0	0	0	0	0	0	2	0	0	0	0	0	0	0	0	0	0	0	0	0	0	0	0	0	0	2	0	0	0	0	6	0
W30	0	0	0	0	0	0	0	0	0	0	2	0	0	0	0	0	0	0	4	0	0	0	0	0	0	0	0	0	0	16	0	0	0	0	0
W31	0	1	0	0	0	0	0	0	0	0	0	0	0	0	0	0	0	0	0	0	0	0	0	0	0	0	0	0	0	0	26	0	0	0	0
W32	0	0	0	0	0	0	0	0	0	0	0	0	0	0	0	0	0	0	0	0	0	0	0	0	0	0	0	0	0	0	0	13	12	0	0
W33	0	0	0	0	0	0	0	0	0	0	0	0	0	0	0	0	0	0	0	0	0	0	0	0	0	0	0	0	0	0	0	0	25	0	0
W34	0	4	0	0	0	0	0	0	0	0	0	0	0	0	0	0	0	0	0	0	0	0	0	0	0	0	0	0	0	0	0	0	0	22	0
W35	0	0	0	0	0	0	1	0	0	0	0	0	0	0	0	0	0	0	0	0	0	0	0	0	0	0	0	0	0	0	0	0	0	0	26

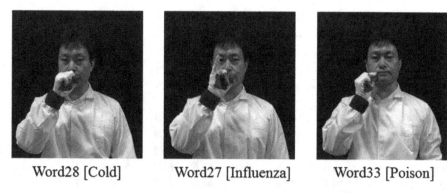

| Word28 [Cold] | Word27 [Influenza] | Word33 [Poison] |

Fig. 8. Representative shots of words where performance was poor (1)

| Word29 | Word32 |
| ["Giri Giri" Painful] | ["Zukin Zukin" Painful] |

Fig. 9. Representative shots of words where performance was poor (2)

only the motion speed of the fingers. Since these motions closely resemble each other, the proposed six feature element cannot classify these words. A new feature element seems to be necessary to discriminate these words.

If we assume that the correct answer is included in the top three rankings for the 35 words, in other words, if we include the third lowest value as a correct result, the result as shown in Table 6 was obtained. The integrated result was raised from 76.2%, shown in Table 5, to 94.1%. This result demonstrates that each recognition result by each feature element and classifier is a good one even though it cannot get the top ranking. This value suggests that the proposed method could be used as a review tool for assessing how well learners have mastered sign language motions.

Table 6. Recognition result by integration of top 3 results

Training data A, B, C, J, K & evaluation data D, E, F, L, M, N

Feature elements	Classifiers						
	HMM	SVM	DT	DA	LCM	KNN	Classifier integrated results
Trajectory	31.0	69.3	61.5	69.6	67.3	65.1	66.9
Position	52.0	77.1	64.3	80.6	68.9	74.5	76.8
Velocity	48.3	70.6	58.9	74.2	71.2	73.1	76.0
Shape	56.8	69.6	66.0	67.6	63.3	66.6	74.5
Direction	39.1	48.4	57.3	46.8	46.3	57.7	57.9
Rotation	35.5	70.8	64.6	70.2	70.2	65.1	72.9
Element integrated results	79.0	90.8	90.1	90.4	88.1	91.5	**94.1**

5   Conclusion

The enhancement of recognition performance for the motions used in sign language is described in this paper. The main feature of the proposed method is integrating the results of six feature elements and six classifiers in order to accurately characterize and discriminate among the sign language motions. The six feature elements were obtained by color extraction from colored gloves and a wrist band. The elements of trajectory, position, and velocity are obtained from the center of gravity of the blue regions of the wrist band. New feature elements were added, and the hand direction was obtained from the angle between each fingertip and the wrist. The hand rotation is calculated from the region size of wrist and palm. Each element is applied to six classifiers to discriminate each motion. The integrated result of six feature elements and six classifiers was 76.2%. In the current investigation, the classification limitation was that six feature elements cannot express the difference in the motions of groups 27, 28 and 33 and groups 29 and 32. However, if we take the top three rankings as a correct result, the integrated success rate for 35 words was increased to 94.2%. This value suggests that the proposed method is a feasible review tool for learners to validate the accuracy of their sign language movements. However, there were four words for which the success rates were approximately 30% or less. The low performance for these words must be resolved if overall performance is to be improved.

References

1. Baatar, B., Tanaka, J.: Comparing sensor based and vision based techniques for dynamic gesture recognition. In: The 10th Asia Pacific Conference on Computer Human Interaction (APCHI), Poster 2P-21 (2012)
2. Zafrulla, Z., Brashear, H., Starner, T., Hamilton, H., Presti, P.: American sign language recognition with the Kinect. In: Proceedings of the 13th International Conference on Multimodal Interfaces, pp. 276–286 (2011)

3. Jitcharoenpory, R., Senechakr, P., Dahlan, M., Suchato, A., Chuangsuwanich, E., Punyabukkana, P.: Recognizing words in Thai Sign Language using flex sensors and gyroscopes. In: i-CREATe2017, 4 p. (2017)
4. Channaiah Chandana, K., Nikhita, K., Nikitha, P., Bhavani, N.K., Sudeep, J.: Hand gestures recognition system for deaf, dumb and blind people. IJIRCCE 5(5), 10058–10062 (2017)
5. Singha, J., Das, K.: Hand gesture recognition based on Karhunen-Loeve transform. In: Mobile & Embedded Technology International Conference 2013, pp. 365–371 (2013)
6. Ozawa, T., Shibata, H., Nishimura, H., Tanaka, H.: Investigation of feature elements and performance improvement for sign language recognition by hidden Markov model. In: Antona, M., Stephanidis, C. (eds.) UAHCI 2017 Part II. LNCS, vol. 10278, pp. 76–88. Springer, Cham (2017). https://doi.org/10.1007/978-3-319-58703-5_6
7. Dietterich, T.G.: Ensemble methods in machine learning. In: Kittler, J., Roli, F. (eds.) MCS 2000. LNCS, vol. 1857, pp. 1–15. Springer, Heidelberg (2000). https://doi.org/10.1007/3-540-45014-9_1
8. Sugaya, T., Suzuki, T., Nishimura, H., Tanaka, H.: Basic investigation into hand shape recognition using colored gloves taking account of the peripheral environment. In: Yamamoto, S. (ed.) HIMI 2013 Part I. LNCS, vol. 8016, pp. 133–142. Springer, Heidelberg (2013). https://doi.org/10.1007/978-3-642-39209-2_16
9. NHK (Japan Broadcasting Corporation), NHK Sign Language CG. http://cgi2.nhk.or.jp/signlanguage/
10. Signing Savvy | ASL Sign Language Video Dictionary. https://www.signingsavvy.com/
11. KCC Corporation, Smart Deaf. http://www.smartdeaf.com/
12. HTK version 3.4.1. http://htk.eng.cam.ac.uk/
13. LIBSVM. https://jp.mathworks.com/help/stats/fitcecoc.html/
14. MATLAB. https://jp.mathworks.com/help/stats/fitcecoc.html/
15. Okayasu, Y., Ozawa, T., Dahlan, M., Nishimura, H., Tanaka, H.: Performance enhancement by combining visual clues to identify sing language motions. IEEE Pacific Rim Conference, 4 p. (2017)

Smart Interaction Device for Advanced Human Robotic Interface (SID)

Rodger Pettitt[1,2(✉)], Glenn Taylor[1,2(✉)], and Linda R. Elliott[1,2(✉)]

[1] Army Research Laboratory, Fort Benning, Fort Benning, GA, USA
{rodger.a.pettitt.civ, linda.r.elliott.civ}@mail.mil,
glenn@soartech.com
[2] Soar Technology, Inc., Ann Arbor, MI, USA

Abstract. Robotic assets used by dismount Soldiers have usually been controlled through continuous and effortful tele-operation; however, more autonomous capabilities have been developed that reduce the need for continuous control of movements. While greater autonomy can make robotic systems more useful, users still need to interact with them, and operator control units (OCUs) for deployed robots still primarily rely on manual controllers such as joysticks. This report describes an evaluation of a multi-modal interface that leverages speech and gesture through a wrist worn device to enable an operator to direct a robotic vehicle using ground guide-inspired or infantry-inspired commands through voice or gesture. A smart watch is the primary interaction device, allowing for spoken input (through the microphone) and gesture input using single-arm gestures.

Keywords: Army robotic systems · Ground robots · Autonomous systems
Gesture commands · Speech commands

1 Introduction

The potential of speech and gesture to enhance Soldier robotic systems is supported at various levels. At the user level, interviews with active duty junior and senior enlisted Soldiers regarding advanced concepts for robotic systems showed high support ratings for more naturalistic controls and autonomous systems and controls [1–3]. A review of gesture-based systems showed significant progress in use of gestures for robot control [4, 5] and concluded that wearable gesture-based systems were more suited to military operations when compared to camera-based systems. Wearable devices have been demonstrated successfully for Soldier use [6, 7]. It was also clear that integration of gesture-based and speech-based controls would likely maximize advantages of each approach. In this effort, a prototype wrist-borne smartwatch that conveys both speech and gesture commands is examined to identify levels of Soldier acceptance and issues for further engineering development.

SoarTech's Smart Interaction Device (SID). SID is a multi-modal human-robot interface designed specifically for the control of robotic mule vehicles that move along with small dismounted units to offload some of the weight carried by Soldiers. SID's

© Springer International Publishing AG, part of Springer Nature 2018
S. Yamamoto and H. Mori (Eds.): HIMI 2018, LNCS 10904, pp. 306–317, 2018.
https://doi.org/10.1007/978-3-319-92043-6_26

supported interactions have focused on ground guide-inspired commands, such as turning in place or moving forward/backward, or infantry-inspired commands, including following a leader or setting rally points. Unlike some other approaches that use sensors on the robot to capture user input, SID employs a smart watch (Fig. 1) as the primary interaction device, allowing for spoken input (through the microphone) and gesture input (using the inertial measurement unit on the watch), while keeping the user's hands free for other tasks. Since it is important to maintain the user's situation awareness, SID also provides multi-modal feedback to the user including visual (on the watch face), speech and vibration, to let the user know the status of commands or of the vehicle.

Fig. 1. Smart Interaction Device (SID) for advanced Human-Robot Interaction (HRI) and Lockheed Martin's Squad Mission Support System (SMSS)

Overview of the Assessment. Twenty-four Soldiers assigned to operational and training units located at Fort Benning, GA. participated in the six-day assessment. The assessment of the SID was based on the evaluation of SID operational commands through each command modality (speech and gesture). Lockheed Martin's Squad Mission Support System (SMSS) (Fig. 1) served as the robotic platform during the evaluation. SMSS uses a number of sensors, including a LIDAR for person tracking and obstacle avoidance. It also hosts a number of computing devices for sensor processing and communications.

Evaluation scenarios were constructed to evaluate command effectiveness within an operational context, and included commands such as Forward, Backward, Turn (left or right), Stop, and Follow me. Each Soldier used both modalities (speech and gesture) separately for robot control. Also, they explored additional speech commands only that explored more autonomous robotic functions such as setting up routes and rally points. After their hands-on sessions, Soldiers participated in after-action reviews and provided feedback through structured ratings regarding ease of use, workload, and operational relevance.

The objective of this effort was to explore options (gesture vs speech) for effective control of a robotic mule using SID. In this preliminary evaluation, data was collected to answer the following questions, to better refine systems for operational use.

- How well does SID recognize inputs from Soldiers?
- Can the Soldiers successfully use SID to control the robotic mule in a variety of maneuver tasks?
- Do the Soldiers find SID useful for controlling the robotic mule?
- What improvements to the system would the Soldiers find useful?

2   Method

Soldiers performed maneuver tasks in three different trials. The task demands were equivalent in Gesture and Speech trials, to allow comparison of task performance. Individual Soldier training was conducted prior to each trial. Training included "hands-on" activity and practice controlling the system. Soldiers performed one trial with the SID speech commands and a trial with gesture commands. Speech and gesture trials were counter balanced. After each Soldier completed both speech and gesture trials, they completed a third trial designed to evaluate additional speech-based commands. Table 1 shows the commands used in each trial. This paper focuses on the comparison of commands used by both speech and gesture.

Table 1. Commands currently implemented in SID for controlling a robotic mule

Command	Gesture	Speech	Additional speech
Forward	X	X	
Backward	X	X	
Turn left	X	X	
Turn right	X	X	
Stop	X	X	
Follow me	X	X	
Start recording route <name>			X
Stop recording route			X
Follow route <name>			X
Set rally point <name>			X
Go to rally point <name>			X
Move forward/backward <x> meters			X
Turn left/right <x> degrees			X

Measures of performance for each trial were recorded by an observer. Measures of performance were based on observations regarding outcomes to each command in terms of correct responses to commands by SMSS. Soldiers completed three trials, one for each condition. Each trial consisted of three sets of the commands. During each set, Soldiers were instructed to give the command up to three times in order to get the correct response from the robot. Based on the robots position and in order to avoid obstacles some commands were attempted more often than others. All correct and incorrect responses were recorded. At the end of each trial, Soldiers were asked to provide verbal (interview) and written (rating scales) feedback with regard to relevance of command, suggestions for improvement, and any operational issues that need to be considered. In addition to objective performance measures, feedback from the Soldiers was collected using 7pt semantic differential rating scales, and verbal/written comments, pertaining to ease of use, comfort, usefulness, perception, workload (NASA TLX), and operational relevance.

3 Results

3.1 Demographics

24 Soldiers agreed to participate in this evaluation. All Soldiers were male, awaiting special training (e.g., Airborne), and were from Infantry (n = 18), Mortar (n = 2), Special Forces (n = 2), or Armor (n = 2) occupational specialties. Their age ranged from 18 to 26 yrs. (mean = 19.8). 6 reported at least some college experience. All had served from four to six months. None reported any injury of any type; all reported they could pass the Army physical fitness test.

All but 3 Soldiers reported wearing their watch on their left hand (1 reported the right hand, 2 did not answer). All reported English as their first language. Only two reported previous experience with robots, regarding (a) manipulator arm on firetruck, and (b) commercial quadcopter. Eighteen reported experience with remote control vehicles. Eight reported experience with speech control capabilities, and reported fair to positive experience regarding effectiveness of use.

3.2 Evaluation of Training

Soldiers provided feedback after all training sessions. Below are mean ratings based on a 7pt scale ranging from 1 = extremely ineffective/negative to 7 = extremely effective/positive. Trainers also rated each trainee on a 7 pt. scale. Trainees were not aware of these trainer ratings. Trainer rating was not correlated to self-ratings of performance (for speech, r = 0.024, p = 0.91; for gesture, r = 0.034, p = 0.87). Neither variable has much variance, the ratings by the trainer ranged from 4 to 5. Ratings of how well each Soldier thought they were going to do in each modality were not significantly correlated with how well they performed; however, their expectations of performance for speech and for gesture were significantly related to each other (r = .771, p < 0.001), suggesting a measure of generalizable confidence and/or humility (e.g., if a Soldier thought he was going to perform very well in speech, he was very likely to think he was going to perform very well in gesture, and vice versa) (Table 2).

Table 2. Quality of training

Aspects	N	Speech mean	Gesture mean	Additional speech mean
Overall effectiveness of training for commands	24	6.17	5.91	6.54
Hands-on training	24	6.38	6.17	6.58
How prepared did you feel for using the device	24	6.25	6.04	6.63
How easy was it to learn and remember commands	24	6.75	6.12	6.42
How well do you think you performed end of training?	24	5.75	5.38	6.33
How well do you expect to perform robot control tasks	24	5.83	5.42	6.29
Trainer rating of trainee	20	4.80	4.38	(N 15) 4.87

3.3 NASA TLX

Soldiers completed NASA TLX rating scales after each performance session. Table 3 shows the mean unweighted ratings for each TLX construct based on a 1 to 10 scale. Five have to do with aspects of workload, while "performance" reflects perception of self-efficacy; therefore, the constructs were not rolled into one overall measure. Table 4 shows paired comparison t-tests for each workload construct and the self-efficacy construct, along with the criteria for Holm's Bonferroni correction for familywise error rates. Results indicate that workload was rated significantly higher for gestures, with regard to mental demand and frustration. Ratings of performance were significantly higher for the speech condition (with or without correction).

Table 3. MEAN NASA TLX for speech, gesture, and additional speech

N = 20	Speech mean	Speech SD	Gesture mean	Gesture SD	Additional speech mean	Additional speech SD
Mental	2.30	1.380	3.81	2.316	3.15	2.033
Physical	1.25	.550	2.48	2.064	1.60	.681
Temporal	2.75	1.552	3.05	1.884	2.00	.795
Effort	2.30	1.750	3.05	2.085	2.65	1.899
Frustration	2.25	1.888	3.95	2.655	1.90	2.033
Performance	8.05	1.432	7.10	1.446	8.50	1.357

TLX ratings of mental demand and frustration were correlated to performance (% correct) in speech and gesture. Correlations were not significant; the only significant correlation was between ratings of frustration for speech and ratings of frustration for gesture, suggesting some individuals experience generalizable levels of frustration, which did not relate to performance.

Table 4. Paired-comparison t–tests comparing speech and gesture, for NASA TLX

Gesture vs speech	t	df	Sig. (2-tailed)	Holm's Bonferroni	
Mental	−2.933	18	.009	.012	*
Physical	−2.549	18	.020	.016	ns
Temporal	−1.035	18	.315	.050	ns
Effort	−1.073	18	.297	.025	ns
Frustration	−2.981	18	.008	.010	*
Performance	2.500	18	.022	.050	*

3.4   Observer-Based Measures

Measures of performance were based on observations regarding outcomes to each command in terms of correct responses to commands by SMSS. Error in response included an incorrect response from the robot when a command was given (i.e. command the robot to go forward and it turns left instead), no response from the robot was when a command was given and the robot acknowledged the command but did not perform the task, and communication error (the robot did not acknowledge receiving the command given by the Soldier).

Errors due to Incorrect Response. Table 5 provides mean number of errors for each command, averaged across each Soldier. The number of attempts varied across Soldiers and commands, as Soldiers were requested to try at least 3 times to get the correct response. Speech commands were interpreted extremely well, and gesture commands were also highly accurate, particularly for "forward" and "stop".

Table 5. Mean number incorrect response errors (N = 24)

Command	Sum speech*	Speech mean (SD)	Sum gesture*	Gesture mean (SD)	Paired t-test (p)
Forward	96	0.13 (0.45)	93	0.33 (0.64)	0.170
Backward	79	0.00 (0.00)	106	3.08 (3.55)	0.000*
Turn left	102	0.00 (0.00)	99	0.71(1.16)	0.007*
Turn right	96	0.00 (0.00)	129	1.88 (2.38)	0.001*
Stop	385	0.00 (0.00)	490	1.42 (2.81)	0.021*
Follow	83	0.00 (0.00)	76	0.42 (0.88)	0.030*

*Differences significant at p </= 0.05
*Total number of command attempts

No Response to Commands. Table 6 provides the mean number of non-responses for each command. It shows relatively high standard deviations for each command, therefore differences among the means were generally not significant, with the exception of "backward".

Table 6. Mean number of no response errors (N = 24)

Command	Sum speech*	Speech mean (SD)	Sum gesture*	Gesture mean (SD)	Paired t-test (p)
Forward	96	0.54 (1.14)	93	0.17 (0.81)	0.224
Backward	79	0.04 (0.20)	106	1.17 (2.01)	0.013*
Turn left	102	1.29 (2.15)	99	0.38 (0.92)	0.059
Turn right	96	0.88 (1.39)	129	0.63 (1.17)	0.485
Stop	385	3.96 (5.78)	490	2.29 (3.85)	0.259
Follow	83	1.38 (1.41)	76	0.79 (1.38)	0.110

*Differences significant at p </= 0.05
*Total number of command attempts

Communication Errors. The mean number of communication errors for each command (Table 7) describes error resulting from the robot not acknowledging having received the command. Means were consistently low, with the exception of no audio responses to speech commands of "stop".

Table 7. Mean number of communication errors (N = 24)

Command	Sum speech*	Speech mean (SD)	Sum gesture*	Gesture mean (SD)	Paired t-test (p)
Forward	96	0.13 (0.45)	93	0.00 (0.00)	0.185
Backward	79	0.13 (0.45)	106	0.17 (0.38)	0.747
Turn left	102	0.13 (0.34)	99	0.04 (0.20)	0.328
Turn right	96	0.17 (0.48)	129	0.13 (0.33)	0.747
Stop	385	1.38 (2.52)	490	0.13 (0.44)	0.025*
Follow	83	0.25 (0.53)	76	0.00 (0.00)	0.031*

*Differences significant at p </= 0.05
*Total number of command attempts

3.5   Post-session Questionnaires

The following summarizes responses to questionnaires administered after performance with the SID controller. Each Soldier used SID, using commands with speech and gesture modalities. Mean ratings are described.

Weight. Weight was rated using a 7pt semantic differential scale ranging from 1 = extremely heavy to 7 = extremely lightweight. The overall mean rating was 6.65 (SD = 0.65).

Comfort. Comfort was rated using a 7pt semantic differential scale ranging from 1 = extremely uncomfortable to 7 = extremely comfortable. The overall mean rating was 6.12 (SD = 0.99).

Ease of use/Difficulty. Ease of use was rated using a 7pt semantic differential scale ranging from 1 = extremely difficult to 7 = extremely easy. Ratings were generally higher for speech commands. For gesture commands, ratings were lowest for "move backward" (mean = 3.92) and highest for "stop" (mean = 6.21). Ratings for "commands in general" were significantly higher for speech (paired-sample t-test = 4.44, df = 23, p < 0.001) (Table 8).

Table 8. Ratings of difficulty of commands, by speech and gesture

Difficulty of command (N = 24)	Speech mean	Speech SD	Gesture mean	Gesture SD
Move forward	6.62	.647	5.92	1.100
Move backward	6.71	.550	3.92	1.909
Left turn	6.12	1.569	5.17	1.579
Right turn	6.42	.929	4.67	1.736
Follow me	6.00	1.348	5.78	1.313
Stop	5.67	1.711	6.21	1.141
Commands in general	6.29	.908	5.29	1.197

Reliability of Gesture Commands. Reliability was rated using a 7pt semantic differential scale ranging from 1 = extremely unreliable to 7 = extremely reliable. Results are summarized in Table 9. Ratings were generally higher for the speech modality. For gesture commands, ratings were lowest for "move backward" (mean = 4.04) and highest for "stop" (mean = 6.21). Ratings for commands in general were significantly higher for speech (paired-sample t-test = 3.042, df = 22, p = 0.006).

Table 9. Reliability of gesture commands

Reliability of command (N = 24)	Speech mean	Speech SD	Gestures mean	Gestures SD
Move forward	6.50	.722	5.96	.999
Move backward	6.50	.722	4.04	2.116
Left turn	6.00	1.445	5.33	1.341
Right turn	6.29	.908	4.79	1.719
Follow me	5.71	1.367	5.67	1.204
Stop	5.46	1.641	6.21	.977
Gestures in general	6.13	.869	5.30	1.295

Opinion Survey. Soldiers used a 7pt Likert scale (1 = strongly disagree to 7 = strongly agree) to respond to statements summarized in Table 10. For positive statements, agreement was generally high for both modalities, with the mean for speech commands (6.11) higher than for gestures (5.65). This difference, using paired comparison t-test, was significant (t = 3.087, df = 23, p = 0.005). For negative statements, the mean for speech commands were lower than for gestures; however, the difference was not significant (t = 0.995, df = 23, p = 0.330). Soldiers agreed that the smartwatch was a good concept for operations.

Table 10. Opinion survey results

Statement	Speech mean	Speech SD	Gesture mean	Gesture SD
Positive statements				
"It was easy to learn the commands for robot control"	6.92	.282	6.46	.779
"It was easy to execute the commands"	6.71	.550	5.83	1.435
"The robot responded effectively to commands"	5.92	.830	5.04	1.398
"Using this modality is a good concept for robot control"	6.26	1.054	5.63	1.555
"Commands helped keep my attention on my surroundings"	6.00	1.022	5.33	1.523
"Commands are useful for 'heads-up' 'eyes-free' performance"	5.67	1.090	5.54	1.285
"I trusted the robot to obey my commands"	5.52	1.123	4.96	1.488
"Using this modality is a good concept for Army dismount operations"	5.92	1.018	5.58	1.472
"The robot was very responsive to commands"	5.63	1.013	4.88	1.624
"The smartwatch was comfortable to wear"	6.46	.932	6.33	1.204
"The smartwatch is a good concept for robot control"	6.21	1.444	6.54	.721
Mean	6.11	0.62	5.65	0.81
Reverse statements				
"It was difficult to maneuver the robot using this modality"	2.54	1.693	3.33	1.949
"Commands would be difficult to remember in operations"	2.33	1.736	2.88	1.825
"The robot was slow to obey"	4.50	1.865	4.17	1.810
Mean	3.12	1.21	3.45	1.25

3.6   4Additional Speech Commands

Soldiers provided feedback on additional speech commands. Ratings for each command are given below. Ratings were based on a 7pt scale ranging from 1 = extremely negative to 7 = extremely positive. Ratings were generally positive, ranging from 5.67 (move faster/slower) to 6.63 (go to rally point) (Table 11).

Table 11. Additional speech commands

Aspect	Rating mean	Rating SD
Move faster/slower	5.67	1.283
Turn left/right X degrees	6.12	.947
Go forward/backward X meters	6.25	1.032
Define route	6.25	.944
Follow route	5.87	1.424
Define rally point	6.46	.721
Go to rally point	6.63	.576

3.7 Overall Experience

The following ratings were based on a 7pt scale ranging from 1 = extremely negative to 7 = extremely positive (Table 12).

Table 12. Overall experience ratings

Aspects N = 24	Mean	SD
Reliability – degree of overall reliability	5.88	.850
Trustworthiness – degree you trust the system overall	5.54	1.021
Ease of use – degree of difficulty	6.37	.576
Comfort/fit – degree system is comfortable over time	6.43	.662
Operational relevance – degree system would be useful in operations	6.04	1.233
Maintaining situation awareness – degree system supports awareness of surroundings	5.75	1.073

End of Experiment Comparative Evaluation Questionnaire. Soldiers were asked to indicate which type of input they preferred for each command; results are summarized in the Table 13.

Table 13. Summary count of system preference regarding command reliability

Reliability	Speech	Gesture	Same
Move forward	11	3	10
Move backward	16	3	5
Left turn	14	3	7
Right turn	15	3	6
Follow me	12	6	6
Stop	9	9	6
All commands in general	14	7	3

Operational relevance. Soldiers were asked, if the system worked perfectly for every command, how useful would the system be in operations, using a 7pt scale, ranging from 1 = extremely useless to 7 = extremely useful. The mean for speech was 6.33, and the mean for gesture was 5.62.

4   Conclusions

Conclusions center on the following questions:

How Well Does SID Recognize Inputs from Soldiers? For errors resulting in the robot performing an incorrect response (i.e. command the robot to go forward and it turns left instead), the overall mean incorrect correct response for speech range was 0.0 for all commands except forward (1.3), for gestures, 0.33 (forward) to 3.08 (backward), indicating that while both systems were fairly accurate, the speech commands were particularly effective. There were also errors resulting from no response from the robot when a command was given, for both types of commands. The overall mean incorrect responses for speech ranged from 0.04 (backward) to 3.96 (stop), for gestures 0.17 (forward) to 2.29 (stop). This suggests further improvements are needed in terms of pure recognition accuracy of the system, both for speech and gesture. Smart watches do not seem to have high quality microphones, and seem to have trouble in noisy environments. This had a noticeable impact on speech recognition. The SMSS's engine revving during turns was problematic for the speech recognizer. The off-the-shelf speech recognizer used in this particular implementation of SID (the Android version of Think-a-Move's SPEAR recognizer) also did not work as well in noisy conditions with the SMSS and a smartwatch as it has in earlier trials with an RGator robotic system and a smart phone microphone [7]. Gesture recognition could also be improved. The gesture recognizer used the native inertial measurement unit (IMU) on the smart watch, which sometimes had sensitivity to ambient arm motion, and generally had too fine a window in which gestures would be successfully recognized.

Can the Soldiers Successfully Use SID to Control the Robotic Mule in a Variety of Maneuver Tasks? Based on Soldier feedback and observations, SID was easy to learn and use for both gesture and speech commands. With few exceptions, all Soldiers successfully maneuvered the robot using all the gesture and speech commands.

Do the Soldiers Find SID Useful for Controlling the Robotic Mule? Overall, Soldiers found SID to be useful for controlling the robotic mule. When Soldiers were asked to indicate which type of input they preferred for each command, more preferred speech than gesture. Also, workload was rated significantly higher for gestures, with regard to mental demand and frustration. Ratings of performance were also significantly higher for the speech condition (with or without correction). For ease of use, ratings were generally higher for speech commands.

What Improvements to the System Would the Soldiers Find Useful? Many Soldiers stated that the gesture command was too sensitive and had to be performed at a very precise speed and motion in order for the robot to recognize the command. Also, several commented that the "follow me command was too similar to the forward command. When asked if the system is usable in its current form, Soldiers noted that the robot not receiving speech commands due to background noise and imperfect gesture recognition leading to either wrong commands or no commands received would be problematic in combat environments.

References

1. Elliott, L., Hill, S.: A vision for future soldier-robot teams. Paper presented at the 7th International Conference on Applied Human Factors and Ergonomics 2016, Orlando (2016)
2. Pettitt, R., Elliott, L., Swiecicki, C.: Soldier-based concepts for squad level autonomous/ intelligent robotic assets. In: Proceedings of the International HCI Human Computer Interface Conference, Vancouver, July 2017
3. Swiecicki, C., Elliott, L., Wooldridge, R.: Squad-level soldier-robot dynamics: exploring future concepts involving intelligent autonomous robots. Technical report No. 7215. Army Research Laboratory Human Research and Engineering Directorate, February 2015
4. Elliott, L., Skinner, A., Vice, J., Walker, A.: Utilizing glove-based gestures and a tactile vest display for covert communications and robotic control. Aberdeen Proving Grounds (MD); Army Research Laboratory (US); Report No. ARL-TR-6971, June 2014
5. Elliott, L., Hill, S., Barnes, M.: Gesture-based controls for robots: overview and implications for use by Soldiers. ARL Technical report 7715. Army Research Laboratory Human Research and Engineering Directorate (2016)
6. Baraniecki, L., Hartnett, G., Elliott, L., Pettitt, R., Vice, J., Riddle, K.: An intuitive wearable concept for robotic control. In: Yamamoto, S. (ed.) HIMI 2017 Part I. LNCS, vol. 10273, pp. 492–503. Springer, Cham (2017). https://doi.org/10.1007/978-3-319-58521-5_38
7. Taylor, G., Quist, M., Lanting, M., Dunham, C., Muench, P.: Multi-modal interaction for robotic mules. Paper presented at the SPIE Defense and Security: Unmanned Systems Technology XIX, Anaheim, CA (2017)

Gestural Transmission of Tasking Information to an Airborne UAV

Alexander Schelle$^{(\boxtimes)}$ and Peter Stütz

University of the Bundeswehr Munich, 85577 Neubiberg, Germany
{alexander.schelle,peter.stuetz}@unibw.de

Abstract. A system is presented that enables an authorized person on ground to transmit mission information to an airborne UAV within line of sight by using gestural expressions of both arms without the need for additional devices on ground. A miniaturized processing board with a discrete GPU is used to detect the body movements via a high resolution onboard camera and to translate them into relevant tasking information. Individual task elements are transmitted consecutively, including numerical and non-numerical information. A context aware gesture recognition approach is implemented to enable the reuse of gestures for different contexts in order to maintain a small gesture set. The system further features a bidirectional communication which allows to dispatch visual feedbacks and to query missing information visually via a LED matrix. Two experiments with different briefing contents in a static and dynamic setup have been conducted to proof the feasibility under real-life conditions.

Keywords: Visual communication · Gesture recognition
Human-UAV-interaction

1 Motivation

Unmanned aerial vehicles (UAV) that are being used for image intelligence purposes, receive their command and guidance information primarily via radio link from a ground control station or handheld devices. Without an adequate device, no access to the UAV is possible. In extraordinary and dangerous situations, like infantryman in unknown terrain or search and rescue personnel in disaster scenarios, this requirement is a disadvantage, since there is no option to transfer the authority of the UAV to a third party via alternative channels these days. New ways of interaction are starting to evolve by using the onboard sensors in combination with gesture recognition to allow a visual communication [1–3]. However, current solutions only utilize this visual channel for low level commands, such as triggering an image capture [4] or telling the UAV to move to a specific direction [5]. High level, mission briefing like communication under real-life conditions has not been demonstrated before. This paper covers a system to address this capability gap.

© Springer International Publishing AG, part of Springer Nature 2018
S. Yamamoto and H. Mori (Eds.): HIMI 2018, LNCS 10904, pp. 318–335, 2018.
https://doi.org/10.1007/978-3-319-92043-6_27

2 System Architecture

The proposed system architecture is designed to allow an authorized person on ground to communicate with an airborne UAV within line of sight by performing a set of gestural expressions to transmit mission relevant details. The UAV can sense the operator's movements and translate them into the needed information components to compose the mission objective on board in real-time. To fulfil this goal, the following requirements have to be met by the system:

- The system shall be robust to a dynamic surrounding, since the flying platform will not be steady at one place at all times.
- The hardware and software components must not exceed the payload weight limitations, however ensuring that the onboard processing power is sufficient for real-time applications.
- The deployed sensor shall be passive with a wide operational range to allow interaction at different distance levels.
- A prompt feedback mechanism shall be included to enable a bidirectional communication between the UAV and the operator. Response time shall not exceed 3 s.

The next chapters cover the relevant system components to satisfy these requirements.

2.1 Communication Model

To map the ongoing processes during the human-to-UAV interaction a communication model has to be designed. To understand the employed model, a few definitions have to be made first.

Operator. A person that is instructed and authorized to send mission details to the UAV is called in the following the *operator*.

Mission. A *mission* is a concatenation of single tasks that are processed by the UAV consecutively to achieve a specific goal (reconnaissance, transport, etc.). These tasks are transferred to the UAV during a *mission briefing*. In the current implementation a briefing starts with a conversation start command (in this case a "Hello" gesture) and always ends with the transmission of a "Return to" task as shown in Fig. 1.

Task. A task is a part of a mission that triggers the activation of a subsystem of the UAV under a specific condition. A "find humans" task for instance activates the onboard sensor system and image processing algorithm to detect persons on the ground once the UAV reaches a defined location. That task consists of two information elements: the command ("find") and an additional non-numerical information "humans". Some tasks can may also require additional numerical information (e.g. "maximum mission time"), where others do not require any additional information at all (e.g. "take off now").

Context. The context is the scope of an interaction. Together with an observation and the already received information this creates a meaning for an observation. If a command requires additional information, it defines the next context for that type of

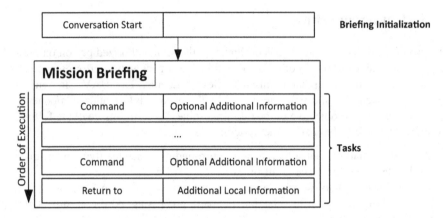

Fig. 1. Layout of a mission briefing

information. Possible contexts are for instance "direction", "location", "numerical information", "tasking" or even "none".

Meaning. A meaning is the interpretation of an observation for a given context. Since the capability for interpretations is ascribed commonly only to humans, this process can be described here as a translation of an observation to a meaning using multiple parameters for the presented system. Meanings can be for example "start conversation", "location of operator" or "count fingers of left hand".

Observation. An observation is the movement or pose of the operator sensed by the UAV. It is the interface between the human operator and the airborne system.

Figure 2 shows the communication model of the proposed system. The operator on ground knows about the type of mission (reconnaissance mission, transport mission, etc.) and the mission specific details. To be able to communicate these details to the UAV, he has to split the mission into multiple single tasks in a way, to represent each task by a set of gestural expression.

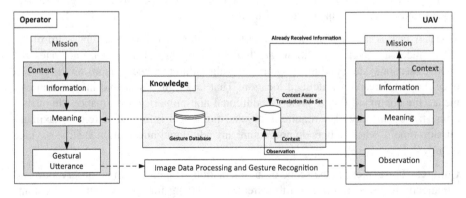

Fig. 2. Communication model

The actual display of the gesture is the gestural utterance, which is visible to the unmanned system and hence can be observed. Both the operator and the gesture recognition system on board the UAV are aware of the possible gestural expressions ("arm up", "arm down", "arm pointing to ground", etc.). This knowledge is implemented in the *Gesture Database* in Fig. 2. Such gesture templates however do not contain any meaning in the first place, instead they can be seen as a means of transportation for the information itself.

The meaning of an observed gesture is assigned in a subsequent step, using the *Context Aware Translation Rule Set*. It takes the current *observation, context,* the *gestural expression* itself as well as the already *received information* as an input and reasons for the *meaning* of the gestural expression under the given circumstances. This way, the same gestural expression can be used for different contexts and the amount of gestures to be memorized by the operator can be reduced.

Once the information is physically sent to the UAV via a gestural utterance, the *image data processing and gesture recognition* onboard transfers these body movements into *observations* which are then translated into *meanings*. If the system detects discrepancies (e.g. expects additional information, but receives instead a different task) it informs the operator about this conflict and asks (again) for the needed information. Once the system has received a valid and complete set of tasks for a mission, it finishes the conversation and starts its mission. Figure 3 illustrates this information transfer process from the operator to the UAV. A programmatic solution for the information processing on UAV side is described in Fig. 15 in the Appendix.

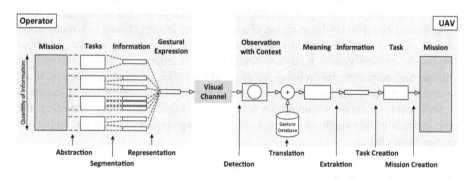

Fig. 3. Information transfer from operator to UAV

2.2 Pose Estimation

A key element for the transmission of gestural information is the detection and localization of the operator's body parts. This process is known as *pose estimation*. In a previous study [6] a depth sensor based approach has been considered to separate the operator from the background reliably and to find extreme points in the depth data to detect the operator's limbs. The results show the advantages of the additional distance information provided by the sensor technology, but also its range limitations, which force the operator to approach the airborne system to an uncomfortable close distance to get within sensor range.

Recent advances in this research domain have brought up a versatile real-time solution [7–9] for this problem, using commonly available 2D image sensors and avoiding the limitations of depth sensors. The open source framework *OpenPose* [10] offers that functionality and supports multiple models with different accuracies and computational demands. However, due to the high degree of parallelism used in that method, a graphics processing unit (GPU) supporting CUDA [11] is required to achieve a real-time processing. Essentially the selected model and the network size determine the performance. The proposed system utilizes the model learned on the MPI dataset [12] using a reduced network size of 288×144 pixel.

2.3   Authentication

As pictured in the introduction of the system architecture, accessing the capabilities of the UAV is only allowed for an authorized person. To include a rudimentary authorization mechanism, a color based approach has been selected for this system. The person that wears a high-visibility vest, like the ones used at construction sites, is assumed to be the operator.

To find the operator in a group of persons, the color of the clothing of each detected person is measured at three points across its upper body part (see Fig. 4). To be robust to changes in illumination each color measurement is converted into the HSV (hue, saturation, value) color space and compared to a reference hue value. If all three measurements are similar to the reference, the person is assumed to be an operator.

More elaborated methods that utilize biometrics for the authorization via face [13], skeleton, body shape [14] or gait recognition [15] can be used for that purpose as well. However, tests with an implementation of a deep metric learning based face recognition [16, 17] showed, that the overall system performance decreases by a factor of about 20% for the chosen system hardware configuration. Furthermore, a face recognition demands high requirements on image quality and resolution which cannot be met in all use cases (especially for long-distance interactions, where the optical magnification is not sufficient to maintain a minimum resolution of the operator's face). Therefore, the computationally lightweight and robust solution using special clothing for authorization has been chosen.

2.4   Gesture Recognition

The pose estimation delivers the detected 2D joint coordinates (Fig. 4) including a confidence value. The angles spanned by the joints *hip-shoulder-elbow* and *shoulder-elbow-wrist* are determined and used as features for each body side. The gesture recognition is divided into two processing chains, a static recognition for pointing and holding type poses and a dynamic recognition for poses that change over time.

Static Gesture Recognition. To detect static gestures, a feature comparison approach is applied. The advantage of this method is the low computational cost (since only two features have to be compared with a reference) and the avoidance of a learning phase, meaning that only reference features have to be defined. Figure 5 visualizes the realized

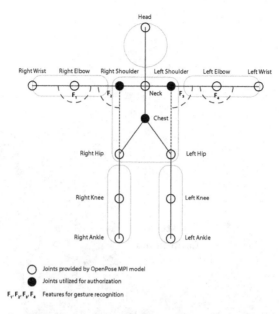

Fig. 4. Provided joints by *OpenPose* and their utilization in the system

gesture set. The inner circle represents the commands that can be followed by one or more additional numerical or non-numerical information, displayed in the second and third ring.

Dynamic Gesture Recognition. To detect dynamic gestures, features have to be analyzed over time. A commonly used and proven approach for this is *Dynamic Time Warping* [18]. This method is suitable for real-time processing and is included in the *Gesture Recognition Toolkit* [19]. Preliminary studies have shown that the achieved frame rate of about 4.5 frames per second (fps) however is not sufficient for the detection of dynamic gestures performed at a common pace (i.e. the operator would need to wave unnaturally slow to get a reliable detection of a "hello" gesture). For that reason, dynamic gestures are not considered in the current state of system implementation. However, dynamic gestures shall be included in the next development iteration of the system once the processing rate can be raised to 10+ fps.

2.5 Finger Detection

For the transmission of numerical information, showing and counting of fingers is a common communication method for humans. Therefore, a hand and finger detection has been included in the system. The implemented method is a modified version of [20], which uses hull curve and convexity defects of a shape to find finger tips in it (see Fig. 6). The location of the hands is estimated based on the wrist information of the pose estimation and the segmentation is performed via skin color thresholding [21].

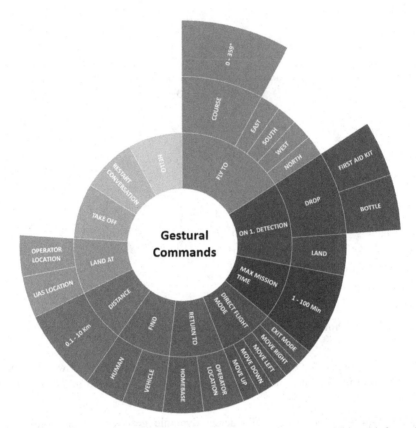

Fig. 5. Supported gestural tasking commands and the task dependent additional information

2.6   Feedback

To provide a feedback mechanism for bidirectional communication, a visual approach has been chosen using a bright LED matrix. To give the operator an acknowledgement for a recognized command, the translated gestural observation is displayed for two seconds followed by a prompt in case additional information is needed or the tasking sequence is not completed. Due to the limited available space for payload, the selected matrix with a width and height of 32 × 8 pixels can display only five letters at a time, hence limiting the allowed word length. Scrolling text horizontally had been tested beforehand and showed multiple disadvantages. Waiting for the information to scroll through extends the communication process. More importantly, it increases the risk for the operator to overlook important parts of the message, if he turns his eyes away from the UAV during the feedback phase. To enhance the readability further, common aeronautical abbreviations (e.g. "CRS" for course), interrogatives or combinations of the needed information followed by a question mark are used (Fig. 7). Experiments under direct sunlight have proven a good readability using green LEDs for distances of up to 75 m with normal vision.

Fig. 6. Visualization of the onboard gesture recognition and finger detection algorithm, from top to bottom line: 1. CNTLR = Count fingers on left and right side, 2. movement status, 3. current context, 4. Number of detected fingers and confidence for each hand, 5. S = detected observation of static gesture recognition, D = result of dynamic gesture recognition (here none, since operator is not moving), yellow circles = detected finger tips (Color figure online)

Fig. 7. Multicopter UAV with stabilized sensor system, processing board and LED matrix for visual feedback

2.7 Hardware Design

A flying platform[1] has been selected that can carry up to 6 kg of payload. It features an autopilot that supports several automated flight modes, such as hovering, automatic take off/landing and waypoint navigation. However, for the experiments described in this paper the automated flight capabilities of the UAV have not been used. Instead, a safety pilot takes over the task of keeping the UAV in an appropriate distance in front

---

[1] DJI Matrice 600 Pro.

of the operator at this stage. The payload includes a stabilized gimbal system[2] that carries a GigE Vision camera[3] with a resolution of 1920 × 1080 pixels and a 20x optical magnification. Furthermore the gimbal hosts the main processing unit, which is a miniaturized multicore system featuring a discrete GPU, the LED matrix[4] and a controller board[5] that drives the matrix. All payload components are powered by the gimbal system, except the LED matrix, which receives its power from the multicopter itself due to the high current demands (Fig. 8). A WLAN connection is used only for the startup of the onboard software modules and for debugging purposes.

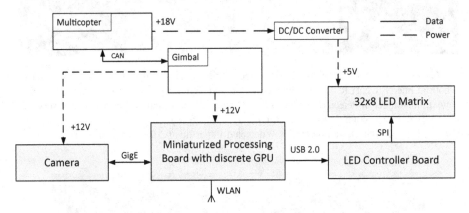

Fig. 8. Involved hardware on board the UAV

2.8   Software Design

The Robot Operating System (ROS) [22] serves as communication infrastructure for all software modules. The software is organized in three major processing nodes (Fig. 9):

Image Acquisition. This node communicates with the GigE Vision camera and provides the received image data to other nodes.

Pose Estimation. This node receives image data and publishes the detected joints to the network. The implementation is a modified version of [23].

VisCom. This is the main node of the visual communication system and is divided into different processing modules. Since the pose estimation node provides its results depending on the order of detection, the *person tracking module* assigns the received skeletal data to each person across frames. The *operator authorization module* determines the authorized person in the image and forwards its joint data to the *gesture recognition module*. This module calculates the features and performs the static and

---

[2] Nvidia Jetson TX2 with Connect Tech Astro carrier and breakout board.

[3] Sony FCB-EH6300.

[4] Adafruit Flexible 8 × 32 NeoPixel RGB LED Matrix.

[5] Arduino Mega 2560.

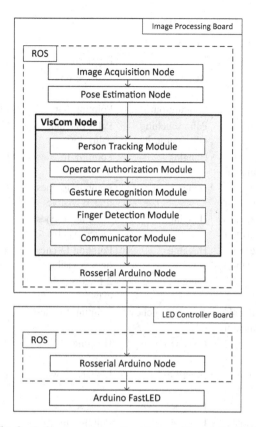

Fig. 9. Software nodes and modules on board the UAV

dynamic gesture recognition. The recognized observations for both body sides are then combined with the results of the *finger detection module* and forwarded to the *communicator module*. This handles the translation of observations, context, available gestures and already received information to a mission relevant meaning. Furthermore, this module controls the visual feedback signals by dispatching display commands to the LED controller board via a *rosserial arduino node*, which acts as an interface between both boards. The actual control of the LED matrix is performed by the *FastLED* library [24].

3 Experiments

Two experiments have been conducted to test the system performance under real life conditions. Multiple missions with different numbers of tasks and objectives were defined for testing with briefing durations from 32 to 209 s. One of them is shown in Table 1. Parts of the communication process are depicted in the image sequences in Figs. 10, 11 and 12.

Table 1. Exemplary mission

Meaning	Context	Observation for arm		Feedback	Question
		Left	Right		
Conversation start	None	Hanging	L-shape up	HELLO	TASK?
Fly	Task	Pointing out	Pointing out	FLY	TO?
Course	Direction	Palm touching other palm, low	Palm touching other palm, low	CRS	C-
1	Numerical information	Hanging	Pointing to chest, 1 finger visible	C 1-	
2	Numerical information	Hanging	Pointing to chest, 2 fingers visible	C 12-	
2	Numerical information	Hanging	Pointing to chest, 2 fingers visible	C 122	DIST?
for 1 km	Numerical information	Hanging	Pointing to chest, 2 fingers visible	1 km	THEN?
Find	Task	Hanging	2 fingers pointing to eyes	FIND	WHAT?
Human	Additional information	Pointing to chest	Pointing to chest	HUMAN	THEN?
On first detection	Task	L-shape up	Hanging	1. DET	DO?
Drop bottle	Additional information	Hanging	L-shape down	DRP B	THEN?
Return to	Task	L-shape down	L-shape down	RET	LOC?
My location	Direction	Hanging	Pointing to ground	URLOC	OK

Fig. 10. Operator starting the mission briefing with airborne UAV, UAV asks for a task

Fig. 11. Operator tasking UAV to fly to a specific direction, UAV responds with question

Fig. 12. Operator tasks UAV to perform an action on the first detection of a human, UAV asks for the action

3.1 Static Performance

The first experiment was conducted with a static platform to test the system performance under optimal conditions without lateral and horizontal movements. For that, the payload had been detached from the flying platform and mounted on the roof of a van in a height of about 2.5 m. The operator was standing in a distance of 20 m away from the van and commanded multiple missions with different numbers of tasks.

3.2 Dynamic Performance

The second experiment included the whole UAV with the stabilized system mounted underneath the flying platform and hovering in a height of 2 to 3 m above the ground to evaluate the influence of the platform movement on the detection and recognition capabilities. The distance between the UAV and the operator was about 22 to 25 m. Fluctuations resulted from windy conditions on that day.

3.3   Results

The experimental results were assessed regarding two major aspects: performance of pose estimation and the response time from the beginning of a gestural movement to the feedback display.

Performance of Pose Estimation. Choosing the appropriate network size for the pose estimation is an accuracy-performance tradeoff (see explanation in chapter 2.2). The larger the size is, the higher the detection rate for small objects gets, but at the same time the processing rate drops significantly. On the other hand, decreasing the network size improves the processing rate, but demands bigger objects in the image frame. The chosen size of 288×144 pixels requires an operator height of about 85% of the frame height for a reliable pose estimation with a sufficient accuracy for the gesture recognition. This can be achieved by a closer hovering next to the operator (not recommended for safety reasons) or a higher optical magnification, that has been selected here. Due to the inference of the convolutional neural network approach, a measured average processing delay of 0.74 s is involved for every frame for the used processing board at maximum CPU and GPU clock rates.

Response Time. To reduce the false positive rate, most of the implemented algorithms use a circular buffer and a confidence value $c \geq 0.8$. Hence, given a buffer size of $b = 5$ and a mean processing performance of $\bar{f} = 4.5$ Hz, a first detection in an ideal use case can be expected after:

$$T = \frac{c \times b}{\bar{f}} = 0.\bar{8} \ s \tag{1}$$

Figures 13 and 14 show the response times for various non-numerical and numerical gestural commands for the static respectively airborne system. All time measurements start in neutral pose, i.e. both arms are hanging, and stop once the LED matrix displays a feedback. So, this measurement includes the transformation from the neutral pose to the final gesture.

The static setup represents the ideal situation with no vehicle movements and an optimal resolution of the operator. Thus, most non-numerical commands are recognized after an average duration of 1.4 s (delay of pose estimation not included), while the numerical commands are decoded with a much greater delay. The reason for this can be found in the implemented color based hand and finger detection, which is sensitive to self-shadowing and skin color-like objects close to the area around both wrists. If these objects occur in that area or shadows cover the hands significantly, the false detection rate tends to rise considerably. The minimum confidence constraint extends then the detection phase and results in a delayed feedback, which misleads the operator to question the correctness of his gestural utterance. For that reason, briefings have been aborted after 15 s of no feedback display.

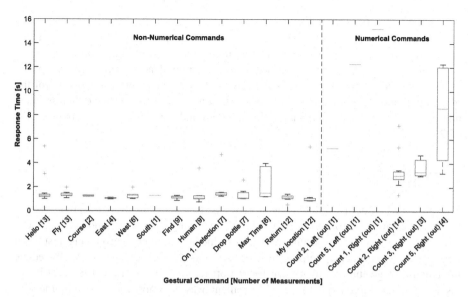

Fig. 13. Response times for gestural commands from first movement to display for static setup, delay of pose estimation not included

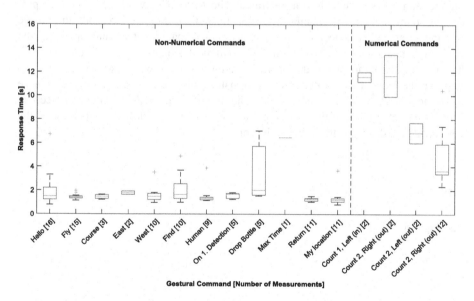

Fig. 14. Response times for gestural commands from first movement to display for airborne setup, delay of pose estimation not included

As expected the airborne setup shows longer average response times. The main reasons are the constant changes in distance to the operator due to manual UAV control and windy conditions, resulting in fluctuations of resolution. Furthermore, the non-optimal positioning of the operator within the frame lead to partial cut offs of the head or the feet of the operator, as the possibilities of the safety pilot to control the multicopter and the gimbal system at the same time were limited. Nevertheless, most non-numerical gestural commands fluctuate around an acceptable response time of 2 s.

4 Conclusion

The presented system allows an authorized person on ground to transmit mission briefing information to an airborne UAV by using his arms and fingers. The developed context aware gesture recognition method delivered promising results in first real-life experiments, helping to keep the false detection rate low and to reduce the amount of gestures to memorize for the operator. Using a bright LED matrix for visual feedback turned out to be a convenient way to improve the communication process and to reduce the uncertainty about the system state. The required robustness against the movement of the flying platform was proven, under the assumption that the operator stays within the image frame. The tradeoff between accuracy and processing rate demands a specific operator resolution. For that reason, the image framing is essential, especially when using optical magnification and operating the UAV in manual mode. These limitations will be taken care of in future steps by handing off the authority for the operator framing to a software tracking module. To improve the recognition of numerical information, more computational power will be added to enable more advanced and light invariant methods for finger detection. Finally, the integration of the autopilot into the system will allow the execution of the commanded tasks from the mission briefing.

Appendix

See Fig. 15.

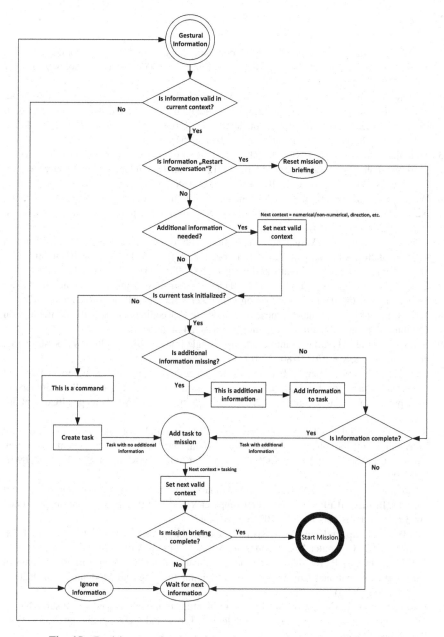

Fig. 15. Decision tree for the information processing on board the UAV

References

1. Schelle, A., Stütz, P.: Modelling visual communication with UAS. In: Hodicky, J. (ed.) MESAS 2016. LNCS, vol. 9991, pp. 81–98. Springer, Cham (2016). https://doi.org/10.1007/978-3-319-47605-6_7

2. Monajjemi, M., Mohaimenianpour, S., Vaughan, R.: UAV, come to me. End-to-end, multi-scale situated HRI with an uninstrumented human and a distant UAV. In: 2016 IEEE/RSJ International Conference on Intelligent Robots and Systems (IROS), pp. 4410–4417 (2016). https://doi.org/10.1109/iros.2016.7759649

3. Monajjemi, V.M., et al.: HRI in the sky: creating and commanding teams of UAVs with a vision-mediated gestural interface. In: 2013 IEEE/RSJ International Conference on Intelligent Robots and Systems (IROS), 03 November 2013, pp. 617–623. IEEE (2013)

4. Wang, T., Wang, M.: Remote control method and terminal Patent US9493232B2 (2016)

5. Nagi, J., et al.: HRI in the sky: controlling UAVs using face poses and hand gestures. In: HRI, pp. 252–253 (2014)

6. Schelle, A., Stütz, P.: Visual communication with UAS: recognizing gestures from an airborne platform. In: Lackey, S., Chen, J. (eds.) VAMR 2017. LNCS, vol. 10280, pp. 173–184. Springer, Cham (2017). https://doi.org/10.1007/978-3-319-57987-0_14

7. Cao, Z., et al.: Realtime multi-person 2D pose estimation using part affinity fields. In: Proceedings of the IEEE Conference on Computer Vision and Pattern Recognition (2017)

8. Wei, S.-E., et al.: Convolutional pose machines. In: Proceedings of the IEEE Conference on Computer Vision and Pattern Recognition, pp. 4724–4732 (2016)

9. Simon, T., et al.: Hand keypoint detection in single images using multiview bootstrapping. arXiv preprint arXiv:1704.07809 (2017)

10. Cao, Z., Simon, T., Wei, S.-E., Sheikh, Y.: OpenPose: real-time multi-person keypoint detection library for body, face, and hands estimation. Carnegie Mellon University, Perceptual Computing Laboratory (2018). https://github.com/CMU-Perceptual-Computing-Lab/openpose

11. Kirk, D.: NVIDIA CUDA software and GPU parallel computing architecture. In: Proceedings of the 6th International Symposium on Memory management, Montreal, Quebec, Canada, pp. 103–104. ACM, New York (2007). https://doi.org/10.1145/1296907.1296909

12. Andriluka, M., et al.: 2D human pose estimation. new benchmark and state of the art analysis. In: 2014 IEEE Conference on Computer Vision and Pattern Recognition, pp. 3686–3693 (2014). https://doi.org/10.1109/cvpr.2014.471

13. Liu, W., et al.: SphereFace. Deep hypersphere embedding for face recognition. In: The IEEE Conference on Computer Vision and Pattern Recognition (CVPR) (2017)

14. Godil, A., Grother, P., Ressler, S.: Human identification from body shape. In: Proceedings of the Fourth International Conference on 3-D Digital Imaging and Modeling, 3DIM 2003, pp. 386–392 (2003)

15. Bouchrika, I.: On using gait biometrics for re-identification in automated visual surveillance. In: Computer Vision: Concepts, Methodologies, Tools, and Applications, pp. 2363–2386. IGI Global (2018)

16. King, D.E.: Dlib-ml: a machine learning toolkit. J. Mach. Learn. Res. **10**, 1755–1758 (2009)

17. Shen, C., Kim, J., Wang, L.: A scalable dual approach to semidefinite metric learning. In: CVPR 2011, Providence, RI, USA, pp. 2601–2608. IEEE (2011)

18. Berndt, D.J., Clifford, J.: Using dynamic time warping to find patterns in time series. In: Proceedings of the 3rd International Conference on Knowledge Discovery and Data Mining, vol. 10, pp. 359–370. AAAI Press, Seattle (1994)

19. Gillian, N.E., Paradiso, J.A.: The gesture recognition toolkit. J. Mach. Learn. Res. **15**(1), 3483–3487 (2014)
20. Yeo, H.-S., Lee, B.-G., Lim, H.: Hand tracking and gesture recognition system for human-computer interaction using low-cost hardware. Multimed. Tools Appl. (2015). https://doi.org/10.1007/s11042-013-1501-1
21. bin Abdul Rahman, N.A., Wei, K.C., See, J.: RGB-H-CBCR skin colour model for human face detection. Faculty of Information Technology, Multimedia University, vol. 4 (2007)
22. Quigley, M., et al.: ROS: an open-source robot operating system. In: ICRA Workshop on Open Source Software (2009)
23. Munaro, M., et al.: OpenPTrack: people tracking for heterogeneous networks of color-depth cameras. In: IAS-13 Workshop Proceedings of the 1st International Workshop on 3D Robot Perception with Point Cloud Library, Padova, Italy, pp. 235–247 (2014)
24. Garcia, D., Kriegsman, M.: FastLED (2015). https://github.com/FastLED/FastLED

A Video Communication System with a Virtual Pupil CG Superimposed on the Partner's Pupil

Yoshihiro Sejima[1]([✉]), Ryosuke Maeda[2], Daichi Hasegawa[2],
Yoichiro Sato[1], and Tomio Watanabe[1]

[1] Faculty of Computer Science and Systems Engineering, Okayama Prefectural
University, Kuboki 111, Soja-shi, Okayama, Japan
sejima@ss.oka-pu.ac.jp
[2] Graduate School of Computer Science and Systems Engineering, Okayama
Prefectural University, Kuboki 111, Soja-shi, Okayama, Japan

Abstract. Pupil response plays an important role in expression of talker's affect. Focusing on the pupil response in human voice communication, we analyzed the pupil response in embodied interaction, and demonstrated that the speaker's pupil was clearly dilated during the burst-pause of utterance. In addition, it was confirmed that the pupil response is effective for enhancing affective conveyance by using the developed system in which an interactive CG character generates the pupil response based on the burst-pause of utterance. In this study, we develop a video communication system with a virtual pupil CG superimposed on the partner's pupil for enhancing affective conveyance. This system generates a virtual pupil response in synchronization of the talker's utterance. The effectiveness of the system is demonstrated by means of sensory evaluations of 12 pairs of subjects in video communication.

Keywords: Human interaction · Nonverbal communication
Affective conveyance · Line-of-sight · Pupil response

1 Introduction

Video communication is a remote communication tool that can support face-to-face communication with talkers using a camera and a microphone in real time. It realizes a better communication environment in comparison with other remote communication tools such as text chat and telephone, because non-verbal behaviors such as body movements, facial expressions and line-of-sight are transmitted through devices [1]. However, the video communication is difficult to provide talkers with an active communication environment where non-verbal behaviors activate and the embodied interaction deepens, because it is not easy to share embodied rhythms in remote communication [2]. Fundamentally, in human face-to-face communication, not only verbal messages but also non-verbal behaviors are rhythmically related and mutually synchronized between talkers. This synchrony of embodied rhythms in communication, called entrainment [3], results in the sharing of embodiment and empathy in human interaction [2].

© Springer International Publishing AG, part of Springer Nature 2018
S. Yamamoto and H. Mori (Eds.): HIMI 2018, LNCS 10904, pp. 336–345, 2018.
https://doi.org/10.1007/978-3-319-92043-6_28

Focusing on the synchrony of embodied rhythms in human interaction and communication, we developed an interactive CG character called "InterActor" that generates the pupil response which is closely related to human affect as well as communicative actions and movements, such as nodding and body movements based on the burst-pause of utterance [4]. In addition, we carried out the communication experiment, and demonstrated that not only body movements but also the pupil response were effective for enhancing affective conveyance and empathy by being synchronized with the timing of response [4]. Therefore, it is expected that a smooth video communication wherein talkers can deepen embodied interaction is provided by visualization of the pupil response.

In this paper, we develop a video communication system which superimposes the partner's pupil CG on the partner's eye position in video image for enhancing affective conveyance and empathy. This system generates a virtual pupil response in synchronization of the talker's utterance. In addition, we carry out an evaluation experiment by using a sensory evaluation. The results demonstrate that the system is effective for enhancing affective conveyance.

2   Video Communication System with a Virtual Pupil CG

2.1   Concept

As the saying goes, "The eyes say more than the mouth," the line-of-sight such as gaze, eye-contact and pupil response is essential to realize the smooth human interaction and communication [5, 6]. In video communication, it was reported that the line-of-sight is important for enhancing embodied interaction [7]. Therefore, in order to develop an advanced video communication system that enhances affective conveyance with line-of-sight, it is effective to express the pupil response that enhances affect as well as the gaze and eye-contact.

The concept of the video communication system is show in Fig. 1. In this system, the pupil CG is superimposed on the partner's eye position in the received video image, and the pupil response is generated in the synchronization to the utterance of speaker [8]. In particular, the system can express a pupil larger or smaller than the human pupil. For example, when a talker's utterance is input to the system, the partner's pupil CG is dilated with the synchronization of utterance. This exaggerated expression encourages and enhances reading the emotional cues and affective conveyance. This in turn causes favorable impressions between talkers. Thus, the system can convey delicate affect using pupil response.

2.2   Development of the System

The setup of the developed system is shown in Fig. 2. This system consists of a desktop PC with Microsoft Windows 10 (CPU: Corei5 2.70 GHz, Memory: 8 GB, Graphics: NVIDIA Geforce GT 730), non-contact motion tracking device (Microsoft Kinect Windows V2), and headset (Logicool H330). The program code is described by using Microsoft Kinect SDK and Face Tracking SDK. The video image is acquired by

Fig. 1. Concept of the system.

the color camera of Kinect. The voice is sampled using 16 bits at 11 kHz. The pupil CG is generated by using image processing library (Itseez OpenCV 3.1). Both image and voice data are transmitted and received between each PC via Ethernet. The frame rate which expresses the pupil CG is 20 fps. Figure 3 shows an example of the developed system in use.

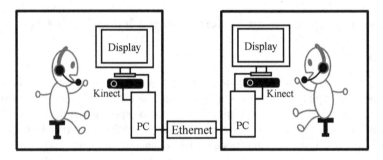

Fig. 2. Setup of the system.

2.3 Superimposing a Pupil CG

In this research, the pupil response is generated by superimposing the pupil CG. However, there is a possibility that a sense of reality may be impaired, because the pupil CG in video image is projected in two dimensions. Therefore, we examine the composition of superimposing the pupil CG to enhance affinity.

The developed pupil CG is shown in Fig. 4(a). Human eyes have a called golden ratio that gives a good impression. The golden ratio is expressed as the ratio of white to iris as 1: 2: 1 as shown in Fig. 4(b) [9]. Therefore, the size of pupil CG is adjusted based on the golden ratio. Here, the diameter of human pupil is known to be about 5 to 6 mm in normal state [10]. Also, the diameter of human iris is known to be about 11 mm [10]. Based on these findings, the size of pupil CG is adjusted by the ratio between the pupil and iris as 5 to 11. The pupil part is expressed with a black circle (Fig. 4(a)). Additionally, the highlight in eyes has the effect that gives a good impression [11]. The highlight part is expressed with a white bar (Fig. 4(a)).

Fig. 3. Example of communication scene.

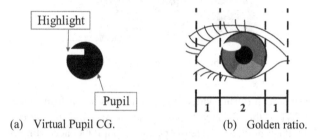

(a)   Virtual Pupil CG.   (b)   Golden ratio.

Fig. 4. Development of virtual pupil CG.

The pupil CG in the developed system is generated as follows. First, in order to superimpose a pupil CG on the partner's pupil, several points of partner's face are detected by using the face tracking function of Kinect. Next, the pupil CG is super-imposed on the eye position of partner. Here, it is necessary to adjust the size of pupil CG for each individual eye. In this research, based on the golden ratio, the size of pupil CG is adjusted from the coordinate values of inner corner and outer corner of eye (Fig. 5). Figure 6 shows an example of screen in which a pupil CG is superimposed.

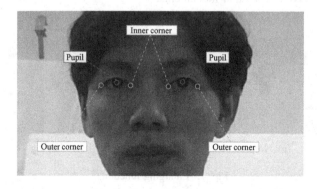

Fig. 5. Face point detection.

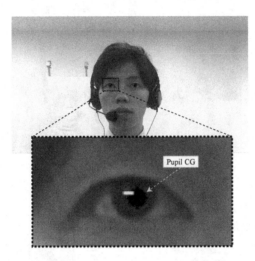

Fig. 6. Superposing a pupil CG.

An example of screen during video communication taken from Kinect's camera is shown in Fig. 7. When the distance between the camera and the talker is far, it has a problem that the talker can't recognize the partners' pupil CG according on the distance (Fig. 7(a)). Therefore, we added two functions to realize smooth video communication. The one function is to zoom the partners' screen. In this function, the green frame is changed with the distance between the camera and the partners' face, and the area surrounded by the frame is enlarged (Fig. 7(b)). The recognition of pupil CG will be enhanced by this function. The other is to apply to various communicative actions and movements. The pupil CG is required to follow various head movements such as nodding in the video communication. In this function, if the partners' head movements are generated within a certain range, the pupil CG is expressed using Kinect's Face Tracking. Figure 8 shows examples of pupil CG by following various head movements. The developed system had the potential to express the pupil CG regardless of any communicative actions and movements.

(a) Normal   (b) Enlarging

Fig. 7. Enlarging function.

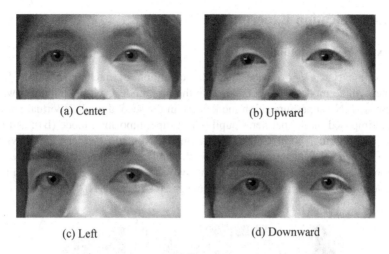

(a) Center   (b) Upward

(c) Left   (d) Downward

Fig. 8. Adjustment of pupil CG.

2.4 Pupil Response Synchronized with Utterance

According to our previous research [4], it was confirmed that the pupillary area enlarges to 1.5 times the normal size by being synchronized with utterance. And, this dilation of pupil causes the enhancement of affective conveyance in embodied interaction and communication. Therefore, the pupil response is generated as follows. When a talker's utterance is input to the system, the absolute value of sound pressure in 30 Hz is calculated. If the value exceeds a threshold value, the speech signal is set as ON. When the speech signal is ON, the area of black circle as the role of pupil increases within 0.5 s. Here, the pupillary area enlarges to 1.5 times the normal size based on the previous research. In addition, if the utterance is continuing, the dilated state of the pupil is maintained. When the utterance is finished, the pupil contracts to the normal size. Thus, the pupil response is generated in synchronization with utterance. An example of dilation in pupil CG is shown in Fig. 9. A distinct pupil response is observed in comparison with the normal size.

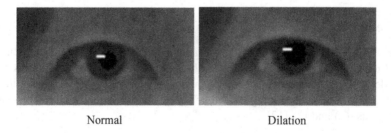

Normal   Dilation

Fig. 9. Dilation of pupil CG.

3 Communication Experiment

3.1 Experimental Method

The experiment was performed under the condition of free conversation. In this experiment, three modes were compared: in the first mode, a virtual pupil CG was not superimposed (No superimposition mode (A)); in the second mode, a virtual pupil CG was superimposed on the partner's pupil (Only superimposition mode (B)); and in the third mode, a virtual pupil CG was superimposed on the partner's pupil with the response to own speech (Superimposition with pupil response mode (C)). We recorded the communication experiment scene using two video cameras as shown in Fig. 10. The subjects were 12 pairs of talkers (12 males and 12 females).

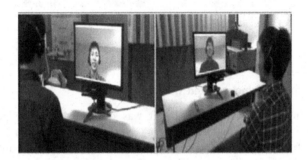

Fig. 10. Example of a communication scene using the system.

The experimental procedure is described as follows. First, the subjects used the system for around 3 min. Next, they were instructed to perform a paired comparison of modes for 2 min. In the paired comparison experiment, based on their preferences, they selected the better mode. Finally, they were urged to talk in a free conversation for 3 min in each mode. The questionnaire used a seven-point bipolar rating scale from −3 (not at all) to 3 (extremely), where a score of 0 denotes "moderately." The conversational topics were not specified in both experiments. Each pair of talkers was presented with the two modes in a random order.

3.2 Result

The results of the paired comparison are summarized in Table 1. In this table, the number of winner is shown. For example, the number of mode (A)'s winner is eleven for mode (B), and the number of total winner is nineteen. Figure 11 shows the calculated results of the evaluation provided in Table 1 based on the Bradley-Terry model given in Eqs. (1) and (2) [12].

$$P_{ij} = \frac{\pi_i}{\pi_i + \pi_j} \tag{1}$$

Table 1. Result of paired comparison.

	(A)	(B)	(C)	Total
(A)		11	8	19
(B)	13		9	22
(C)	16	15		31

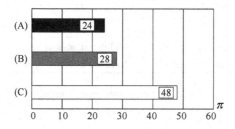

Fig. 11. Comparison of the preference π based on the Bradley-Terry model.

$$\sum_i \pi_i = const.(= 100) \tag{2}$$

π_i : Intensity of i

p_{ij} : probability of judgment that i is better than j

The consistency of mode matching was confirmed by performing a goodness of fit test ($x^2(1, 0.05) = 3.84 > x_0^2 = 0.00$) and a likelihood ratio test ($x^2(1, 0.05) = 3.84 > x_0^2 = 0.00$). The proposed mode (C), superimposition with pupil response, was evaluated as the best; followed by mode (B), only superimposition; and mode (A), no superimposition.

The questionnaire results are shown in Fig. 12. From the results of the Friedman signed-rank test and the Wilcoxon signed rank test, "(f) Enthusiasm," and "(g) Interest" had a significance level of 1% between modes (A) and (C). In addition, "(e) Familiarity" was at the significant level of 1%, and "(g) Interest" was at 5% between modes (B) and (C). In both experiments, mode (C) of the proposed method was evaluated as the best for video communication. These results indicate the effectiveness of the developed system.

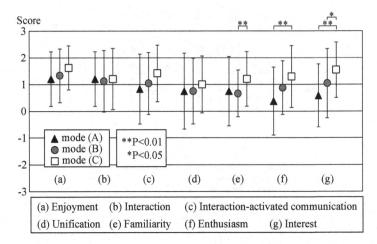

Fig. 12. Seven-points bipolar rating.

4 Conclusion

In this paper, we designed a media representation method for enhancing line-of-sight expression and developed an advanced video communication system. The developed system can express the partner's pupil CG by superimposing on the partner's eye position in video image. The pupil CG can be adjusted based on the features of the face detected by a non-contact motion tracking device, and the pupil response is generated by being synchronized with utterance. In addition, we performed a communication experiment to evaluate the system. The effectiveness of the system was demonstrated by means of sensory evaluations in video communication.

In future research, we will evaluate the system with various parameters such as pupillary size, rapidity and frequency for deepening the close relationship.

Acknowledgments. This work was supported by JSPS KAKENHI Grant Numbers JP16K01560, JP26280077.

References

1. Schoeffmann, K., Hudelist, M.A., Huber, J.: Video interaction tools: a survey of recent work. ACM Comput. Surv. **48**(1), 14:1–14:34, (2015)
2. Watanabe, T.: Human-Entrained Embodied Interaction and Communication Technology, Emotional Engineering, pp. 161–177. Springer, London (2011). https://doi.org/10.1007/978-1-84996-423-4_9
3. Condon, W.S., Sander, L.W.: Neonate movement is synchronized with adult speech. Science **183**, 99–101 (1974)
4. Sejima, Y., Watanabe, T., Jindai, M.: Speech-driven embodied entrainment character system with pupillary response. Bull. JSME Mech. Eng. J. **3**(4), 832–837 (2016). Paper No. 15-00314

5. Argyle, M., Dean, J.: Eye contact, distance and affiliation. Sociometry **41**(3), 289–304 (1965)
6. Hess, E.H.: Attitude and pupil size. Sci. Am. **212**(4), 46–54 (1965)
7. Vertgaal, R., Veer, V.G., Vons, H.: Effects of gaze on multiparty mediated communication. In: Proceedings of Graphic Interface 2000, pp. 95–102 (2000)
8. Maeda, R., Egawa, S., Sejima, Y., Sato, Y., Watanabe, T.: Development of a video chat system superimposing a pupil CG synchronized with utterance on the partner's pupil. In: Proceedings of 23rd International Symposium on Artificial Life and Robotics (AROB2018), pp. 236–239 (2018)
9. Johnson & Johnson Co, Ltd.: The "golden ratio of pupil" of white eyes and black eyes is "1: 2: 1" (2014). http://acuvue.jnj.co.jp/corp/press/pdf/p0138.pdf. Accessed 8 Feb 2018. (in Japanese)
10. Oyama, S., Imai, S., Wake, T.: Handbook of Sense and Perception Psychology, p. 895. Sei Shin Shobo, Tokyo (1994). (in Japanese)
11. Kawatani, H., Kashiwazaki, H., Takai Y., Takai, N.: Feature evaluation by moe-factor of ANIME character image and its application. In: IEICE Technical Report of Image Engineering, vol. 34, no. 6, pp. 113–118 (2010). (in Japanese)
12. Luce, R.D.: Individual Choice Behavior: A Theoretical Analysis. Wiley, New York (1959)

bRIGHT – Workstations of the Future and Leveraging Contextual Models

Rukman Senanayake[✉], Grit Denker, and Patrick Lincoln

SRI International, 333 Ravenswood Ave, Menlo Park, CA, USA
{rukman.senanayake,grit.denker,
patrick.lincoln}@sri.com

Abstract. Experimenting with futuristic computer workstation design and specifically tailored application models can yield useful insights and result in exciting ways to increase efficiency, effectiveness, and satisfaction for computer users. Designing and building a computer workstation that can track a user's gaze; sense proximity to the touch surface; and support multi-touch, face recognition etc meant overcoming some unique technological challenges. Coupled with extensions to commonly used applications to report user interactions in a meaningful way, the workstation will allow the development of a rich contextual user model that is accurate enough to enable benefits, such as contextual filtering, task automation, contextual auto-fill, and improved understanding of team collaborations. SRI's bRIGHT workstation was designed and built to explore these research avenues and investigate how such a context model can be built, identify the key implications in designing an application model that best serves these goals, and discover other related factors. This paper conjectures future research that would support the development of a collaborative context model that could leverage similar benefits for groups of users.

Keywords: Contextual model · Cognitive model · Task automation
Multimodal input · Gaze tracking

1 Introduction

1.1 Motivation

What features and capabilities would future computer workstations have? How would they impact our productivity and ability to solve complex problems? Will they help us collaborate more effectively? Is it possible to achieve a significant leap in performance without changing the application models being used today? These are some of the questions we asked ourselves when we set out to experiment on futuristic workstation design. We were motivated by the ready availability of unused computational power (most PCs have CPUs that do not function to full capacity under normal use), rapid growth in the availability of RAM and hard drive storage, the wide availability of inexpensive sensor systems, and the clear understanding that we do not use all these resources in PCs a vast majority of the time. We were also interested in seeing if we could raise microprocessor features such as instruction pipelining and branch prediction

© Springer International Publishing AG, part of Springer Nature 2018
S. Yamamoto and H. Mori (Eds.): HIMI 2018, LNCS 10904, pp. 346–357, 2018.
https://doi.org/10.1007/978-3-319-92043-6_29

to the user level tasks, what the system requirements would be to achieve that, and whether these features would significantly increase a user's efficiency and effectiveness.

1.2 Design Rationale

From the very start of this project, we were interested in experimenting with multi-touch as a primary input model. We wanted the main input area of the workstation to be fairly large so that more than one person could work together (to support possible future experiments with multiple users simultaneously engaging with the system). We selected gaze-tracking as one of the key input models to the new workstation because of its ability to help us understand the user's context. We wanted to experiment with building a system that could put screen controls (like Ok or Cancel buttons) directly under a user's hands, shifting the traditional paradigm of reaching or searching for the controls. This line of investigation revealed that proximity detection (detecting the nearness of the user's hands to the main input surface) would be a valuable capability.

We also determined that capturing user-level semantics during interactions with the system would significantly improve the usefulness of information related to gaze tracking as well as interactions on the touch screen. Therefore, we extended the applications used for testing so that they can report accurately about on-screen items of interest with which the user engages. For areas of research such as task automation and contextual/cognitive auto-fill, we introduced a semantically enhanced interaction model for the applications being used to enable the capture of a user's workflows while preserving user-level operational semantics. For example, if the user pressed the Send button in a mail client, our system would respond, "The user executed the send mail function in Thunderbird with the following parameters..." instead of reporting a less meaningful output such as "Send button pressed".

To facilitate research about team dynamics and methods for improving collaboration, we decided to build a centralized server modeled on massively multi-player online role playing (MMORPG) game systems. This approach would enable us to capture the context of all users irrespective of their location in real time, reason about it, and then adjust the user's experience based on the analysis results.

2 Related Work

The development of bRIGHT involved R&D in various areas of HCI. In this section we will be looking at some of those areas such as multimodal input, gaze tracking and context modeling. bRIGHT utilizes semantic interaction and visualization models to capture the end user's context in a very rich fashion, see [1] for a detailed discussion on how this is achieved.

Gaze Tracking (GT) is the process of detecting gaze coordinates on a display or the point where a person is looking at. For bRIGHT it was fundamentally critical that we support gaze tracking in a fairly unusual configuration. Since the bRIGHT workstation forces an end user to stand about four feet away from the vertically mounted primary screen they are looking at, we found it important to design a system with various features that are not in typical GT systems. For example, since the user is standing in front of a large screen, we felt it was critical that our GT system allows the user to turn their head naturally when looking at the edge of the screen. This also meant the 'head box' (the area within which a user's gaze is effectively tracked) is much larger in our system than conventional approaches. Gaze tracking recently has been making important contributions in research in conventional HCI [3–5] as well as human cognitive studies [6–11].

Many explorations have been made into developing both nonintrusive [12–20] and intrusive GT techniques [21–23] in order to develop more accurate, efficient and user friendly systems. Nonintrusive methods have been much preferred over its intrusive counterpart due to the potential comfort when performing EGT. Compared to intrusive techniques, image/video processing lies at the core of nonintrusive methods where users are not prodded with electromechanical hardware (e.g. head-mounted cameras, lenses, etc.).

Generally, the process of GT involves four main components. They are (1) head pose detection, (2) eye location detection, (3) gaze estimation (GE), and (4) area of interest analysis. In the case of bRIGHT steps (1–3) were implemented very similar to conventional approaches that are based on active illumination using Infrared. Step (4) on the other hand was radically different to most common approaches due to our ability to access the semantic visualization model of the application being used by the end user (discussed later).

In bRIGHT we grounded our approach based on the work in [24] as one of our team members was Zhiwei Zhu, who designed and built the first generation of GT for this workstation. These were then further improved upon by increasing the accuracy and updating the IR cameras to much higher fidelity.

bRIGHT is an extensible multimodal input system. In essence we have a workstation design where new input modalities can be added as needed by end user context and the systems contextual modeling and other features can dynamically adapt to these. The reason for this approach is grounded in benefits of multimodal input. Various research efforts in the past have shown that multimodal systems are preferred by users over their unimodal alternatives, resulting in higher flexibility and reliability while better meeting the needs of a variety of users [25, 26]. Such systems provide several advantages, including improved efficiency, alternating interaction techniques, and accommodation of individual differences, such as permanent or temporary handicaps [27, 28].

Industrial robot programming using markerless gesture recognition and augmented reality is evaluated in [29]. It allows the user to draw poses and trajectories into the workspace, we believe our semantic interaction models will give us the ability to extend such modalities into various domains where seamless gesture recognition would be of great value. The authors state that such systems show a significant reduction of required teach-in time. A combination of human intention and language recognition for task-based dual-arm robot programming is evaluated in [30]. They conclude that using multimodal input reduces the programming complexity and allows non-experts to move a robot easily. User group differences in modality preferences for interaction with a smart-home system are presented in [31]. Their study shows that female participants prefer touch and voice over gestures. Male subjects prefer gesture over voice. Hinckley and Oviatt evaluate different input modalities based on various aspects like input speed or user feedback possibilities [32, 33]. By using multiple modalities, the system can make up for the shortcomings of other modalities.

3 Methodology

This section describes the various processes we used to develop both the hardware and software components of the bRIGHT workstation and framework.

3.1 Hardware Components and Sensor Platform

Multi-touch and Proximity Detection

We built all the multi-touch surfaces used in bRIGHT based on the technique pioneered by Han [2]. We selected Frustrated Total Internal Reflection (FTIR) as the multi-touch basis because our experiments showed that this method of vision-based multi-touch can be extended to support proximity detection. Being able to approximately detect where the user's hands are along the surface normal of the multi-touch plane is a key requirement of bRIGHT's ability to position controls under a user's hands. Using fairly cheap infrared emitters (near Infrared in the 825-nm to 920-nm range) and widely available machine vision cameras, allowed us to test and confirm the practical viability of this approach.

Figure 2 shows a screenshot of the raw signal processing software developed for this task. As can be seen in the image we stitch the 4 Infrared (IR) camera's inputs into a single video feed, and then apply background removal. After that OpenCV based blob detection is used to detect both touch and proximity. We use two different blob detectors for these tasks. Proximity detection requires that we configure a blob detector to be highly sensitive to IR data with a high value for maximum blob size.

Our approach significantly differs from the technique described in [2] in one significant way. When it comes to how the Infrared emitters are attached to the side of the acrylic surface which forms the main multi-touch area, we made sure that the angle of contact was such that Infrared would not only flood the acrylic but also the area above it. This creates the effect of having an Infrared light field above the multi-touch surface which in turn is useful for proximity detection using the same Infrared sensors that detect touch on the surface. Then we created a second instance of the same blob detection algorithm in OpenCV and configured the parameters to detect the user's hands as they hover close to the multi-touch surface. When configuring the blob detection algorithm we had to setup the parameters to track two fairly large blobs that had very low threshold of Infrared being reflected. This was achieved by increasing the values for amplification, lowering the values for IR threshold, increasing the values for highpass blur and increasing the values for the smallest blob detected to a size similar to a palm.

Our multi-touch software then converts detected blobs into TUIO (Tactile User Input/Output) output. We then developed a windows driver that registers our device as a USB Human Interface Device (HID) with MS Windows operating systems. This allows us to pass on the TUIO output to windows operating systems so they can be dealt with the same way as key strokes or mouse pointer movements. This approach was selected because we wanted the multi touch aspect of our development to be useful for any application within MS Windows. And the TUIO processing is then the same as any other Windows base user interaction handling.

Gaze Tracking

The gaze-tracking system we developed has characteristics that differentiate it from most commercial gaze trackers. We were interested in developing a gaze tracker that support a user's significant head movements; i.e., the head box for the gaze tracker had to be significantly larger in volume than usual. The gaze tracker also had to track the user's gaze on a large TV instead of on a regular PC monitor. In our case, we used a 60" LCD TV (150 cm measured diagonally) as the primary display. This meant that the user was significantly further away (about 115 cm) from the surface on which the system would track gaze movements. (Typical gaze trackers work on PC monitors, which are about 75 cm measured diagonally, and the user is a maximum of about 70 cm away. Our technique is based on active infrared illumination and built on the approach discussed in [2].

Face Recognition

The active illuminated infrared system for gaze tracking also detects and recognizes the user's face. We used the OpenCV library and its support for face detection and recognition for this feature. Face recognition is the primary authentication method in bRIGHT and is how the MMORPG backend server distinguishes individual users.

Speech/Speaker Recognition

We initially incorporated speech recognition support for bRIGHT using SRI's STAR Lab's DynaSpeak technology. In many of the application domains where bRIGHT will be useful (such as Battle Management Command and Control, Intelligence analysis etc), we believe voice commands and speech input will have a significant role to play. But at present our focus is on interaction models associated with the multi touch surface and proximity detection. Therefore speech and speaker recognition features are not being currently utilized or experimented with, but may be used in the future if a need arises.

3.2   Software Components

Application Models and Extensions

Maximizing the benefits of integrated gaze tracking in this workstation required a technique that would allow us to identify and label (mark up) all the on-screen con-structs of an application that may be of interest to a user. To satisfy this requirement, we developed a Semantic Visualization Model (SVM) for an application. The SVM is a collection of markup that captures user-level semantics of on-screen constructs. This is done only for controls or screen components that are of interest and are used frequently. For example, we do not markup controls such as a scroll bar in a user interface.

To experiment with application modeling approaches and ways to capture user's workflows at a highly detailed level, we needed an approach that would record inter-actions without losing user-level operational semantics. To support this requirement, we introduced the concept of a Semantic Interaction Model (SIM) for an application.

The SIM is a collection of markup that describes the operational semantics of possible user interactions while using the application. Refer to [1] for a more in depth discussion of this topic and examples of how this is used in a cyber-security context.

MMORPG-Based Design

For selecting support for multiple users, we based our application model and backend server design on techniques used in MMORPGs such as World of Warcraft. The MMORPG model will allow us to experiment in the future with team dynamics and investigate ways to effectively study collaboration in geographically disparate groups. We collaborated with a commercial game server developer Electrotank (www. electrotank.com) and studied the design of their ElectroServer product, which is used in commercial MMORPG style games. We used this design as the base for the design of bRIGHT's backend server.

The backend server handles every single user's input to bRIGHT systems (shown in Fig. 1[b]). The input gathered from users are instances of the SVM and SIM that pertain to the applications being used by the end user. The server then forwards these updates to the Contextual Model (CM), which attempts to build a model of the user's current context based on these inputs.

Fig. 1. bRIGHT workstation and its main components: (a) multi-touch surface, (b) backend server, (C) gaze tracker, (D) Contextual Model

For every single user, that utilizes a bRIGHT system, the backend server creates an individual instance of that user in the bRIGHT universe, and maintains 2 types of information: (a) pre-cognitive information (information that the bRIGHT workstation has already been exposed to but the user has not, (b) SIM and SVM instances generated by the user engaging with the bRIGHT system. An example of pre-cognitive information might be an email the user has received, but has not yet read, but the bRIGHT system has already read and processed this (since it has access to all the data feeds to the user). In such cases, the information is the pre-cognitive information is evaluated to see if it is relevant to the current user's context. If it is deemed so, then the bRIGHT system can take steps to alert the user to such information.

Contextual Model

The Contextual Model (CM) in bRIGHT, for now, is built per individual user, based on her gaze (SVM instances) and interaction (SIM instances) with the system. The CM applies rules and weights to data elements to which the user has been exposed. Figure 2 shows a sample CM after a user has been performing a few tasks in two cyber security related applications.

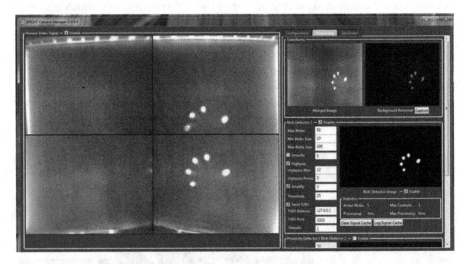

Fig. 2. FTIR based multi-touch processing software developed using OpenCV

The CM can grow extremely large, spanning hundreds to thousands of components due to the volume of information a typical user engages in. The size of an individual node represents how the rules and weights in the CM evaluate the significance of a particular item at this time in the user's context. In Fig. 2 the IP address

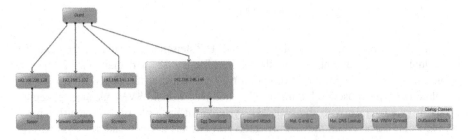

Fig. 3. Sample Contextual Model after a few actions from the user

192.168.248.146 has much more significance in the current context than other items in the CM. Refer to [3] for a more in-depth look at how the CM is built and its impact (Fig. 3).

The contextual model is built primarily based on what we refer to as "contextual indexes". Contextual indexes are references that tie various data elements in a user's context together based on relationships reported due to application characteristics or known properties in the ontological definition of the SIM or SVM instance associated with the data. As can be imagined as a user goes through a multitude of applications and vast amount of information during the course of a regular work day, the contextual index set generated by them tends to be extremely large. This causes the contextual model to be a highly complex map (connected component) that has a vast array of data and attributes being bound together in various relationships reflecting the end user's context.

4   Discussion

4.1   Current Focus of Research

Figure 4 lays out our envisioned technology development road map for bRIGHT. We are currently experimenting with the CM and working on contextual auto-fill.

We are no longer pursuing research in Learning by Demonstration or in developing a bRIGHT desktop. The learning by demonstration approach yielded results that were not flexible enough to support highly dynamic use cases such as cyber security operations, where the same task may be performed in several similar but different ways. We abandoned the experiments on building a desktop that could leverage bRIGHT's features when we realized that the CM should ideally be used as a basis for any future desktop. The lessons learned point us in the direction of a contextual desktop where the user's current context or mental model is the primary interface of the system. Therefore, we have halted the work on incremental desktop design until the CM and its implications and potential are fully realized.

Since the CM is built to reflect the user's context and figure out the significant items of interest within that context, the CM can be used to build a highly accurate auto-fill feature for most applications. This is similar to the auto filling most users experience when they use a web search engine such as Google. In the case of bRIGHT, however, auto fill can be done for any data field in any bRIGHT-enabled application. In Fig. 4 the contextual auto-fill popup (A) appears when the user selects the Get IP Reputation field in the Infected World application, since this field requires the user to input an IP address. Given the user's current context, the CM informs the bRIGHT framework when a user needs an item of type IP address to use this collection of IP addresses and also sends additional information, such as where the user has interacted with these IP address before (From column). If the user has not yet seen this item, it will be marked as pre- cognitive (as in the group with the green check mark). It is possible for bRIGHT to include items that the user has not yet seen; for example, the user receives an email that is not yet open but has been processed by bRIGHT. In such a circumstance, if there is a relevant item that bRIGHT can link to current context (by

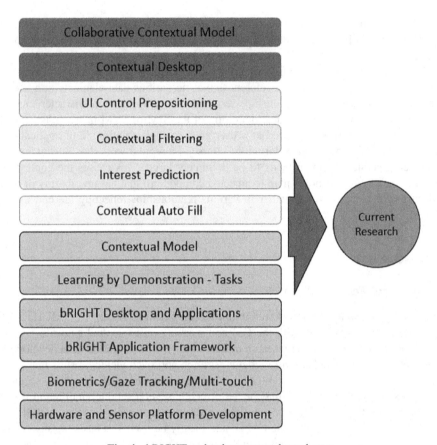

Fig. 4. bRIGHT technology research road map

type or other contextual link), bRIGHT will extract that and place it in the contextual auto-fill popup and mark it as pre-cognitive.

We are also investigating how to improve the semantic markup process. Writing plug-ins to applications or extending their code base to support bRIGHT is a significant hurdle for wider adoption of our approach. Therefore we are at present investigating how to automate this process significantly by developing tools that would allow us to markup an application by linking the semantic interaction model and semantic visualization model constructs to the applications on screen controls and widgets. This reduces the overhead associated with deploying applications on bRIGHT systems and getting them to interact correctly with the backend server and other signal generating components. It also reduces the engineering development overhead associated with adding new tools or software to the platform.

4.2   Conclusion

bRIGHT as an experimental HCI platform that has been an exciting development so far. The features we are experimenting with at present, such as contextual auto-fill and the ability to successfully implement them encourages us to pursue more ambitious research goals such as contextual filtering. Contextual filtering is a great way to manage an end user's cognitive load when they are performing critical tasks. But the risk of such an approach is that if we filter out data that the user needs to solve a problem in the current context, then we will add to the challenges they are facing rather than alleviate them. Whether that can be successfully achieved depends to a great degree on the accuracy of the contextual model we can build and the inferences we can derive from it. At present, in terms of interactions with the system, bRIGHT does not lose any vital information when recording the user's context. There is a degree of uncertainty introduced into our modeling approach due to gaze tracking and how we interpret that in the user's context but we are confident that we can compensate for that and still achieve the level of accuracy we strive for such that the contextual model can be used to increase the user's effectiveness, efficiency and reduce human errors.

References

1. Senanayake, R., Denker, G.: Towards more effective cyber operator interfaces through semantic modeling of user context. In: Nicholson, D. (ed.) Advances in Human Factors in Cybersecurity, vol. 501, pp. 19–31. Springer, Cham (2016). https://doi.org/10.1007/978-3-319-41932-9_3
2. Han, J.Y.: Low-cost multi-touch sensing through frustrated total internal reflection. In: Proceedings of the 18th Annual ACM Symposium on User Interface Software and Technology. ACM (2005)
3. Bulling, A., Gellersen, H.: Toward mobile eye-based human computer interaction. IEEE Pervasive Comput. 9(4), 8–12 (2010)
4. Rantanen, V., Vanhala, T., Tuisku, O., Niemenlehto, P.-H., Verho, J., Surakka, V., Juhola, M., Lekkala, J.: A wearable, wireless gaze tracker with integrated selection command source for human-computer interaction. IEEE Trans. Inf Technol. Biomed. 15(5), 795–801 (2011)
5. Corcoran, P.M., Nanu, F., Petrescu, S., Bigioi, P.: Real-time eye gaze tracking for gaming design and consumer electronics systems, pp. 347–355 (2012)
6. Bolmont, M., Cacioppo, J.T., Cacioppo, S.: Love is in the gaze: an eye-tracking study of love and sexual desire. Psychol. Sci. 25, 1748–1756 (2014)
7. Judd, T., Ehinger, K., Torralba, A.: Learning to predict where humans look. In: ICCV, pp. 2106–2113 (2009)
8. Senju, A., Johnson, M.H.: The eye contact effect: mechanisms and development. Trends Cogn. Sci. 13(3), 127–134 (2009)
9. Tylén, K., Allen, M., Hunter, B.K., Roepstorff, A.: Interaction vs. observation: distinctive modes of social cognition in human brain and behavior? A combined fMRI and eye-tracking study. Front. Hum. Neurosci. 6, 331 (2012)
10. Kochukhova, O., Gredeba, G.: Preverbal infants anticipate that food will be brought to the mouth: an eye tracking study of manual feeding and flying spoons. Child Dev. 81(6), 1729–1738 (2010)

11. Kano, F., Tomonaga, M.: How chimpanzees look at pictures: a comparative eye-tracking study. Proc. Biol. Sci. **276**(1664), 1949–1955 (2009)
12. Bergasa, L.M., Nuevo, J., Sotelo, M.A., Barea, R., Lopez, M.E.: Real-time system for monitoring driver vigilance. IEEE Trans. Intell. Transp. Syst. **7**(1), 63–77 (2006)
13. Qi, Y., Wang, Z., Huang, Y.: A non-contact eye-gaze tracking system for human computer interaction. In: 2007 International Conference on Wavelet Analysis Pattern Recognition, pp. 68–72, November 2007
14. Hennessey, C., Lawrence, P.: Noncontact binocular eye-gaze tracking for point-of-gaze estimation in three dimensions. IEEE Trans. Biomed. Eng. **56**(3), 790–799 (2009)
15. Iqbal, N., Lee, H., Lee, S.-Y.: Smart user interface for mobile consumer devices using model-based eye-gaze estimation. IEEE Trans. Consum. Electron. **59**(1), 161–166 (2013)
16. Guestrin, E.D., Eizenman, M.: General theory of remote gaze estimation using the pupil center and corneal reflections. IEEE Trans. Biomed. Eng. **53**(6), 1124–1133 (2006)
17. Hennessey, C., Noureddin, B., Lawrence, P.: Fixation precision in high-speed noncontact eye-gaze tracking. IEEE Trans. Syst. Man. Cybern. Part B (Cybern.) **38**(2), 289–298 (2008)
18. Nawaz, T., Mian, M., Habib, H.: Infotainment devices control by eye gaze and gesture recognition fusion. IEEE Trans. Consum. Electron. **54**(2), 277–282 (2008)
19. Asteriadis, S., Karpouzis, K., Kollias, S.: Visual focus of attention in non-calibrated environments using gaze estimation. Int. J. Comput. Vis. **107**(3), 293–316 (2013)
20. Zhu, Z., Ji, Q.: Novel eye gaze tracking techniques under natural head movement. IEEE Trans. Biomed. Eng. **54**(12), 2246–2260 (2007)
21. Xia, D., Ruan, Z.: IR image based eye gaze estimation. In: Eighth ACIS International Conference Software Engineering Artificial Intelligence Networking, Parallel/Distributed Computing (SNPD 2007), vol. 1, pp. 220–224, July 2007
22. Nguyen, Q.X., Jo, S.: Electric wheelchair control using head pose free eye-gaze tracker. Electron. Lett. **48**(13), 750 (2012)
23. Rae, J.P., Steptoe, W., Roberts, D.J.: Some implications of eye gaze behavior and perception for the design of immersive telecommunication systems. In: 2011 IEEE/ACM 15th International Symposium on Distributed Simulation Real Time Applications, pp. 108–114, September 2011
24. Panwar, P., Sarcar, S., Samanta, D.: EyeBoard: a fast and accurate eye gaze-based text entry system. In: 2012 4th International Conference on Intelligent Human Computer Interaction, pp. 1–8, December 2012
25. Ji, Q., Zhu, Z.: Eye and gaze tracking for interactive graphic display. In: Proceedings of the 2nd International Symposium on Smart Graphics. ACM (2002)
26. Dahlback, N., Jönsson, A., Ahrenberg, L.: Wizard of Oz studies - why and how. Knowl.-Based Syst. **6**(4), 258–266 (1993)
27. Cohen, P.R., Johnston, M., McGee, D., Oviatt, S.: The efficiency of multimodal interaction: a case study. In: International Conference on Spoken Language Processing (ICSLP), Sydney, Australia (1998)
28. Oviatt, S., Lunsford, R., Coulston, R.: Individual differences in multimodal integration patterns: what are they and why do they exist? In: Conference on Human Factors in Computing Systems (CHI), New York, USA (2005)
29. Ruiz, N., Chen, F., Oviatt, S.: Multimodal input. In: Thiran, J.-P., Marques, F., Bourlard, H. (eds.) Multimodal Signal Processing, pp. 231–255 (2010). Chapter 12
30. Turk, M.: Multimodal interaction: a review. Pattern Recogn. Lett. **36**, 189–195 (2014)
31. Perzylo, A., Somani, N., Profanter, S., Rickert, M., Knoll, A.: Toward efficient robot teach-in and semantic process descriptions for small lot sizes. In: Robotics: Science and Systems (RSS), Workshop on Combining AI Reasoning and Cognitive Science with Robotics, Rome, Italy (2015)

32. Akan, B., Ameri, A., Cürüklü, B., Asplund, L.: Intuitive industrial robot programming through incremental multimodal language and augmented reality. In: International Conference on Robotics and Automation (ICRA), Shanghai, China (2011)
33. Stenmark, M., Nugues, P.: Natural language programming of industrial robots. In: International Symposium on Robotics (ISR), Seoul, Korea (2013)
34. Stenmark, M., Malec, J.: Describing constraint-based assembly tasks in unstructured natural language. In: World Congress of the International Federation of Automatic Control (IFAC) (2014)

Development of Frame for SPIDAR Tablet on Windows and Evaluation of System-Presented Geographical Information

Yuki Tasaka[1(✉)], Kazukiyo Yamada[1], Yasuna Kubo[1],
Masanobu Saeki[1], Sakae Yamamoto[1], Takehiko Yamaguchi[2],
Makoto Sato[3], and Tetsuya Harada[1]

[1] Tokyo University of Science, 6-3-1, Niijuku, Katsushika, Tokyo, Japan
8114063@ed.tus.ac.jp
[2] Tokyo University of Science, Suwa, 5000-1, Toyohira, Chino, Nagano, Japan
[3] Tokyo Metropolitan University, 6-6 Asahigaoka, Hino, Tokyo, Japan

Abstract. When viewing a map, we understand the terrain by the symbols marking the buildings, roads, and landmarks. However, these pieces of information are in planar form. The original road has a slope and is irregular in shape. In the event of disaster, the ways in which people can safely evacuate must be carefully considered, so terrain characteristics must be well understood. In this study, we use not only visual information but also information from other senses. To present information for the other senses, we used a force-sense presentation device designed for tablet PCs, known as the SPIDAR tablet. We developed an application that can display maps on the tablet screen and present sensory information regarding the slope when the user traces the road on the map with a finger. Then, we evaluated the amount of road information that can be understood and which sensory presentation was most effective. The subjects of the evaluation were adults and children who completed a questionnaire regarding their degree of comprehension. The children participants were third graders at the Aijitsu elementary school. The results of the questionnaire reveal noticeable differences in the comprehension of adults and children with respect to sensory information presented. Based on this result, we plan to present more helpful information in future work. Moreover, we identified the need to thoroughly consider the modality of the sensory information presented.

Keywords: Haptic device · Map · Presenting geographical information
Understanding geographical information · Force sense · Auditory sense
Visual sense

1 Introduction

1.1 Background

When travelling to a destination, we look at a map, determine which route to proceed along, and proceed to the destination using a route with the shortest possible distance to that destination. Decisions regarding the route are based on the roads, intersections, and landmarks shown on the map. However, in the event of a disaster, evacuation routes

© Springer International Publishing AG, part of Springer Nature 2018
S. Yamamoto and H. Mori (Eds.): HIMI 2018, LNCS 10904, pp. 358–368, 2018.
https://doi.org/10.1007/978-3-319-92043-6_30

must be selected based on their being safe and efficient. Therefore, it is important to understand the road conditions at each location. Furthermore, road information must be provided about whether they are physically difficult to use and situations that are evolving in the event of secondary disasters. For example, elderly people should avoid roads with major disruptions even if they offer the shortest evacuation routes. It is possible to provide guidance for optimum evacuation decisions as follows:

(1) Optimum route search based on geographical features
(2) Presentation method providing easily understandable information

In this research, we focused on the second factor: the information presentation method. Buildings, roads, and landmarks are represented on a planar map, but, to understand road conditions such as road height differences, maps are somewhat limited. Therefore, to better convey geographical information, in this study we tested the possibility of offering sensory information and focused on the provision of geographic altitude information, which can be obtained from contour lines. Although contour lines are taught as part of elementary school education, they are presented only as a way to distinguish between mountains and lowlands. That is, urban altitudes would not be expressed using contour lines. We propose the use of sensory information by physically touching the map to identify road height differences that are typically perceived by the five senses. As our presentation method, we present geographical information on a SPIDAR tablet, a force-sense presentation device for tablet PCs described in the next section. Today, tablet PCs are also widely adopted in the education field. In this work, we focused on fundamental research for the SPIDAR tablet using a tablet PC.

1.2   SPIDAR-Tablet

In recent years, the usage rates of portable devices, such as tablet PCs and smartphones, have skyrocketed. Regarding this popularization trend, Tamura et al. [1] conducted a study using a SPIDAR tablet with a tablet PC connected to a haptic device, the SPIDAR mouse, which adds a sense of force to the PC mouse operation. The SPIDAR tablet is a device that enables users to experience on their finger a force sense on a two-dimensional plane parallel to the screen when touching the screen and operating the tablet PC. Force sense is achieved by controlling a motor and winding a thread around the pulley of the motor. Figure 1 shows a SPIDAR mouse and Fig. 2 shows a SPIDAR tablet.

The frame consists of a motor, thread, pulley for winding the thread, a surface fastener for attaching a finger, and a control circuit using a peripheral interface controller (PIC). A pulley is attached to the shaft of the motor and the thread is wound around the pulley.

However, since the SPIDAR tablet itself is a device that is not portable, it cannot take full advantage of the features of portable tablet terminals. Therefore, a new SPIDAR tablet frame must be developed so that the SPIDAR tablet can be carried around and used.

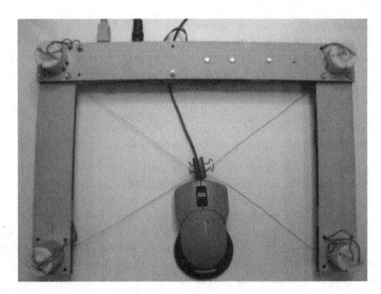

Fig. 1. SPIDAR mouse

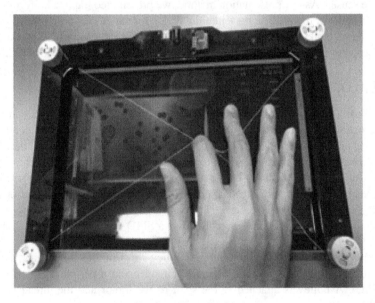

Fig. 2. SPIDAR tablet

1.3   Purpose

We developed a frame that makes the SPIDAR tablet portable. Furthermore, to convey height differences, we developed a map application that presents sensory stimuli on the

SPIDAR tablet. Using this novel map application, the user's understanding of altitude information increases due to the addition sensory stimulus on the map.

2   Proposed Device and Application Configuration

2.1   Configuration of Proposed Device

We created a new frame for the SPIDAR tablet that is portable, which was not possible with the conventional SPIDAR tablet. To do so, we combined the tablet PC and frame of the new SPIDAR tablet. In addition, we wanted user evaluations from third–grade students, so it was necessary to consider material and shape to ensure safety and durability. It was also necessary to devise a design that allows the tablet PCs and electronic devices to be easily removed from the frame when the proposed device fails.

For the frame material, we used ABS HG [2]. For the tablet PC, we constructed a frame suitable for the size of the device, based on the Surface Pro 4 made by Microsoft Corporation. Figure 3 shows the manufactured frame of the SPIDAR tablet.

Fig. 3. Manufactured frame of the SPIDAR tablet

We designed the motor, the PIC board, and the connectors connecting the leads to be hidden to prevent them from being easily touched. We then placed this frame on a commercially available tablet PC cover, as shown in Fig. 4, and sandwiched the frame between the tablet PC and the cover to integrate the SPIDAR-tablet frame and the tablet PC. For the tablet PC cover, we used a PDA-TAB 13 manufactured by Sanwa Supply Co., Ltd. Figure 5 shows the integration of these parts.

Fig. 4. Commercially available tablet PC cover

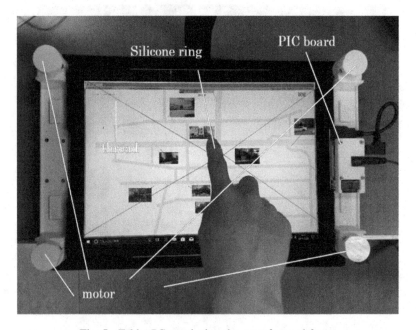

Fig. 5. Tablet PC attached to the manufactured frame

When operating the SPIDAR tablet, the screen is touched by the index finger, so next, we considered the versatility of the ring attached to the finger with respect to the finger thickness and decided to use a silicon ring. For the power supply to the motor, we used a mobile battery "DE-M 04 L-3015" manufactured by ELECOM Co., Ltd., as shown in Fig. 6. Table 1 lists the specifications of the mobile battery.

Fig. 6. Mobile battery

Table 1. Mobile battery specifications

Output current [A]	1.5
Output voltage [V]	5.0
Rated battery capacity [mAh]	3000

Table 2 shows the specifications of the motor we used, which was a motor with brush RF-300FA-12350 [3] manufactured by Mabuchi Motor Co., Ltd.

Table 2. Motor specifications

Rated voltage [V]	3
Rated torque [mNm]	0.48
Stall torque [mNm]	2.51
Weight [g]	22

2.2 Application Configuration

We developed a map application that indicates elevation height differences in the force, visual, and auditory senses by tracing with a finger a road on the map displayed on the tablet PC screen. With respect to the cross-platform development and rendering, we used the Unity comprehensive development environment on this application. Figure 7 shows the information flow during the operations of the SPIDAR tablet and map application.

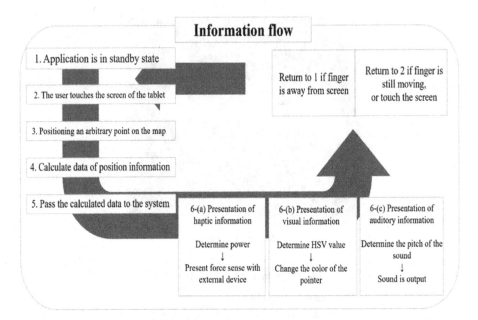

Fig. 7. Information flow between application and device

2.2.1   System for Calculating Sensory Information Output

(1) Force sense expression

To express elevation height differences, a downward force of the slope is always presented to the user's finger. In this way, the difficulty of climbing an uphill slope and relative ease of descending a downhill slope is conveyed. This force is expressed by Eq. (2). For altitude data, we used the numerical map with 5-m mesh data issued by Geospatial Information Authority of Japan [5]. Based on this data, we created nodes at intersections and corners, as well as in the middle of every road five meters wide or more. We also incorporated the node graph data in the application.

$$\theta_i = \tan^{-1} \frac{h_i}{d_i} \tag{1}$$

$$\vec{F_i} = \vec{R_i} \theta_i^2 \alpha \tag{2}$$

$$d_i \leq 5 \qquad i \in N \tag{3}$$

θ_i: Inclination angle at local point [rad]
d_i: Actual distance between two nodes forming the road of the local point [m]
h_i: Difference in elevation between two nodes forming the road of the local point[m]

$\overrightarrow{R_i}$:   Unit vector from the higher to the lower elevation, between two nodes forming the road of the local point

$\overrightarrow{F_i}$:   Force value at an arbitrary point [N]

α: Parameters for determining force

N: Natural number indicating the route number on a map

i: Route number of node graph

(2) Auditory sense expression

By changing the pitch of the sound with altitude, geographical height differences can be expressed. For the sound source, we used a copyright-free wind sound [6]. We conveyed geographical height differences by the pitch of the wind sound, as expressed by Eq. (4).

$$f = \frac{f_{max} - f_{min}}{H_{max} - H_{min}} h - \frac{f_{max} H_{min} - f_{min} H_{max}}{H_{max} - H_{min}} \tag{4}$$

f: Frequency representing the pitch of the sound source to be output [Hz]

h: Altitude of local point [m]

f_{max}, f_{min}:   Maximum and minimum values of the audible range of the frequency representing the pitch of the sound to be output [Hz]

H_{max}, H_{min}:   Maximum and minimum elevations at all points of the displayed map information.

(3) Visual sense information

By changing the color of the pointer indicating the position of the user's finger, we conveyed elevation height differences. To change the color of the pointer, we changed the V value (0 to 100) of the HSV value (H: hue, S: saturation, V: lightness) based on altitude. The color of the pointer changes in gradations between white and black. The changing V value is expressed by Eq. (5).

$$V = \frac{100}{H_{max} - H_{min}} h - \frac{100 H_{min}}{H_{max} - H_{min}} \tag{5}$$

V: Changing V value

h: Altitude of local point [m]

H_{max}, H_{min}:   Maximum and minimum elevations at all points of the displayed map information

3 User Evaluation Experiment

3.1 Experimental Method

We asked study participants to follow a road in the map application using their dominant index finger. Then, we presented the sensation expressing a slope using the following procedure.

(1) Only force sense
(2) Force sense and the sound of wind
(3) Force sense and changing the pointer color (gradation between white and black)
(4) Force sense, the sound of wind, and changing the pointer color

Regarding steps (1)–(4), we conducted questionnaires in which users were asked to verbally rate on a five–point scale the extent to which they could understand the information about the slope. The five-point evaluation is shown below:

5: I understood very well.
4: I understood well.
3: I understood.
2: I understood a little.
1: I didn't understand.

The subject included 10 adults (22 to 25 years old) and 45 children (8 and 9 years old). The children were elementary-school third graders at the Aijitsu elementary school in Shinjuku, Tokyo. This school allotted time for VR lessons for its students.

3.2 Experimental Results and Discussion

In the evaluations obtained from the adults and children, there is differences in how people feel, such as force sense. Because of this, the obtained results were normalized and expressed as a proportional scale. The average values of standardized ones are shown in Table 3. Similarly, for children among men and women, average values of standardized ones are shown in Table 4. The higher the value, the higher the degree of comprehension, and the lower the value, the lower the degree of comprehension.

Table 3. Results of questionnaire for all users

User	Sensation			
	Force sense	Force sense and sound	Force sense and changing the color of pointer	Force sense, sound and changing the color of pointer
Children	−0.022	−0.11	−0.17	0.64
Adults	1.14	−0.12	−0.73	−0.59

Table 4. Questionnaire results of children for boys and girls

User	Sensation			
	Force sense	Force sense and sound	Force sense and changing the color of pointer	Force sense, sound and changing the color of pointer
Boys	0.10	−0.20	−0.09	0.41
Girls	−0.24	0.05	−0.31	1.04

In Table 3, we can see that when only force sense was presented, adults could gain a better understanding than children. However, when the force sense and changing the pointer color were presented, children could understand better than adults. When force sense, sound, and changing the pointer color were all presented, the results was the same. We discuss this finding below.

First, when trying to understand something, people recognize and consider irritation, which is a sensation hinting about something. It is thought that human understanding will deepen based on various experiences and thoughts as one becomes an adult. However, since children have had fewer experiences and thoughts than adults, it may be more difficult to understand some information. As such, adults and children have different levels of experience in understanding things.

Regarding changing the pointer color, it is not obvious that shades of white and black would correspond to elevation height differences. Because of this, it seems that it was difficult to understand for adults. Therefore, regarding the experience of moving up or down a slope, we found that when only force sense was presented, adults found this to be the most intuitive and easy to understand.

From Table 4, we can consider the differences in perception of sensory presentations between boys and girls. In a study conducted by 6. Toi [6], the authors studied a combination of visual–tactile stimulus and visual–auditory stimulation modalities for sensory presentation. They found women to be more aware of auditory than tactile information, and men to be more awareness of tactile than auditory information. Therefore, in our study, boys more easily recognized uphill and downhill slopes depending on the force sense, whereas girls were aware of the uphill and downhill slopes depending on the sound of the wind. So, our results differed by gender. When force sense and changing the pointer color were both presented, the degree of comprehension was negative for both boys and girls. Therefore, we can conclude that both sexes have difficulty identifying uphill and downhill slopes by a change in the pointer color. Finally, when force sense, the sound of wind, and changing the pointer color were all presented, we found the decision about uphill and downhill slopes to depend on the wind sound rather than the pointer color change.

Also, as visual information was not included in the experiment, we asked subjects whether altitude can be recognized by the illustration shown in Fig. 8 regarding uphill and downhill slopes. Regardless of age, most subjects could easily understand this geographical information.

Fig. 8. Person going up and down a slope

4   Conclusions and Future Prospects

In this research, we designed a portable frame for the SPIDAR tablet and combined the frame with a tablet PC. Our questionnaire results reveal that the proposed system promotes understanding of geographical map information by presenting information regarding slope by the use of force, auditory, and visual senses.

However, we found there to be a difference in people's sensory perceptions. Therefore, we propose the need for a system for users of the SPIDAR tablet that accommodates these individual differences to the degree possible. Regarding the force sense, the optimal size of the presented force must be determined through preliminary experiments.

Also, the sensory stimuli considered in this study were the force, visual, and auditory senses. In the future, we hope to enable the search for the shortest map route with respect to geographical features by adding other sensory information, which will be useful for disaster prevention and evacuation route searches.

References

1. Tamura, A., Murayama, J., Hirata, Y., Sato, M., Harada, T.: Development of haptic device for touch panel-SPIDAR-tablet-and its haptic computational methods. Virtual Real. Soc. Jpn. **16** (3), 363–366 (2011)
2. MUTOH, 3D printer. https://www.mutoh.co.jp/products/MagiX/supply_mf.html. Accessed 25 Jan 2018
3. Mabuchi Motor. https://product.mabuchi-motor.co.jp/detail.html?id=74. Accessed 25 Jan 2018
4. Geographical Survey Institute download service. https://fgd.gsi.go.jp/download/menu.php. Accessed 25 Jan 2018
5. Hurt record. http://www.hurtrecord.com/. Accessed 25 Jan 2018
6. Toi, C.: Visual-auditory stimulus during out-of-body illusion can induce the sense of self as well as visual-tactile stimulus. J. Grad. Sch. Hum. Sci. **19**, 215–224 (2017)

Information in Virtual and Augmented Reality

The Lessons of Google Glass: Aligning Key Benefits and Sociability

Leo Kim[(⊠)]

Ars Praxia, Seoul Finance Center 21F, Seoul, South Korea
Leo_kim@arspraxia.com

Abstract. This article presents a case study of the user experience of Google Glass when it was initially introduced in 2013. By applying the combined methods of on-line data research, semantic network analysis and field research, it is argued that awkwardness of form factor and use, and failures of Google Glass's user interface explain the low acceptability of the device. From a methodological perspective that combines big data analysis and qualitative research, this article discusses the user needs and preferences that should inform development of new tech.

Keywords: Google Glass · Big data research · Semantic network analysis
UX

1 Introduction

Predicting what kind of new device will be accepted in the market is a difficult task. The difficulty is compounded when a novel type of technology and function is introduced. Google Glass, introduced and initially distributed to a limited number of tech users in 2013, was the typical case. Even though there was public concern that the attached video camera could cause privacy issues (Guardian, 6 March 2013), neither technicians nor market experts at the time knew if social concerns could hamper its proliferation in the market. On the other hand, there were reasons to imagine that new functions, with its technological charm, could deliver unprecedented benefits to the users (MIT Technology Review, 1 July 2013).

This study was conducted as market research in 2013 to investigate the key factors that might determine user acceptance. Although the privacy concern appeared as a salient one, I argue there were subtler aspects that hindered the product's success. As user experience (UX) is 'dynamic and continually modified over time in response to changing circumstances and innovations' [1], the researcher should look into what kind of invariability might exist in accordance to the changing circumstances and innovations.

This research demonstrates that the low acceptability of Google Glass was in part affected by sociological factors relating to sociability and control, and the social-psychological ego-alter relationship. By utilizing on-line data research and text-mining methods to extract key variables (concepts), this article also attempts to present a process to link them with existing qualitative methods.

© Springer International Publishing AG, part of Springer Nature 2018
S. Yamamoto and H. Mori (Eds.): HIMI 2018, LNCS 10904, pp. 371–380, 2018.
https://doi.org/10.1007/978-3-319-92043-6_31

2   Methodology

2.1   On-line Data Research

The research began with the collection of data from Twitter. In 2013, Twitter was one of the most widely used social media services where users communicated on a variety of issues. To concentrate on posts that gained a certain degree of agreement and/or resonance, the researcher applied the search word "google glass" in English and collected 758 original posts that were retweeted at least 10 times between 6th and 21st May 2013. The data was manually tagged in 4 categories: positive, neutral, negative and informative; categories are further defined by related keywords (Fig. 1).

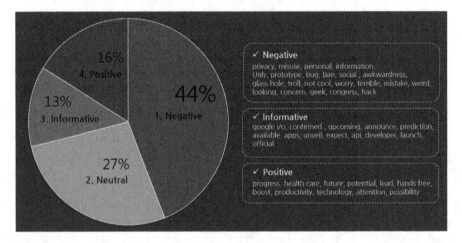

Fig. 1. Opinion of Google Glass in Twitter (Original postings between 6–21 May 2013)

While the contents tagged as "neutral" were slight (insignificant) expressions that contained no value judgement, those tagged "informative" were mostly linked news or blog contents that reported the announcement of Google Glass and experts' general comments. By collecting these data sets, the researcher can assess overall on-line sentiment regarding the device, but also parse responses to specific aspects of the technology.

Keywords for negative contents include "privacy", "misuse", "personal", "information", "ugly", "prototype", "bug", "ban", "social", "awkwardness", "glasshole", "troll", "not cool", "worry", "terrible", "mistake", "weird", "looking", "concern", "geek", etc. For the positive contents, expressions like "progress", "health care", "future", "potential", "lead", "hands free", "boost", "productivity", "technology", "attention", and "possibility" were frequently used.

2.2   Qualitative Research

Structure of Twitter Contents

Based on the grounded theory approach [2], the qualitative approach adopted here reflects an effort to derive meaningful categories from the acquired data. The four categories of Twitter content were more specifically classified along their converging themes. As shown in Fig. 2, the themes represent the additional layer of (1) privacy and sociability, (2) fashion and design, (3) function and technology (data processing and desired features), (4) UI and UX (user interface and user preference), and (5) market (price and acceptability, applications and ecosystem). Among them, the three key positive themes include "new value and capability", "practical advantage", and "potential of Google Glass" and the four negative themes containing "privacy concerns", "deficiency as fashion item (or accessory)", "overall concern about emerging technology", and "difficulty in actual usage"; these require further scrutiny.

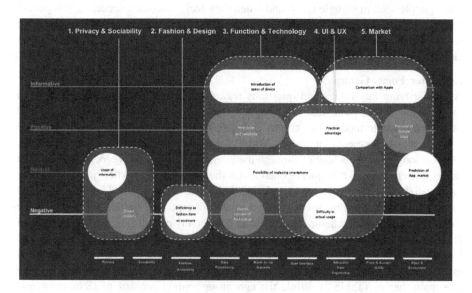

Fig. 2. Emerging thematic categories with sentiments

In-depth Interview of Experts

After deriving the five key themes, my research team conducted in-depth interviews with experts relevant in each theme (Table 1). The main purpose of the interview was to grasp guiding concepts that were important in understanding privacy and sociability, fashion and design, function and technology of Google Glass, related UI and UX, and market forecast. Those guiding concepts in turn clarified what questions to pose for the users and potential consumers.

For example, the interview with Dr. Judith Donath (Director of social media group in MIT Media Lab) clarified that public concern about privacy was more significantly related to the "fear of losing control of information" rather than the exposure of

Table 1. Interviewed experts

Domain	Name	Affiliation
Privacy & sociability	Judith Donath	MIT Media Lab
Fashion & design	Jeff Salazar	Lunar Design
Function & technology	Danny Roa	Wedding Party co.
UI & UX	David Witt	Symphony Teleca co.
Market	Jaekwon Sohn	Maeil Economy Newspaper

him/herself to the data space such as social media. Jeff Salazar (vice president of Lunar Design) pointed out there was difference between stylish object and fashionable item: if the former is judged by experts with respect to the product's design, the latter is recognized publicly by social value. These kinds of comments from the domain experts lead the research team to develop specific questions such as 'what kind of controllability people seek in Google Glass and when they feel secure or insecure?', 'Do people feel that Google Glass is fashionable? If not, why?' This kind of specific questioning was applied in field research: interviews with lead users and frequent observers of Google Glass in the Bay Area of San Francisco.

Ego-Alter Focus Groups

After clarifying key issues and questions regarding Google Glass, the research team recruited a lead user group and an observer group each composed of 6 people. The lead user group was termed "ego" group, and the observer group "alter". Following the logic of social representation theory [3], it is supposed that a common opinion of an object (Google Glass) is formed through the interaction of the user (ego, or self) and the others (alter) who make frequent contact with the user. The Ego-Alter 'jointly generate their social reality - objects of knowledge, beliefs or images' [4]. Unlike a common market segmentation, this social-psychological concept was utilized to build up the focus group interview scheme. As there was not yet any established market segment of the new device, catching early signals of social acceptability of a new technology and its usage required a more social interaction-oriented design of interview. The interview questions were semi-structured and posed separately to the two groups regarding the five main themes (Table 2). While the ego group mainly consists of technicians and app developers, the alter group included various kind of lay people like the ego group member's family, friend, colleague, restaurant owner, journalist, colleague, etc.

Semantic Network Analysis of Transcription

After gathering transcripts from the focus groups, a fully automated method of semantic network analysis was utilized to derive the structural pattern of both ego and alter group's key needs and concerns. Although the adoption of semantic network analysis is not new to UX research [5, 6], there is still no consensus as to what kind of features to focus on when interpreting the key relational concepts. A lack of theoretical underpinning makes meaningful interpretation difficult.

Literature in communication research [7, 8] offers theoretical frameworks to visualize human desires and concerns, in terms of the connectivity of expressed concepts, and helps to operationalize content analysis through semantic network analysis.

Table 2. Ego-alter focus group interviews

Domain	Name	Affiliation
Privacy & sociability	Ego	• Were there any concerns that Google Glass might affect others' privacy before purchasing it? • How have people responded when you wear Google Glass? • How did you feel and what was your reaction? • What is the awkward situation that hampers social interaction? • If you were to keep using a glass type device that might not necessarily be a Google Glass, how could the product be better designed to improve social acceptance?
	Alter	• How did you feel about people using Google Glass around you? What is the main reason? • Do you feel that you want to use some wearable device such as Google Glass? If so, why? If not, why not? • How have you reacted to people using Google Glass? • Do you think Google Glass will become acceptable to the majority of society? What could be main issues and how they might be addressed? • If some new glass type device comes out and it does not look like Google Glass, how would you imagine the product could be better designed, to improve social acceptance?
Fashion & design	Ego	• How do you evaluate the design of Google Glass? How do you think it can be improved? • If some new glass type device comes out and it does not look like Google Glass, how would you imagine the product could be better designed?
	Alter	• How do you evaluate the design of Google Glass? How do you think it can be improved? • If some new glass type device comes out and it does not look like Google Glass, how would you imagine the product could be better designed?
Function & technology + UI & UX	Ego	• From a technical perspective, can you tell your experience using Google Glass? • What are the good things and bad things? • Based on your experience, how do you feel the device can be improved?
	Alter	• Is there some potential downside when interacting with a friend using Google Glass? • Based on your experience, how do you think the device can be improved to ease your interaction with a counterpart using a Glass type device?
Consumer expectation (Market)		• What do you believe future Google Glass should improve and provide more benefits?

(continued)

Table 2. (*continued*)

Domain	Name	Affiliation
		• If a new form of glass type device comes out, what do you imagine it would look like and how could it provide better satisfaction? Would you be willing to purchase it?
		• What do you believe future Google Glass should improve and provide more benefits? • If a new form of glass type device comes out, what do you imagine it would look like and how could it provide better satisfaction to a potential consumer or an audience?

If classical content analysis [9] focuses on the frequency of keywords in the text, semantic network analysis elucidates the relation of the keywords based on a co-occurrence matrix [10]. In this research, we utilized a commercial semantic network analysis tool Optimind (http://arspraxia.com/product) that automatically codes text, extracts core features, and visualizes them as a network. The procedure of analysis goes through:

1. Preprocessing:
   a. Checking words and listing context-specific thesauri of synonyms
   b. Automatic lemmatization of variable words (transformation into basic form) based on the English natural language processing (NLP) library and system
   c. Automatic deletion of functional words such as articles and adverbs
2. Processing:
   a. Transformation of the remaining text into an adjacency matrix of keywords, with the window size of every paragraph
   b. Applying a backbone extraction model threshold that extracts a core set of substantive keywords falling around 200
3. Visualization and Interpretation:
   a. Visualization of a network focusing on the adjacent keywords of the key concept, using the k-neighbor algorithm
   b. Interpretation of the represented semantic network.

3   Results of Semantic Network Analysis

The transcribed focus group interviews were categorized into two parts: (1) current evaluation of Google Glass and (2) future market expectation. Figure 3 below depicts the extracted frames of ego users and alter observers.

For the ego part (top half of figure), the most important frame consists of 3 keywords: "information", "interesting" and "social". Linked words emerge after passing a backbone extraction model's threshold [11] and are extracted by the k-neighbor [12] algorithm to represent directly connected words. If the users are interested ("interesting") in the new device, key binding values turn out to be "information" and sociability ("social" attached to situations and interactions). In particular, information should be

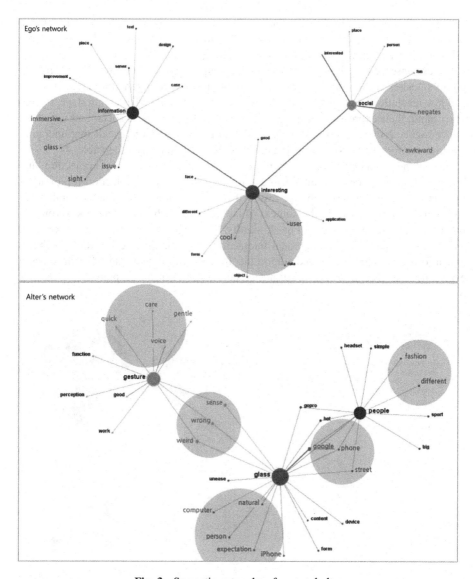

Fig. 3. Semantic networks of ego and alter

delivered with a key "issue" within the limited screen size and low resolution of Google Glass, and be "immersive". However, if social interactions are made "awkward", this "negates" the intended function and value of the technology.

While the ego group (users) point to the problem of social interaction, alters clarify the specifics of the problem: the wrongness of gesture while using Google Glass. Keywords related to preferred social gestures are: "quick", "gentle", "care", and "voice" while gestures elicited by Google Glass are considered "weird" and "wrong". Users think the introduction of Google Glass is as natural as the adoption of a "computer" or an "i-phone". Nevertheless, the new device should offer a distinct value from smartphones ("phone") in the "street", and Google Glass can be fashionable ("fashion") only if it can address people's "different" tastes.

These observed needs of the ego and alter group elucidate their differences. The ego group's desires are focused on curated information, considering the limited screen size and low resolution of the device, that is adapted to the Google Glass's interface. On the other hand, their expressions that Google Glass is in some way a-social and weird are rather vague. Semantic network analysis of alter group's opinion reveals that operating gestures are judged important in making Google Glass socially acceptable.

From the perspective of social representation theory, one may conjecture that interaction between ego and alter will eventually form a stable projection (opinion) of Google Glass. In this research, the result of the interaction is analyzed separately by each ego and alter group. In the context of the individual, the categorized keywords in relation to gesture, depicted in Fig. 4, signify that finger swiping is often inaccurate, and seems "wrong" for interpersonal interaction. A minimalist design for operability and information feed is deemed important. In a social context, the blinking gesture especially was thought to be weird, not graceful, and even unacceptable.

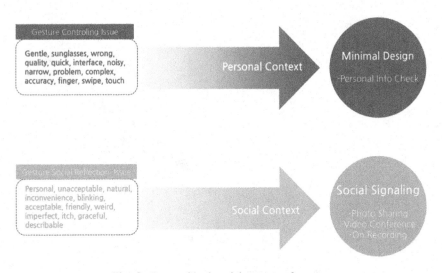

Fig. 4. Personal and social context of gesture

Beyond the evaluation reported here, there were expectations of the marketability of Google Glass for various usages in the future from the focus groups. Figure 5 demonstrates that the future market is closely related to the novel usage of data. If the non-user group of alter (left side of the figure) mainly expect health related utilization of data (e.g. blood, health, calorie), the existing user group of alter (right side of the figure) expect new services for sharable vision, photography, object recognition, and entertainment.

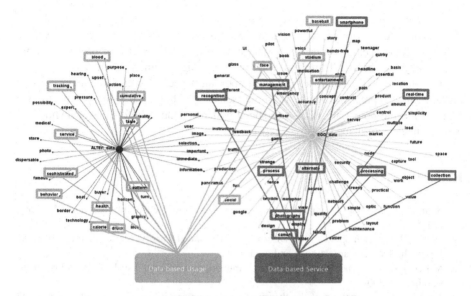

Fig. 5. Data-related market: ego and alter

While researching potential consumers' thoughts in the Bay Area, our research team also interviewed one of the developers of Google Glass to compare our results with the provider. The developer said that Google's team was fully aware of complex social issues that would determine Google Glass's success in the market. Therefore, the company had invested in a sizable expenditure to hire a market research agency. When asked what kind of killer usage would cause Google Glass to proliferate, the developer replied that providing services for music would make users more hands-free, mentioning people riding bicycles while wearing Google Glasses and listening to music, and this will soon create a boom in the market, according to the marketing research results (Interview with Mr. Ryan (alias) on 15 May 2013). This expectation was contradictory to our research outcomes; the provider did not seriously consider the issue of Google Glass being a-social, and did not know what part of UX was truly at stake.

4   Conclusion

Our research results were consistent with what had actually happened with Google Glass. When Google's social network service, Google Plus, received criticisms, Sergey Brin, the co-founder of Google, admitted that he was "not a very social person" (The Verge, 28 May 2014). Perhaps this scant attention to address social expectations and needs led to the failure of Google Glass. Google has practically ended sales of the device as of 16 January 2015.

My proposed research procedure: on-line data analysis, expert interview, focus group interview with ego and alter groups, and finally application of semantic network analysis, proved useful in deriving key UX issues, clarifying users' reservations about

Google Glass, especially that they deemed it to be socially jarring. The social psychological concepts of ego and alter and the study of their interaction, reflected by each groups's separate expressions of experiences, seems to be useful in projecting what kind of key variables will turn out to be important in the emerging market.

The study of UX should reflect the important aspect of social interaction. Using methods that represent sociological and social psychological concepts in market research has the potential to embrace social dynamics and therefore a worthwhile and perhaps neglected element of user experience.

References

1. Shin, D.: User value design for cloud courseware system. Behav. Inf. Technol. **34**(5), 508 (2015)
2. Glaser, B.G., Strauss, A.L.: The Discovery of Grounded Theory: Strategies for Qualitative Research. Aldine Transaction, London (1967)
3. Moscovici, S.: Social Representations: Explorations in Social Psychology. Polity Press, Cambridge (2000)
4. Marková, I.: The making of the theory of social representations. Cadernos de Pesquisa **47** (163), 368 (2017)
5. Rhie, Y., Lim, J., Yun, M.: Evaluating representativeness of qualitative text data in identifying UX issues. Int. J. Hum.-Comput. Interact. **33**(11), 868–881 (2017)
6. Chung, J., Nah, K., Kim, S.: Study on design research using semantic network analysis. J. Ergon. Soc. Korea **34**(4), 563–581 (2015)
7. Kim, L., Kim, N.: Connecting opinion, belief and value: semantic network analysis of a UK public survey on embryonic stem cell research. J. Sci. Commun. 14(1) (2015)
8. Kim, L.: Denotation and connotation in public representation: semantic network analysis of Hwang supporters' internet dialogues. Pub. Underst. Sci. **22**(3), 335–350 (2013)
9. Krippendorff, K.: Content Analysis: An Introduction to its Methodology. Sage, Thousand Oaks (2004)
10. Carley, K.: Coding choices for textual analysis: a comparison of content analysis and map analysis. Sociol. Methodol. **23**, 75–126 (1993)
11. De Nooy, W., Mrvar, A., Batagelj, V.: Exploratory Social Network Analysis with Pajek. Cambridge University Press, New York (2005)
12. Serrano, M.A., Boguñá, M., Vespignani, A.: Extracting the multiscale backbone of complex weighted networks. Proc. Natl. Acad. Sci. U.S.A. **106**(16), 6483–6488 (2009)

Study of Virtual Reality Performance Based on Sense of Agency

Daiji Kobayashi$^{(\boxtimes)}$ and Yusuke Shinya

Chitose Institute of Science and Technology, Chitose, Hokkaido, Japan
{d-kobaya, b2140760}@photon.chitose.ac.jp

Abstract. In recent years, virtual reality (VR) technology has been applied to various needs and problems. However, there are as yet few guidelines for providing VR from user's characteristics. Therefore, we aimed to make design guideline for VR from an ergonomics viewpoint and by observing the characteristics of the user's performance in a virtual environment during a task. Thus, the task was designed to be performed in both in reality and a virtual environment. Using the task, we observed undesirable performances by the user which could be affected by aspects of the virtual environment. First, 15 participants completed the task in reality and their performance was measured based on the movement of their hand and a surface electromyogram on their fifth finger and lower arm. We therefore obtained the characteristics of the task. Second, seven participants performed the task in a virtual environment and their performance was observed. When analyzing the results, we found undesirable performance by the participants. We therefore consider the unusual phenomena related to aspects of our VR to be based on the concept of a sense of agency (SoA). Consequently, we estimated that knowledge or predefined significance is limited in a virtual environment and it is not useful to perform tasks in VR and keep SoA to some extent. In this context, we confirmed that introducing the concept of SoA is useful when explaining performance in VR. However, our conceptual consideration should be confirmed in further research.

Keywords: Virtual reality · Sense of agency · Haptics · Electromyogram

1 Introduction

Recently, virtual reality (VR) technology has been applied to an increasing variety of human–computer interactions, offering realistic an operating experience for training, amusement, and other uses. VR includes visual and audial user interfaces, as well as haptic devices such as data gloves stimulating our haptic or kinesthetic senses. These user interfaces offer a greater sense of realism and feeling of immersion. The VR technology produced by high-specification computers used to be appropriate for every task. However, designing cost-effective VR appropriate to an objective is now a necessary aspect of VR design. Furthermore, VR should be coordinated based on not only system performance but also from a user's viewpoint. However, there are as yet few guidelines for providing VR from an ergonomic viewpoint. With this in mind, we set out to produce ergonomics-based VR design guidelines. However, the characteristics of

© Springer International Publishing AG, part of Springer Nature 2018
S. Yamamoto and H. Mori (Eds.): HIMI 2018, LNCS 10904, pp. 381–394, 2018.
https://doi.org/10.1007/978-3-319-92043-6_32

a user's cognitive and physical performance in recent virtual environments were not clear. Consequently, concepts to explain and describe a user's performance had to be considered.

In VR, the avatar is the projection of the user and performs or manipulates objects in the virtual environment; however, the user experiences and learns skills through performing tasks in the virtual environment rather than the avatar. In this context, Forbes-Pitt [1] argued the context of playing a 3D game from the perspective of agency theory, and proposed that the ego agent becomes enhanced and is inseparable from the machine and that the ago agent feels whole in the world of the game. The ego agent, in every human's social world, refers to the self-knowledge attendant in the awareness of one's own mental abilities. As well as when playing 3D games, therefore, the ego agent of a user in a virtual environment could be inseparable from VR technology and may feel whole in the virtual environment. This viewpoint suggests that it is appropriate to regard users who move their bodies to manipulate objects in virtual environments rather than those who interact with VR.

In case of evaluating VR, including user interfaces stimulating our senses and computer systems creating and controlling virtual environments, the sense of immersion is an understandable concept. Adams (2014) classified the sense of immersion into three types [2]. However, immersion is a higher-level concept and quantitative measure or metrics for evaluation of the sense of immersion into virtual environment has not yet been confirmed.

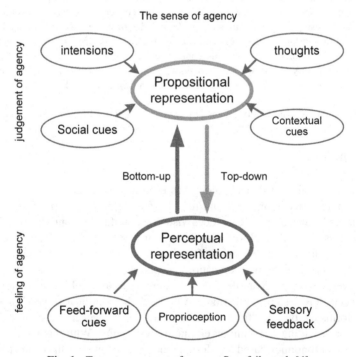

Fig. 1. Two-step account of agency Synofzik et al. [4].

The sense of Agency (SoA), a concept related to the sense of immersion, has been focused on by Frith [3] in the field of the neuropsychology of schizophrenia. Synofzik et al. [4] also discussed whether the empirical comparator model explaining the sensorimotor control mechanism and SoA and concluded that a framework of a "two-step account of agency" could allow investigation of the concept of SoA when understanding SoA and its disruptions in schizophrenia, as shown in Fig. 1.

Synofzik et al. [4] explain that the extent to which the feeling and judgement of agency contribute to the overall SoA depends on the context and task requirements, respectively. The feeling of agency is produced by a sub-personal weighting process of different action-related perceptual and motor cues, and forms the overall judgement of agency.

Action-related authorship indicators, such as feed-forward cues, proprioception, and sensory feedback produce which feeling of agency contributing to the overall SoA, depend on the context and the task requirements. In this sense, a SoA deficit could be the result of incongruence between action-related authorship indicators when the user acts, due to an imperfect virtual environment. However, SoA relates not only to the feeling of agency but also to the judgement of agency. A mismatch between different authorship indicators triggers (i) a primary basic feeling of not being the initiator of an event and (ii) a second interpretative mechanism which looks for the best explanation, resulting in a specific belief formation about the origins of the change in perception. In this sense, a user who is confused by an unexplainable situation in the virtual environment has a reduced SoA until the user finds the best explanation. However, we have little knowledge of the characteristics of the influence on performance of a reduced SoA in virtual environments. Thus, we tried to observe user's performance as characterized by the SoA in the virtual environment.

Focusing on the SoA relating to human–computer interaction (HCI), Sato and Yasuda [5] defined a sense of self-agency as the sense that I am the one who is generating an action, and focused on the degree of discrepancy resulting from comparison between predicted and actual sensory feedback through experiment. Limerick et al. [6] propose that SoA refers to the experience of controlling both one's body and the external environment and considered human–computer interaction (HCI) using the concept of SoA.

Limerick's study [6] also reviews some previous empirical studies which examine HCI issues. In these studies, the conditions of audial, visual, or tactile stimuli were assigned as independent variables and the participant's SoA-related performance was measured as the dependent variable under proper laboratory conditions. However, when considering the design of VR from a user's viewpoint, we need to know the user's psychological condition when their SoA becomes impaired while executing the task in the virtual environment, as this unusual condition commonly occurs due to either malfunction of VR or system design problems.

In this study, we designed an experimental task for users in order to observe the user's conditions and tried to consider the user's performances based on the concept of SoA.

2   Creating Rod Tracking Task for Different Environments

To observe user performance in a virtual environment, we first created a task named the rod tracking task (RTT) and set up the similar experimental environment in reality and a virtual environment to compare performance in these environments.

2.1   Design Concept for Making Task in VR

It was assumed that the user can look for the best explanation even if the user is inexperienced and cannot perform as intended in reality; however, it is hard to find an appropriate explanation when the user faces an unexperienced situation in the virtual environment. In this case, it could occur that the judgement of agency relating to SoA is difficult. To observe this phenomenon, we created the Rod Tracking Task (RTT) and observed the user's performance. When executing the RTT, the user is required to grasp a rod by the right hand and effort to move the rod back and forth without contact from end to end in a slit panel. The slit panel is installed in front of the user and was rotated anticlockwise by 45 degrees with respect to the user, as shown in Fig. 4.

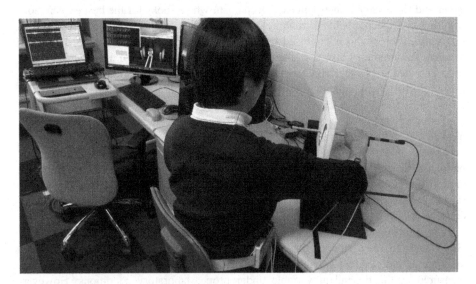

Fig. 2. Experimental scene of a participant executing the RTT in reality. Two electrodes for the EMG are attached to the participants' lower arm and near the base of their fifth finger.

The RTT was designed considering the following requirements: firstly, the users' performance must be different between normal situations and under unusual conditions. In this regard, the user could be affected by various factors arising from the design of VR. Therefore, the performance under normal situation refers to the performance in reality and must be compared to the performance in the virtual environment. We then set up VR for the RTT, as well as apparatus for a similar task in reality, as shown in

Fig. 3. Secondly, the experimental task should immerse oneself because the user's performance depends on the attention to the user's own behavior. Therefore, successful performance of the RTT requires a certain level of skill and the task must interest participants to some extent.

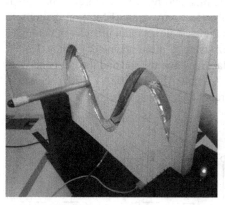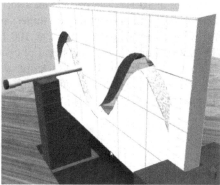

Fig. 3. The real slit panel and the rod for the RTT in reality (left hand side) and its equivalent virtual scene in VR (right hand side).

The slit width and the size of the slit panel were decided based on the difficulty of the task (see Fig. 4). Regarding the difficulty, we concretely determined that the RTT needs a level of difficulty such that an inexperienced participant was required to repeatedly practice around ten times before achieving the task in reality. The sine-curved slit and the direction of installation of the slit panel was also chosen from several patterns based on the difficulty of the RTT.

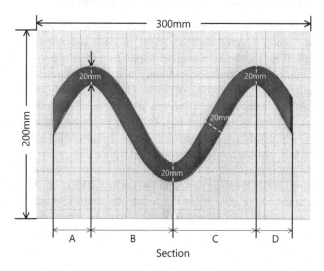

Fig. 4. Specification of the slit panel. The width of the slit panel is 20 mm and the track are divided into four sections (A–D) for convenience.

2.2   Experimental Equipment for Reality

To observe the participant's performance, the experimental equipment for the RTT was composed of the several devices as shown in Fig. 5. To record the movement of the right hand, a motion sensor (Leap Motion) connected to a personal computer (DELL XPS 8700) was used. The movement was sampled and recorded as three-dimensional coordinates at 10 Hz. The participant's surface electromyogram (EMG) on the muscle abductor digiti minimi (on the fifth finger) and muscle flexor carpi ulnaris (on the lower arm) were sampled and recorded at 100 Hz using a multi-telemeter (Nihon Koden WEB-9500). Furthermore, the signal of contact between the rod and the slit panel was sent to the PC via an USB I/O terminal (Contec AIO-160802AY-USB) and to a multi-telemeter, as well as illuminating a red LED indicator for the benefit of the participant.

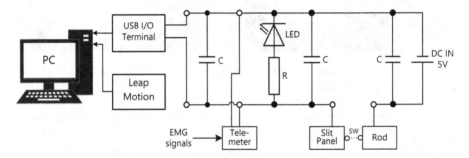

Fig. 5. Connection diagram of the devices recording the participant's performance in reality.

2.3   Experimental Equipment for Our Virtual Environment

The VR environment we made for investigating human performance consisted of two user interfaces including a haptic interface (SPIDER-HS) and a head-mounted display (HTC VIVE). SPIDER-HS, the human-sized haptic user interface for VR, consisted of controllers and eight motor modules including a motor, a threaded pulley, and an encoder for reading the yarn winding of each. The motor modules also presented a force sense as well as detecting the positions and angle of end effectors i.e. the rod moved by the participant, as shown in Fig. 6. Other experimental conditions and the specifications of SPIDER-HS were shown in Table 1 and further details of the specifications of SPIDER system are introduced in reference [7].

In the case of the experiment in the virtual environment, the movement of the rod was recorded using SPIDER-HS system and the surface EMGs on the same two body parts described above were also recorded using the multi-telemeter (Nihon Koden WEB-9500).

Table 1. Specifications of the SPIDER-HR used for the RTT.

Details	Specifications
Stick length (length of the rod in VR)	400 mm
Stick diameter (width of the rod in VR)	10 mm
Path size (size of the slit panel in VR)	$200 \times 300 \times 20$ mm
Path thickness (width of the slit in VR)	20 mm
Required force for moving vertical upward	Over 0.42 N

Characteristics of Our Virtual Environment. The physics of the rod and the slit panel was similar to that of the real object, especially visually, but somewhat different in terms of haptic sense because of being in a VR environment. For instance, the rod as the end effector of the SPIDER-HS did not fall due to gravity during the task.

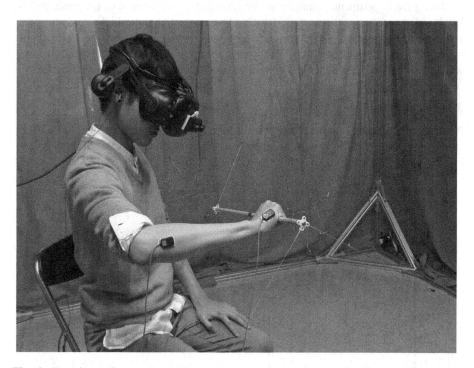

Fig. 6. Experimental scene shows the participant grasps a rod on eight strings to eight motor modules of SPIDER-HS. The head-mounted display presents the scene shown on the right-hand side of Fig. 3.

3 Evaluation of the RTT in Reality

The objective of evaluating the RTT was to clarify the characteristics of the RTT in reality using the real experimental equipment described above and then to evaluate the performance for the RTT in the virtual environment was based on the knowledge of the characteristics of the RTT.

3.1   Method

Participants. We selected 15 male students ranging from 21 to 23 years of age and participants gave their informed consent for participation in advance. All participants were right handed and had no experienced of the RTT.

Experimental Task and Procedure. A trial of the RTT in this experiment was to pull the rod from the end of the slit to the end of the near side along the track; after that, the participant had to push the rod from the end of near side to the far side and had to avoid contact of the rod with the slit panel except at both ends of the slit. The speed at which the participant moved the rod was chosen the respective participants, as long as the participant held the head as steady as possible during the trial, as shown in Fig. 2. The respective participants repeated the trial ten times; however, the participants could take a rest and relax for a while if necessary.

Using the experimental equipment, we recorded the movement of the participant's hand and the participant's surface EMGs. The contact signal which was generated when the rod and slit panel were contacted was recorded by both the PC and the multi-telemeter. Furthermore, the participants' opinions, especially about which section was most difficult, were recorded in an interview after all the trials were complete.

3.2   Result

Number of Contacts. As the result of trials, the participants accomplished the RTT in around five attempts. However, the average number of contacts in each section (see Fig. 3) throughout all trials is shown in Fig. 7.

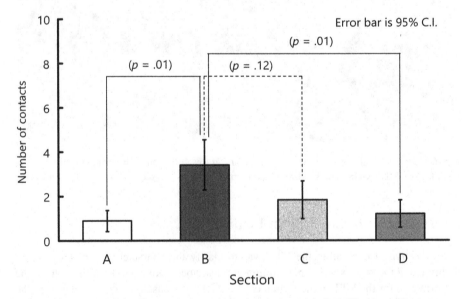

Fig. 7. Number of contacts in each section during trials in reality (n = 15).

Figure 7 represents that it is significantly more difficult for participants to avoid contact in section-B. However, we also confirmed that the relation between the number of contacts and the direction of movement of the rod was insignificant ($p =.45$). We can therefore infer that the section of the slit panel was the main factor affecting the number of contacts.

Participants' Opinions. The results of interviewing the participants indicated that the section subjectively judged most difficult was section-B, as shown in Table 2. Almost of participants found that section B was difficult, and the section-C was easy, despite the length of the slit at both section B and C was the same. Participants said that the width of the slit at section-C was perceivable, but that it was hard to see the slit at section-B and it was therefore difficult to judge the appropriate course when moving the rod. (see the left hand side of Fig. 4)

Table 2. Number of subjects who found the section was difficult or easy (n = 15).

Section	Moving to near side			Moving to far side		
	Easy	Difficult	N/A	Easy	Difficult	N/A
A	1	1	13	1	0	14
B	0	13	2	0	11	4
C	12	2	1	9	4	2
D	4	1	10	1	0	14

Relation Between EMG and Skill at the RTT The envelope of the surface EMG signal at the fifth finger and lower arm were calculated by root-mean-square (RMS) and the envelopes of each were compared between the case of inexperienced trial and the trial during which participants accomplished the RTT. Figure 8 shows that the difference between the two situations for each surface EMG is insignificant. However, the results suggest that participants manipulated the rod with their fifth finger rather than by using their lower arm.

Time for the RTT. The speed at which the participant moved the rod was at the discretion of the respective participants. thus, we observed that they were tried to accomplish the RTT in any way they wished. The time taken to finish the trial did not shorten but lengthened significantly ($p = .02$).

3.3   Discussion

The above results show that characteristics of the RTT. In particular, it is difficult to avoid contact in section-B because it is hard to perceive the slit and consequently hard to discern the position of the rod. In other words, there were inexperienced participants who could not obtain feed-forward visual cues to make a perceptual representation. However, the main point to note is that all participants could achieve the RTT in

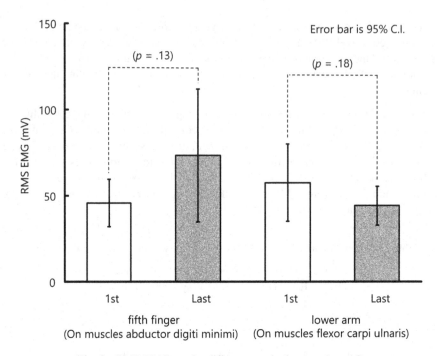

Fig. 8. RMS EMGs at the different two body parts (n = 15).

reality. According to the framework of two-step account of agency [7], this result explains that the feeling of agency has been produced, judgement of agency has formed and therefore that a sense of agency was stable until the end of all trials through repeated practice.

4   Observing the Performance of the RTT in VR

The aim of this study was to observe several performances during unusual situations in the virtual environment and to consider the performance based on a SoA. We set up the virtual environment for the RTT and conducted an experiment using seven participants and the equipment for VR described above.

4.1   Method

Participants. We selected seven male students ranging from 21 to 23 years of age as participants. All participants were right handed and had no experience in doing the RTT in reality but had experienced performing other tasks in a virtual environment using the SPIDER system and a head-mounted display. In this regard, the participants were used to performing experimental tasks in virtual environments and we assumed they had accrued sufficient skill to perceive the virtual environment to some extent.

Experimental Task and Procedure. A trial of the RTT in this experiment was the same as in the prior experiment in reality. That is, they pulled the end effector like the rod from the end of the slit to the near side along the track and then pushed the rod from the near side to the far side. During both sections of the task they had to avoid contact of the rod with the slit panel except as both ends. The speed at which the participant moved the rod was decided by the respective participants. However, the participant was required to hold their head as steady as possible during the trial in the virtual environment. The respective participants repeated the trial until their performance was improved up to a maximum of ten trials, and the participants could take a rest and relax as required.

Using the experimental equipment, we recorded the changing position and angle of the end effector as well as the participant's surface EMGs. The contact signal, which was generated by PC when the rod and slit panel made contact, were also recorded by the multi-telemeter. Furthermore, the participants' opinions, especially about which section was most difficult, were recorded in an interview after all the trials were completed.

4.2   Results

All participants had no experience of the RTT in reality; therefore, the different conditions in reality and in the virtual environment were not particularly noticed by any of the participants. In this sense, they could concentrate on the RTT in the virtual environment.

Number of Contacts. No participants accomplished the RTT within 10 trials. According to the average number of contacts in each section, the difference between the number of contacts in each section is insignificant; however, it is similar to the result from the experiment in reality that the number of contact in section-B was the highest of all sections (see Fig. 9). In other words, to move the rod through section-B was the most difficult for the participants in the virtual environment as well as in reality, though, the number of contacts in each section was higher in VR than in reality.

Participants' Opinions. The results of interviewing the participants indicated that five out of seven participants found that section B was difficult. A participant pointed out not only the visual problems in section B mentioned by participants in the reality experiment, but also several issues relating to haptic sense; e.g. a participant said that they could not perceive and understand which surface contacted to the rod from the visual and the haptic sense. The other participants said that they felt unnatural force in section B. In this sense, we assumed they experienced abnormal perception in section B.

Deviation of RMS EMG Before and After Contact at Section B. From the participants' opinions, we investigated the performance at section-B in all trials and found that there were twelve cases in which the rod continued to contact with the surface of slit at section-B regardless of the participant. In these cases, the standard deviation (SD) of RMS EMG at the surface of two body part before and after contact were significantly different, as shown in Fig. 10.

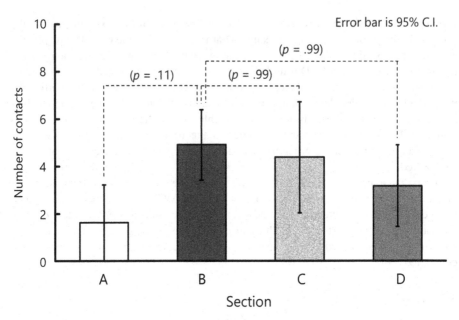

Fig. 9. Number of contacts at each section during a trial in the virtual environment (n = 7).

4.3   Discussion

Trying to explain the observed undesirable performance based on the framework of SoA shown in Fig. 1, the results suggest that the participants made a strong effort to control the rod continuously when the participants faced those cases. According to their opinions, we assumed that the participant tried to create a perceptual representation mainly based on haptic sensory feedback and attempted to move the rod in various directions because they could not obtain visual feed-forward cues and produce a perceptual representation. They could not produce a feeling of agency, and consequently the bottom-up mechanism was disabled, and concrete judgement of agency was not formed. Thus, the extent of SoA in those situations was lower than before and the lower SoA led to undesirable performance.

We assumed that there were several factors affecting feeling of agency in VR, such as incompleteness of visual feedback and haptic feedback despite participants having knowledge about their characteristics, which differed from those in the real environment. In this regard, it was estimated that forming propositional representation took more time than producing a perceptual representation because perceptual representation from unintentional sensory information was produced more quickly than processing conceptual capacities and attitudes such as beliefs or desires shown in Fig. 1. This indicates that knowledge or predefined significance was limited in virtual environment and could not be useful to perform tasks in VR and consequently retain a SoA to some extent.

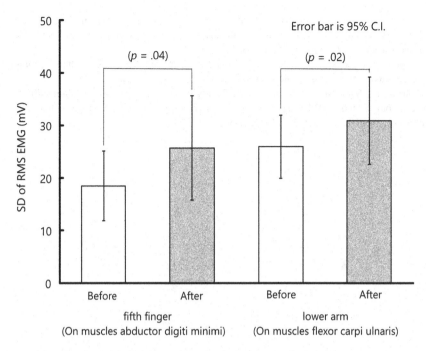

Fig. 10. Standard deviation of RMS EMG before and after contact (n = 7).

5 Conclusion

In this study, we designed a VR environment and a task named the RTT. The RTT was performed experimentally in reality and in VR, and we investigated the problem concerning user performance in VR. The results of our investigation were considered based on a framework of SoA, and we found several points to consider when designing VR. Therefore, we concluded that the framework of SoA could be useful when determining design principles for VR. However, our conceptual considerations should be confirmed in further research.

References

1. Forbes-Pitt, K.: The Assumption of Agency Theory. Routledge, Oxfordshire (2011)
2. Adams, E.: The designer's notebook: Postmodernism and the 3 types of immersion. https://www.gamasutra.com/view/feature/130531/the_designers_notebook_.php. Accessed 20 Feb 2018
3. Frith, C.D., Blakemore, S.J., Wolpert, D.M.: Abnormalities in the awareness and control of action. Philos. Trans. R. Soc. B. Biol. Sci. **355**(1404), 1771–1788 (2000)
4. Synofzik, M., Vosgerau, G., Newen, A.: Beyond the comparator model: a multi-factorial two-step account of agency. Conscious. Cogn. **17**(1), 219–239 (2008)

5. Sato, A., Yasuda, A.: Illusion of sense of self-agency: discrepancy between the predicted and actual sensory consequences of actions modulates the sense of self-agency, but not the sense of self-ownership. Cognition **94**(3), 241–255 (2005)
6. Limerick, H., Coyle, D., Moore, J.W.: The experience of agency in human-computer interactions a review. Front. Hum. Neurosci. **8**(643), 1–10 (2014)
7. Akahane, K., Sato, M.: Research on high fidelity haptic interface based on biofeedback. In: Yamamoto, S. (ed.) HIMI 2017. LNCS, vol. 10273, pp. 481–491. Springer, Cham (2017). https://doi.org/10.1007/978-3-319-58521-5_37

Airflow for Body Motion Virtual Reality

Masato Kurosawa[1], Yasushi Ikei[1(✉)], Yujin Suzuki[1],
Tomohiro Amemiya[2], Koichi Hirota[3], and Michiteru Kitazaki[4]

[1] Tokyo Metropolitan University, Tokyo 1910065, Japan
{kurosawa,ikei,suzuki}@vr.sd.tmu.ac.jp
[2] NTT Communication Science Laboratories, Atsugi, Kanagawa 2430198, Japan
amemiya.tomohiro@lab.ntt.co.jp
[3] University of Electro-Communications, Tokyo 1828585, Japan
hirota@vogue.is.uec.ac.jp
[4] Toyohashi University of Technology, Toyohashi, Aichi 441-8580, Japan
mich@tut.jp

Abstract. The present study investigates the characteristics of cutaneous sensation evoked by airflow to the face of the seated and standing user during the real and virtual walking motion. The effect of airflow on enhancement of a virtual reality walk was demonstrated. The stimulus condition provided in the evaluation involved the airflow, the visual, and the vestibular presentations, and the treadmill and walk-in-place real motions. The result suggested that the cutaneous sensation of air flow was suppressed while the movement was performed actively with visual information provided. The equivalent speed of air flow for the participants was 5 $\sim$ 29% lowered from the air flow speed in the real walk.

Keywords: Airflow · Cutaneous sensation · Virtual walk

1 Introduction

Airflow in a virtual space has been developed to augment the quality of experience of a simulated environment [1–3]. It has also been used to impart tactile sensation to the hand [4–8] and to deliver scent to the nose [9, 10]. The sensation of airflow (sensation of wind) is one of the information sources for acquiring spatial experience in which we can perceive both the motion of the self-body and the condition of environmental air [11]. The airflow has a compelling effect on reality of translation or ground speed direction in the simulation of vehicles [12–14] or the experience of flight as a bird [15]. Airflow induces and augments vection when it is used with other modality presentation [16, 17]. These features are quite attractive when they are applied to a virtual environment in which the user is immersed, and specifically the sensation of motion is needed. Although many aspects have been investigated as shown above, the relation between the cutaneous sensation by airflow and the body motion state of the user has not been discussed in detail.

In the present research, we investigated basic characteristics of airflow perception on the face during the user is moving in the real and the virtual space. We intend to use airflow to replicate spatial experience of others who moved in a variety of interesting

© Springer International Publishing AG, part of Springer Nature 2018
S. Yamamoto and H. Mori (Eds.): HIMI 2018, LNCS 10904, pp. 395–402, 2018.
https://doi.org/10.1007/978-3-319-92043-6_33

places of the world or who had a high level physical skill that could be very hard to pass to others. To transfer experience accurately, multisensory information needs to be reproduced as the past person received and felt. A multisensory display [18] has many components that interact with each other where the airflow specifically contributes to the perception of dynamic motion of the body [16, 17]. In order to design such a multisensory system, the relation between the perception of body motion and the airflow display effect in modalities needs to be elucidated.

In this paper, first we measure the walking sensation and the cutaneous sensation of airflow under visual and vestibular stimuli provided to a sitting participant. Since the cutaneous sensation of airflow changes by the existence of other stimuli, the equivalent airflow speed in those conditions was measured. Then, we show the airflow characteristics observed when the user is walking/walking in place on the treadmill and receives the visual stimulus moving through a corridor. We consider the results could provide the data for the optimal design of the airflow display.

2   Cutaneous Sensation, Walking Sensation and Air Velocity in a Seated Virtual Reality Walk

2.1   Objective and Participants

Airflow presentation characteristics in reliving of virtual reality walk for a seated user were investigated. The extent to which walking sensation and cutaneous sensation depend on airflow condition was measured. The subjectively equivalent airflow speed was adjusted to the real walk.

The participants of sensation intensity experiments were ten (under)graduate students of mean age of 23.1 years. Eight students measured the airflow speed of subjective equality.

2.2   Stimulus

Airflow was presented to the face of the participant by a fan display shown in Fig. 1a 80 cm in front of the face. The airflow speed was 0.92 m/s that was also the real walk speed. Three degree-of-freedom (lifting, pitching, rolling) motion seat built in-house presented vestibular stimulus. The motion of the seat was optimized beforehand to provide a walking sensation. An omnidirectional view in a virtual corridor translating at 0.92 m/s was presented by the head-mounted display (Vive, HTC Inc.). Three factors of cutaneous, vestibular and visual stimuli in two levels (presence or absence) were presented in a random order. In the equality adjustment experiment, the initial airflow speed was randomized within 0 to 3.2 m/s at each trial to cancel perceptual hysteresis.

2.3   Procedure

The participant wore the HMD and sat on the 3D motion seat as shown in Fig. 1b and c. The intensity of walking sensation and cutaneous sensation was evaluated by a visual analogue scale whose ends were indicated as 'no sensation' and 'real walk equivalence

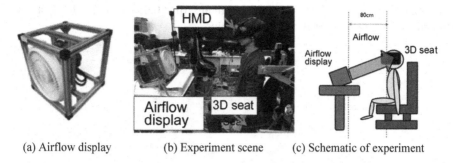

(a) Airflow display   (b) Experiment scene   (c) Schematic of experiment

Fig. 1. Airflow display and the experimental setup.

level'. A white noise was provided by earphones. The equivalent airflow speed was adjusted by the participant using a wheel on a mouse.

2.4   Results

Figure 2 shows the sensation of walking. The rating was highest when all the stimuli of airflow, visual and vestibular stimuli were provided. The main effect of all the three factors was significant ($p < 0.001$) based on the ANOVA. The interaction between the airflow and vestibular stimuli was observed. A simple main effect was significant when the vestibular stimulus was presented.

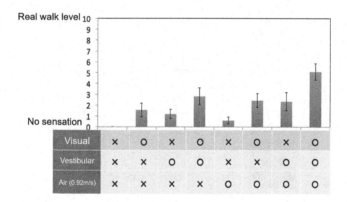

Fig. 2. Sensation of walking

Figure 3 shows the intensity of cutaneous sensation. The cutaneous sensation was larger in this virtual reality walk than the real walk. This is considered due to the sensory suppression that is activated in the case of the real walk. When the visual and vestibular stimuli were added, the cutaneous sensation decreased most to close to that in the real walk. The main effect of both the airflow and the vestibular stimulus was significant ($p < 0.001$). The interactions between the airflow and the visual stimulus and also between the airflow and the vestibular stimulus were observed. The simple main effect was detected for the visual stimulus ($p < 0.05$), and for the vestibular stimulus ($p < 0.001$) under the airflow stimulus.

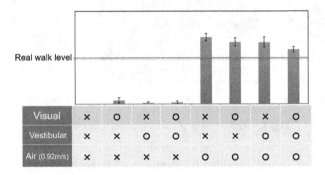

Fig. 3. Intensity of cutaneous sensation

Figure 4 shows the equivalent airflow speed. It depends on stimulus condition, and is lower than the real airflow speed, around 0.7 to 0.85 m/s. As the main effect of vestibular stimulus was significant ($p < 0.05$), it is suggested that the adjusted speed came closer to the airflow speed of the real walk when the vestibular stimulus was included.

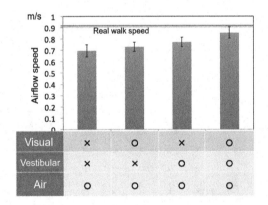

Fig. 4. Equivalent airflow speed for a seated participant

These results provided the optimal airflow speed in the experience of the walk virtual reality. It was suggested that the cutaneous sensation was suppressed when the visual and the vestibular stimulus were involved.

3 Cutaneous Sensation in a Standing Virtual Reality Walk

3.1 Objective and Participants

The intensity of cutaneous sensation and environmental reality were evaluated when the standing user received the airflow and visual stimuli during walking and walking in place on a treadmill. The cutaneous sensation was compared with that in the real walk. The airflow speed equivalent to the real walk was adjusted by the participant.

Nine (under)graduate students 22.8 years old in average participated in the evaluation. Nine students of the average age of 23.1 years performed the adjustment of subjective equality.

3.2 Stimulus

Two airflow displays were placed at the both sides of a monitor screen (80 inches) as shown in Fig. 5 where the airflow comes from the screen. The two airflows converge at 200 mm front of the face, and the airflow is directed normal to the face. The speed of the airflow was 0.92 m/s that was equal to the real walk speed. The cycle time of walk (in place) steps on the treadmill was 0.7 s. The visual stimulus was a monocular movie clip of a first person view going down a campus corridor at 0.92 m/s. The initial speed of airflow in the adjustment was randomly set between 0 to 3.2 m/s.

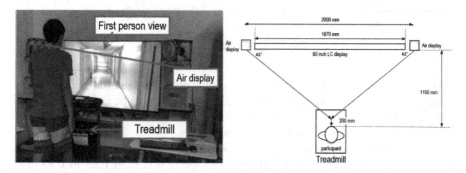

Fig. 5. Apparatus for standing on a treadmill

Twelve test stimuli from four factors of airflow, visual, walking in place and walking on the treadmill with two levels each of them were presented to the participant in a random order. The treadmill walk and the walk in place are mutually exclusive.

3.3 Procedure

The cutaneous sensation and the reality of environment were evaluated. The anchor of the VAS for cutaneous sensation was from 'no sensation' to 'equivalent to real walk', while for environmental reality from 'sensation on the treadmill in the laboratory' to 'sensation of walking down the corridor in the campus'. The sound of a metronome with 0.7 s period was provided via earphones to muffle the sound of the treadmill. In the adjustment session, the participant modified the airflow speed by a wheel on a computer mouse.

3.4   Results

Figure 6 shows the cutaneous sensation. It came to most close to that of the real walk in the condition with the airflow, the visual stimuli and the treadmill walk. The closest condition was statistically different with the condition of airflow and visual stimuli. This indicates that the active motion (treadmill walk) suppressed cutaneous sensation.

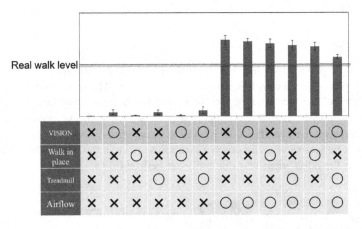

Fig. 6. Cutaneous sensation

Figure 7 shows the reality of the environment. The reality rating of walking down the actual corridor was the highest at the stimulation level where the airflow, the vision, and the treadmill walk were involved. It is the same as the cutaneous sensation. It looks that the reality was increased with the airflow.

Figure 8 shows the equivalent airflow speed obtained by the adjustment by participants regarding the stimulus conditions. The speed was from 0.65 to 0.87 m/s, although it depends on the stimulus condition. These values are smaller by 5 to 29% than the airflow speed of the real walk. The equivalent airflow speed at the condition of the airflow, the visual, and the treadmill walk was significantly ($p < 0.001$) higher than the airflow single stimulus.

This result is consistent with the fact that the cutaneous sensation was suppressed during active motion shown in Fig. 6. The result provided us with the optimum airflow speed for conditions including a walk VR, and the suppression of cutaneous sensation during active motion was confirmed.

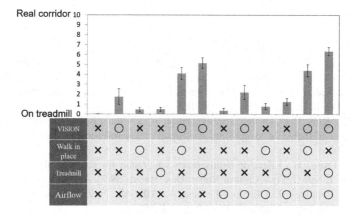

Fig. 7. Reality of environment

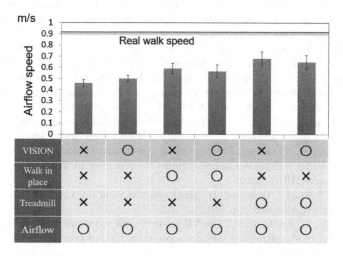

Fig. 8. Optimum airflow speed in sitting posture

4 Conclusion

In the present research, the effect conveyed by the airflow in a virtual space was investigated. It was suggested that the airflow to the face of the participant had a positive effect in providing the walking sensation and the environmental reality, although the conditions used are limited. In addition, the suppression of the cutaneous sensation during an active walk was observed in the experiment.

Acknowledgments. This research was supported by SCOPE project 141203019 at MIC and JSPS KAKENHI Grant Number JP26240029, and a past funding of NICT in Japan.

References

1. Heilig, M.L.: Sensorama Simulator. US Patent 3,050,870 (1962)
2. Kulkarni, S., Fisher, C., Pardyjak, E., Minor, M., Hollerbach, J.: Wind display device for locomotion interface in a virtual environment. In: Proceedings - 3rd Joint EuroHaptics Conference and Symposium on Haptic Interfaces for Virtual Environment and Teleoperator Systems, World Haptics 2009, pp. 184–189 (2009)
3. Moon, T., Kim, G.J.: Design and evaluation of a wind display for virtual reality. In: Proceedings of the ACM symposium on Virtual reality software and technology, pp. 122–128 (2004)
4. Ogi, T., Hirose, M.: Effect of multisensory integration for scientific data sensualization. JSME Int. J. Ser. C: Dyn. Control Robot. Des. Manuf. **39**(2), 411–417 (1996)
5. Sodhi, R., Poupyrev, I., Glisson, M., Israr, A.: AIREAL: interactive tactile experiences in free air. ACM Trans. Graph. (TOG) **32**(4), 134 (2013)
6. Gupta, S., Morris, D., Patel, S., Tan, D.: AirWave: non-contact haptic feedback using air vortex rings. In: Ubicomp 2013, pp. 1–11 (2013)
7. Tsalamlal, M.Y., Issartel, P., Ouarti, N., Ammi, M.: HAIR: HAptic feedback with a mobile AIR jet. In: 2014 IEEE International Conference on Robotics and Automation (ICRA), pp. 2699–2706 (2014)
8. Tsalamlal, M.Y., Ouarti, N., Ammi, M.: Psychophysical study of air jet based tactile stimulation. In: 2013 World Haptics Conference, WHC 2013, pp. 639–644 (2013)
9. Yanagida, Y., Kawato, S., Noma, H., Tomono, A., Tetsutani, N.: Air cannon design for projection-based olfactory display. In: Proceedings of the 13th International Conference on Artificial Reality and Teleexistence, ICAT 2003, pp. 136–142 (2003)
10. Matsukura, H., Nihei, T., Ishida, H.: Multi-sensorial field display: presenting spatial distribution of airflow and odor. In: Proceedings – IEEE Virtual Reality, pp. 119–122 (2011)
11. Feng, M., Dey, A., Lindeman, R.W.: An initial exploration of a multi-sensory design space: tactile support for walking in immersive virtual environments. In: Proceedings of IEEE 3D User Interfaces (3DUI) (2016)
12. Cardin, S., Vexo, F., Thalmann, D.: Head mounted wind. In: Computer Animation and Social Agents (CASA2007), pp. 101–108 (2007)
13. Deligiannidis, L., Jacob, R.J.K.: The VR scooter: wind and tactile feedback improve user performance. In: Proceedings of IEEE Symposium on 3D User Interfaces 2006, 3DUI 2006, pp. 143–150 (2006)
14. Nakano, T., Saji, S., Yanagida, Y.: Indicating wind direction using a fan-based wind display. In: Isokoski, P., Springare, J. (eds.) EuroHaptics 2012. LNCS, vol. 7283, pp. 97–102. Springer, Heidelberg (2012). https://doi.org/10.1007/978-3-642-31404-9_17
15. Rheiner, M.: Birdly an attempt to fly. In: ACM SIGGRAPH 2014 Emerging Technologies (SIGGRAPH 2014), vol. 3, pp. 1. ACM, New York (2014)
16. Seno, T., Ogawa, M., Ito, H., Sunaga, S.: Consistent air flow to the face facilitates vection. Perception **40**, 1237–1240 (2011)
17. Murata, K., Seno, T., Ozawa, Y., Ichihara, S.: Self-motion perception induced by cutaneous sensation caused by constant wind. Psychology **5**, 1777–1782 (2014)
18. Ikei, Y., Abe, K., Hirota, K., Amemiya, T.: A multisensory VR system exploring the ultra-reality. In: Proceedings of 18th International Conference on Virtual Systems and Multimedia (VSMM2012), pp. 71–78 (2012)

Designing Augmented Sports: Merging Physical Sports and Virtual World Game Concept

Takuya Nojima[1(✉)], Kadri Rebane[1], Ryota Shijo[1], Tim Schewe[1],
Shota Azuma[1], Yo Inoue[1], Takahiro Kai[1], Naoki Endo[1],
and Yohei Yanase[2]

[1] University of Electro-Communications, Tokyo, Japan
tnojima@nojilab.org, {kadri,shijo,timschewe,sazuma,
yoiknoue,kai,naoki.e}@vogue.is.uec.ac.jp
[2] Unity Technologies Japan G.K., Tokyo, Japan
yoheiy@unity3d.com

Abstract. It is important to encourage people to play sports and physical activities to keep their own health and well-being. However, not many people can keep playing sports regularly. In addition, people in today tend to become physically inactive. Thus, a novel way to motivate people to become physically active, by playing sports, is desired. Augmented sports are novel sports that integrates concepts of computer games into existing physical sports. Physical sports are played in our physical, real world. Thus, physical law limits the methods. On the other hand, such methods for computer games are limitless. Augmented sports are novel sports that integrates various methods to fill or reduce the unwanted gap between humans, to make sports enjoyable regardless of their physical skill or conditions. It will contribute every sports player to feel fun and enjoyment more, that should also lead to motivate people to play sports more. In this paper, detail concept of augmented sports is described. Then, we developed an augmented dodgeball, a proof-of-concept of augmented sports. The detail of the system is also described.

Keywords: Augmented sports · Augmented dodgeball · Sports
Exertion games · Video gaming

1 Introduction

The word "sports" is deemed to be etymological in Latin [18]. This word has the meaning of de (= away) and portare (= carry), distractions from what is indispensable for survival, distraction, rest and enjoyment. In other words, sports is basically a leisure activity, it is an activity that strongly includes elements as play. However, it is not perfectly equivalent to "play". As a sociologist A. Guttmann tries to define sports by incorporating concepts such as play, structuring, and competition. The relationship between play and sports [19] is illustrated in Fig. 1. For example, fun playing just like playing the ball somehow or throwing it on the ground is classified as "Spontaneous play". When this becomes a catch ball, some rules are applied such as the ball must not

© Springer International Publishing AG, part of Springer Nature 2018
S. Yamamoto and H. Mori (Eds.): HIMI 2018, LNCS 10904, pp. 403–414, 2018.
https://doi.org/10.1007/978-3-319-92043-6_34

be dropped, throwing the ball so that others can easily receive it. This is classified as "Organized play". Catch ball does not have any competitive elements, but baseball has. Since baseball is a competitive play with physical activity, it can be classified as sports.

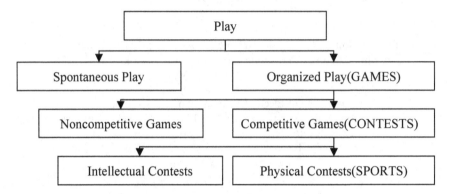

Fig. 1. Definition of sports(re-edited)[33]

One of the big question is from when human play sports. Sports have long history. About 2100 BC Gilgamesh epic poems have descriptions that thought to be about wrestling. The number of kinds of sports has remarkably increased during the 4000 years from that description. For example, in the first Ancient Olympics held in 776 BC, there were at most only a few sports events. On the other hand, at the Rio de Janeiro Olympic Games in 2016, 306 events of 28 races are being conducted. It is well known that there are lots of sports that are not adopted to the official Olympic Games. All sports currently being enjoyed were created by people of the past. In other words, sports can be created with our own hands. This seems to be obvious, but for many people today, sports is not the area of creation but the area of choosing. However, if we strongly recognize that "we can create sports by ourselves", it will contribute enlarging the possibility of new sports. In addition, the progress of technology is expected to accelerate this movement. Motor sports and e-sports can be typical examples.

Why do people play sports? According to the study in the UK, fun, enjoyment and social interaction were reported to be the key factors whether or not to get involved in the sport [2]. It is well known that playing sports contribute keeping one's health. This also can be a reason to play sports, but not the key factors. When considering improve one's health, playing sports are highly recommended. However, it is also known that not many people can keep playing sports regularly. In addition, people in today tend to become physically inactive. In a U.S. report, the percentage of people who work for a job requiring physical activities decrease from almost 50% to less than 20% in these 50 years [5]. As this kind of change in the human behavior has resulted in negative consequences in people's health and well-being, it is important to encourage people to play sports. Thus, a novel way to motivate people to become physically active, by playing sports, is desired. To comply with this issue, computer games may be helpful when considering the reason why people play sports. The Olympic Movement has also acknowledged that – "The Olympic Movement should strengthen its partnership with the computer game industry in order to explore opportunities to encourage physical activity and the practice and understanding of sport among the diverse population of computer game users" [1].

Augmented sports are novel sports that integrates concepts of computer games into existing physical sports. Competitiveness, is one of the key factor of sports as we stated in Fig. 1. To achieve this, rules must be fair. But fairness must be discussed carefully. Physical abilities of human differ from person to person. Gender, height, weight, age, presence or absence of disability, and degree of disability, of course, are different from each other. Rules of sports are required to fill or reduce the unwanted gap in an appropriate form. Most existing sports have sophisticated rules on this point, and those rules are often updated if necessary. For example, weight classification in fighting sports is a rule to reduce the impact of difference of player's weight which greatly affects result of the match. This way of thinking is not limit to physical sports. Games (including computer games) also have the same way of thinking. A lot of effort have been done to develop enjoyable games for both novice and skilled players. The difference between computer games and physical sports is limitation of methods can be used to fulfill the purpose. Physical sports are played in our physical, real world. Thus, physical law limits the methods. On the other hand, such methods for computer games are limitless. Augmented sports are novel sports that integrates various methods to fill or reduce the unwanted gap between humans, to make sports enjoyable regardless of their physical skill or conditions. It will contribute every sports player to feel fun and enjoyment more, that should also lead to motivate people to play sports more.

2 Related Works

From the point of the view of computer games, sports are important theme fields to create new games. In addition, several computer game products that have novel interface like Nintendo Wii [16] and Kinect [12] enables people to play such games with a certain amount of physical activity. It is the same for sports area. Computer games are getting focused as a technology to motivate people to play sports. Research have been done [9, 10] on how to add exertion elements to otherwise button controlled video games. It is believed that a certain technology can induce motivation for people to become physically active. Richards et al. developed a card game on screen to motivate people who do repetitive exercises that can easily get boring [17]. Kajastila et al. developed a novel trampoline game, that uses trampoline activity as a game input [8]. This makes trampoline jumps into jumping in the game world. In the game world, it has graphical obstacles to give trainer's feedback. Adding social elements is considered to contribute to improving enjoyment of sports. Mueller and Muirhead developed a jogging companion robot that stays with a jogger [14]. They also developed a system that enables people in remote places to play sports together [11, 15]. Such systems are intended to improve social elements when playing sports under a certain situation. Swimtrain [4] is a system that enhances a group fitness training like swimming with social elements, to motivate people to put more effort on the exercise.

It is hard to enjoy playing sports with players who have too much different skill level. Balancing among different players an important factor for being the sports more enjoyable. Altimira et al. proposed a novel balancing way between in table tennis by limiting table size according to the player's skill level [3]. Thanks to the big interest to the field of movement based games, guidelines have also been published [7], to give game designers insights of what things have been proved to work and which things not.

In this research work, we apply concept of augmented sports to dodgeball to make augmented dodgeball as a proof of concept. It is expected to make the dodgeball more balanced, enjoyable sports. There are also many versions of dodgeball that have been developed to make the classical version more versatile, strategic and/or give weaker players some advantages. Some versions of them include giving the weaker players more "lives" so that they can play longer as an infield player. There is also a version using a "buddy" system which means that the players are also divided into pairs and when one player gets hit, actually the other one in that pair will go out of the game [13]. Some variations also use the time restriction on how long time the players can hold the ball, other ones have a doctor in the team which means that the players who have got hit can return to the game when the doctor has healed them and the game finishes when the doctor has got hit etc.

Our research focuses on using virtual parameters known from computer games. This method is considered to enable players with different skill levels would be able to play together and enjoy their activity. In addition, this method is expected to used as more sophisticated way of balancing, by adjusting parameters. In the next chapter, detail of our approach and developed system is described.

3 Augmented Dodgeball-The Proof of Concept Augmented Sports

3.1 Basic Concept of Augmented Sports

Our research group has focused on developing augmented sports which could be enjoyable for everyone, regardless of their physical skills and abilities. People tend to feel the sports as enjoyable if it provides right amount of exertion, feeling of contribution to the outcome of the game. It is often difficult to maintain this condition when people with too much different physical abilities play together. Traditionally, analogical handicapped methods are pursuit to balance players, such as adding extra scores to the player or team that assumed to have lower abilities before playing the game. Such kind of explicit handicap methods can be thought to be an unfavorable label that the player/team is not skillful.

The basic concept of the augmented sports is making use of virtual parameters and abilities that are common among computer games. People plays (existing) physical sports that have rules modified to use those virtual parameters. Those parameters are affected by the physical event such as hitting and/or a physical ball, running to/from a certain place, etc. Then, a certain kind of special effects are generated based on those events and parameters, to support and/or entertain players.

Those virtual parameters and abilities are also used to fill or reduce the unwanted gap between humans. At the same time, such virtual parameters never harm the fairness of the game. The fairness is the key factor to make the sports enjoyable. In augmented sports, a person's playing ability depends on the person's actual physical playing ability and virtual parameters and abilities. In other words, although a person has low physical ability, the person's ability for play can be augmented by the effect of virtual parameters and abilities. This is a kind of implicit handicapping method to improve competitiveness of sports. In addition, such virtual parameters and abilities can be used to increase fun for play.

To realize the augmented sports, major research topic will be (a) rule design to motivate people with different physical abilities to play while keeping fairness of game, (b) systems to detect physical events during play to support the rule, which is necessary to make a certain effect to virtual parameters, (c) feedback systems to show the status of virtual parameters and abilities for players and audience.

3.2   Development of Augmented Dodgeball

In this section, detail of augmented dodgeball (Fig. 2) is described. This is a proof of concept of the augmented sports.

Fig. 2. Playing augmented dodgeball

Dodgeball is a ball game mostly played in elementary and junior high schools all over the world. Although detail of rules may differ from place to place, the basic rule is common: "throwing a ball against opponents to hit". Dodgeball is not only a sports of personal physical abilities, but requires a cooperation with their team members and a certain strategy to win. One of the typical strategy is keep passing the ball between infield players and outfield players to keep moving opponents in their field. While passing the ball, they try to find a target and throw a ball against the player. When throwing a ball, the thrower often assesses physical abilities of thrower's own and that of opponent's who is about to be thrown against. In this condition, a physically weak player often has difficulty continue playing as an infield player. By making existing dodgeball as a kind of augmented sports, it is expected that all players could enjoy playing "the modified version of dodgeball" without depending on their own physical abilities. We call this "modified version of dodgeball" as "Augmented dodgeball". Augmented dodgeball is designed to use virtual parameters which is mainly used for balancing the physical skill levels between different players. Following subsections show how to modify the existing dodgeball into augmented dodgeball.

Rule design

Firstly, rules must be modified to use virtual parameters while playing dodgeball. To achieve this, we added a virtual point named "life points" to each player to indicate their virtual life status. This is a well-known concept in computer games. When a player gets hit in the Augmented Dodgeball game, they will lose a certain amount of points from their life points. This enables all the players to keep being an infield player until the player loses all life points. In addition, players are also assigned virtual parameters named "attack points" and "defense points". Attack points indicates the level of offence, and defense points indicates the level of defense. The Eq. (1) shows the points to be reduced from the player's life points when they got hit. In this equation, m, n, u, v are natural numbers to identify players and teams. T_m denotes a team, P_u denotes a player. $AP\left(T_{n(\neq m)}, P_{v(\neq u)}\right)$ denotes an attack points of the thrower, $DP(T_m, P_u)$ is a defence point of a player who is hit by the ball. If a player P_u in team T_m got hit by a ball thrown by the player $P_{v(\neq u)}$ in team $T_{n(\neq m)}$, the life point of the player P_u in team T_m, which is denoted as $L(T_m, P_u)$ will be updated as follows:

$$L_d(T_m, P_u) = \frac{AP\left(T_{n(\neq m)}, P_{v(\neq u)}\right) - DP(T_m, P_u)}{2} \tag{1}$$

$$L_{new}(T_m, P_u) = L_{prev}(T_m, P_u) - L_d(T_m, P_u) \tag{2}$$

$L_{new}(T_m, P_u)$ in the Eq. (2) denotes the life points of the player P_u in team T_m after being hit, $L_{prev}(T_m, P_u)$ denotes that of before being hit. To use this concept as a balancing method, three player roles who have different virtual points are designed: Attacker, Defender and Balanced. Attacker players have greater attack power but lower defence power than other player roles. On the other hand, defender players have greater defence power but lower attack power. Balanced players have moderate attack and defence power. All the player roles have the same life points. Then, based on a computational game simulation, detail virtual points of each roles are defined to achieve fair and competitive game as shown in Table 1. Those roles are assigned only to infield players. If the player becomes outfield player, the role will be changed to balanced type automatically. The reason behind developing these roles was to give players to act as those roles such as aggressive attacker players or less aggressive defence players. Such method is expected to add enjoyment of "playing the assigned role" in the game.

Table 1. Roles and parameters used in the augmented dodgeball game

Player role	Points		
	Life points	Attack power	Defence power
Attacker	120	140	120
Balanced	120	120	160
Defender	120	110	180

System Development

In the augmented dodgeball, life points must be updated in a real time. However, the rules behind the game is too complicated to calculate mentally in a real time. Thus, a supportive system is required.

To make virtual parameter rules effective, the supportive system should equip functions as follows: (I) detecting a player who throws the ball, (II) detecting a player who got hit by the ball, (III) updating life points. The main goal of this paper is to prove the effectiveness of the concept of augmented sports through developing augmented dodgeball. Thus, a system with minimum functions (I) and (III) are developed.

In this subsection, we describe a helmet type system to register the thrower, and a point management application. When playing the augmented dodgeball, all the players wear a helmet system as shown in Fig. 3 left. Then, players "register" the ball to the point management application through the helmet just before throwing against an opponent. The occurrence of the hit and identifying who got hit is done by a human referee. If appropriate hit occurs, the referee input the information of the player who got hit into the point management application. Then life points of that player will be updated.

Fig. 3. Left: The helmet system for players to register the thrower. Right: The ball covered with RFID tags.

The helmet system is equipped with a thrower registering system. The system consists of an RFID tag reader, an Arduino UNO board, an XBee [8] wireless module, a small speaker and a battery. The helmet system has unique ID to identify the player who wears it. The tag reader is installed in front of the helmet to make it easier for the players to register the ball that the player has. The ball used in the augmented dodgeball is a sponge ball (ɸ 160 mm) covered with RFID tags as shown in Fig. 3 right. To register the ball, players hang it on the RFID tag reader area of the helmet for a second until beep sound is generated as shown in Fig. 4. In the world of augmented dodgeball, we explain this action to players as "pouring special force into the ball". This registering event and the helmet ID are transmitted to the point management application wirelessly. The whole system is powered with a 9 V battery.

Fig. 4. Registering action

The point management application receives the helmet ID, which represents a player who holds the ball. Then, if the human referee inputs player ID who got hit, the life points of the player will be updated based on Eqs. (1) and (2).

3.3   Feedback Systems

Virtual parameters must be shown all to players and spectators. Otherwise no one can understand the current status of the game. At this moment, a scoreboard is placed in the center of the play field to show life points of all players as shown in Fig. 5.

Although all players and spectators could perceive all virtual parameter information through this display, it is difficult to check actual game situation and the scoreboard simultaneously. Thus, appropriate information display system is necessary for both players and spectators. For the first step of display system for players, we developed a prototype of wearable parameter indicator as shown in Fig. 6. This indicator consists of four LED bars, two for each side of the body, an XBee for data communication, an Arduino UNO R3 as an LED controller. The color of LED bars changes according to the life points that the player has, which is transmitted from the point management application. Figure 7 shows the relationship between color of LED bars and life points.

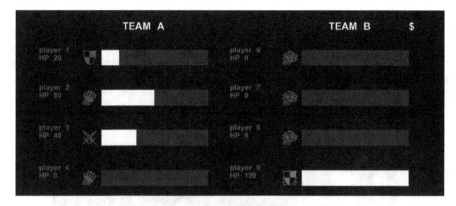

Fig. 5. The scoreboard of the augmented dodgeball

Fig. 6. A prototype of wearable parameter indicator (Color figure online)

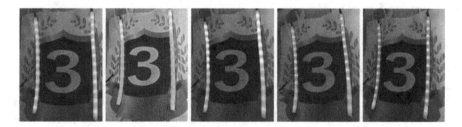

Fig. 7. Relationship between color and life points (Left: all green indicates maximum life points, Right: most of LEDs turns red to indicate exhausted life points) (Color figure online)

Basically, sports cannot live through without considering the existence of spectators. Spectators are potential provider of psychological, material and financial support to maintain the environment for play. In addition, they are also potential players in the future. Thus, it is important designing enjoyment for spectators.

As mentioned above, the scoreboard system can be used as a supplemental information display, but not enough. For the first step to comply with this issue, we developed an AR display system. In this system, players wear specially designed bibs as shown in Fig. 8 left, which act as markers for developed AR display. Figure 8 right shows example screenshot what spectators watch. As shown in the figure, player role icon and current life points are overlaid on each players.

Fig. 8. Left: example designs of bibs, Right: prototype AR image for spectators

To achieve this system, position information of all players are necessary. In this prototype system, a web camera is placed outside of the playfield to recognize AR markers designed as bibs worn by players. Vuforia (version 6.2.10) is used to recognize the AR markers. Then, the player position information and the status information of each player are sent to the AR display device worn by spectators. Here, Microsoft HoloLens is used as the AR display device. This application was developed using Unity(version 5.6.3p4).

3.4   Preliminary Evaluation

To evaluate the concept of augmented dodgeball, a preliminary evaluation was conducted. 16 participants joined this evaluation session. Age of participants were between 20 and 26, average was 23. None of the participants played dodgeball regularly. In this evaluation, the wearable parameter indicator for players and the AR display system for spectators were not used. The scoreboard system is the only medium to indicate the status of virtual parameters. In this evaluation, two sessions were conducted. One or two games of regular dodgeball was played in the first session. Two augmented dodgeball game was played in the second session. One team consists of four players. Participants organized the team by themselves. After finishing the second session, a survey was conducted. The survey consists of 26 questionnaires to be answered by a five point Likert scale.

From the result of the questionnaire, all of the answered participants stated that they had fun playing the augmented dodgeball. 60% even strongly agreed to that claim. In addition, 93% answered that they would play the game again. The augmented dodgeball has complex rules than those of existing regular dodgeball. However, 64% of participants answered that the rule was always clear to play. Although 29% stated that they had some trouble with that.

4   Conclusions

In this paper, we propose the concept of augmented sports, which intended to be sports enjoyable for everyone, regardless of their physical skills and abilities. To achieve this concept, we claim that integrating computer game elements into existing physical sports. Then a proof of concept system was developed named augmented dodgeball, which is a re-invented version of dodgeball by using the concept of augmented sports. The augmented dodgeball utilizes virtual parameters that are well known in computer games to indicate characters' status in that world. In the augmented dodgeball, players are assigned different player roles that have different virtual parameters. Then, by throwing a physical ball against opponents, those parameters are changed according to the event in the physical world. To support this, a wearable system, point management application, display feedback system were developed. In addition, a preliminary evaluation was conducted. Through the evaluation, it is agreed that the augmented dodgeball is enjoyable sports.

At this moment, the functions equipped with the wearable system is not enough unfortunately. More detail, a function for detecting a player who got hit by the ball should be equipped to eliminate human referee. In addition, more sophisticated mechanism for detecting the player who throws the ball is strongly required. In the future, this augmented dodgeball supporting system is expected to be updated as a generally applicable system to make various kinds of ball sports into augmented sports.

Acknowledgement. This work was supported by JSPS KAKENHI Grant Number JP16H01741.

References

1. XII Olympic Congress: The Olympic Movement in Society Recommendations (2009). www.olympic.org/Documents/Conferences_Forums_and_Events/2009_Olympic_Congress/Olympic_Congress_Recommendations.pdf Accessed 1 Feb 2018
2. Allender, S., Cowburn, G., Foster, C.: Understanding participation in sport and physical activity among children and adults: a review of qualitative studies. Health Educ. Res. 21(6), 826–835 (2006)
3. Altimira, D., Mueller, F.F., Clarke, J., Lee, G., Billinghurst, M., Bartneck, C.: Digitally augmenting sports: an opportunity for exploring and understanding novel balancing techniques. In: Proceedings of the 2016 CHI Conference on Human Factors in Computing Systems, pp. 1681–1691 (2016)

4. Choi, C., Oh, J., Edge, D., Kim, J., Lee, U.: SwimTrain: exploring exergame design for group fitness swimming. In: Proceedings of the 2016 CHI Conference on Human Factors in Computing Systems, pp. 1692–1704 (2016)

5. Church, T.S., Thomas, D.M., Tudor-Locke, C., Katzmarzyk, P.T., Earnest, C.P., Rodarte, R. Q., Martin, C.K., Blair, S.N., Bouchard, C.: Trends over 5 decades in US occupation-related physical activity and their associations with obesity. PLoS ONE 6(5), e19657 (2011)

6. FunAndGames.org: Dodge Ball Games. http://funandgames.org/tag/dodgeball/ Accessed 1 Feb 2018

7. Isbister, K., Mueller, F.F.: Guidelines for the design of movement-based games and their relevance to HCI. Hum. Comput. Interact. 30(4), 366–399 (2015)

8. Kajastila, R., Holsti, L., Hämäläinen, P.: Empowering the exercise: a body-controlled trampoline training game. Int. J. Comput. Sci. Sport (Int. Assoc. Comput. Sci. Sport) 13(1), 6–23 (2014)

9. Ketcheson, M., Walker, L., Nicholas Graham, T.C.: Thighrim and Calf-Life: a study of the conversion of off-the-shelf video games into exergames. In: Proceedings of the 2016 CHI Conference on Human Factors in Computing Systems, pp. 2681–2692 (2016)

10. Ketcheson, M., Ye, Z., Nicholas Graham, T.C.: Designing for exertion: how heart-rate power-ups increase physical activity in exergames. In: Proceedings of the 2015 Annual Symposium on Computer-Human Interaction in Play (CHI PLAY 2015), pp. 79–89 (2015)

11. Jensen, M.M., Rasmussen, M.K., Mueller, F.F., Grønbæk, K.: Keepin' it real:challenges when designing sports-training games. In: Proceedings of the 33rd Annual ACM Conference on Human Factors in Computing Systems (CHI 2015), pp. 2003–2012 (2015)

12. Microsoft: Kinect for Xbox One. www.xbox.com/en-US/xbox-one/accessories/kinect-for-xbox-one Accessed 1 Feb 2018

13. Moritake, T.: Different versions of dodgeball. homepage1.nifty.com/moritake/taiiku/bo-ru/dozzi-ball.htm Accessed 1 Dec 2016

14. Mueller, F.F., Muirhead, M.: Jogging with a Quadcopter. In: Proceedings of the 33rd Annual ACM Conference on Human Factors in Computing Systems (CHI 2015), pp. 2023–2032 (2015)

15. Mueller, F.F., O'Brien, S., Thorogood, A.: Jogging over a distance: supporting a jogging together experience although being apart. In: CHI 2007 Extended Abstracts on Human Factors in Computing Systems, pp. 1989–1994 (2007)

16. Nintendo: Wii. wii.com Accessed 1 Feb 2018

17. Richards, C., Nicholas Graham, T.C.: Developing compelling repetitive-motion exergames by balancing player agency with the constraints of exercise. In: Proceedings of the 2016 ACM Conference on Designing Interactive Systems, pp. 911–923 (2016)

18. Penjak, A., et al.: Sport and literature: an overview of the wrestling combats in the early literary texts. J. Hum. Soc. Sci. 3(5), 49–55 (2013)

19. Guttmann, A.: From Ritual to Record: The Nature of Modern Sports. Columbia University Press, New York (1979)

Comparison of Electromyogram During Ball Catching Task in Haptic VR and Real Environment

Issei Ohashi[1(✉)], Kentaro Kotani[1], Satoshi Suzuki[1], Takafumi Asao[1], and Tetsuya Harada[2]

[1] Department of Mechanical Engineering, Kansai University, Osaka, Japan
{k526812,kotani,ssuzuki,asao}@kansai-u.ac.jp
[2] Department of Applied Electronics,
Tokyo University of Science, Tokyo, Japan

Abstract. The objective of this study was to construct systems for haptic virtual reality (VR) environment and to conduct an experiment to compare muscular activity during ball catching tasks in real and VR environments, where the level of the presence was evaluated. A ball catching task was demonstrated in two environments, where head-mounted display and SPIDAR-HS, the haptic presentation device using tensile force of the wire, were applied for constructing VR environment. As an index of dynamic muscular activity, forearm EMG signals were measured in the time course of a ball catching task. Average peak RMS value for forearm EMG in VR environment was 45.2% smaller than that in real environment. This difference was apparent because the amount of force generated by SPIDAR-HS was relatively lower than that made by the gravity force of the ball. On the other hand, the trends in dynamic muscular activities were similar for both environment, indicating that two tasks were fairly unique regardless the type of environments. It was concluded that the presence of VR was observable by the dynamic muscular changes during VR tasks with further adjustment of force levels required for the task in VR environment.

Keywords: Virtual reality · Haptics · Presence · Electromyogram

1 Introduction

Virtual reality (VR) is a computer-based environment that simulates realistic experiences [1]. As the progress for the development of VR, the market for head-mounted displays (HMD) like HTC Vive [2], PlayStation VR [3] and Oculus Rift [4] has grown rapidly. HMD is a wearable device attached to the head and supports users' head movement tracking. This makes it easier for the user to look around virtual scenes along with the head movement. Although HMD offers high reality in visual circumstances, other modalities such as haptics play a key role for apprehending the environment [5]. Haptic display devices such as SPIDAR [6] and Geomagic Phantom [7] have been developed to present high-reality haptics in VR environment.

According to Tachi et al. [1], there are three essential elements to establish the appropriate VR environment.

© Springer International Publishing AG, part of Springer Nature 2018
S. Yamamoto and H. Mori (Eds.): HIMI 2018, LNCS 10904, pp. 415–425, 2018.
https://doi.org/10.1007/978-3-319-92043-6_35

- Autonomy: building a natural three-dimensional space
- Interaction: immediate and prompt reflection for the users' behavior
- Presence: feeling like you are present inside the VR environment

The autonomy and the interaction can be determined by the performance of computers to construct the VR. In contrast, a method for quantifying the presence has not been established because it varies depending on human senses and physiological responses. In VR environment where the presence is limited, users may feel stress, yielding frequent operational errors. For instance, in the telerobotic surgery, if an operator does not perceive force feedback correctly through the haptic display, this can cause fatal medical accidents [8]. Therefore, it is important to quantify the presence for preventing users' excessive stress and the potential accident due to lack of accurate presence.

The presence has attracted a lot of attention and research regarding the presence has flourished in recent years [9, 10]. In addition to the subjective questionnaire, evaluation for the presence using physiological indices such as electromyogram (EMG) and heart rate have attracted attention [11, 12]. We focused on EMG as a physiological index to evaluate the presence. In the previous study, EMGs when catching a free-falling ball in the haptic VR environment have been evaluated [13]. They focused on the start time of activation of the muscles in the forearm before catching a ball. However, they did not compare EMGs in VR environment and that in real environment. If the presence decreases with inappropriate haptic presentation, different muscular activities from the real exercise may be generated. Therefore, it is speculated that the temporal characteristics on muscular activities were the potential index for evaluating the presence quantitatively.

Therefore, the objective of this study was to construct the system for demonstrating haptic VR environment and to conduct comparison of EMGs during catching a ball in VR and real environments.

2 Methods

2.1 Participants

A total of 3 males participated in this study. Two were in early their twenties and one was in his early fifties. They were right-handed. They have experienced VR several times, but not regularly.

2.2 Experimental Setup

Figure 1 shows the overview of the experimental set up. The experimental setup was composed of VR environment presentation equipment and EMG measuring instrument. HTC Vive and space interface device for artificial reality – human scale (SPIDAR-HS) were used for constructing VR environment.

HTC Vive is shown in Fig. 2. The frame rate of HTC Vive was 90 frames per second. HTC Vive has a field of view of 110 degrees, slightly less than typical viewing angle of the human eye (approximately 120 degrees). The resolution of HTC Vive is 1080×1200 per eye [2]. HTC Vive wearer can see the VR space created by Unity.

SPIDAR-HS is derived from the original SPIDAR which was introduced in 1991 by Sato and colleagues [6]. After that, Scaleable-SPIDAR [14] was developed as a predecessor of SPIDAR-HS in early 2000. SPIDAR-HS is 2.5 m × 2.5 m × 2.5 m in size. Within this space, different kinds of force feedback senses associated with weight, contact and inertia can be displayed to the user's hand by tensioned wire. The device used tensioned wire techniques to track hand position as well as to provide haptic feedback sensations. The force feedback felt on the user's hand was the same as the resultant force of tension generated by the collection of eight wires equipped at each corner of the frame. In order to control the tension and length of each wire, one end was connected to an end-effector and the other end was wound around a pulley, which was driven by a DC motor. By controlling the power applied to the motor, the system was able to create designated tensions all the time. A rotary encoder was attached to the DC motor to detect the wire's length variation. The updating frequency of the motor was set to 500 Hz. Maximum force was 6.30 N. A half-sphere-shaped end-effector, shown in Fig. 3, was made with a 3D printer. The set of DC motor, pulley and encoder controlling each wire was fixed on the frame. The frame was made of aluminum (HF S6-3030-2500). Figure 3 shows the structure for SPIDAR-HS.

Muscular activity was measured with a bioamplifier (BA1104 m EMG, Nihon Santeku, high-pass frequency of 1000 Hz, TC: 0.03 s). EMG signals were recorded with a sampling rate of 1000 Hz and stored at the data logger (GL900, GRAPHTEC). In this study, we used the surface electrodes (BA-U410 m, Nihon Santeku) for measurement of EMG signals. Also, we used an electrode (M-00-Sm, Ambu) and a cord (YCE116 m, Nihon Santeku) as a body ground for the measurement of EMG.

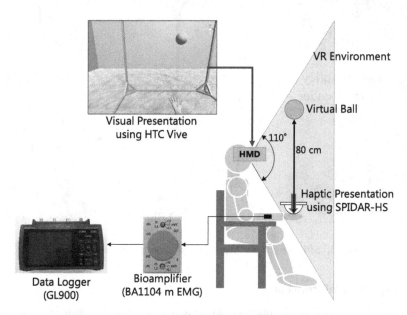

Fig. 1. Overview of experimental setup.

Fig. 2. HTC Vive for visual presentation in VR environment.

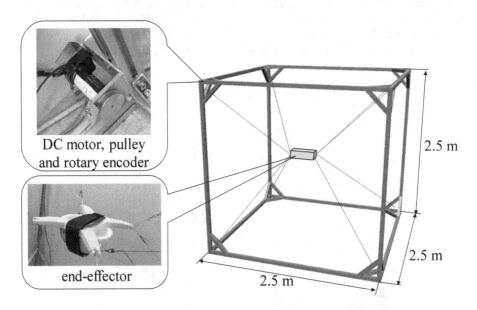

Fig. 3. SPIDAR-HS for haptic presentation in VR environment

2.3 Experimental Protocol

Throughout the experiments, the participant was sitting on the chair with the right forearm placed on the armrest with the right palm facing upwards. The muscular loads for two muscles in the right forearm areas were registered by electromyography.

Participant was applied electrodes to the palmaris longus muscle (PLM) and the extensor carpi radialis longus muscle (ECRLM) in the right forearm (Fig. 4). PLM manipulates the palmar flexion of the wrist and ECRLM manipulates the wrist to move the hand away from the palmar side [15]. The electrode for the body ground was attached to the right elbow. EMGs were measured while participant was at rest, during ball catching task in real environment (Real task) and during ball catching task in VR environment (VR task).

Measurement of EMG at Rest. Participant's EMGs were measured for 15 s at rest. We instructed participant not to exert on his right arm. This task was executed once before the Real task. If it was confirmed that the muscles were activated during this measurement, the measurement was retried.

Real Task. In Real task, participants were asked to perform catching free-falling ball sequence in real environment. The initial position of the ball was 80 cm above the right palm. EMG measurement was started when the ball began to fall. When catching a ball, participant was instructed not to grasp the ball with fingers but to catch as much to support it with the palm as possible. This task was carried out three times. Figure 5 shows the ball used at the Real task. The ball was 220 g and its diameter was 80 mm.

VR Task. Participant attached HTC Vive to the head. The initial position of the virtual ball on the HMD's display was 80 cm above the right palm. EMG measurement was started when the virtual ball began to fall. SPIDAR-HS presented haptic information to the hand at the moment the virtual ball reached to the palm (see Fig. 6). When catching the virtual ball, participant was instructed not to grasp the object with fingers but to catch as much to support it with the palm as possible. This task was carried out three times.

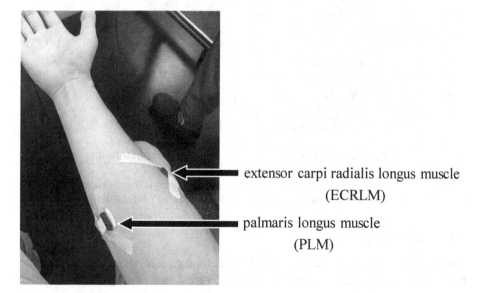

Fig. 4. Surface EMG electrode positions. Palmaris longus muscle (PLM) and extensor carpi radialis longus muscle (ECRLM) are in forearm

Fig. 5. Ball used at Real task. The weight was 220 g and diameter was 80 mm.

Fig. 6. Integration of vision and haptics in VR environment. Free-falling ball was projected on HTC Vive display. Haptics as ball impact force was presented by SPIDAR-HS.

2.4   Data Analysis

LabVIEW2017 (National Instruments) was used for data analysis. For the analysis of EMG data, band-stop filtering 49 Hz – 51 Hz and band-pass filtering 10 Hz – 350 Hz were used, followed by root mean square (RMS) processing.

The moment when the ball began to fall was taken as zero second. Catching Onset Time is the start time of activation of muscles in the forearm before catching a ball. Catching Onset Time was determined by rectified signals exceeded the maximum value at rest. Time when the ball contacted the palm (Contact Time) was calculated from Eq. (1) that was based on Newton's law of motion. Let the time when the ball contacted the palm be t, gravitational acceleration be g and the height of the ball from the palm be h.

$$t = \sqrt{\frac{2h}{g}} \tag{1}$$

3   Results and Discussion

3.1   Results of Experiments

Figure 7 shows the EMGs during the ball catching task in VR and real environments for three participants. In the figure, the up arrow shows Catching Onset Time and the down arrow shows Contact Time. It was not successful to measure ECRLM activity of participant C at VR task due to excessive motor noises contaminated to EMG signals.

3.2   Tendency of Muscular Activity

Figure 8 shows correlation scatter diagrams of the EMG for each Real VR tasks from Catching Onset Time to Contact Time. Correlation coefficient is also shown in each diagram. The high correlation coefficient means that the tendency of the muscular activity at VR task and at Real task is similar.

High correlation coefficients ($r > 0.9$) were found in PLM of participant B and C, and in ECRLM of participant A and B. Relatively high correlation indicated that the tendency of the muscular activity at VR task and at Real task was similar in most cases. ECRLM of participant C showed low correlation coefficient ($r = 0.334$) due to incomplete measurement at VR task. PLM of participant A showed slightly lower correlation coefficient ($r = 0.770$) than the other conditions. It was interpreted that when Catching Onset Time was determined, time delay was occurred between VR and Real tasks. The method of establishing Catching Onset Time in this experiment was the moment of exceeded the maximum rectified EMG signals at rest. According to Ishida et al. [16], there were three methods to establish Catching Onset Time and further analysis should be applied with specific onset time estimation methods.

3.3   Amount of Muscular Activity

Average peak RMS value for forearm EMG at VR task was 45.2% smaller than that at Real task. There were two probable causes for this difference. First, the amount of force generated by SPIDAR-HS at VR task was relatively lower than that made by the gravity force of the ball at Real task. Therefore, less muscular strength of the forearm

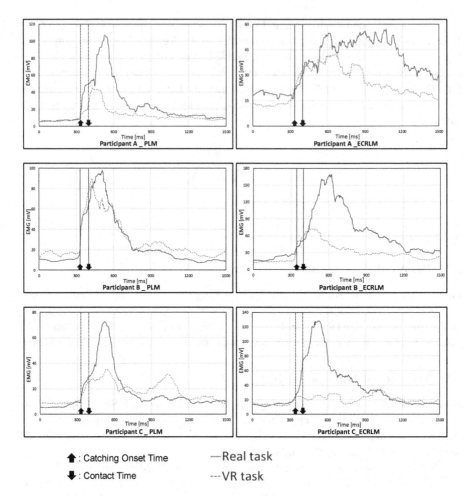

**: Catching Onset Time   —Real task

**: Contact Time   ---VR task

Fig. 7. EMGs during ball caching at VR and Real task for three participants. The vertical line shows the muscular activity and the horizontal one shows elapsed time from the ball began to fall. Left: PLM, Right; ECRLM.

was required for VR task compared to Real task. Second, participants seemed not to flex the forearm very well at VR task because they did not require as much effort to catch the ball as the Real task. For instance, even if they caught the ball with edge of the palm, the virtual ball stayed securely at the palm regardless of the precision of hand location. On the other hand, unless they performed the catching task precise enough such that their palm stayed exactly beneath the center of mass of the falling ball, the ball dropped to the floor at Real task. Therefore, muscle of the forearm was not strong enough at VR task compared to Real task.

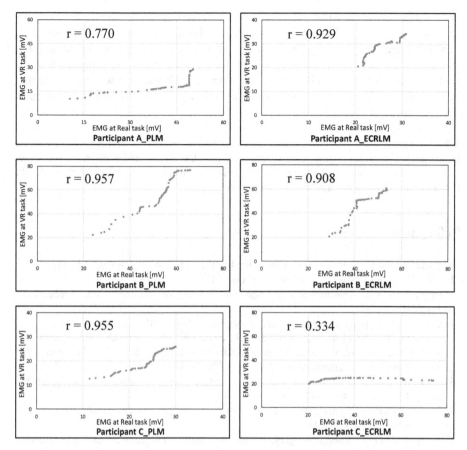

Fig. 8. Correlation of the EMG signals between real task and VR task from catching onset time to contact time. Left: PLM, Right; ECRLM. The high correlation coefficient (r) means that the tendency of the muscular activity at VR task and at real task is similar.

As the amount of muscular activity changed depending on the VR environment, it was indicated that it can be used as an evaluation index of the presence. It was necessary to adjust the amount of force generated by SPIDAR-HS and to correct the ball-falling program for further evaluation of the presence using VR task.

4 Conclusion

In this study, we constructed the haptic VR environment whose visual presentation by HTC Vive, and haptic presentation by SPIDAR-HS. Also, we compared the dynamic muscular activities of the forearm during the ball catching task in VR and real environments, and investigated whether dynamic muscular activity would be an evaluation index of the presence of VR. Dynamic changes in the amount of muscular activity confirmed that the average peak RMS value for forearm EMG at VR task was 45.2%

smaller than that at Real task. The difference in force magnitude was present between the tasks in VR and real environments. SPIDAR-HS could only generate a force less than the gravity of the ball. The tendency of muscular activity confirmed that the muscular loads from the preliminary motion for ball catching to the completion of the catch was similar between VR and real environments. Therefore, it was concluded that the presence of VR environment was observable by changes in dynamic muscular activities during VR tasks.

Acknowledgements. Part of the present study was funded by Environmental control based on human environment interaction research group Kansai University, and Kakenhi of the Japan Society for the Promotion of Science (17H01782). The authors would like to thank Ryuki Tsukikawa and Kanata Nozawa during the data collection.

References

1. Tachi, S., Sato, M., Hirose, M.: Virtual Reality Gaku. Corona Publishing, Japan (2011)
2. HTC Vive. https://www.vive.com/us/ Accessed 18 Jan 2018
3. PlayStation VR. https://www.playstation.com/en-us/explore/playstation-vr/. Accessed 18 Jan 2018
4. Oculus Rift. https://www.oculus.com/rift/ Accessed 18 Jan 2018
5. Tzovaras, D., Nikolakis, G., Fergadis, G., Malasiotis, S., Stavrakis, M.: Design and implementation of haptic visual environments for the training of the visually impaired. IEEE Trans. Neural Syst. Rehabil. Eng. **12**(2), 266–278 (2004)
6. Sato, M., Hirata, Y., Kawarada, H.: Space interface device for artificial reality-SPIDAR-. The IEICE Trans. **74**(7), 887–894 (1991)
7. 3D SYSTEMS 3D Systems Touch Haptic Device. https://ja.3dsystems.com/haptics-devices/touch Accessed 18 Jan 2018
8. Ballantyne, G.H.: Robotic surgery, telerobotic surgery, telepresence, and telementoring review of early clinical results. Surg. Endosc. **16**(10), 1389–1402 (2002)
9. Sanchez-Vives, M.V., Slater, M.: From presence to consciousness through virtual reality. Nat. Rev. Neurosci. **6**, 332–339 (2005)
10. Schuemie, M.J., Van Der Straaten, P., Krijn, M., Van Der Mast, C.: Research on presence in virtual reality: a survey. CyberPsychol. Behav. **4**(2), 183–201 (2004)
11. Egan, D., Brennan, S., Barrett, J., Qiao, Y., Timmerer, C., Murray, N.: An evaluation of heart rate and electrodermal activity as an objective QoE evaluation method for immersive virtual reality environments. In: 2016 Eighth International Conference on Quality of Multimedia Experience (2016)
12. Watanabe, S., Nagano, Y., Okanoya, K., Kawai, N.: A proposal for quantification of immersion in the virtual space - Do the smoothness and delay of the body-movement in the virtual space degrade immersion measured by subjective scales and physiological responses?. The 31st Annual Meeting of the Japanese Cognitive Science Society, pp. 92–95 (2014)
13. Hong, S., Kim, J., Sato, M., Koike, Y.: A research of human's time-to-contact prediction model for ball catching task. Inst. Electron. Inf. Commun. Eng. **88**(7), 1246–1256 (2005)

14. Buoguila, L., Ishii, M., Sato, M.: A large workspace haptic device for human-scale virtual environments. In: First International Workshop on Haptic Human-computer Interaction, pp. 86–91 (2000)
15. Ishii, N., Sa, M., Yamaguchi, N.: Color zukai kinniku no shikumi hataraki jiten. SEITOSHA, Japan (2010)
16. Ishida, M., Tsushima, E.: Comparison with standards judging an onset of EMG activity for measurement of reaction time at gastrocnemius. Japanese J. Rehabil. Med. **29**(9), 843–849 (2001)

A Virtual Kitchen for Cognitive Rehabilitation of Alzheimer Patients

Paul Richard[1]([✉]), Déborah Foloppe[2], and Philippe Allain[2]

[1] Laboratoire d'Ingénierie des Systèmes Automatisés (LARIS),
University of Angers, Angers, France
Paul.Richard@univ-angers.fr
[2] Laboratoire de Psychologie des Pays de la Loire (LPPL),
University of Angers, Angers, France

Abstract. This article presents an innovative interactive tool that has been designed and developed in the context of the preventive treatment of Alzheimer's disease. This tool allows simulating different cooking tasks that the patient has to perform with the computer mouse. The virtual environment is visualized on a simple computer screen. Gradual assistance is provided to the patient so that he/she trains and learns to perform the tasks requested. In order for the training to be relevant and effective, no errors are allowed by the system.

Keywords: Virtual environment · Activity of daily living
Cognitive rehabilitation · Errorless learning · Alzheimer disease

1 Introduction

Cooking and shopping are examples of Instrumental Activities of Daily Living (IADL) which contribute to independent living. Although mainly automatized, IADL can be disturbed by various cognitive impairments (attentional processes, executive functions, short memory), as in Alzheimer's disease [1–3]. During the last decade, research has focused on developing non-pharmacological solutions likely to maintain or enhance independence of people with Alzheimer's disease in everyday activities. In this context, some studies have underlined the importance of Alzheimer's disease patients staying cognitively active to prevent functional deterioration over time [4, 5]. However, to efficiently improve daily functioning, interventions in Alzheimer's disease should (i) employ specific functional tasks involving different ecological contexts, (ii) target simple goals, (iii) structure training tasks, sessions and intervention, (iv) use feedback to keep patients engaged, and (v) tune the difficulty to appropriate levels [6, 7]. Indeed, for instance, interventions using decontextualized tasks to restore cognitive processes have failed to show any effects on everyday life of trained people, as underlined by several studies involving 11,430 participants [8–10], in elderly people with mild cognitive impairments [11, 12]. Alternatively, functional rehabilitation-based interventions directly focus on improving autonomy on specific IADL [13, 14]. In such interventions, patients are confronted to complex situations that they could encountered in their everyday life.

© Springer International Publishing AG, part of Springer Nature 2018
S. Yamamoto and H. Mori (Eds.): HIMI 2018, LNCS 10904, pp. 426–435, 2018.
https://doi.org/10.1007/978-3-319-92043-6_36

- to choose stimulus parameters (e.g., their nature, intensity) and virtual objects (e.g., their appearance, their behaviors), their mode of presentation (e.g., progressive or brutal, accompanied or not by narration);
- manage different levels of difficulty in order to accompany the person in their progression, by choosing
- sufficient difficulty to maintain interest, but reasonable enough not to discourage the person;
- to amplify the sensory or multisensory returns in such a way that they are very perceptible;
- to place the person as the main actor and co-creator of the experienced situation;
- to generate ecologically valid behaviors, soliciting together the components of the functioning;
- to involve the patient in his training (motivation to engage, to repeat successively and to start over another day), because you have to interact to move forward in the scenario, with tasks that are rewarding or enjoyable and verbal encouragement;
- to practice cognitive rehabilitation techniques (e.g., spaced-out recovery, learning method without error, fading technique).

The paper is organized around the following sections (based on [15]). In Sect. 2, we present a review of the approaches proposed in the real world and then we present different approaches and numerical tools developed for the evaluation and rehabilitation of cognitive disorders related to Alzheimer's disease. In Sect. 3 we present our innovative tool based on a virtual kitchen environment. The article ends with a conclusion and proposes ways of improvement.

2   Related Works

Several learning techniques have been used, independently or combined [16–20]. These techniques can be classified as either errorless learning approach or self-generation based approach. Errorless learning approach gathers methods aiming to limit the encoding of non-pertinent information. For this objective, patients receive helps during the training sessions. In addition, they can receive positive feedback which reinforces the encoding of their correct actions. Among these methods, based on implicit learning processes, one can cite "pure" errorless learning method [21], forward or backward chaining methods, the classical vanishing-cues method [22], and spaced-retrieval methods with fixed or increasing time intervals [23]. Contrary to the errorless learning approach, the approach with auto-generation focuses on the solicitation of cognitive processes which underline the formulation of answers, while errors limitation is not a primary goal. It gathers methods aiming to provide the minimum cues which are necessary for the patient to execute the required actions. Thus, patients are first left free to autonomously produce actions and receive feedback in case of difficulties. Trial and error based methods and reversed vanishing-cue methods are examples inscribed in the approach with auto-generation. The errorless approaches with auto-generation have also been proposed [24].

In the context of cognitive rehabilitation of the (I)ADLs in AD, the errorless learning methods have often been naturally combined with specific (i.e., for each step) direct instructions [25, 26]). The principle is based on the limitation of error production during the learning process, strategies to gain commitment of the patients, and a guided practice to promote learning. In practice, the activities to train are firstly structured in a series of actions, to obtain simple and concrete standardized instructions. During the learning process, these instructions can be used to engage, reinforce the correct answer, guide, and/or reduce error production. They can also be of different natures (e.g., oral, physical). In this context, Padilla [27] has recommended adapting the cognitive demand during the activities according to the severity of the functional impairment. Thus, he has proposed the following levels of cueing: a neutral instruction (e.g., it's noon), an oral general direct instruction (e.g., now, prepare a salad with corn, tomato, salad) or an oral specific instruction (e.g., put salad leaves in the bowl), followed by help via gestures, such as showing the target object or the action to perform, and physical priming (e.g., touching the bowl while repeating the specific instruction). Similar hierarchies according to an increasing level of assistance — from the most cognitive help to the most physical help — can be found in the context of the ecological assessment of daily functioning with Alzheimer's disease [28]. Several studies support that the errorless approach and cueing can successfully improve the performance of patients with mild to moderate Alzheimer's disease on various IADLs. Lancioni et al. [29] have obtained positive results by using mainly specific oral instructions and some physical assistance during the patient's performance, if needed. Simard and Grand-maison obtained positive results, maintained over five weeks, using a learning method that combined the spaced retrieval method and three levels of assistance: a full demonstration of the task, prompts with specific oral instructions, and a small measure of guidance if needed when the patient could name and carry out each step [30]. Following a successful trial, the level of assistance decreased and retrieval delay increased for the next trial, whereas if an error occurred, the level of assistance increased and retrieval delay decreased for the next trial. Dechamps et al. [31] obtained positive results maintained over one month using errorless learning and learning by modeling. The authors showed the effectiveness of these two methods over a trial-and-error method. In their errorless learning, oral and specific pictured plus written instructions were given and then hidden. Then, if patients did not perform the action after five seconds, they received as prompting the same pictured plus written instructions. Physical help was given if needed. In learning by modeling, the task was firstly demonstrated with specific oral instructions. Then, if an error occurred, the instruction was repeated, and if, despite this cue, patients were unable to continue, the sequence was shown again.

In the aim to develop functional support, neuropsychology has focused on the contributions of Virtual Reality (VR) [32–34]. In addition to integrating the interests of classic functional rehabilitation offers additional possibilities. In particular, VR allows:

- to propose situations that are difficult to achieve in real life (e.g., to fly under the storm);
- to choose the parameters to modify the emotional, motor and cognitive experience of the real situation simulated;

- to provide complex, appropriate, and controlled multisensory feedback, according to the patient's actions, abilities or level of performance;
- to provide sensory feedback close to those delivered in everyday life (e.g., sound of a coffee maker on the way) or not (e.g., score display, sounds in case of errors);
- to select behavioral interfaces adapted to people's sensory and motor abilities;
- to tune the level of interactivity and difficulty;
- to repeat tasks in a standardized way, while minimizing the environmental variability;
- to place the patient in a secure environment, the consequences of possible errors remaining only virtual (e.g., forget to extinguish the hob);
- to provide portable systems, allowing the therapist to easily intervene in several places, including the patients' home;
- to provide inexpensive, space-saving and autonomous systems to offer remote care, accessible to patients' homes;
- to integrate social aspects through group work (with other sick people or virtual characters) or through interactions with the therapist (real or virtual);
- to collect performance and behavioral data in real time and in a non-intrusive way;
- to interrupt the session at any time if necessary (e.g., external event, comment), to resume training where the person was, or again from the beginning;
- to free the therapist of various practical tasks, allowing him/her to focus on the situation, and make good decisions about the support provided;
- to involve the patient in the training using various motivational techniques, such as verbal encouragement, score display, or a virtual coach.

These benefits have spurred the development of many VR systems aimed at improving autonomy, in various domains and clinical populations. The following parts are intended to illustrate this variety, by the presentation of some systems. The latter place themselves in the desire to improve the daily functioning, in people with deficiencies of origin acquired or innate, but target certain particular behaviors.

3 System Description

We developed a VR based system which allows patients-users to perform a familiarization task and 10 cooking tasks in a learning condition adapted to their disease. We used Unity 3D (http://unity3d.com/, version 4.1.3) as game engine.

3.1 Tasks Design

In order to increase the opportunity for patients to choose the tasks they want to train (as recommended by [35]), we selected 10 cooking activities likely to be useful for pleasant for the French elderly population (Table 1). In addition, the activities were selected to require similar number of utensils and ingredients. To create simple verbal instructions for our error-reduction training method, we broke down each activity into a sequence of 12 or 13 motor steps. The number of steps and also the length and the syntax of instructions were closed between tasks. To ensure that the sequences corresponded to what people are used to doing, we have submitted the scripts of tasks to 10 healthy subjects. This work leads us to rephrase several actions to make the instructions more natural and understandable. Oral instructions were synthetic voices generated from texts and recorded in .mp3 format. We also designed a generic familiarization task which proposes to the patients-users to execute the different basic interactions by manipulating colored geometries in 8 instructions-steps.

Table 1. Short names of activities and corresponding general instructions

Short name	General instruction
Table	Set the table for two people
Coffee	Prepare a cup of coffee with the coffee machine and serve with a sugar piece
Espresso	Prepare a cup of coffee with the espresso machine and serve with a sugar piece
Tea	Prepare a cup of sweet tea using the kettle and serve with a sugar piece and a teaspoon
Breakfast	Toast bread slides for two people and serve with butter and jam on the plates
Vegetables	Prepare frozen vegetables with a stove and serve in the dish
Salad	Prepare a salad-corn-tomato salad with seasoning in the salad bowl
Soup	Heat up a soup in a microwave et serve two ladles into the soup plate
Sandwich	Prepare a sandwich ham-butter-pickle with a napkin around
Cake	Prepare a chocolate cake baked with the electric oven

3.2 Virtual Environment

Based on our tasks design, we established a technical documentation to list:

- all the necessary machines, perishable items and utensils;
- the real dimensions of objects
- the visual feedback potentially used in real-life situations to follow the task progression.

For example, the coffee machine has a water gauge, to check whether there is water in the tank. All the actions were associated with visual and/or auditory feedback to ease immersion, and to inform the user about the state of the task (Fig. 1).

We created the 3D models (e.g., drawer, salad leaf, colliders) using 3D Studio Max (2013/2014). The models have been optimized for real-time interaction. PhotoFiltre Studio X (version 10.7.3) was used to create the textures, either entirely or from photographs of real objects taken in a white box. The models were then exported in .fbx format to be imported into Unity 3D. We also added sounds to provide feedback of actions performed (e.g., opening the salad spinner, heating butter, closing the tap, setting up filter in the filter holder, spoon placed in a cup). To provide a more realistic rendering than in previous versions, we modeled a new kitchen environment, based on real therapeutic cooking. In particular, we have taken over his furniture (dimensions, textures, and relative provisions).

Fig. 1. Kitchen model in 3D Studio Max (left), and kitchen model rendered in Unity 3D (right).

The phase of modeling first required documentation on the actual measurements of objects, as well as on their appearance. Then, the objects were modeled and mapped under 3D Studio Max and then imported into Unity 3D.

3.3 General Software Architecture

Our virtual kitchen consists of three main software components: (i) a menu for configuration, (ii) a tasks management system, (iii) and a data (user's actions) saving manager (see Fig. 2). It was mainly developed using the C# programming language.

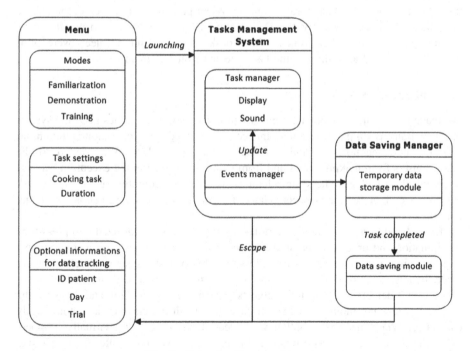

Fig. 2. General software architecture.

The menu allows therapists-users to choose the mode in which the task will be launched.

There are three available modes (i) familiarization, (ii) demonstration, and (iii) training. In the demonstration mode, the task is performed automatically. In the modes familiarization and training, the task can be performed by the actions of patients-users. For the demonstration and training modes, the therapists-users can precise which cooking task she/he wants to run among a list of ten cooking task (the first is selected by default). The setting duration indicates the time given to patients-users to perform actions before the software launches assistance during the training mode. The familiarization mode launches a familiarization task. Finally, the menu allows therapists-users to specify different types of information aiming to facilitate the identification of the recorded data for a subsequent analysis.

The tasks management system instantiates the 3D objects (i.e., kitchen environment, objects required for the launched task), activates the "task manager" script corresponding to the selected Mode, and set the duration parameters if necessary. The task manager indicates the generic task's behaviors, including the ending conditions or the management of data storage and saving. Interactive objects are associated to generic scripts which allow managing the object's behavior in function of some objects properties (i.e., is a lid, is a container, is draggable).

The event manager triggers the modifications corresponding to a performed action. Actions are user actions on a virtual object via the computer mouse or the interactions between two virtual objects. Moreover, at each step of the task in the training mode, the event manager sends information (including, assistances given to the patients-users, time) to the temporary data storage module. When patients-users have completed the task, those data are saved by the Data saving in a CSV file in a row. It allows therapists-users to read data from the local desktop. The required objects were automatically displayed according to the task selected from the main menu.

3.4   Interaction System

We made some objects interactive only when it was required and designed the system of interactions as simple as possible. Indeed, the targeted user population is unfamiliar with the new technologies and the use of the mouse [36]. We wanted to the patients-users from making errors and from being interrupted in the execution because of technical difficulties. The system focuses more on the cognitive aspect of the task performance than on the aspects related to sensory-motor coordination or spatial control.

Each step described in the phase of tasks design can be categorized into press-type manipulation and drag-type manipulation. For the simple actions with a single object (e.g., opening a lid, closing the tap, pressing a button), the system detects a trigger when the left button of the computer mouse was pressed down while the mouse cursor rolled over a detection area. Virtual objects were moved by clicking and dragging the object. Because the mouse is a 2D interface, the depth dimension was automatically calculated. When the mouse button was released, the object was virtually dropped. When the object arrived at a destination of interest (e.g., when the coffee filter arrived at the port filter), the action was performed (e.g., the filter holder was placed in the filter holder). To ease the manipulation, the objects' trajectory was constrained along invisible Bezier curves, which were recalculated in real time from the current position of the object (to be able to release the object mid-way and resume).

4   Conclusion and Future Work

We presented the design and development of an innovative interactive tool based on a virtual kitchen environment that allows performing different tasks. This tool has been proposed for the preventive treatment of Alzheimer's disease. The cooking tasks are performed using computer mouse. Gradual assistance is provided to the patient so that he/she trains and learns to perform the tasks requested. In order for the training to be relevant and effective, no errors are allowed by the system. Preliminary results showed that our tool allowed patients to learn the proposed tasks and to maintain this knowledge over time. In the future work we plan to investigate the use of a tactile surface in order to make easier for the patient to select grad and drop the objects.

References

1. Mahurin, R.K., DeBettignies, B.H., Pirozzolo, F.J.: Structured assessment of independent living skills: preliminary report of a performance measure of functional abilities in dementia. J. Gerontol. **46**(2), 58–66 (1991)
2. Marshall, G.A., Amariglio, R.E., Sperling, R.A., Rentz, D.M.: Activities of daily living: where do they fit in the diagnosis of Alzheimer's disease? Neurodegener. Dis. Manag. **2**(5), 483–491 (2012)
3. Perneczky, R., Pohl, C., Sorg, C., Hartmann, J., Tosic, N., Grimmer, T., Kurz, A.: Impairment of activities of daily living requiring memory or complex reasoning as part of the MCI syndrome. Int. J. Geriatr. Psychiatry **21**(2), 158–162 (2006)
4. Friedland, R.P., Fritsch, T., Smyth, K.A., Koss, E., Lerner, A.J., Chen, C.H., Petot, G.J., Debanne, S.M.: Patients with Alzheimer's disease have reduced activities in midlife compared with healthy control-group members. Proc. Natl. Acad. Sci. USA **98**(6), 3440–3445 (2001)
5. Wilson, R.S., De Leon, C.F.M., Barnes, L.L., Schneider, J.A., Bienias, J.L., Evans, D.A., Bennett, D.A.: Participation in cognitively stimulating activities and risk of incident Alzheimer disease. JAMA **287**(6), 742–748 (2002)
6. Holden, M.K.: Virtual environment for motor rehabilitation: review. Cyber-Pschol. Behav. **8**(3), 187–211 (2005)
7. Sveistrup, H.: Motor rehabilitation using virtual reality. J. Neuroeng. Rehabil. **1**(1), 10 (2004)
8. Owen, A.M., Hampshire, A., Grahn, J.A., Stenton, R., Dajani, S., Burns, A.S., Ballard, C.: Putting brain training to the test. Nature **465**(7299), 775–778 (2010)
9. Plautz, E.J., Milliken, G.W., Nudo, R.J.: Effects of repetitive motor training on movement representations in adult squirrel monkeys: role of use versus learning. Neurobiol. Learn. Mem. **74**(1), 27–55 (2000)
10. Wykes, T., Huddy, V., Cellard, C., McGurk, S.R., Czobor, P.: A meta-analysis of cognitive remediation for schizophrenia: methodology and effect sizes. Am. J. Psychiatry **168**(5), 472–485 (2011)

11. Man, D.W.K., Chung, J.C.C., Lee, G.Y.Y.: Evaluation of a virtual reality-based memory training program for Hong Kong Chinese older adults with questionable dementia: a pilot study. Int. J. Geriatr. Psychiatry **27**(5), 513–520 (2012)
12. Galante, E., Venturini, G., Fiaccadori, C.: Computer-based cognitive intervention for dementia: preliminary results of a randomized clinical trial. Giornale Italiano Medicina del Lavoro ed Ergon. **29**(3), 26–32 (2007)
13. Klinger, E., Chemin, I., Lebreton, S., Marié, R.-M.: Virtual action planning in Parkinson's disease: a control study. CyberPsychol. Behav. **9**(3), 342–347 (2006)
14. Le Gall, D., Allain, P.: Applications des techniques de réalité virtuelle à la neuropsychologie clinique. Champ Psychosomat. **22**(2), 25–38 (2001)
15. Foloppe, D.A.: Evaluation et entraînement des activités de la vie quotidienne dans la maladie d'Alzheimer: Intérêt de la réalité virtuelle (Assessment and training on activities of daily living in Alzheimer disease: Interest of virtual reality). Doctoral thesis, University of Angers (2017). www.theses.fr/s137644
16. Bertens, D., Fasotti, L., Boelen, D.H.E., Kessels, R.P.C.: A randomized controlled trial on errorless learning in goal management training: study rationale and protocol. BMC Neurol. **13**(1), 64 (2013)
17. Clare, L., Wilson, B.A., Breen, K., Hodges, J.R.: Errorless learning of face-name associations in early Alzheimer's disease. Neurocase **5**(1), 37–46 (1999)
18. Dechamps, A., Fasotti, L., Jungheim, J., Leone, E., Dood, E., Allioux, A., Kessels, R.P.C.: Effects of different learning methods for instrumental activities of daily living in patients with Alzheimer's dementia: a pilot study. Am. J. Alzheimer's Dis. Other Dement. **26**(4), 273–281 (2011)
19. Kessels, R.P.C., de Haan, E.H.F.: Implicit learning in memory rehabilitation: a meta-analysis on errorless learning and vanishing cues methods. J. Clin. Exp. Neuropsychol. **25**, 805–814 (2003)
20. Wu, H.S., Lin, L.C., Wu, S.C., Lin, K.N., Liu, H.C.: The effectiveness of spaced retrieval combined with Montessori-based activities in improving the eating ability of residents with dementia. J. Adv. Nurs. **70**(8), 1891–1901 (2014)
21. Baddeley, A.D., Wilson, B.A.: When implicit learning fails: amnesia and the problem of error elimination. Neuropsychologia **32**(1), 53–68 (1994)
22. Glisky, E.L., Schacter, D.L.: Extending the limits of complex learning in organic amnesia: computer training in a vocational domain. Neuropsychologia **27**(1), 107–120 (1989)
23. Camp, C.J., Bird, M.B., Cherry, K.E.: Retrieval strategies as a rehabilitation aid for cognitive loss in pathological agin. In: Hill, R.D., Bäckman, L., Neely, A.S. (eds.) Cognitive Rehabilitation in Old Age, pp. 224–248. Oxford University Press, New York (2000)
24. Tailby, R., Haslam, C.: An investigation of errorless learning in memory-impaired patients: improving the technique and clarifying theory. Neuropsychologia **41**(9), 1230–1240 (2003)
25. Sohlberg, M.M., Turkstra, L.S.: Optimizing Cognitive Rehabilitation: Effective Instructional Methods. Guilford Press, New York (2011)
26. Van der Linden, M., Juillerat, A.-C., Delbeuck, X.: La prise en charge des troubles de la mémoire dans la maladie d'Alzheimer. In: Belin, C., Ergis, A.M., Moreaud, O. (eds.) Actualités sur les démences: Aspects cliniques et neuropsychologiques, pp. 167–197 (2006)
27. Padilla, R.: Effectiveness of interventions designed to modify the activity demands of the occupations of self-care and leisure for people with Alzheimer's disease and related dementias. Am. J. Occup. Ther. **65**(5), 523–531 (2011)
28. Anselme, P., Poncelet, M., Bouwens, S., Knips, S., Lekeu, F., Olivier, C., Majerus, S.: Profinteg: a tool for real-life assessment of activities of daily living in patients with cognitive impairment. Psychol. Belgica **53**(1), 3–22 (2013)

29. Lancioni, G.E., Pinto, K., La Martire, M.L., Tota, A., Rigante, V., Tatulli, E., Oliva, D.: Helping persons with mild or moderate Alzheimer's disease recapture basic daily activities through the use of an instruction strategy. Disabil. Rehabil. **31**(3), 211–219 (2009)

30. Thivierge, S., Simard, M., Jean, L., Grandmaison, E.: Errorless learning and spaced retrieval techniques to relearn instrumental activities of daily living in mild Alzheimer's disease: a case report study. Neuropsychiatr. Dis. Treat. **4**(5), 987–990 (2008)

31. Dechamps, A., Fasotti, L., Jungheim, J., Leone, E., Dood, E., Allioux, A., Robert, P.H., Gervais, X., Maubourguet, N., Olde Rikkert, M.G., Kessels, R.P.: Effects of different learning methods for instrumental activities of daily living in patients with Alzheimer's dementia: a pilot study. Am. J. Alzheimers Dis. Other Dement. **26**(4), 273–281 (2011)

32. De Mauro, A.: Virtual reality based rehabilitation and game technology. In: Blandford, A., De Pietro, G., Gallo, L., Gimblett, A., Oladimeji, P., Thimbleby, H. (eds.) Proceedings of the 1st International Workshop on Engineering Interactive Computing Systems for Medicine and Health Care. CEUR-WS, Pisa, vol. 727, pp. 48–52 (2011)

33. Garcia-Betances, R.I., Jimenez-Mixco, V., Arredondo, M.T., Cabrera-Umpiérrez, M.F.: Using virtual reality for cognitive training of the elderly. Am. J. Alzheimer's Dis. Other Dement. **30**(1), 49–54 (2015)

34. Levac, D.E., Galvin, J.: When is virtual reality «therapy»? Arch. Phys. Med. Rehabil. **94**(4), 795–798 (2013)

35. Choi, J., Twamley, E.: Cognitive rehabilitation therapies for Alzheimer's disease: a review of methods to improve treatment engagement and self-efficacy. Neuropsychol. Rev. **23**(1), 48–62 (2013)

36. Dickinson, A., Arnott, J., Prior, S.: Methods for human-computer interaction research with older people. Behav. Inf. Technol. **26**(4), 343–352 (2007)

Emotion Hacking VR: Amplifying Scary VR Experience by Accelerating Actual Heart Rate

Ryoko Ueoka[1,2(✉)] and Ali AlMutawa[1,2]

[1] Faculty of Design, Kyushu University, 4-9-1 Shiobaru Minami ku,
Fukuoka City, Japan
`r-ueoka@design.kyushu-u.ac.jp`
[2] Graduate School of Design, Kyushu University, 4-9-1 Shiobaru Minami ku,
Fukuoka City, Japan
`zerokizer@gmail.com`

Abstract. An emotion hacking virtual reality (EH-VR) system is an interactive system that hacks one's heartbeat and controls it to accelerate a scary VR experience. The EH-VR system provides vibrotactile biofeedback, which resembles a heartbeat, from the footrest. The system determines a false heartbeat frequency by detecting the user's heart rate in real time. The calculated false heart rate is higher than the user's actual heart rate. This calculation is based on a quadric equation mode that we created. Using this system, we demonstrated at emerging technologies at Siggraph Asia 2016. Approximately 100 people experienced the system and we observed that for all participants, the heart rate was more elevated than in the beginning. Additional experiments endorsed that this effect was possibly caused by the presentation of a false heartbeat by the EH-VR.

Keywords: Emotion · Virtual reality · Scare · False heart rate

1 Introduction

Virtual reality has long been pursued to impart a "sense of reality" by generating various stimulations artificially. Recently it has been argued that the incorporation of emotional feeling is also effective for enhancing an immersive virtual experience. An emotional state is considered to be ambiguous; however, in the cognitive field, it is known that emotion can be evoked as a result of physiological changes in the body induced by the autonomic nervous system [1]. This theory endorses that emotions may be amplified or enhanced by a system that controls the physiological state externally. In the past, various technologies to generate artificial emotions have been proposed [2–4]. These previous studies added a sensory modality to deceive a user's emotions through an external control. Specifically, Fukushima and Kajimoto argued that haptic displays

© Springer International Publishing AG, part of Springer Nature 2018
S. Yamamoto and H. Mori (Eds.): HIMI 2018, LNCS 10904, pp. 436–445, 2018.
https://doi.org/10.1007/978-3-319-92043-6_37

are effective in altering emotions via somatosensory areas that represent emotional body images [5]. The EH-VR system is an interactive system that hacks one's heartbeat and controls it to accelerate a scary VR experience. The system provides vibrotactile stimulus, which resembles a heartbeat, from the floor. It determines a false heartbeat frequency by detecting the user's heart rate in real time and calculates a false heart rate, which is faster than the one observed. By providing the vibrotactile biofeedback of a false heartbeat while measuring the user's heart rate in real time and adjusting the acceleration rate accordingly, we observed that the real heart rate synchronized with the false one more significantly than the step-wise acceleration rate in our previous experiment [6]. Based on this research outcome, we developed an EH-VR walk-through system that provides a virtual experience using vibrotactile biofeedback. We demonstrated the EH-VR system at emerging technologies at Siggraph Asia 2016 held in Macau, China [7]. Approximately 100 people experienced the system and we observed that for all participants, the heart rate was more elevated than in the beginning. Additional experiments endorsed that this effect was possibly caused by the presentation of a false heartbeat by the EH-VR.

2   Related Works

2.1   Effect of False Biofeedback in Terms of Emotion

According to the study by Valins in 1967, false heart sounds had an effect the emotional user's impression of slides of seminude females [8]. Bogus heart sounds, which were faster or slower than a subject's real heat rate, were played while showing slides of seminude females. Subjects believed the heart sound played was their real heart sound, and the faster the false heart rate became, the higher the preference the subject indicated for the slide. Though this early work was a tentative analysis, it suggests that external information may affect a person's emotional state. Nishimura et al. developed a tactile feedback device to provide a false heartbeat vibration on the chest as well as a tactile feedback device implemented in a cushion to provide a false heart rate unobtrusively [9]. They carried on Valins's experiment to evaluate whether a false tactile heartbeat influenced the evaluation of attractiveness of female/male photographs. They reported that false heartbeats could alter physiological or emotional states. Costa et al. evaluated the effect of false vibrotactile biofeedback of the heart rate on the wrist to regulate anxiety [10]. The research clarified the fact that believing that the vibration represented their heart rate, the interoceptive awareness generated by the vibration would affect the regulation of their emotion. These related studies endorse our research to amplify fright by false biofeedback generated using a vibrotactile sensation. However, the purpose and system design are completely different from the one we propose. In terms of software design, their system provided a simple false heartbeat, either increased heart rate or decreased heart rate, and did not calibrate to a subject's heart rate when generating false heartbeats. Our research purpose is to discover an effective way of deriving a false vibrotactile heart rate from a real heart rate, which is not yet certain. Therefore, we develop a vibrotactile feedback system to accelerate a user's real heart rate by referring to a subject's heart rate in real time and employing an acceleration equation model to calculate a false one. Finally, the goal is to discern an effective algorithm for deriving a false vibrotactile heart-beat from a real heartbeat to artificially add the sense of subjective reality to VR content.

2.2   Relationship Between Physiological State and Subjective Fright

Brosschot and Thayer indicated in [11] that the responses of heart rate associated with negative emotions prolonged the recovery of heart rate compared to responses associate with positive emotions. This implies that physiological response and emotional valence are interrelated. In our previous research [12], we concluded that an enclosed environment changed the physiological state as well as the emotional impression of a movie. We found that watching a horror movie in an enclosed space increased the heart rate more significantly rather than watching it in an open space. We analyzed the data collected in a previous experiment with respect to the relationship between physiological change and emotion. The result is summarized in Fig. 1. The graph shows the difference in the mean heart rate for each movie scene on the basis of a pre-test mean heart rate. The line plot shows the mean ratings of subjective assessment of fear for each scene (4 categories on the Likert scale: 1 (not scary) to 4 (extremely scary)) reported by 9 subjects. This suggests that the acceleration of heart rate and the increased ratings of subjective assessment of fear are related. In order to analyze the relationship between the two variables, we utilized single regression analysis. The heart rate value x is an explanatory variable of rating of subjective assessment of fear y; the equation is described as Eq. 1 (p < .01). Although we require further investigation to bring this to a conclusion, this result is a positive endorsement of the fact that the acceleration of heart rate drives the increase in fright. Previous related research suggests that emotion and physiological changes are well-related and thus biofeedback will be effective in changing the emotional state based on external stimulus.

$$y = 0:04437x + 0:47759 \tag{1}$$

3   EH-VR System

3.1   EH-VR Hardware

Figure 2 shows the system. We implemented a cylinder type wireless speaker (JBL Charge2) on the wooden plate set for the footrest. By putting the user's feet on the footrest while wearing an Oculus Rift head-mounted display, a false vibrotactile heart beat is produced through the footrest at sole of the feet while gradually accelerating the heartbeat frequency. In order to measure a user's heart rate to calculate a false heartbeat frequency, we used an ear worn photodiode

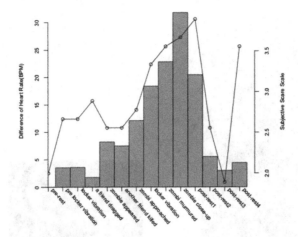

Fig. 1. Mean heart rate and subjective assessment of scare

pulse sensor which measures heart rate variability (HRV). An interactive walk-through in EH-VR is controlled by a wheel chair interface which detects a wheel's speed and direction by the switch of a game controller. As a virtual body can be an important part of a presence-generating experience [13], an avatar that was gender matched and calibrated to the height of the user increased immersion citelin2013. We intentionally used a wheel chair as an interface because sitting on a wheel chair cancels the height difference of each participant, thereby unifying the eye-level position. In addition, we draw a part of a user's legs and a part of the wheel chair from the first-person perspective as Fig. 3 shows.

3.2 EH-VR Software

We recorded a heartbeat sound using a stethoscope and edited it as a sampling sound of a cycle of a heartbeat in order to provide vibrotactile feedback. We developed software that measures the interval time records of the heartbeat of a participant, calculates a false heart rate, and outputs a vibration pattern of the heartbeat through a spealer. In terms of determining the frequency of the false heart rate, we created an acceleration model of a heart rate, which is approximated by a quadratic equation. We derived this assumption from the analysis of data collected in the previous experiment [12]. The line plot in the Fig. 4 shows the state of the mean heart rate aligned as an increase over the baseline heart rate. The blue line illustrates the resulting analysis of 9 subjects (excluded one subject's data from the result of the Cochran's C test) who watched a 4-min horror short movie in a closed locker in an upright position. It shows that the biological boundary of fluctuation of heart rate in the upright position is about 28 bpm faster than the baseline and the outline of the increase obtains an approximate curve of the quadratic equation ($R^2 = 0.9$). Based on this result, we conducted an experiment and determined that calibrating the heart rate

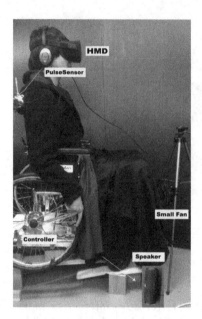

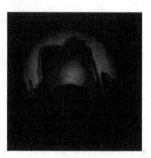

Fig. 2. Emotion hacking VR system

Fig. 3. Sitting posture from the first-person point of view

every 10 s to correct the coefficient of the quadratic equation more effectively accelerated the real heart rate than calculating the false heart rate with a fixed coefficient [6]. In our EH-VR system, a participant maintains a seated posture during the experience. It is known that the biological boundary of fluctuation of heart rate in a seated posture is 30 % lower than that of an upright position [15]. Therefore, we revised the quadratic equation based on the acceleration model shown in Fig. 4 to account for the difference in heart rate of the upright position. In this EH-VR walk-through system, as a person controls walk-through speed by its own pace, we predetermined the fixed position where starting to present false heart rate in the CG world. Figure 4 shows an overview of walk-through space and each alphabet describes the position of either change of the scene or heart-beat feedback. From starting point of A, real heart rate feedback was presented and when it arrived the point B, a false heart rate feedback was presented until the point E.

3.3   EH-VR Contents

With our EH-VR system, we provide a walk-through of the VR experience through artwork called the "Pressure of the unknown" demonstrating a fear metaphor. Figure 6 shows an image of the artwork. The concept of the artwork is as follows:

The "unknown" is a well-known cause of fear but sources of the 'unknown' vary. People are one of these sources, and the larger the number of people staring at a

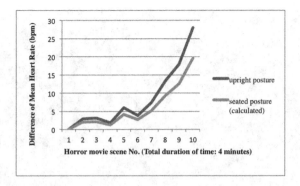

Fig. 4. Heart rate increase of two postures (Color figure online)

Fig. 5. Overall image of the walk-through CG space

person the more'unknowns' are produced in that person's brain. This is similar to what happens during a public speech, where the audience analyzes everything about the presenter and therefore, become a source of fear (Fig. 5).

In this artwork, the viewers go into a closed unfamiliar organic space, constructed with human body parts. A large number of eyes stare at the viewer, trying to imitate the pressure induced by an audience when a presenter is on stage. In addition, attempting to amplify the feeling of fear, viewers sit in a wheelchair to provide a sense of vulnerability and constrained movement.

4   Demonstration at Siggraph Asia 2016

EH-VR was demonstrated at Siggraph Asia 2016 held in Macau, China from Dec 6th to 8th in 2016. The duration of a demonstration for one person was about five to seven minutes. During three days of demonstration, 125 people experienced the contents. Comments from the participants were as follows: "It was a terrifying space", "The vibration of tactile sensation felt like a natural heartbeat", "I was surprised that my heart rate was rising before I noticed". Therefore, it is thought that the EH-VR system was able to induce fearful emotions. Figure 7 shows a person experiencing the EH-VR. The contents of EH-VR was not intended to surprise people with ghosts and zombies but it induces the imagination about the inside of the human body so the tactile stimulation

Fig. 6. A work image of EHVR

emulating the heartbeat sound was effective in enhancing the feeling of walking through the inside of a human body. In addition, we calculated the mean rate of increase in heart rate among 121 participants (4 participants were excluded because of a failure of the recording data). Based on the average heart rate of 74.7 bpm of 121 people before the experience, it increased by 21% on average after the experience. From the perspective of physiological response, as everyone had a rise in heart rate, it was shown that EH-VR amplified fearful emotions. Figure 8 shows one of the examples of a participant who shows an increase in heart rate close to the mean value. It has been shown that the heart rate rises gradually from the beginning of the experience and participants maintain a relatively high heart rate.

Fig. 7. Experience scene at Siggraph Asia 2016

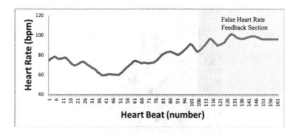

Fig. 8. Change in heart rate of a person close to average increase rate

5 Verification of Heart Rate Rise Effect of EH-VR

In the walkthrough system of EH-VR, the rotation of the wheel of the wheelchair was detected and used as a space movement input such that the user could move through the space arbitrarily. In the system developed in the previous study used for the generation of a false heartbeat [6], it was observed that the subject was in a stationary state and the heart rate increased due to the presentation of false vibrotactile feedback. In the current EH-VR system, since the exercise of rowing a wheelchair was included as a part of the interaction, there is a possibility that the heart rate rose as a result of the exercise. It may also be possible that there was no influence caused by the visual and haptic feedback presented by the system at all. Therefore, a verification experiment was carried out to confirm whether the rise in heart rate occurred in the seated position as well according to the sensory stimulation presentation. Eight subjects performed an EH-VR walk-through experience using the fame controller with all other conditions the same as the conventional EH-VR as Fig. 9 shows. Based on the average heart rate at the beginning, all subjects increased their heart rate by 8% on average (the mean at the beginning was 82.3 BPM, the mean at the end was 89.1 BPM). Figure 10 indicates the heart rate of one subject during the experiment. It shows that the heart rate remained high and was unchanged during the presentation of the false heartbeat as observed in previous studies. Although the number of subjects in this experiment is different from the number of participants at the exhibition and it is difficult to compare, an increase in heart rate was observed without the effect of exercise. Therefore, it can be assumed that the sensory stimulus presentation generated by EH-VR had some influence on the physiological response.

6 Conclusion and Future Works

In this research, we created VR experience content called Emotion Hacking VR (EH-VR) for the purpose of amplifying a scary VR experience. The system was showcased at Siggraph Asia 2016. In a previous study, EH-VR produced a false heartbeat on the sole of a user as a tactile sensation while the user watched

Fig. 9. Subject experiencing EH-VR walk-through system using a game controller

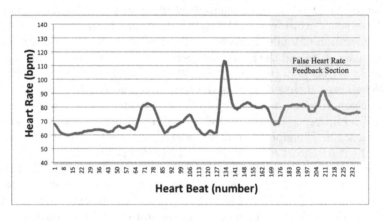

Fig. 10. Change in heart rate of a subject

3D horror content in a standing position. However, in this study we applied a method to raise and maintain the heart rate of a user in a seated position. At the exhibition, all users experienced an increased heart rate within a certain range. It appeared that the system allowed users to enjoy the VR-based content. Based on the additional experiment, it was observed that there was a change in the physiological response via the presentation of the tactile stimulus. As a part of a future study, a method that can generate the experience along with artificial emotions will be developed using a combination of multiple senses.

Acknowledgements. This work was supported by JSPS KAKENHI Grant Number JP16K12514.

References

1. William, J.: What is an emotion? Mind **9**, 188–205 (1884)
2. Fukushima, S., Kajimoto, H.: Facilitating a surprised feeling by artificial control of piloerection on the forearm. In: Proceedings of 3rd Augmented Human International Conference, Article No. 8 (2012)
3. Yoshida, S., Sakurai, S., Narumi, T., Tanikawa, T., Hirose, M.: Incendiary reflection: evoking emotion through deformed facial feedback. In: Proceedings of SIG-GRAPH 2013 Emerging Technologies, Article No. 8 (2013)
4. Ando, H., Watanabe, J., Sato, M.: Empathetic heartbeat exhibited in 2010. http://www.junji.org/eh/
5. Fukushima, S., Kajimoto, H.: Chilly chair: facilitating an emotional feeling with artificial piloerection. In: Proceedings of SIGGRAPH 2012 Emerging Technologies, Article No. 5 (2012)
6. Ueoka, R., Ishigaki, K.: Development of the horror emotion amplification system by means of biofeedback method. In: Yamamoto, S. (ed.) HCI 2015. LNCS, vol. 9173, pp. 657–665. Springer, Cham (2015). https://doi.org/10.1007/978-3-319-20618-9_64
7. Ueoka, R., Almutawa, A.: Emotion hacking VR (EH-VR): amplifying scary VR experience by accelerating real heart rate using false vibrotactile biofeedback. In: Proceedings of Siggraph Asia 2016, Article No. 7 (2016)
8. Valins, S.: Emotionality and information concerning internal reactions. J. Person. Soc. Psychol. **6**(4), 458–463 (1967)
9. Nishimura, N., Ishi, A., Sato, M., Fukushima, S., Kajimoto, H.: Facilitation of affection by tactile feedback of false heartbeat. In: Extended Abstracts of the ACM SIGCHI 2012, pp. 2321–2326 (2012)
10. Costa, J., Adams, A.T., Jung, M.E., Guimbretiere, F., Choudhury, T.: EmotionCheck: leveraging bodily signals and false feedback to regulate our emotions. In: Proceedings of UBICOMP 16, pp. 758–769 (2016)
11. Brosschot, J.F., Thayer, J.F.: Heart rate response is longer after negative emotions than after positive emotions. Int. J. Psychophysiol. **50**, 181–187 (2003)
12. Omori, N., Tsutsui, M., Ueoka, R.: A method of viewing 3D horror contents for amplifying horror experience. In: Yamamoto, S. (ed.) HIMI 2013. LNCS, vol. 8018, pp. 228–237. Springer, Heidelberg (2013). https://doi.org/10.1007/978-3-642-39226-9_26
13. Usoh, M., Arthur, K., Whitton, M.C., Bastos, R., Steed, A., Slater, M., Brooks Jr., F.P.: Walking > walking-in-place > flying, in virtual environments. In: Proceedings of the 26th Annual Conference on Computer Graphics and Interactive Techniques, SIGGRAPH 1999, pp. 359–364 (1999)
14. Lin, Q., Rieser, J.J., Bodenheimer, B.: Stepping off a ledge in an HMD-based immersive virtual environment. In: Proceedings of the ACM Symposium on Applied Perceptin, SAP 2013, pp. 107–110 (2013)
15. Schmidt, R.R.: Human Physiology. Springer-Verlag, Berlin (1989)

The Nature of Difference in User Behavior Between Real and Virtual Environment: A Preliminary Study

Takehiko Yamaguchi[1]([✉]), Hiroki Iwadare[1], Kazuya Kamijo[1],
Daiji Kobayashi[2], Tetsuya Harada[3], Makoto Sato[4],
and Sakae Yamamoto[3]

[1] Tokyo University of Science, Suwa, 5000-1 Toyohira,
Chino-City, Nagano, Japan
tk-ymgch@rs.tus.ac.jp
[2] Chitose Institute of Science and Technology,
758-65 Bibi, Chitose, Hokkaido, Japan
[3] Tokyo University of Science, 6-3-1 Niijuku, Katsushika-ku, Tokyo, Japan
[4] Tokyo Metropolitan University, 6-6 Asahigaoka, Hino-shi, Tokyo, Japan

Abstract. In this study, we examined the effect of different types of behavioral strategy on performance as well as on behavior in three types of different information representation method such as real task environment, VR-based task environment, and MR-based task environment in order to identify some features that enable to be applied for performance-based/behavioral-based measurement for the characterization of the SoE and its sub-components. As the results, we found that there was a significant difference in task performance such as time completion time, and parameter of time-to-collision distribution, as well as on user behavior such as decomposed motion data.

Keywords: The sense of embodiment · Virtual reality
Singular value decomposition method

1 Introduction

Virtual reality (VR) is a technology that enables to create a fully three-dimensional computer-generated environment in which a person can move around as well as interact as if he/she actually were in the virtual space [1]. VR has made great strides in the past 20 years, having great potential as sites for research in social, behavioral and economic sciences, as well as in human-centered computer science [2]. Because VR enables to afford a user to walk through a computer-generated environment that can be controlled to assess hypotheses that are hard to examine systematically in the real world [3].

According to the recent technology, it is possible to develop highly immersive and presence evoking environment; that is, a head-mounted display (HMD) based virtual environment. In this immersive virtual environment, user's viewpoint is fixed on the eyes of a virtual body which substitute a user's own biological body with synchronous

© Springer International Publishing AG, part of Springer Nature 2018
S. Yamamoto and H. Mori (Eds.): HIMI 2018, LNCS 10904, pp. 446–462, 2018.
https://doi.org/10.1007/978-3-319-92043-6_38

vasomotor feedback, so that when the user moves in the real world, his/her virtual body moves in the virtual world in real time and synchronously [4].

1.1 The Sense of Embodiment in Virtual Reality

The sense of embodiment (SoE) is an essential component of user experience in immersive virtual environments through the embodied virtual body. The SoE consists of three subcomponents such as the sense of self-location, the sense of agency, and the sense of ownership [5]. The sense of self-location is a determinate volume in space where a user feels located. The self-location, as well as body-space normally coincide in the sense that a user feels self-located inside the user's own biological body [6]. The sense of agency is a sensation defined as *"global motor control, including the subjective experience of action, control, intention, motor selection and the conscious experience of will"* [7]. The sense of ownership defined as a user's self-attribution of a body, influencing by morphological similarities as well as by spatial correlation of the body and so on [8].

To enhance the SoE would be to enhance each of its three subcomponents in order to design more immersive experience with an embodied virtual body in the virtual environment. In this line of the enhancement, it is essential to measure the effect of these subcomponents on different factors.

1.2 The Measurement of the Sense of Embodiment

The measurement of the SoE including its subcomponents usually relies on questionnaires or physiological responses [6, 9, 10]. However, there is no explicit measure of the SoE as well as its subcomponents for the moment, including performance-based as well as behavioral-based measurement.

This study aims to explore the nature of difference on performance as well as on behavior in a different kind of information representation methods such as real environment, VR environment, and Mixed Reality environment to identify some features that enable to be applied for performance-based/behavioral-based measurement for the characterization of the SoE and its subcomponents.

2 Methods

2.1 Participants

Eleven students were recruited from our University. All of the participants were male whose mean age was 21 years ($SD = 0.31$). All participants had little or no experience with Virtual Reality as well as Mixed Reality application.

2.2 Apparatus

Visual Display/Rendering. HTC Vive head-mounted display was used for the experimental task. The Vive consists of a headset, two controllers, and two infrared laser emitter units. The headset covers a nominal field of view about 110° through two 1080 × 1200 pixel display that is updated at 90 Hz. As for the visual rendering, Unity3D was employed, rendering the graphics at 60 Hz.

3D Motion Tracker. HTC Vive tracker was used for the experimental task. The Vive tracker is a motion tracker that enables to measure its 3D position as well as orientation in real time. The field of view of the tracker is 270°. The Tracker weighs is 85 g and the size is 99.65 mm (Diameter) × 42.27 mm (Height).

Data Acquisition. AIO-160802AY-USB was used for the data acquisition in the experiment. AIO-16080AY-USB is a high-precision analog I/O terminal which has analog input (16 bit, 8ch), and analog output (16 bit, 2ch). The maximum conversion speed of the terminal is 10 μs.

Real Task Environment. A custom experimental task called a "rod tracking task" was developed for this study. The developed experimental environment had a square wave-shaped slit on a paper board made by grid paper to provide a path for a rod tracking by a participant (See Fig. 1). The system captured data regarding trial time; whether the user's rod was touching a wall along the slit; the vive tracker position as the rod location in x, y, and z coordinates.

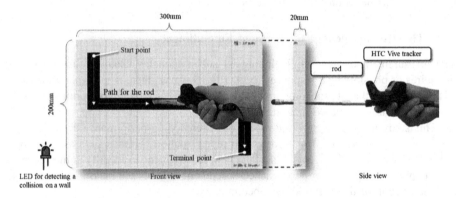

Fig. 1. A rod tracking task in a real environment. A LED light turns on when the rod collides with a wall along the slit. The applied voltage was measured on the AIO-16080AY-USB.

Virtual Reality-Based Task Environment. A rod tracking task using VR representation was developed based on the real task environment, utilizing HTC Vive head mounted display (See Fig. 2). The real task environment was 3D reconstructed in a virtual environment. The virtual environment was visually represented through the HMD. However, the paper board in the virtual environment was not haptically represented. The system captured data regarding trial time; whether the user's rod was touching a wall along the slit in the virtual environment; the vive tracker position as the rod location in x, y, and z coordinates.

Mixed Reality-Based Task Environment. A rod tracking task using MR representation was developed based on the real task environment, utilizing HMD as well as the

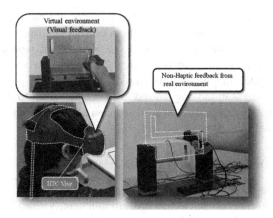

Fig. 2. A rod tracking task in a VR-based environment.

paper board in the real task environment (See Fig. 3). The real task environment was 3D reconstructed in a virtual environment. The position of the paper board in the real environment was calibrated to the position of the paper board in the virtual environment; that is, the paper board was visually represented in the virtual environment through the HMD, as well as was haptically represented in the real environment. The system captured data regarding trial time; whether the user's rod was touching a wall along the slit in the virtual/real environment; the vive tracker position as the rod location in x, y, and z coordinates.

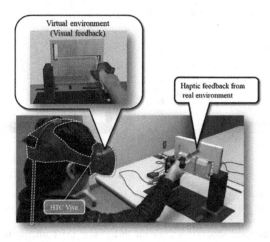

Fig. 3. A rod tracking task in a MR-based environment.

2.3   Experimental Task

The protocol for the experimental task was identical for all conditions. A start point on the left top of the path on the paper board is touched with a rod grabbing by a participant for three seconds. After counting three seconds, the rod would be moved by the participant to the terminal point on the right bottom of the path on the paper board (trial 1). When it reaches the terminal point, the participant would stay for three seconds, then move back to the start point (trial 2). These trials would be repeatedly conducted for three times; that is, six trials would be performed by participants in total. Participants were tasked with moving the rod from the start point to the terminal point as quickly and accurately as possible, as well as much as possible without touching the wall along the slit with the rod.

2.4   Experiment Design and Independent Variables

To examine the effect of behavioral strategy on performance as well as on behavior in the custom task environments such as real task environment, VR task environment, and MR task environment, a balanced 2×3 within subjects factorial experiment design was used, which result in 6 experimental tasks. A summary of the independent variables is shown in Table 1.

Behavioral Strategy. Behavioral strategy for the experimental task was designed to assist two purposes: (1) to control the participant's intention to move the rod accurately on the center line of the path, relating body-image construction regarding the correct physical motion, (2) not to provide any cues for controlling the participant's intention regarding physical motion (non-body-image construction) so that participants are required to conduct the task in standard manner of the experimental task; that is, with moving the rod from the start point to the terminal point as quickly and accurately as possible, as well as much as possible without touching the wall along the slit with the rod. These concepts were illustrated in Fig. 4.

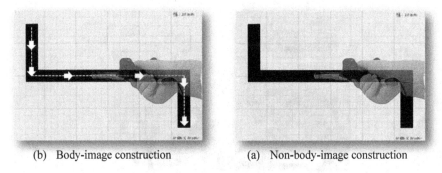

(b)   Body-image construction   (a)   Non-body-image construction

Fig. 4. The two behavioral strategy applied in this study; (a) Participants were required to imagine the straight line on the center of the path, as well as to move on the line as accurately as possible. (b) There is no cue regarding the body-image construction.

Information Representation Method. The three distinct representation method was applied such as Real task environment, VR-based task environment, and MR-based task environment.

Table 1. Summary of Independent variables.

Factor	Level	Definition
Behavioral strategy	None body image	There is no cue regarding the body-image construction
	Body image	Participants were required to imagine the straight line on the center of the path, as well as to move on the line as accurately as possible.
Information representation method	Real task environment	Visual/haptic feedback from real environment
	VR task environment	Visual feedback from virtual environment, No haptic feedback
	MR task environment	Visual feedback from virtual environment, haptic feedback from real environment

2.5 Dependent Variables

To systematically investigate the effect of behavioral strategy in the custom task environments, this study used several dependent measures, which can be categorized into two types of variables: (a) task performance including task completion time and, a probability distribution of time-to-collision; (b) user behavior.

Task Performance. Task performance includes task completion time, and a probability distribution of time-to-collision. Task completion time was defined as the time from when participants start the first trial to the moment they finish the last trial except for the time for the three seconds counting on each start/terminal point. A probability distribution of time-to-collision is a probability distribution of the time to collision with the wall on the path, which was characterized based on the Gamma distribution (See Fig. 5). The distribution can be represented by the Eq. (1).

$$f(x) = \frac{1}{\Gamma(k)\theta^k} x^{k-1} e^{-\frac{x}{\theta}} \quad (x > 0) \tag{1}$$

The gamma distribution is continuous probability distribution which has two parameters such as a shape parameter k, as well as scale parameter θ. These parameters were estimated using the maximum likelihood estimation method, as well as was used as dependent variables.

User Behavior. A user's behavior interacting with the task was analyzed using Singular value decomposition (SVD) method. $x_n^i \in R^N$ is a time series data of the position of the rod in nth trial. A set of trial data for subject i is represented by X^i described as

$$X^i = \left(\{x_1^i\}^T \{x_2^i\}^T \cdots \{x_n^i\}^T \right)^T \tag{2}$$

where n is the number of trials in the experimental task. A set of behavior data represented by X^i can be formed as matrix D (See Eq. (3)), which represents matrix for subject group. The number of the data point of each X^i was standardized using interpolation technique.

$$D = \left(X^1 X^2 \cdots X^M\right) \tag{3}$$

The singular value decomposition (SVD) method was applied to the matrix D in order to extract similarities of behaviors within the subject group. As a result of the SVD, the matrix D can be decomposed as follows:

$$D = U \Sigma V^T \tag{4}$$

Where U is a unitary matrix which has left singular vector $u_i \in R^{n \cdot N}$ as its elements. The matrix V is a unitary matrix which has right singular vector $v_i \in R^M$ as its elements. The matrix Σ is a diagonal matrix which has a set of singular values $\sigma_i (i = 1, 2, \cdots, M)$ as its diagonal elements. The value of similarities was decomposed as the σ_i so that the behavior of individual X^i can be represented using left singular vector u_i, and right singular vector v_i as follows:

$$X^i = \sum_{j=1}^{n} \sigma_j v_{i,j} u_i \tag{5}$$

If the value of 1st singular value σ_1 is high, it indicates that decomposed time series data are similar; that is, participants did almost same physical motion in all of the trials. The similarity based on the singular value was defined as a dependent variable.

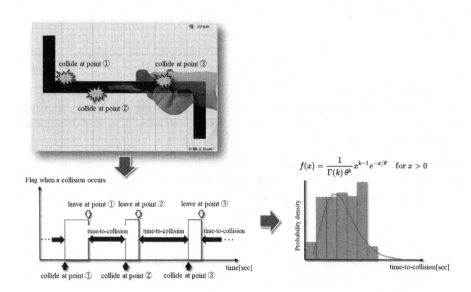

Fig. 5. The concept of the probability distribution of time-to-collision.

2.6   Procedure

Participants were required to read and signed an informed consent. After completion of the introduction paperwork, short training was conducted to familiarize participants with interacting with the custom experimental environment. Participants were informed of their goal to move as quickly and accurately in each trial.

Following the training session, the experimental tasks began. Each participant completed 6 tasks in a randomized order, with each task consisting of 6 trials.

2.7   Research Question

This study seeks to examine the effect of different types of behavioral strategy on performance as well as on behavior in three types of different information representation method such as real task environment, VR-based task environment, and MR-based task environment. We assumed that the result from the experiment in the real environment could be defined as the normal state of the SoE. The result from the all the other data in the virtual environment would be compared to the result from the real environment in order to identify some features that enable to be applied for performance-based/behavioral-based measurement for the characterization of the SoE and its subcomponents.

3   Results

3.1   Task Performance

Task Completion Time. A 2 × 3 ANOVA with Behavioral Strategy (Body-image, None body-image) and Representation method (RL, VR, MR) as within-subject factors revealed a main effect on Representation method, $F(2, 20) = 5.017, p < .038, \eta_p^2 = .334$. A post hoc comparison (Ryan's method, $\alpha = .05$) indicated that there is significant difference between RL ($M = 84.76$ s, $SD = 8.09$ s) and MR ($M = 97.98$ s, $SD = 2.96$ s) condition. However, there is no significant difference between VR and MR condition, as well as between RL and VR condition. The main effect was not qualified by an interaction between Representation method and Behavioral strategy, $F(2, 20) = 1.001, p < .355, \eta_p^2 = .091$. The results showed in Table 2.

Table 3 showed the result of descriptive statistics for the task completion time.

The descriptive plot for the time completion time was illustrated in Fig. 6.

Shape Parameter of the Gamma Distribution. A 2 × 3 ANOVA with Behavioral Strategy (Body-image, None body-image) and Representation method (RL, VR, MR) as within-subject factors revealed a main effect on Representation method, $F(2, 20) = 6.327, p < .025, \eta_p^2 = .388$. A post hoc comparison (Ryan's method, $\alpha = .05$) indicated that there is significant difference between VR ($M = 10.52$, $SD = 0.95$) and MR ($M = 2.19$, $SD = 0.31$) condition, as well as between RL ($M = 2.63$, $SD = 1.49$) and VR ($M = 10.52$, $SD = 0.95$) condition. However, there is no significant difference between RL and MR condition. The main effect was not qualified by

Table 2. The result of the ANOVA on the task completion time.

	Sphericity correction	Sum of squares	df	Mean square	F	p	η^2	η_p^2	ω^2
Representation method	None	2070.6	2	1035.3	5.017	0.017	0.334	0.334	0.259
	Greenhouse-Geisser	2070.6	1.252	1653.3	5.017	0.038	0.334	0.334	0.259
Residual	None	4127.6	20	206.4					
	Greenhouse-Geisser	4127.6	12.525	329.6					
Behavioral strategy	None	366.2	1	366.2	1.389	0.266	0.122	0.122	0.031
	Greenhouse-Geisser	366.2	1	366.2	1.389	0.266	0.122	0.122	0.031
Residual	None	2636.5	10	263.7					
	Greenhouse-Geisser	2636.5	10	263.7					
Representation method * behavioral strategy	None	462.7	2	231.3	1.001	0.385	0.091	0.091	0
	Greenhouse-Geisser	462.7	1.237	374.2	1.001	0.355	0.091	0.091	0
Residual	None	4620.1	20	231					
	Greenhouse-Geisser	4620.1	12.365	373.6					

Table 3. The descriptive statistics for the task completion time.

Behavioral strategy	Information representation method			Mean	SD
	RL (N = 11)	VR (N = 11)	MR (N = 11)		
Body-image	90.49	93.81	100.07	94.79	4.87
None body-image	79.04	95.31	95.89	90.08	9.56
Mean	84.76	94.56	97.98		
SD	8.09	1.06	2.96		

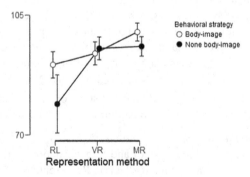

Fig. 6. The descriptive plot for the time completion time. The error bars represent standard error of means.

an interaction between Representation method and Behavioral strategy, $F(2, 20) = .277, p < .648, \eta_p^2 = .027$. The results showed in Table 4.

Table 4. The result of the ANOVA on the shape parameter of the Gamma distribution.

	Sphericity correction	Sum of squares	df	Mean square	F	P	η^2	η_p^2	ω^2
Representation method	None	965.238	2	482.619	6.327	0.007	0.388	0.388	0.317
	Greenhouse-Geisser	965.238	1.128	855.463	6.327	0.025	0.388	0.388	0.317
Residual	None	1525.5	20	76.275					
	Greenhouse-Geisser	1525.5	11.283	135.201					
Behavioral strategy	None	2.735	1	2.735	0.041	0.843	0.004	0.004	0
	Greenhouse-Geisser	2.735	1	2.735	0.041	0.843	0.004	0.004	0
Residual	None	664.654	10	66.465					
	Greenhouse-Geisser	664.654	10	66.465					
Representation method * behavioral strategy	None	32.919	2	16.46	0.277	0.761	0.027	0.027	0
	Greenhouse-Geisser	32.919	1.184	27.795	0.277	0.648	0.027	0.027	0
Residual	None	1189.623	20	59.481					
	Greenhouse-Geisser	1189.623	11.844	100.443					

Table 5 showed the result of descriptive statistics for the shape parameter of the Gamma distribution.

Table 5. The descriptive statistics for the shape parameter of the Gamma distribution.

Behavioral strategy	Information representation method			Mean	SD
	RL (N = 11)	VR (N = 11)	MR (N = 11)		
Body-image	1.58	11. 19	1.98	4.92	5.44
None body-image	3.69	9.85	2.42	5.32	3.97
Mean	2.63	10.52	2.20		
SD	1.50	0.95	0.31		

The descriptive plot for the shape parameter of the Gamma distribution was illustrated in Fig. 7.

Scale Parameter of the Gamma Distribution. A 2×3 ANOVA with Behavioral Strategy (Body-image, None body-image) and Representation method (RL, VR, MR) as within-subject factors revealed a main effect on Representation method, $F(2, 20) = 4.808, p < .027, \eta_p^2 = .325$. A post hoc comparison (Ryan's method, $\alpha = .05$)

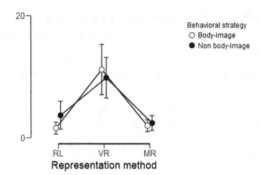

Fig. 7. The descriptive plot for the shape parameter of the Gamma distribution. The error bars represent standard error of means.

indicated that there is significant difference between VR (M = 3.19, SD = 0.54) and MR (M = 5.25, SD = 0.39) condition, as well as between RL (M = 5.14, SD = 0.59) and VR (M = 3.19, SD = 0.54) condition. However, there is no significant difference between RL and MR condition. The main effect was not qualified by an interaction between Representation method and Behavioral strategy, $F(2, 20)$ = 1.456, $p <$ .259 η_p^2 = .127. The results showed in Table 6.

Table 6. The result of the ANOVA on the scale parameter of the Gamma distribution.

	Sphericity correction	Sum of squares	df	Mean square	F	P	η^2	η_p^2	ω^2
Representation method	None	58.652	2	29.326	4.808	0.02	0.325	0.325	0.249
	Greenhouse-Geisser	58.652	1.667	35.186	4.808	0.027	0.325	0.325	0.249
Residual	None	121.998	20	6.1					
	Greenhouse-Geisser	121.998	16.669	7.319					
Behavioral strategy	None	0.746	1	0.746	0.091	0.769	0.009	0.009	0
	Greenhouse-Geisser	0.746	1	0.746	0.091	0.769	0.009	0.009	0
Residual	None	81.67	10	8.167					
	Greenhouse-Geisser	81.67	10	8.167					
Representation method * behavioral strategy	None	7.971	2	3.985	1.456	0.257	0.127	0.127	0.038
	Greenhouse-Geisser	7.971	1.53	5.209	1.456	0.259	0.127	0.127	0.038
Residual	None	54.745	20	2.737					
	Greenhouse-Geisser	54.745	15.301	3.578					

Table 7 showed the result of descriptive statistics for scale parameter of the Gamma distribution.

Table 7. The descriptive statistics for the scale parameter of the Gamma distribution.

Behavioral strategy	Information representation method			Mean	SD
	RL (N = 11)	VR (N = 11)	MR (N = 11)		
Body-image	5.55	2.82	5.53	4.63	1.57
None body-image	4.72	3.57	4.97	4.42	0.74
Mean	5.14	3.20	5.25		
SD	0.59	0.54	0.40		

The descriptive plot for the scale parameter of the Gamma distribution was illustrated in Fig. 8.

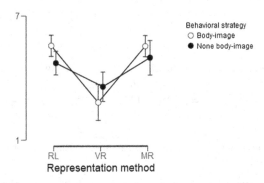

Fig. 8. The descriptive plot for the scale parameter of the Gamma distribution. The error bars represent standard error of means.

3.2   User Behavior

Cumulative Contribution Ratio for Singular Value (Mode 1). A 2 × 3 ANOVA with Behavioral Strategy (Body-image, None body-image) and Representation method (RL, VR, MR) as within-subject factors revealed no significant difference. The results showed in Table 8. The ANOVA was performed using nine participants data since there was missing data on two participants out of eleven.
Table 9 showed the result of descriptive statistics for the cumulative contribution ratio for singular value (mode 1).
   The descriptive plot for the scale parameter of the cumulative contribution ratio for singular value (mode 1) in Fig. 9.

Cumulative Contribution Ratio for Singular Value (Mode 1 and Mode 2). A 2 × 3 ANOVA with Behavioral Strategy (Body-image, None body-image) and Representation method (RL, VR, MR) as within-subject factors revealed a main effect on Behavioral strategy, $F(1, 8) = 9.171$, $p < .016$, $\eta_p^2 = .534$. The main effect was not qualified by an interaction between Representation method and Behavioral strategy, $F(2, 16) = .134$, $p < .813$ $\eta_p^2 = .017$. The results showed in Table 10. The ANOVA

Table 8. The result of the ANOVA on the cumulative contribution ratio for singular value (mode 1)

	Sphericity correction	Sum of squares	df	Mean square	F	P	η^2	η_p^2	ω^2
Representation method	None	22.551	2	11.276	1.75	0.205	0.18	0.18	0.073
	Greenhouse-Geisser	22.551	1.613	13.985	1.75	0.213	0.18	0.18	0.073
Residual	None	103.08	16	6.443					
	Greenhouse-Geisser	103.08	12.901	7.99					
Behavioral strategy	None	4.1	1	4.1	0.924	0.365	0.104	0.104	0
	Greenhouse-Geisser	4.1	1	4.1	0.924	0.365	0.104	0.104	0
Residual	None	35.495	8	4.437					
	Greenhouse-Geisser	35.495	8	4.437					
Representation method * behavioral strategy	None	0.176	2	0.088	0.019	0.981	0.002	0.002	0
	Greenhouse-Geisser	0.176	1.567	0.112	0.019	0.96	0.002	0.002	0
Residual	None	72.665	16	4.542					
	Greenhouse-Geisser	72.665	12.539	5.795					

Table 9. The descriptive statistics for the scale parameter of the cumulative contribution ratio for singular value (mode 1)

Behavioral strategy	Information representation method			Mean	SD
	RL (N = 9)	VR (N = 9)	MR (N = 9)		
Body-image	58.19	59.81	59.38	59.13	0.84
None body-image	57.78	59.26	58.69	58.57	0.75
Mean	57.98	59.53	59.03		
SD	0.29	0.39	0.49		

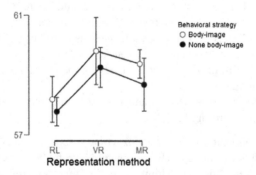

Fig. 9. The descriptive plot for the scale parameter of the cumulative contribution ratio for singular value (mode 1). The error bars represent standard error of means.

was performed using nine participants data since there was missing data on two participants out of eleven.

Table 11 showed the result of descriptive statistics for the cumulative contribution ratio for singular value (mode 1 and mode 2).

The descriptive plot for the scale parameter of the cumulative contribution ratio for singular value (mode 1 and mode 2) in Fig. 10.

Table 10. The result of the ANOVA on the cumulative contribution ratio for singular value (mode 1 and mode 2)

	Sphericity correction	Sum of squares	df	Mean square	F	P	η^2	η_p^2	ω^2
Representation method	None	7.548	2	3.774	1.225	0.32	0.133	0.133	0.023
	Greenhouse-Geisser	7.548	1.874	4.027	1.225	0.319	0.133	0.133	0.023
Residual	None	49.309	16	3.082					
	Greenhouse-Geisser	49.309	14.995	3.288					
Behavioral strategy	None	16.861	1	16.861	9.171	0.016	0.534	0.534	0.45
	Greenhouse-Geisser	16.861	1	16.861	9.171	0.016	0.534	0.534	0.45
Residual	None	14.707	8	1.838					
	Greenhouse-Geisser	14.707	8	1.838					
Representation method * behavioral strategy	None	0.661	2	0.331	0.134	0.875	0.017	0.017	0
	Greenhouse-Geisser	0.661	1.472	0.449	0.134	0.813	0.017	0.017	0
Residual	None	39.388	16	2.462					
	Greenhouse-Geisser	39.388	11.775	3.345					

Table 11. The result of the ANOVA on the cumulative contribution ratio for singular value (mode 1 and mode 2)

Behavioral strategy	Information representation method			Mean	SD
	RL (N = 9)	VR (N = 9)	MR (N = 9)		
Body-image	92.89	92.80	93.73	93.14	0.52
None body-image	91.60	91.99	92.47	92.02	0.44
Mean	92.24	92.40	93.10		
SD	0.91	0.57	0.89		

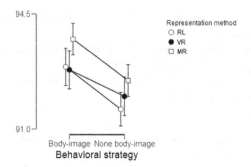

Fig. 10. The descriptive plot for the scale parameter of the cumulative contribution ratio for singular value (mode 1 and mode 2). The error bars represent standard error of means.

4 Discussion

4.1 Task Performance

Task Completion Time. Research in multimodal feedback suggest improvements in performance when additional modalities are used for user activities [11]. In this case, performance improvements were expected with additional modality, namely haptic feedback. However, there was no significant difference between VR and MR condition. In addition, although the haptic feedback in RL and MR conditions was totally same, there was a significant difference between RL and MR condition. These results indicate that the visual feedback affects the performance; that is, the consistency of visual feedback must be considered.

Shape and Scale Parameter of the Gamma Distribution. Scale parameter θ has the effect of stretching or compressing the range (time-to-collision) of the Gamma distribution. Whereas, shape parameter k controls the shape of the family of distributions. The fundamental shapes of the Gamma distribution are characterized by values of k; (1) When $k < 1$: the Gamma distribution can be exponentially formed as well as asymptotic to both the vertical and horizontal axes, (2) When $k = 1$: A Gamma distribution with shape parameter $k = 1$ as well as scale parameter θ is the same as an exponential distribution of scale parameter θ, (3) When $k > 1$: the Gamma distribution assumes a unimodal, namely skewed shape. The skewness reduces when k increases.

The result of these parameters indicates that the Gamma distribution on VR condition assumes more skewed than that of RL and MR conditions; that is, a collision to the wall during each trial on VR task occurs much more constantly than all the other task. According to the participants' impression when they play the task, they reported they could conduct the task easily when there is haptic feedback.

4.2   User Behavior

Cumulative Contribution Ratio for Singular Value (Mode 1). The result indicates that extracted motion pattern has similarity in 1st mode singular value. However, the cumulative contribution ratio for 1st mode was about between 53.54–66.35% in all conditions so that it would not be possible to reconstruct original motion data only using 1st mode data. The remaining about 40% data would be expected to spread out in other mode singular value; that is, feature difference between conditions would be shown in later mode value.

Cumulative Contribution Ratio for Singular Value (Mode 1 and Mode 2). The result indicate that decomposed motion pattern has the difference in Behavioral strategy factor. We performed ANOVA on the other combination in cumulative contribution ratio, however, there was no significant difference on all of factors; that is, feature difference was spread out by 2nd mode singular value (88.68 ∼ 95.88%). We did not perform deeper analysis on difference in this study. However, these behavioral differences must be considered as a feature to identify the quality of the SoE as well as its subcomponents.

5   Conclusion

In this study, we examined the effect of different types of behavioral strategy on performance as well as behavior in three types of different information representation method such as real task environment, VR-based task environment, and MR-based task environment in order to identify some features that enable to be applied for performance-based/behavioral-based measurement for the characterization of the SoE and its sub-components. Especially, we focused to explore what kind of difference it will reveal regarding performance as well as behavior data.

As the results, we found that there was a significant difference in task performance such as time completion time, and parameter of time-to-collision distribution, as well as on user behavior such as decomposed motion data. However, since this study was conducted as a preliminary study, more large number of samples should be analyzed as well as compared to the result of traditional questionnaire measurement of the SoE and its subcomponents as future work.

Acknowledgements. This work was supported by the Grant-in-Aid for Scientific Research (B) from Ministry of Education, Japan, Grant Number: 17H01782.

References

1. Satava, R.M.: Virtual reality surgical simulator. Surg. Endosc. **7**, 203–205 (1993)
2. Bainbridge, W.S.: The scientific research potential of virtual worlds. Science **317**, 472–476 (2007)
3. Fink, P.W., Foo, P.S., Warren, W.H.: Catching fly balls in virtual reality: a critical test of the outfielder problem. J. Vis. **9**, 14 (2009)

4. Banakou, D., Groten, R., Slater, M.: Illusory ownership of a virtual child body causes overestimation of object sizes and implicit attitude changes. Proc. Natl. Acad. Sci. **110**, 12846–12851 (2013)
5. Kilteni, K., Groten, R., Slater, M.: The sense of embodiment in virtual reality. Presence: Teleoper. Virtual Environ. **21**, 373–387 (2012)
6. Lenggenhager, B., Mouthon, M., Blanke, O.: Spatial aspects of bodily self-consciousness. Conscious. Cogn. **18**, 110–117 (2009)
7. Blanke, O., Metzinger, T.: Full-body illusions and minimal phenomenal selfhood. Trends Cogn. Sci. **13**, 7–13 (2009)
8. Argelaguet, F., Hoyet, L., Trico, M., Lécuyer, A.: The role of interaction in virtual embodiment: effects of the virtual hand representation. In: 2016 IEEE Virtual Reality (VR), pp. 3–10. IEEE (2016)
9. Longo, M.R., Schüür, F., Kammers, M.P.M., Tsakiris, M., Haggard, P.: What is embodiment? A psychometric approach. Cognition **107**, 978–998 (2008)
10. Aspell, J.E., Lenggenhager, B., Blanke, O.: Keeping in touch with one's self: multisensory mechanisms of self-consciousness. PLoS ONE **4**, e6488 (2009)
11. Lee, J.-H., Poliakoff, E., Spence, C.: The effect of multimodal feedback presented via a touch screen on the performance of older adults. In: Altinsoy, M.E., Jekosch, U., Brewster, S. (eds.) HAID 2009. LNCS, vol. 5763, pp. 128–135. Springer, Heidelberg (2009). https://doi.org/10.1007/978-3-642-04076-4_14

A Fingertip Glove with Motor Rotational Acceleration Enables Stiffness Perception When Grasping a Virtual Object

Vibol Yem$^{(\boxtimes)}$ and Hiroyuki Kajimoto

The University of Electro-Communications, Chofu, Tokyo, Japan
{yem,kajimoto}@kaji-lab.jp

Abstract. We developed a 3D virtual reality system comprising two fingertip gloves and a finger-motion capture device to deliver a force feedback sensation when grasping a virtual object. Each glove provides a pseudo-force sensation to a fingertip via asymmetric vibration of a DC motor. In this paper, we describe our algorithms for providing this illusionary force feedback, as well as visual feedback, which involved deforming the shape of a virtual object. We also conducted an experiment to investigate whether presenting pseudo-force sensation to the tip of the thumb and index finger during grasping enabled participants to interpret the material stiffness of a virtual object. We changed the initial vibration amplitude, which represents the reaction force when the thumb and the index finger initially contact the surface of an object, and asked participants to match each haptic feedback condition with a visual feedback condition. We found that most participants chose the rubber or wood material (task 1) and highly deformable material (task 2) when the initial vibration was weak, and chose the wood or aluminum (task 1) and non-deformable material (task 2) when the initial vibration was strong.

Keywords: Fingertip glove · Stiffness perception
Motor-rotational acceleration · VR interaction

1 Introduction

Recent technological advances in computer graphics and head mounted displays have enabled the active development of various high quality virtual reality (VR) systems, which are increasingly being adapted to new applications. Such systems can enable users to view, move through, and even physically interact with objects in fully immersive virtual environments. Realistic simulation of the experience of physically touching or grasping a virtual object requires new technology, designed to present physical force feedback sensations that communicate the stiffness of the object. Additionally, for use in free-space virtual environments, devices that present force feedback sensation must be wearable.

Stiffness is an important material property that is generally sensed when grasping an object. To reproduce such a sensation, a haptic device must produce a force sensation that simulates backward extension of the fingertip. Various studies have used a mechanical actuator to provide a grasping feedback sensation in which the fingertips

© Springer International Publishing AG, part of Springer Nature 2018
S. Yamamoto and H. Mori (Eds.): HIMI 2018, LNCS 10904, pp. 463–473, 2018.
https://doi.org/10.1007/978-3-319-92043-6_39

are pushed in a backward-extension movement [1, 2]. Some techniques involve grounded actuators and others comprised wearable robotic mechanisms [3, 4]. However, devices that use physical force are often large, heavy, and rely on complicated mechanisms.

To overcome these issues, we developed a technique to deliver pseudo-force sensation to the fingertips. We previously reported that a DC motor could produce an illusionary rotational force sensation when the input voltage was asymmetric (i.e. saw tooth waveform) [5]. We also mounted DC motors to the backside of the thumb and index finger and confirmed that the pseudo-force sensation occurred in multiple fingers simultaneously [6]. Here, we applied this technique to the presentation of a feedback sensation during grasping of a virtual object. Our haptic feedback device, shown in Fig. 1, is simple in mechanism, compact, and lightweight. We also developed a 3D virtual reality system that enables users to perceive stiffness sensations via both haptic and visual feedback.

Fig. 1. A 3D virtual reality system that uses pseudo-force perception produced by DC motor rotational acceleration to present stiffness feedback sensations to the tips of the thumb and index finger during grasping of a virtual tube. We modulated the visual feedback by changing the cross-sectional shape of the tube from a circle to an ellipse.

In this paper, we describe the algorithms used to produce haptic feedback via our VR glove and visual feedback via deformation of the shape of a virtual object. We also conducted an experiment to investigate whether participants could interpret the material stiffness of a virtual object during grasping. We modulated the initial vibration amplitude, which represented the reaction force when the thumb and index finger contacted the surface of the object, and asked participants to match each haptic feedback condition to a visual feedback condition. Our results showed that a stronger initial vibration represented materials that were harder and more rigid.

2 Related Work

2.1 Exploring Virtual Objects via Haptic Feedback Delivered to the Fingertip

Several studies have examined the efficacy of vibration feedback during exploration of the surface or shape of a virtual object [7–9]. However, vibration actuators cannot deliver force sensations, and thus cannot be used to communicate the stiffness properties of an object. Other haptic technologies have used grounded actuators to produce force feedback [3, 4, 10]. A representation of the stiffness of a virtual object, from very soft to hard, can be communicated to the user by outputting a force through an end effector. Such devices are tool based, and limit the movement of the user's fingers. Further, while some grounded actuators allow finger movements with a higher degree of freedom, these devices are relatively large [11, 12].

Several researchers have developed wearable devices that can deliver grasping-force feedback to the fingertips without imposing workspace limitations [1, 2]. Although these devices, which typically include an exoskeleton, can deliver a reaction force to the fingertips, they are often large and heavy because they contain robotic mechanisms. Other devices provide tactile feedback to the finger pads via skin deformation [13–15]. Although skin deformation has been reported to enhance the sensation of stiffness, cutaneous stimulation devices cannot present the sensation of a force that extends the finger backward. Therefore, reproduction of the sensation of grasping an object is limited in virtual reality.

2.2 Delivering Pseudo-force to a Fingertip via a Vibration Actuator

Previous studies have used physical force to present grasping feedback sensations to the fingertips. However, because such methods generally require an actuator with a high power output and a relatively large size and weight, we sought to use different technology to deliver an illusionary sensation instead of a physical force. Previous studies on human perception have reported that reciprocating asymmetric vibrations with different accelerations can elicit the sense of being pulled in a particular direction [16–20]. This is called a pseudo-force sensation. Existing linear vibration actuators that have been found to produce an illusionary force sensation include the voice coil [21], Hapuator (Tactile Labs Inc.) [22], and Force Reactor (Alps Electric Co.) [23].

Similarly, we previously reported that when driven by a saw tooth waveform, a DC motor can engage in asymmetrical motor rotation, producing a rotational pseudo-force sensation [5]. In addition, we developed a fingertip glove in which a DC motor is mounted onto the tip of the index finger. We found that an illusionary force could be felt in the finger even when not gripping the vibration actuator. This illusionary force can induce fingertip forward-flexion or backward-extension, and so we considered it to be useful in the development of a VR glove. Previously, our experiment revealed that the illusionary force could be compared to the reaction force of grasping a real object. The equivalent physical force generated by this illusion ranged from 10 to 30 grams [6].

3 System

Figure 1 shows the system used in our study, which consisted of (1) two fingertip gloves to present pseudo-force sensation to the fingertips, (2) a virtual environment to provide visual feedback, and (3) a motion capture device (Leapmotion Inc.) to measure finger movements.

3.1 Fingertip Glove and Hardware

The fingertip glove comprises a DC motor (Maxon, 118396) and a motor-fixing attachment. The attachment for fixing the DC motor to the fingertip was made from titanium using a three-dimensional printer. It can be mounted such that the motor shaft rotates in the pitch direction with respect to the fingertip. The size of the glove can be adjusted to the size of the user's finger. The pseudo-force sensation was presented in the direction of the pitch axis, and the user felt a force in their finger moving from the finger pad to the fingernail (Fig. 2).

The vibration waveform for driving the DC motors was generated by a micro-controller (mbed LPC1768, NXP Semiconductors) and amplified with an amplifier (OPA2544T, Texas Instruments, Inc.). Figure 3 shows the control hardware and asymmetric waveform used to produce the pseudo-force sensation. The microcontroller shared information with a computer via serial communication. The vibration amplitude could be adjusted from the computer. The outgoing data was renewed every 20 ms according to the frame rate of the visual feedback.

Fig. 2. A fingertip glove without DC motor (left) and two gloves attached to the tips of the thumb and index finger (right).

3.2 Visual Feedback and Algorithm

The movement of a user's virtual fingers in a virtual environment should follow the position of their fingers in the real world, as measured via a motion capture device. However, the user's virtual fingers can easily move inside the virtual rigid body of an

 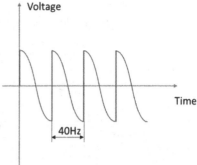

Fig. 3. Hardware for driving the DC motors (left) and asymmetric vibration waveform (right).

object when the user attempts to touch or grasp it if there is no physical force to resist the fingers (Fig. 4 [left]). This is a common issue for every wearable haptic device. In the current study, we developed an algorithm to address this issue as follows. When the user contacts a virtual object using one finger, this is classified as a pushing state, and the object moves in the direction of the pushing movement (Fig. 4 [right]). In contrast, when the user contacts the object using two fingers (i.e. thumb and index finger), this is classified as a grasping state, and the object follows the moment of the palm (Fig. 5 [left]). We made the virtual fingers invisible when they moved inside the virtual rigid body of the object, and showed a copy of the fingers grasping the surface of the object (Fig. 5 [right]).

We used the following equation to determine the strength of the force feedback on the fingertips according to the vibration amplitude of the input voltage.

$$V = \begin{cases} k_{haptic}\Delta x + V_0 & (if\ V < V_{max}) \\ V_{max} & (if\ V \geq V_{max}) \end{cases} \tag{1}$$

where V is the asymmetric vibration amplitude of the input voltage, k_{haptic} is a constant that represents the spring coefficients for haptic feedback, and V_0 is the initial vibration amplitude for presenting force feedback when the thumb and index finger initially contact the surface of the object. Δx is the total distance that the thumb and index finger move inside the object when grasping it (Fig. 5). V_{max} is a constant that limits the voltage of the vibration amplitude.

The surface of the virtual object starts to deform when the thumb and index finger apply a grasping force to the object. The amount of deformation of the virtual surface is proportional to the total distance that the thumb and index finger move inside the object. It is expressed by the following equation.

$$\Delta d = d_{visual}\Delta x \tag{2}$$

where Δd is the total deformation distance between two contact points on the skin of the thumb and index finger, and d_{visual} is a factor value that represents the deformation property of the visual feedback. The virtual object becomes a non-deformable rigid body when d_{visual} is equal to zero. Figure 6 shows the invisible fingers and the finger copies when the deformable tube is pressed in our virtual environment.

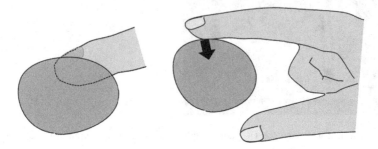

Fig. 4. Representation of a common issue in which the virtual finger moves inside the virtual rigid body (left), and the pushing state prior to grasping the object (right).

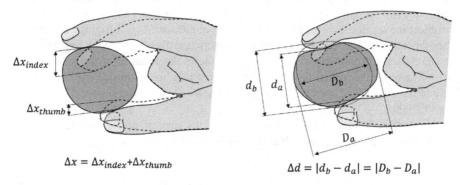

$$\Delta x = \Delta x_{index} + \Delta x_{thumb} \qquad \Delta d = |d_b - d_a| = |D_b - D_a|$$

Fig. 5. The proposed algorithm for maintaining thumb and index finger contact with the surface of the virtual object (left) and deformation of the virtual object when grasping (right).

Fig. 6. When the user's virtual fingers entered the rigid virtual object in the virtual environment (left), we made them invisible and showed copy fingers grasping the surface of the object (right).

4 Experiment

We conducted an experiment in which we asked participants to complete a matching task where they paired a haptic feedback condition with a visual feedback condition. Our goal was to determine whether participants could interpret the type of material that a virtual object was made of when they were exposed to each corresponding haptic feedback condition. There were two tasks in this experiment. In the first task, we investigated whether pseudo-force provided sufficient information about the material of the object, such as whether it was made of rubber or metal. In the second task, we investigated the requirement of virtual object deformation with respect to the initial vibration amplitude V_0.

4.1 Design

In first task, there were two initial vibration amplitudes ($V_0 = 0$, $V_0 = V_{max}$) and virtual cylindrical objects made of three types of material (rubber, wood, aluminum) (Fig. 7). In the second task, there were three initial vibration amplitudes ($V_0 = 0$, $V_0 = 0.5V_{max}$, $V_0 = V_{max}$) and three factor deformation conditions for a virtual tube ($d_{visual} = 0.0$, $d_{visual} = 0.2$, $d_{visual} = 0.4$) (Fig. 8). $d_{visual} = 0.0$ was a condition in which the virtual tube was non-deformable. The value of k_{haptic} was fixed for all haptic feedback conditions.

Fig. 7. Task 1: Two asymmetric vibration amplitudes (left) and cylindrical virtual objects made of three types of material (right).

Fig. 8. Task 2: Three asymmetric vibration amplitudes (left) and three deformation factors for a virtual tube (right).

4.2   Participants and Procedure

Six participants took part in this experiment: five males and one female, ranging in age from 22 to 25 years. All participants were right-handed.

In the beginning of the experiment, participants were asked to sit on a chair where we attached each of the two finger gloves to the thumb and index finger of their right hand. As shown in Fig. 1, the participant moved their right hand in range of the motion capture system (Leapmotion) to grasp a virtual object shown on a monitor. The participants were instructed to grasp the virtual object with the tip of their thumb and index finger. We asked the participants to grasp the object one or two times in each condition of haptic feedback condition, and then in each visual feedback condition. Thus, the participants had experienced all of the haptic and visual feedback conditions in each task before matching the associated conditions. For matching, we asked them to choose the visual feedback that they considered to be most appropriate match for each haptic feedback that had been presented. We explained that there was no correct answer and asked them to just follow their perception. We allowed them to adjust their answer if they felt they had made a mistake.

4.3   Result

Figures 9 and 10 show the results for tasks 1 and 2. The horizontal axis in each figure shows the haptic feedback conditions in terms of initial vibration amplitude. The vertical axis in Fig. 9 shows the participant responses in terms of the three kinds of material (rubber = 1, wood = 2, and aluminum = 3). The vertical axis in Fig. 10 shows the participant responses in terms of the three deformation factors d_{visual}.

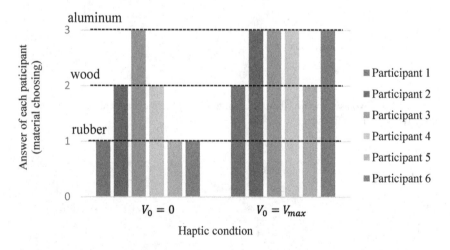

Fig. 9. Results for task 1: matching each virtual material to each haptic feedback condition.

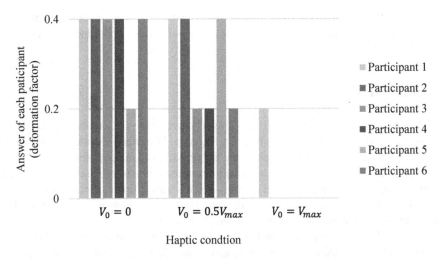

Fig. 10. Results for task 2: matching each deformation factor to each haptic feedback condition.

5 Discussion

The results of task 1 show that when the initial vibration amplitude V_0 was zero, most participants matched this haptic feedback condition to the rubber or wood. In contrast, when V_0 was equal to V_{max}, most of the participants matched this condition to wood or aluminum. In this task, the virtual object did not undergo deformation when touched, and so the participants only received visual feedback regarding the color of the material. Our data indicate that the participants interpreted the stimulation as resulting from touching a harder material when the initial vibration amplitude was higher. Thus, even though we did not provide visual feedback regarding object deformation, the participants clearly interpreted the aluminum object to be harder than the wood or rubber objects.

In task 2, we presented a virtual object that could be deformed when grasped with the thumb and index finger. The virtual object was a tube that changed in shape from a circle to an ellipse. When the initial vibration amplitude V_0 was zero, most of the participants paired this condition with that with the highest amount of deformation. When the initial vibration amplitude V_0 was equal to V_{max}, most participants matched this with the non-deformable object. Thus, a lower initial vibration amplitude elicited a perception of deformable and softer material.

The results of tasks 1 and 2 showed that, when the amplitude of the initial vibration was low (or even zero), participants engaged in material comparison chose the softer object (rubber or wood), and those in the shape deformation condition chose the more highly deformable object. Therefore, certain haptic feedback stimuli may be useful in multiple stiffness perception conditions, according to the visual feedback presented.

In this experiment, all participants interpreted a higher initial vibration amplitude to represent harder material. This initial vibration amplitude provides important information about the strength of the pseudo-force sensation when the thumb and index finger initially contact the surface of the object. Thus, the reaction force during the initial contact is important for presenting the stiffness (softness or hardness) property of the material.

6 Conclusion

We developed a system in which two finger gloves with DC motors delivered a pseudo-force sensation to the thumb and a fingertip. We examined the utility of this system in presenting grasping feedback and the sensation of material stiffness. We also developed a 3D virtual reality system that allows users to grasp any object in the virtual environment. To address a common issue in which a virtual finger enters a virtual rigid body, we proposed an algorithm where the finger inside the object is rendered invisible and is replaced by a copy that moves on the surface of the object.

We tested our system using two tasks. Our results showed that, when the initial vibration was weak, most participants in task 1 interpreted the object to be made of rubber or wood, and most participants in task 2 interpreted the material to be highly deformable. When the initial vibration was strong, they interpreted the material as wood or aluminum and non-deformable in tasks 1 and 2, respectively. In future work, we plan to examine the cross-modal relationship between haptic and visual feedback.

Acknowledgement. This work was partly supported by JSPS KAKENHI Grant Number JP17F17351, JP15H05923 (Grant-in-Aid for Scientific Research on Innovative Areas, "Innovative SHITSUKSAN Science and Technology"), and the JST-ACCEL Embodied Media Project.

References

1. Ma, K.Z., Ben-Tzvi, P.: RML Glove - an exoskeleton glove mechanism with haptics feedback. IEEE/ASME Trans. Mechatron. **20**(2), 641–652 (2015)
2. Choi, I., Hawkes, E.W., Christensen, D.L., Ploch, C.J., Follmer, S.: Wolverine: a wearable haptic interface for grasping in virtual reality. In: Proceedings of the IEEE/RSJ Intelligent Robots and Systems (IROS), pp. 986–993 (2016)
3. Liu, J., Song, A., Zhang, H.: Research on stiffness display perception of virtual soft object. In: Proceedings of the IEEE International Conference on Information Acquisition (ICIA), pp. 558–562 (2007)
4. Nojima, T., Sekiguchi, D., Inami, M., Tachi, S.: The SmartTool: A system for augmented reality of haptics. In: Proceedings of the IEEE Virtual Reality (VR), pp. 67–72 (2002)
5. Yem, V., Okazaki R., Kajimoto, H.: Vibrotactile and pseudo force presentation using motor rotational acceleration. In: Proceedings of the Haptics Symposium, pp. 47–51 (2016)
6. Sakuragi, R., Yem, V., Kajimoto, H.: Pseudo force presentation to multiple fingers by asymmetric rotational vibration using a motor: consideration in grasping posture. In: Proceedings of the World Haptics Conference, pp. 305–309 (2017)
7. Martínez, J., García, A., Oliver, M., Molina, J.P., González, P.: Identifying virtual 3D geometric shapes with a vibrotactile glove. IEEE Comput. Graphics Appl. **36**(1), 42–51 (2016)

8. Muramatsu, Y., Niitsuma, M., Thomessen, T.: Perception of tactile sensation using vibrotactile glove interface. In: IEEE Cognitive Infocommunications (CogInfoCom), pp. 621–626 (2012)

9. Murray, A.M., Klatzky, R.L., Khosla, P.K.: Psychophysical characterization and testbed validation of a wearable vibrotactile glove for telemanipulation. Presence Teleoperators Virtual Environ. **12**(2), 156–182 (2003)

10. McNeely, W.A., Puterbaugh, K.D., Troy, J.J.: Six degree-of-freedom haptic rendering using voxel sampling. Proceedings of the ACM SIGGRAPH, pp. 401–408 (1999)

11. Kim, S., Hasegawa, S., Koike, Y., Sato, M.: Tension based 7-DOF force feedback device: SPIDAR-G. In: Transactions on Control, Automation, and Systems Engineering, vol. 4, no. 1, pp. 9–16 (2002)

12. Endo, T., Kawaski, H., Nouri, T., Doi, Y., Yoshida, T., Ishigure, Y., Shimomura, H., Matsumura, M., Koketsu, K.: Five-fingered haptic interface robot: HIRO III. In: Proceedings of the Eurohaptics Conference and Symposium on Haptic Interfaces for Virtual Environment and Teleoperator Systems, pp. 458–463 (2009)

13. Minamizawa, K., Kajimoto, H., Kawakami, N., Tachi, S.: Wearable haptic display to present gravity sensation—preliminary observations and device design. In: Proceedings of the IEEE World Haptics Conference (2007)

14. Leonardis, D., Solazzi, M., Bortone, I., Frisoli, A.: A wearable fingertip haptic device with 3 DoF asymmetric 3-RSR kinematics. In: Proceedings of the IEEE World Haptics Conference, pp. 388–393 (2015)

15. Tsetserukou, D., Hosokawa, S., Terashima, K.: LinkTouch: a wearable haptic device with five-bar linkage mechanism for presentation of two-DOF force feedback at the fingerpad. Proceedings of the IEEE Haptics Symposium, pp. 307–312 (2014)

16. Amemiya, T., Gomi, H.: Distinct pseudo-attraction force sensation by a thumb-sized vibration that oscillates asymmetrically. In: Proceedings of the EuroHaptics, Part II, pp. 88–95 (2014)

17. Rekimoto, J.: Traxion: a tactile interaction device with virtual force sensation. In: Proceedings of the ACM Symposium User Interface Software and Technology (UIST 2013), pp. 427–432 (2013)

18. Amemiya, T., Ando, H., Maeda, T.: Virtual force display: direction guidance using asymmetric acceleration via periodic translational motion. In: Proceedings of the World Haptics Conference, pp. 619–622 (2005)

19. Tanabe, T., Yano, H., Iwata, H.: Properties of proprioceptive sensation with a vibration speaker-type non-grounded haptic interface. In: Proceedings of the Haptics Symposium, pp. 21–26 (2016)

20. Culbertson, H., Walker, J.M., Okamura, A.M.: Modeling and design of asymmetric vibrations to induce ungrounded pulling sensation through asymmetric skin displacement. In: Proceedings of the Haptics Symposium, pp. 27–33 (2016)

21. Yao, H.Y., Haywad, V.: Design and analysis of a recoil-type vibrotactile transducer. J. Acoust. Soc. Am. **128**, 619–627 (2010)

22. Tactile Labs Inc. http://tactilelabs.com/. Accessed 9 Feb 2018

23. Alps Electric Co. http://www.alps.com/e/. Accessed 9 Feb 2018

Information and Vision

A Study for Correlation Identification in Human-Computer Interface Based on HSB Color Model

Yikang Dai, Chengqi Xue[(⊠)], and Qi Guo

School of Mechanical Engineering, Southeast University,
211189 Nanjing, China
ipd_xcq@seu.edu.cn

Abstract. In recent years, visual perception has been paid more attention by many researchers in the field of data visualization. The study of visual perception has become another research hot spot in the research of visualization. As an important tool for visualization, this paper focuses on the scatterplots. Series of scatterplot were generated by programming and then used in the experiment. The results of experiments indicate that the influence of the color, amount and correlation of the interference points on the reaction time is significant under the white background and suitable combination is found which is important for designing the scatterplots.

Keywords: Visual perception · Scatterplots · Correlation identification

1 Introduction

Human-computer interface which is used to transfer and exchange information is the medium and dialogue interface between human and computers. It is an important part of the human-computer system which converts the internal form of information into an acceptable one to users. Status of the system can be understood by users so that they can monitor the output of it while running and adjust it at any time due to the human-computer interface.

With the development of computer and network technology, the amount of information in human-computer interface increases sharply. Massive amounts of data are generated by various kinds of devices, sensors, electronic websites and social networks every day, triggering an explosive growth in data size. As a result, the concept of 'Big data' appears. Lately, the term 'Big Data' tends to refer to the use of predictive analytics, user behavior analytics, or other advanced data analytics methods that extract value from data, and seldom to a particular size of data set. Big Data has 5 features (named 5 V): Volume, Variety, Velocity, Veracity and Value of high density, it brings new opportunities and challenges for human. Nowadays, big data has become a hot topic in academic research and is considered to be another revolutionary information technology after cloud computing and the Internet of Things.

© Springer International Publishing AG, part of Springer Nature 2018
S. Yamamoto and H. Mori (Eds.): HIMI 2018, LNCS 10904, pp. 477–489, 2018.
https://doi.org/10.1007/978-3-319-92043-6_40

2 Background

Scatterplot which has also been called scatter diagram, scatter gram, and scatter graph was first introduced in the 18th century, during the boom of statistics graphics [1]. There are several variations and extensions, including animated scatterplot, scatterplot matrix [2], and glyph-based scatterplot [3]. Scatterplots can be used for observing correlation, clusters, and outliers [4, 5].

As one of the visualization tool, scatterplots have many advantages in terms of displaying data sets in high dimension: It can quickly preview and analyze data sets, a variety of data points are applicable. Both continuous data set and discrete data points can be visualized by using it; In scatterplots, users can observe the distribution of data points and obtain the overall information through the way of dimensionality reduction of data. Moreover, they can also estimate the correlation among data plots to make judgment in order to provide better service; Scatterplots can easily visualize the data displayed to get useful information. This visualization technology can quickly and easily help users distinguish the abnormal points from all plots. Also, scatterplots play an important role in ensuring the correctness of data. In summary, scatterplots visualization is a valuable tool for data visualization.

Cartesian coordinates are used in scatterplots to present multi-dimensional data in a two-dimensional plane. Information which implied in the data can be clearly revealed in this way. Linear association in the scatter-plots is the most basic and simple relationship between variables, and at the same time, statistical analysis uses correlation to quantify the strength and direction of such bivariate linear associations [9]. Throughout the method above, not only can the relationship between variables display visually, but users can observe the overall information of the data. As a result, using scatterplots has become an effective way to visualize data.

If the graphical representation is designed well, data can be analyzed rapidly, accurately, and precisely. In such situations, the way of analyst's visual system perceiving structure in a dataset is the same way as it perceiving structure in the real world. Therefore, the perception of such graphical representations has considerable potential to help researchers investigate various aspects of our visual intelligence [6–8].

In recent years, visual perception has been paid more attention by many researchers in the field of data visualization. The study of visual perception has become another research hot spot on visualization.

In statistical analysis, R (Pearson's product-moment coefficient, PPMC) is commonly used to define the correlation between variable [9]. In a specific data set, R equals to

$$R = \frac{S_{xy}}{S_x S_y} \tag{1}$$

For x and y are the sample data for the two variables in this data set. Where Sx and Sy are the sample standard deviations of x and y, Sxy is the sample covariance between x and y. As shown above, it can be seen that r ranges from 0 which means the perfect negative correlation to 1 which means perfect positive correlation [10]. Figure 1(a) and

(b) show the sampling distributions of R is 0.3 and 0.5, respectively, Fig. 1(c) and (d) show the sampling distributions of R is 0.7 and 0.9, respectively.

There are a number of cognitive influences which can affect the human perception of correlation. These may include gestalt laws of grouping, shape interpretation, learned knowledge about statistical measures such as Pearson's product-moment correlation coefficient (PPMCC). In recent years, there has been growing interest in studying the underlying models of human perception of correlation.

Sher et al. [11] found that for different PPMCC values (R value), data distribution had an impact on the average offsets which meant the differences between the estimated and actual PPMCC), and result showed that only large variations in density caused a statistically significant impact.

Li [12] studied the symbol size perception of scatterplots in the context of analytic tasks which required size discrimination. They conducted an experiment in three visual analysis tasks represented by a circle, divided into three groups with 8 linearly varying radii. 24 subjects were participated in this experiment. The result showed that approximate uniformity of size perception existed in complex tasks and could be described by power-law transformation with an index of 0.4.

Li et al. [9] observed correlation as a function of the sample correlation under different visualization methods, sample sizes and observation time. In the study, they introduced a discriminating index to characterize performance under different conditions. Furthermore, they came to a conclusion that users could reliably differentiate two different degrees of relevance when using scatterplots and using PCP.

Micallef et al. [13] in order to automatically set parameters to enhance the visual quality of the scatterplot, they studied the use of perceptual models and quality metrics. They took the construction of cost function as the key consideration to capture several relevant aspects of the human visual system and they used different experiments to test different data analysis task in a scatterplot design.

Gleicher et al. [14] used a realistic task which evaluated the difference in class means in a scatterplot. And in the end, they explored the assemble judgment in visualizations using.

Stenholt et al. [15] explored the visualization of 3D scatterplots in an immersive virtual environment. The results showed that CVA glyphs did better understand shapes in 3D scatterplots than regular perspective glyphs, especially when there were large numbers of clutter. In addition, their assessment showed that the perception of structure in the 3D scatterplot was affected by the volumetric density of the glyphs in the figure significantly.

Bertini et al. [16] in order to reduce cluttering in the scatterplots, they presented a strategy which relied on a combination of non-uniform sampling and pixel displacement. The paper mentioned that it could also be used to define the precise quality measurement which allowed for validating their approach.

Rensink [17] used four different data point distributions to test. They found that JND was a linear function of distance from r = 1, and the perceived relative magnitude was a logarithmic function of that amount. Then the other three cases were checked and the performance was found to be similar in all situations. The results showed that the basis of the correlation perception was not in a geometric structure, such as the shape of a point cloud, but rather the shape of the probability distribution of points that might be inferred by a collective encoding.

Doherty et al. [18] studied experiments on four different scatter-related perceptions and found that except for the error variance, all graphical attributed the affect subjective but non-objective correlations (R) remained the same.

Hasse and and Kaczmarek [19] compared the visual perception of auditory and scatterplots and found the similar correlation estimation performance in both modes. Their results showed that electrical tactile complexity of the graphics and also provided useful information to improve the future version of the tactile display.

3   Experiment

3.1   Method and Materials

In this paper, the stimuli are scatterplots that contain 100, 200 and 500 normally-distributed points (including inherent points and interference points) which are generated by Python using pseudo-random numbers. The Pseudo-random numbers which are taken from a Gaussian distribution data set that the mean and the standard deviation was set to 0.5 and 0.2 are chosen to build correlated pairs (x, y') and are then used to generate scatterplots. The x-coordinate is the first number chosen from the Gaussian distribution. Y value is then created and transformed using Eq. (2) to create y'.

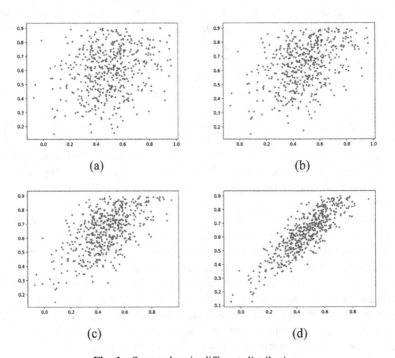

Fig. 1. Scatterplots in different distributions

$$y' = \frac{\lambda x + (1-\lambda)y}{\sqrt{\lambda^2 + (1-\lambda)^2}}, \text{ where } \lambda = \frac{r^2 - \sqrt{r^2 - r^4}}{2r^2 - 1} \tag{2}$$

In order to avoid points outside the range of the graph, any points which are greater than 2 standard deviations from the mean are eliminated and then generated a new point to take its place.

Figure 2 shows 200 inherent points which R value is 0.5 and interference points which R value is 0.5 in the scatterplot. The different colors of interference points are shown in the Table 1.

Table 1. The different color of interference points

Component	Value					
H	30°	60°	90°	120°	150°	180°
S	100%	100%	100%	100%	100%	100%
B	100%	100%	100%	100%	100%	100%
Sample	●	●	●	●	●	●

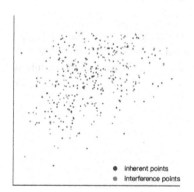

Fig. 2. Scatterplot with 200 inherent points (r = 0.5) and interference points(r = 0.5)

3.2 Experimental Design

In this experiment, there are two scatter plots presented in the screen with different R value which size are both 464 × 464 pixels. Participants need to observe the trends of these two scatterplots and decide which scatterplot has higher R value. Press 'f' key in keyboard to response if the left one is higher, press 'j' key in keyboard to response if the right one is higher.

The experiment is divided into two parts: the training session and the formal session. At the beginning of the experiment, the computer screen shows experimental guidance. After reading the guidance, participants enter the experimental stage by pressing any key. The cross-visual guidance appears in the center of screen for 500 ms at first and then participants need to observe the trends of these two scatterplots and decided which scatterplot has higher R value. Press 'f' key in keyboard to response if the left one is higher, press 'j' key in keyboard to response if the right one is higher. In this experiment, the presentation time of each task images is set to infinite. The cross-visual guidance appears after each image. Procedure of the experiment is shown in the Fig. 3.

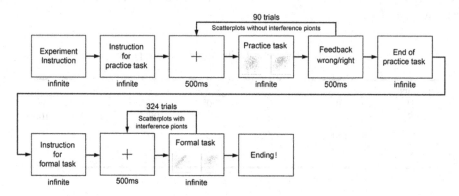

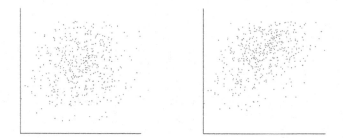

Fig. 3. The procedure of experiment

The interference points are not included in the scatterplots presented in the practice task. Training session is used to familiarize with the task for all participants. The training session consists of 90 trials. Figure 4 shows scatterplots in the training session.

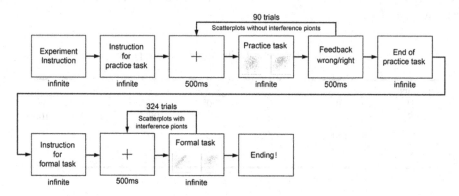

Fig. 4. Scatterplots in the training session (Color figure online)

Interference points are added in the scatterplots of the formal session. Participants are asked to make judgments based on the R value of red points (inherent points, H = 0°, S = 100%, B = 100%) on which scatterplot has higher R value. Press 'f' key in keyboard to response if the left one is higher, press 'j' key in keyboard to response if the right one is higher. Figure 5 shows scatterplots in the formal session. For instances, in Fig. 5, interference points which are blue (H = H = 180°, S = 100%, B = 100%) are added in the scatterplots, participants should make decisions based on the red points. The scatterplot on the left has higher R value, so they are supposed to press 'f' key in the keyboard to response. In this experiment, there are 50% pictures on the left which have higher R value, 50% pictures on the right which have higher R value. The reaction time and response of each participants are recorded through E-prime.

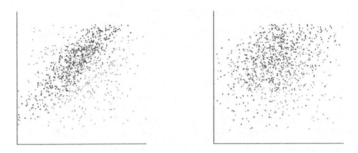

Fig. 5. Scatterplots in the formal session

25 participants attend this experiment, and each seats 64 cm from a screen with the resolution of 1366 × 768-pixel laptop. All participants are tested on both training and formal sessions. Average age of participants is 24 years old who has at least some experience with scatterplots; most have made and used scatterplots on several occasions. They are given as much time as needed to complete each task and they are also mentioned that accuracy is important.

4 Result

The independent variables of this experiment are as follow: (1) The amount of interference point (Amount): 3 levels, 100, 200, 500; (2) The color of interference point (Color): 6 levels, H = 30°, H = 60°, H = 90°, H = 120°, H = 150°, H = 180°; (3) The correlation levels of interference point (Rvalue): 3 levels, 0.1 (low correlation), 0.5 (general correlation), 0.9 (high correlation).

The dependent variable of this experiment is the reaction time (RT) of each subject. There are 25 subjects participate in the experiment. After data analysis by using SPSS, the 12th and 24th subjects' reaction time (RT) are eliminated because of the abnormal deviation and too many singular values in reaction time.

The analysis of variance (ANOVA) is performed on RT and ACC. The results are shown in Table 2. As shown in the Table 2, the main effect of Color (F = 5.981, P = 0.000, P ≤ 0.01), the main effect of Rvalue (F = 34.864, P = 0.000, P ≤ 0.01) and the main effect of Amount (F = 32.103, P = 0.000, P ≤ 0.01) on reaction time are significant and the main effect of the second-order interaction between factors: Rvalue*Color, F = 1.076, P = 0.377, P > 0.01; Amount*Color, F = 1.085, P = 0.370, P > 0.05; Amount*Rvalue F = 3.144, P = 0.014, P > 0.01 are not significant. The main effect of third-order interaction is not significant (F = 0.897, P = 0.591, P > 0.05). It can be seen that these three independent variables (the color of the interference point, the amount of the interference points and the correlation of the interference points) all have significant influence on the reaction time(RT) of participants. As shown in Tables 3, 4 and 5, the multiple post-test multiple comparison of the least significant difference is then performed.

Table 2. Tests of between-subjects effects

Source	Type III sum of squares	df	F	Sig.	Partial Eta squared
Rvalue	15424064.531	2	34.864	.000	.055
Color	6614853.091	5	5.981	.000	.025
Amount	14202908.242	2	32.103	.000	.051
Rvalue* color	2381025.137	10	1.076	.377	.009
Rvalue* amount	2782197.923	4	3.144	.014	.010
Color* amount	2400687.825	10	1.085	.370	.009
Rvalue* color* amount	3968684.949	20	.897	.591	.015

The results of multiple comparisons in Table 3 show that there are significant differences between the mean reaction time (RT) of H = 30° and the reaction time (RT) of H = 60°, H = 90°, H = 120°, H = 150°, H = 180° which indicates that if H > 30°, the reaction time(RT) is not influenced by the H value changes of the interference points. As the HSB value for inherent points in scatterplots is H = 0°, B = 100%, H = 100%, the ΔH between inherent and interference points is 30 which means the difference between colors is not significant. As a result, it causes a huge interference which makes test images difficult to identify for the correlation of participants' perception.

According to the result of multiple comparisons for the amount of interference point in Table 4, it can be seen that there are significant differences between the mean reaction time (RT) of 100, 200 and 500. It can be indicated that the number of interference points has a huge interference on the correlation of participants' perception.

It can be seen from multiple comparisons for the correlation of interference point (Rvalue) in Table 5 that there are significant differences between the mean reaction time (RT) of R = 0.1 and R = 0.5, R = 0.9. It can be indicated that when R > 0.1, the mean reaction time (RT) is not influenced by the changes of interference points' correlation. The reason is that when R = 0.1, the interference points are in low correlation which means the distribution for interference points is straggling and makes test images difficult to identify.

After each reaction time is compared, Fig. 6a–c are line charts which show the influence of different combinations of variables on mean reaction time.

Figure 6a exposes the relationship between color, correlation of the interference points and the mean reaction time of each subject. As it shown in the graph, when the

Table 3. Multiple comparisons (color)

(I) Color	(J) Color	Mean difference (I-J)	Std. error	Sig.
30	60	240.0427*	46.23047	.000
	90	132.7625*	46.23047	.004
	120	131.7375*	46.23047	.004
	150	180.5395*	46.23047	.000
	180	164.4614*	46.23047	.000
60	30	−240.0427*	46.23047	.000
	90	−107.2802	46.23047	.020
	120	−108.3052	46.23047	.019
	150	−59.5032	46.23047	.198
	180	−75.5813	46.23047	.102
90	30	−132.7625*	46.23047	.004
	60	107.2802	46.23047	.020
	120	−1.0250	46.23047	.982
	150	47.7770	46.23047	.302
	180	31.6989	46.23047	.493
120	30	−131.7375*	46.23047	.004
	60	108.3052	46.23047	.019
	90	1.0250	46.23047	.982
	150	48.8019	46.23047	.291
	180	32.7238	46.23047	.479
150	30	−180.5395*	46.23047	.000
	60	59.5032	46.23047	.198
	90	−47.7770	46.23047	.302
	120	−48.8019	46.23047	.291
	180	−16.0781	46.23047	.728
180	30	−164.4614*	46.23047	.000
	60	75.5813	46.23047	.102
	90	−31.6989	46.23047	.493
	120	−32.7238	46.23047	.479
	150	16.0781	46.23047	.728

Table 4. Multiple comparisons (amount)

(I) Amount	(J) Amount	Mean difference (I-J)	Std. error	Sig.
100	200	156.1486*	32.68988	.000
	500	260.2089*	32.68988	.000
200	100	−156.1486*	32.68988	.000
	500	104.0604*	32.68988	.001
500	100	−260.2089*	32.68988	.000
	200	−104.0604*	32.68988	.001

Table 5. Multiple comparisons (Rvalue)

(I) Rvalue	(J) Rvalue	Mean difference (I-J)	Std. error	Sig.
.10	.50	232.5958*	32.68988	.000
	.90	240.0262*	32.68988	.000
.50,	.10	−232.5958*	32.68988	.000
	.90	7.4304	32.68988	.820
.90	.10	−240.0262*	32.68988	.000
	.50	−7.4304	32.68988	.820

interference point color is 30°(H = 30°, S = 100°, B = 100°), the mean reaction time of subjects is the longest in different correlation levels; when the interference point color is 60°(H = 60°, S = 100°, B = 100°), the mean reaction time of subjects is the shortest during these three correlation levels; when the interference point correlation level is low (R = 0.1), the mean reaction time is higher than the other two correlation levels; in all combinations of variables, the mean reaction time is the shortest when h = 60° and the correlation level is general (R = 0.5).

Figure 6b shows the relationship between the color, amount of interference points and the mean reaction time. It can be concluded that when the H component of the interference point located at 30° in the color wheel (H = 30°, S = 100°, B = 100°), the mean reaction time is the longest; when the H component of the interference point located at 60° in the color wheel (H = 60°, S = 100°, B = 100°), the mean reaction time is the shortest compared to the other five locations of H component; when the interference point number is 100, the mean reaction time is higher than the other two levels; In all the combinations of variables, the mean reaction time of subjects is the shortest when H component of the interference point located at 60° (H = 60°, S = 100°, B = 100°) in the color wheel and the amount of the interference points is 500.

Figure 6c shows the relationship between the amount, correlation of interference points and the mean reaction time of subjects. It can be seen from the figure that when the number of interference points is 100, the mean reaction time is the longest compared to 200 and 500. When the number of interference points is 500, the mean reaction time is the shortest in different correlation levels; when the correction level of the interference points is low (r = 0.1), the mean reaction time is longer than the other two levels of 0.5 and 0.9; When the correlation level is general (r = 0.5) and the amount is 500, the mean reaction time is the shortest among the all combinations of variables.

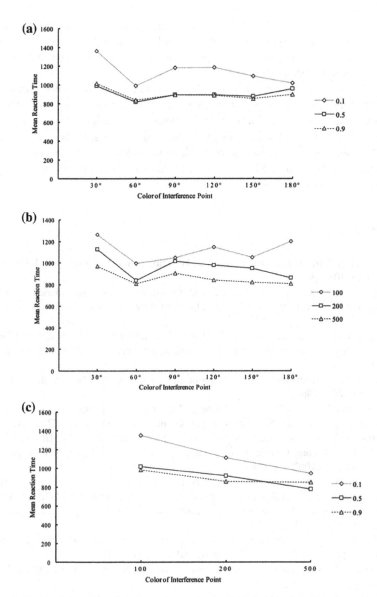

Fig. 6. a. The mean reaction time for color and correlation of the interference points, b. The mean reaction time for color and amount of the interference points, c. The mean reaction time for amount and correlation of the interference points

5   Conclusion

The effect of the color of interference points, amount of interference points and correlation of interference points on correlation recognition are analyzed in this paper according to the correlation of scatter plot correlation test. What's more, the degree of interference is compared. The result shows that the advantages and disadvantages of the scatter plot visualization can be judged according to correlation recognition of the users. That is to say, the faster the subjects recognize the correlation, the more reasonable scatter plot visualization is.

The results of main effect indicate that the influence of the color, amount and correlation of the interference points on the reaction time is significant under the white background; the result of post-test multiple comparison analysis shows the different mean reaction time between different combinations of variables. In the end, the shortest reaction time among the combination of the three variables is concluded; the method of quantifying color based on HSB color mode in this paper can intuitively represent the color for experimental research.

References

1. Friendly, M., Denis, D.: The early origins and development of the scatterplot. J. Hist. Behav. Sci. **41**(2), 103–130 (2005)
2. Elmqvist, N., Dragicevic, P., Fekete, J.-D.: Rolling the dice: multidimensional visual exploration using scatterplot matrix navigation. IEEE Trans. Visual. Comput. Graph. **14**(6), 1141–1148 (2008)
3. Chung, D.H.S., Legg, P.A., Parry, M.L., Bown, R., Griffiths, I.W., Laramee, R.S., Chen, M.: Glyph sorting: interactive visualization for multi-dimensional data. Inf. Visual. **14**(1), 76–90 (2013)
4. Lewandowsky, S., Spence, I.: The perception of statistical graphs. Sociol. Methods Res. **18** (2–3), 200–242 (1989)
5. Kanjanabose, R., Abdul-Rahman, A., Chen, M.: A multi-task comparative study on scatter plots and parallel coordinates plots. Comput. Graph. Forum **34**(3), 261–270 (2015)
6. Rensink, R.A.: On the prospects for a science of visualization. In: Huang, W. (ed.) Handbook of Human Centric Visualization, pp. 147–175. Springer, New York (2014). https://doi.org/10.1007/978-1-4614-7485-2_6
7. Cleveland, W.S., McGill, R.: Graphical perception: the visual decoding of quantitative information on graphical displays of data. J. Roy. Stat. Soc. **150**(3), 192–229 (1987)
8. Meyer, J., Taieb, M., Flascher, I.: Correlation estimates as perceptual judgments. J. Exp. Psychol. Appl. **3**(1), 3–20 (1997)
9. Li, J., Martens, J.B., Wijk, J.J.V.: Judging correlation from scatterplots and parallel coordinate plots. Inf. Vis. **9**, 13–30 (2010). Palgrave Macmillan
10. Carpenter, M.: The new statistical analysis of data. J. Am. Stat. Assoc. **42**(2), 205–206 (1996)
11. Sher, V., Bemis, K.G., Liccardi, I.: An empirical study on the reliability of perceiving correlation indices using scatterplots. Comput. Graph. Forum **36**(3), 61–72 (2017)
12. Li, J., Martens, J.B., Van Wijk, J.J.: A model of symbol size discrimination in scatterplots (2010)
13. Micallef, L., Palmas, G., Oulasvirta, A.: Towards perceptual optimization of the visual design of scatterplots. IEEE Trans. Vis. Comput. Graph. **23**(6), 1588–1599 (2017)
14. Gleicher, M., Correll, M., Nothelfer, C.: Perception of average value in multiclass scatterplots. IEEE Trans. Vis. Comput. Graph. **19**(12), 2316–2325 (2013)
15. Stenholt, R., Madsen, C.B.: Shape perception in 3-D scatterplots using constant visual angle glyphs. In: Virtual Reality Short Papers and Posters (2012)

16. Bertini, E., Santucci, G.: Improving 2D scatterplots effectiveness through sampling, displacement, and user perception. In: International Conference on Information Visualisation. IEEE Computer Society (2005)
17. Rensink, R.A.: The nature of correlation perception in scatterplots. Psychon. Bull. Rev. **24** (3), 1–22 (2017)
18. Doherty, M.E., Anderson, R.B., Angott, A.M.: The perception of scatterplots. Percept. Psychophys. **69**(7), 1261–1272 (2007)
19. Haase, S.J., Kaczmarek, K.A.: Electrotactile perception of scatterplots on the fingertips and abdomen. Med. Biol. Eng. Comput. **43**(2), 283 (2005)

Investigating Effects of Users'
Background in Analyzing Long-Term
Images from a Stationary Camera

Koshi Ikegawa[(✉)], Akira Ishii, Kazunori Okamura, Buntarou Shizuki,
and Shin Takahashi

University of Tsukuba, Tsukuba, Japan
{ikegawa,ishii,kokamura,shizuki,shin}@iplab.cs.tsukuba.ac.jp

Abstract. Images recorded over a long term using a stationary camera have the potential for revealing various facts regarding the recorded target. We have been developing an analyzing system with a heatmap-based interface designed for visual analytics of long-term images from a stationary camera. In our previous study, we experimented with participants who were recorded in the images (*recorded participants*). In this study, we conducted a further experiment with participants who are not recorded in the images (*unrecorded participants*) to reveal the discoveries that participants obtain. By comparing the results of participants with different backgrounds, we investigated the difference between discoveries, functions used, and analysis process. The comparison suggests that *unrecorded participants* could discover many facts about environment, and *recorded participants* could discover many facts about people. Moreover, the comparison also suggests that *unrecorded participants* could discover many facts comparable to *recorded participants*.

Keywords: Data visualization · Big data management
Information presentation · Heatmap · Surveillance system
Visual analytics · Lifelog

1 Introduction

Images recorded over a long term using a stationary camera have the potential for revealing various facts regarding the recorded target. For example, if a department store installs a stationary camera that produces aerial images of a floor, the recorded images could contain useful data for evaluating the layout of the floor. However, it is difficult to view such images in their entirety, because the speed at which the images are replayed must be sufficiently slow for the user to comprehend them, and thus it is difficult to obtain valuable information from the images quickly.

To address this problem, we developed an analyzing system [7], which has a heatmap-based interface designed for visual analytics [11,13] of long-term images

© Springer International Publishing AG, part of Springer Nature 2018
S. Yamamoto and H. Mori (Eds.): HIMI 2018, LNCS 10904, pp. 490–504, 2018.
https://doi.org/10.1007/978-3-319-92043-6_41

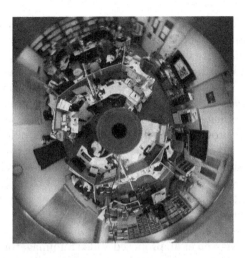

Fig. 1. An image from our omni-directional camera.

from a stationary camera (Fig. 1). This system provides heatmaps, each of which summarizes the movement of people and objects during a specific term, allowing the user to analyze the recorded target by comparing heatmaps of two different intervals. In our previous study, we experimented with participants who were recorded in the images (*recorded participants* [4]).

In this study, we conducted a further experiment with participants who are not recorded in the images (*unrecorded participants*) to reveal the discoveries that participants obtain. Furthermore, we compared the results with those of our previous study. The comparison suggests that *unrecorded participants* could discover many facts about environment, and *recorded participants* could discover many facts about people. Moreover, the comparison also suggests that *unrecorded participants* could discover many facts comparable to *recorded participants*.

2   Related Work

Heatmap-based visualization for analyzing images from a stationary camera have been explored. Romero et al. proposed Viz-A-Vis [8] and evaluated it [9]. While their visualization is different, we use their evaluating method as a reference. Heatmaps were also used in Viz-A-Vis, which are 3D heatmaps in which time is represented in the third axis to provide a spatial and temporal abstraction of a video. Kubbat et al. proposed TotalRecall [5] that focused on transcribing and adding annotations based on audio and video recorded at the same time for a 100,000 h. While the visualization of the above systems is similar to ours, our system focuses on discovering by comparing two different terms using 2D heatmaps, each of which summarizes the movement of people and objects during a specified term.

Various systems for analyzing images from a stationary camera have explored visualization techniques other than heatmaps. DeCamp et al. proposed House-Fly [2], a system that visualizes the entire floor, consisting of several rooms, as one 3D representation by mapping the image of the camera attached to the ceiling of each room as a texture on the floor plan. In addition to video browsing, this system serves as a platform for visualizing patterns of a multi-modal data over time, including person tracking and speech transcripts. Chiang and Yang [1] proposed a browsing system to help the users quickly locate desired targets in surveillance videos. Their basic idea is to collect all moving objects, which carry the most significant information in surveillance videos to construct a corresponding compact video. Shin et al. proposed VizKid [10] that visualizes the position and orientation of an adult and a child as they interact with one another in one graph.

Many studies have explored interfaces for browsing videos from non-stationary cameras. Nguyen et al. proposed Video Summagator [6] that visualizes a video in 3D, allowing a user to look into the video cube for rapid navigation. Tompkin et al. [12] proposed a video browsing system that embeds videos into a panorama, allowing the users to comprehend the videos within its panoramic contexts across both the temporal and spatial dimensions. Higuchi et al. proposed EgoScanning [3], a video fast-forwarding interface that helps users find important events from lengthy first-person videos continuously recorded with wearable cameras.

In contrast to the above work, our visualization is designed to help the users analyze the images from one stationary camera on the ceiling by providing the summarization of the movement of people and objects in the images as a heatmap.

3   Implementation

This section describes the system used in this experiment. Our system consists of a recording system and an analyzing system. The recording system obtains images from a stationary camera that was mounted on the ceiling of the authors' laboratory room, preprocesses the images for the generation of a heatmap, and stores the images to a network attached storage (NAS). The analyzing system generates heatmaps using the images stored in the NAS and presents the heatmaps to users.

3.1   Recording System

Our recording system stores the images, with a 608 × 608 pixel spatial resolution at 1 fps, from an omni-directional camera (Sharp Semiconductor LZ0P3551) mounted on the ceiling of our laboratory as shown in Fig. 2. This frame rate is the one frequently used in the video archives of surveillance systems and produces 86,400 frames per day. The images are stored in a NAS (QNAP TS-859 and TS-859 Pro+). The recording system (Fig. 3) runs on a laptop computer (MacBook Pro 13-in. Late 2011 and MacBook Pro Retina 15-in. Early 2013).

We installed two sets of the above system in our laboratory's two rooms (Room-A and Room-B) with the sizes of approximately 7.50 m × 7.75 m (58 m²) and 7.5 m × 15.0 m (113 m²), and heights of approximately 2.5 m and 2.7 m, respectively.

3.2   Analyzing System

Our analyzing system generates heatmaps using the images stored on NAS and presents the heatmaps to users. Figure 4 shows our analyzing system. It consists of Image-Presenting Panel, Time-Operation Panel, and Heatmap-Operation Panel.

Image-Presenting Panel displays a camera image (Fig. 4A). The users can select a part (Fig. 4B) of the image for further analysis. Time-Operation Panel allows the users to specify the date (Fig. 4C) and time (Fig. 4D) of the camera image. The system colors the calendar (Fig. 4C) and the date slider (Fig. 4D) blue with the depth which shows the amount of the movement in the selected part. Using this function, the users can identify the terms to be analyzed. Heamap-Operation Panel controls two colors of heatmaps; our system can display two different heatmaps with two different colors, each of which can be turned on/off using the two checkboxes (Fig. 4E) on Heatmap-Operation Panel. Turned-on heatmaps are overlaid on the camera image. The users can specify the terms for the two heatmaps using the term selectors (Fig. 4F).

Our heatmap summarizes the movement of people and objects in the specified term based on the pixel changes of the camera images: the more movement there

Fig. 2. Omni-directional camera mounted on the ceiling of our laboratory.

Fig. 3. Recording system.

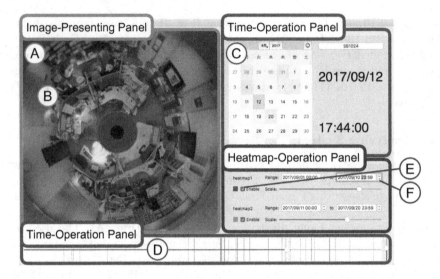

Fig. 4. Analyzing system using heatmaps. (Color figure online)

is within the specified term, the more densely the pixel is colored. Therefore, the heatmap allows the users to recognize areas with little movement and areas with much movement within the specified term at a glance. Moreover, the users can compare movements in different terms by using two heatmaps.

4   Experiment

We conducted an experiment to examine what and how users obtain discoveries by each function provided by our analyzing system. In addition, we used the experimental results of *recorded participants* in the previous study for comparison.

4.1   Participants

Four new *unrecorded participants* (three males, one female) aged 18 to 22 years were recruited for the experiment. The age of *recorded participants* (three males, one female) in the previous study was 22 to 23 years. None of the participants had previously used our system, nor did they have prior knowledge regarding our system.

4.2   Experimental Environment

We used a MacBook Air 13-in. Mid 2011 (CPU: 1.7 GHz Intel Core i5, RAM: 4 GB, OS: macOS 10.12.6) as the computer for running our analyzing system. For high-speed communication, the computer and the NAS connected to the local network using a wired LAN cable. The experiment was conducted in a calm office environment. We recorded the whole experiment with a video camera, voice recorder, and screen capture software (QuickTime Player 10.4). In the experiment, two authors were sitting near the participants. One author recorded the participants' remarks and the other author addressed the question of the participants.

4.3 Procedure

First, we informed the participants of the purpose and the procedure of the experiment. We then informed the participants that the reward for participation in the experiment included not only a basic reward (820 JPY, 7.78 USD) but also a bonus depending on the number of discoveries (30 JPY per discovery). After the explanation, we explained the use of our analyzing system to the participants. We then asked them to engage with the system until they felt that they completely understood how to use it, as a practice. We used the Room-A images for the practice, in which *recorded participants* were not recorded. For reference, we showed the seating chart of the two rooms to the participants (Fig. 5).

In the analyzing task, we asked the participants to use our analyzing system for 30 min, during which they should attempt to make as many discoveries as possible and inform the experimenters of each of these using think aloud protocol. In addition, we asked the participants to inform the experimenters of the facts that led them to each discovery (e.g., the color of the heatmap is dense in certain areas, as described in Sect. 5.3). In this analyzing task, we used the images of Room-B that were recorded over six months (July 1, 2014 to December 31, 2014; 4,416 h; approximately 32 million images) using the stationary camera.

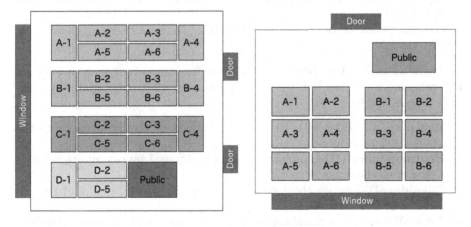

Fig. 5. Seating chart: (left) Room-A, (right) Room-B.

After the analyzing task was completed, we asked the participants to answer a questionnaire related to our system. The experiment took approximately 60 min in total.

5 Results

In this section, we show the results of the experiment and discuss them. First, we classify discoveries by five properties and five functions and compare them based on participants' background. Next, we examine the results of questionnaires. Finally, we compare the analysis process based on participants' background.

5.1 Discoveries and Classification

As a result, *unrecorded participants* (P1–P4) had 32.25 discoveries on average. In the previous study, *recorded participants* (P5–P8) had 24.00 discoveries on average. As the normalities of the results were satisfied with Shapiro-Wilk tests, we compared the number of discoveries using t-test. As a result of t-test, there was no significant difference ($t = 1.02$; $df = 6$; $p = 0.38 > 0.05$). This result suggests that *unrecorded participants* could discover many facts comparable to *recorded participants* when using our system.

We classified the discoveries by the following five properties (Figs. 6 and 7):

1. **Overviewing.** Discoveries obtained by paying attention to the entire image.
2. **People.** Discoveries obtained by paying attention to the people in recorded images.
3. **Environment.** Discoveries obtained by paying attention to the environment (e.g., objects in recorded images and changes in the appearance of the room).
4. **Suggestion.** Opinions/discoveries that are related to the analyzing system (e.g., requests to extend functionality, proposals of a new function, and ideas to improve the system).
5. **Other.** Additional opinions/discoveries (e.g., suggestions for applications of our system).

We also classified the discoveries by the five functions of our analyzing system by examining which function the participants used when they obtained discoveries (Figs. 8 and 9):

1. **Heatmap/all (HM/all).** Discoveries obtained by paying attention to the whole heatmap (Fig. 4A).
2. **Heatmap/part (HM/part).** Discoveries obtained by paying attention to a part of the heatmap (Fig. 4B).
3. **Calendar.** Discoveries obtained by paying attention to the color of the calendar (Fig. 4C).
4. **Time slider.** Discoveries obtained by paying attention to the color of the time slider or comparing different images by operating the time slider (Fig. 4E).
5. **Camera image.** Discoveries obtained by paying attention to the camera image (Fig. 4A).

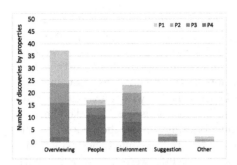

Fig. 6. Results of classifying discoveries of *unrecorded participants* by properties.

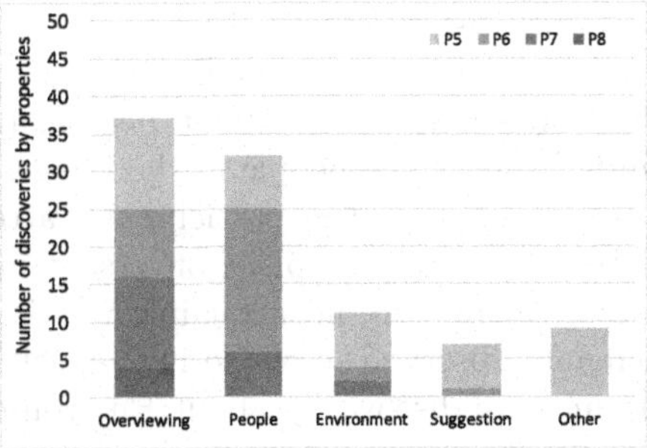

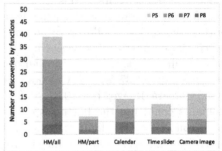

Fig. 7. Results of classifying discoveries of *recorded participants* by properties.

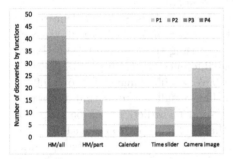

Fig. 8. Results of classifying discoveries of *unrecorded participants* by functions (breakdown by participants).

Fig. 9. Results of classifying discoveries of *recorded participants* by functions (breakdown by participants).

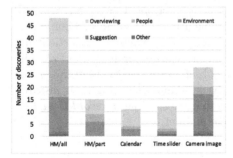

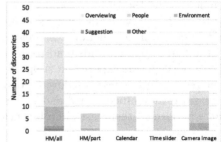

Fig. 10. Results of classifying discoveries of *unrecorded participants* by functions (breakdown by properties).

Fig. 11. Results of classifying discoveries of *recorded participants* by functions (breakdown by properties).

Comparison between Figs. 6 and 7 shows that *unrecorded participants* discovered many facts about Environment. On the contrary, *recorded participants* discovered many facts about People. In the experiment with *recorded participants*, most of the people in the images were acquaintances of the participants. This would draw the participants' attention to the people, resulting in more discoveries about People. Conversely, in the experiment with *unrecorded participants*, there were no acquaintances in the images. This would make the participants observe the whole images without focusing on specific people, resulting in more discoveries about Environment. We also examined the result using statistical analysis. As the normalities of the results were not satisfied with Shapiro-Wilk tests, we compared the number of discoveries by properties using Mann-Whitney U test. However, there was no significant difference (Overviewing: $p = 0.8 > 0.05$; People: $p = 0.69 > 0.05$; Environment: $p = 0.09 > 0.05$; Suggestion: $p = 1 > 0.05$; Other: $p = 1 > 0.05$); this would be because the number of participants in this experiment was small.

Moreover, there is not much difference between Figs. 8 and 9. Due to the normalities of the results were not satisfied with Shapiro-Wilk tests, we compared the number of discoveries by functions using the Mann-Whitney U test. As a result of test, there was no significant difference (HM/all: $p = 0.74 > 0.05$; HM/part: $p = 0.37 > 0.05$; Calendar: $p = 0.89 > 0.05$; Time slider: $p = 0.97 > 0.05$; Camera Image: $p = 0.31 > 0.05$). This result suggests that both participants could use the functions of our system regardless of the participants' background. The two figures also suggest that usage proportion of each function of our system has the same tendency regardless of the participants' background. Because the usage proportion has the same tendency, the cause of the difference in the previous paragraph might be the difference in participants' background.

As shown in Figs. 10 and 11, both participants discovered many facts about Overviewing, People, and Environment using the HM/all function. Compared to *recorded participants*, *unrecorded participants* discovered many facts about Environment using the HM/part and Camera Image functions. Conversely, *recorded participants* discovered many facts about People using the Calendar, Time slider, and Camera Image functions. Therefore, these results suggest that participants with different backgrounds tend to discover facts with different properties. The videos of the whole experiment support this observation. We found in the videos that *recorded participants* enjoyed looking back on and analyzing their research days since they often smiled and laughed sometimes, while *unrecorded participants* performed their tasks in a businesslike manner.

5.2   Questionnaire Results

We asked all participants to answer a questionnaire that consisted of two questions: *"did you use our system with ease?"* (Q1) and *"do you want to use our system in the future?"* (Q2). Each question included a five-point Likert scale form (1 = strongly disagree, 5 = strongly agree) and a comment form. The results of Q1 and Q2 for *recorded participants* and *unrecorded participants* are shown in Figs. 12, 13, 14 and 15.

As shown in Figs. 13 and 15, the scores provided by P6 were lower than the others. While P6 skillfully used all functions, he stated that "the analyzing was fun for me, but I cannot imagine application examples" in the questionnaire. Furthermore, P6 provided many requests for extending the functionality, proposals for new functions, and ideas to improve our system.

As shown in Figs. 12 and 13, *unrecorded participants* had similar results compared to *recorded participants*; this means that participants in both backgrounds were neutral to the usability of the analyzing system. In addition, P3 and P5 mentioned complaints about the processing speed of the analyzing system. Therefore, if the processing speed improves, the usability would improve.

As shown in Figs. 14 and 15, we found that there were many participants who wanted to use our system in the future. In addition, P2, who is a college student studying biology, commented that our system could be used to observe and analyze places where animals gather.

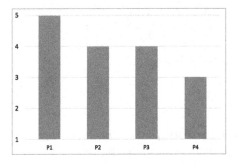

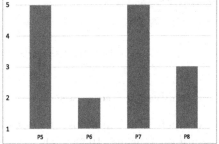

Fig. 12. Answer to the question *"did you use our system with ease?"* by *unrecorded participants* (mean = 3.50; SD = 0.50).

Fig. 13. Answer to the question *"did you use our system with ease?"* by *recorded participants* (mean = 4.00; SD = 0.71).

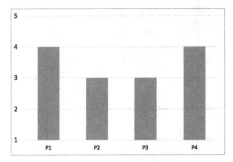

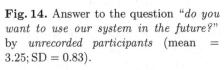

Fig. 14. Answer to the question *"do you want to use our system in the future?"* by *unrecorded participants* (mean = 3.25; SD = 0.83).

Fig. 15. Answer to the question *"do you want to use our system in the future?"* by *recorded participants* (mean = 3.75; SD = 1.30).

Fig. 16. Discovery about desktop monitor by P2.

Fig. 17. Discovery about large screen monitor by P2.

Fig. 18. Discovery about dartboard by P3.

5.3 Analyzing Processes

We compared the analyzing processes of *unrecorded participants* with those of *recorded participants*. To examine each participant's analyzing processes, we analyzed the screen captures. As a result, we found that the processes of both participants were similar: all the participants first browsed the images recorded in July, and then browsed the images recorded in August and later in the sequence.

Examining the analyzing processes of *unrecorded participants* with those of *recorded participants* gives us a suggestion for future possible improvement of our system. P1, P2, and P4 discovered the facts about monitors. For example, P2

Fig. 19. Discovery about kite by P4.   **Fig. 20.** Discovery about 3D printer by P5.

discovered that there was a movement behind the seat B-6 using the HM/all function (Fig. 16). Then, P2 used the Camera Image function and guessed there was a monitor; however, P2 could not identify what was displayed on that monitor. By contrast, P5, P6, and P7, who discovered similar facts about the monitors, could identify what the monitor displays as a clock since they had prior knowledge of Room-B. Moreover, we found another suggestive result. P2 discovered that the large monitor installed in the public space was used less frequently (Fig. 17 Left). Then, P2 used the HM/part function and discovered the detailed date and time that the large monitor was used. In addition, P2 discovered that an animation had been displayed on the large monitor (Fig. 17 Right). These results suggest that *unrecorded participants* would be able to discover more detailed facts if the resolution of the camera image is improved, because P2 succeeded in identifying the content displayed on the monitor since the monitor was large.

We also found how *unrecorded participants* discovered many facts about Environment while *recorded participants* discovered many facts about People when we observed the analyzing processes of both participants. P3, P4, and P5 discovered the facts about objects installed in Room-B. For example, P3 discovered that the dartboard was removed on October 12th using the Camera Image function (Fig. 18). P4 discovered that there was a kite in the public space (Fig. 19). In addition, P4 discovered that there was a movement around the kite using the HM/all function and guessed that someone was touching it. However, P4 stopped the analysis at this moment. By contrast, after P5 discovered that the 3D printer was installed on the shelf on October 16th by operating Time-Operation Panel (Fig. 20), P5 continued the analysis and discovered a person who installed the 3D printer using the Camera Image function. From the above, the focus of the analysis might depend on the background of the participants. Namely, it seems that P3 and P4 (*unrecorded participants*) focused on the object itself and thus discovered the fact about the object; the fact's property becomes Environment. Conversely, it seems that P5 (*recorded participants*) focused on the person and thus discovered the fact about the object; the fact's property becomes People.

In summary, even *unrecorded participants* can discover many facts. Furthermore, by improving the resolution of the camera image, *unrecorded participants* would be able to discover more detailed facts.

6 Discussion

In our experiment, we found that *unrecorded participants* can discover many facts using our system even if they had no prior knowledge of the room. This means that it is possible to crowdsource the task of analyzing images, which will lead to finding many facts in the images quickly.

We could not identify any difference in the tendency of functions to be used due to the difference in the background of participants. However, properties of the discovered facts tended differently depending on the background of the participants. Compared to *recorded participants*, *unrecorded participants* discovered several facts about Overviewing and Environment and few discoveries on People. However, this does not mean that *unrecorded participants* cannot discover facts about People; we are interested in how the discoveries of *unrecorded participants* will change by instructing *unrecorded participants* explicitly to try their best to discover facts about People before the analysis task.

7 Future Work

Currently, the recording system runs on a high-performance laptop PC (MacBook Pro). Therefore, we need to use a UPS with a large capacity battery in case of long-term power failure. We plan to modify our recording system to run on a low-power computer such as Raspberry Pi. Furthermore, we plan to set up another recording system with an omni-directional camera with higher resolution to examine whether *unrecorded participants* can discover more detailed facts.

There was an opinion that the processing speed of the analyzing system from participants was slow. We think that this problem is influenced by transferring images from NAS. We plan to implement an image prefetching algorithm based on the locality of reference. In addition, we will change the transfer protocol and compress image data before transfer.

In this experiment, we installed a recording system in our laboratory and used the images obtained from that system. In the future, we will use our system in a place other than our laboratory and conduct experiments with the images obtained from it. In addition, we will investigate the adaptability of our system for the observation of animals, plants, and natural phenomena.

8   Conclusions

In this study, we conducted a further experiment using the images in which the participants have not been recorded (*unrecorded participants*) to reveal the discoveries that participants obtain. By comparing the results of users with different backgrounds, we investigated the difference between discoveries, functions used, and analysis process. The comparison suggests that *unrecorded participants* could discover many facts about Environment, and *recorded participants* could discover many facts about People. Moreover, the comparison also suggests that *unrecorded participants* could discover many facts comparable to *recorded participants*.

References

1. Chiang, C.C., Yang, H.F.: Quick browsing and retrieval for surveillance videos. Multimedia Tools Appl. **74**(9), 2861–2877 (2015)
2. DeCamp, P., Shaw, G., Kubat, R., Roy, D.: An immersive system for browsing and visualizing surveillance video. In: Proceedings of the International Conference on Multimedia, MM 2010, pp. 371–380. ACM, New York (2010)
3. Higuchi, K., Yonetani, R., Sato, Y.: EgoScanning: quickly scanning first-person videos with egocentric elastic timelines. In: Proceedings of the 2017 CHI Conference on Human Factors in Computing Systems, CHI 2017, pp. 6536–6546. ACM, New York (2017). http://doi.acm.org/10.1145/3025453.3025821
4. Ishii, A., Abe, T., Hakoda, H., Shizuki, B., Tanaka, J.: Evaluation of a system to analyze long-term images from a stationary camera. In: Yamamoto, S. (ed.) HIMI 2016. LNCS, vol. 9734, pp. 275–286. Springer, Cham (2016). https://doi.org/10.1007/978-3-319-40349-6_26
5. Kubat, R., DeCamp, P., Roy, B.: TotalRecall: visualization and semi-automatic annotation of very large audio-visual corpora. In: Proceedings of the 9th International Conference on Multimodal Interfaces, ICMI 2007, pp. 208–215. ACM, New York (2007)
6. Nguyen, C., Niu, Y., Liu, F.: Video summagator: an interface for video summarization and navigation. In: Proceedings of the SIGCHI Conference on Human Factors in Computing Systems, CHI 2012, pp. 647–650. ACM (2012)
7. Nogami, R., Shizuki, B., Hosobe, H., Tanaka, J.: An exploratory analysis tool for a long-term video from a stationary camera. In: Proceedings of the 5th IEEE International Symposium on Monitoring & Surveillance Research, ISMSR 2012, vol. 2, pp. 32–37 (2012)
8. Romero, M., Summet, J., Stasko, J., Abowd, G.: Viz-A-Vis: toward visualizing video through computer vision. IEEE Vis. Comput. Graph. **14**(6), 1261–1268 (2008)
9. Romero, M., Vialard, A., Peponis, J., Stasko, J., Abowd, G.: Evaluating video visualizations of human behavior. In: Proceedings of the SIGCHI Conference on Human Factors in Computing Systems, CHI 2011, pp. 1441–1450. ACM, New York (2011)
10. Shin, G., Choi, T., Rozga, A., Romero, M.: VizKid: a behavior capture and visualization system of adult-child interaction. In: Salvendy, G., Smith, M.J. (eds.) Human Interface 2011. LNCS, vol. 6772, pp. 190–198. Springer, Heidelberg (2011). https://doi.org/10.1007/978-3-642-21669-5_23

11. Thomas, J.J., Cook, K., et al.: A visual analytics agenda. IEEE Comput. Graph. Appl. **26**(1), 10–13 (2006)
12. Tompkin, J., Pece, F., Shah, R., Izadi, S., Kautz, J., Theobalt, C.: Video collections in panoramic contexts. In: Proceedings of the 26th Annual ACM Symposium on User Interface Software and Technology, UIST 2013, pp. 131–140. ACM (2013)
13. Wong, P.C., Thomas, J.: Visual analytics. IEEE Comput. Graph. Appl. **24**(5), 20–21 (2004)

Decreasing Occlusion and Increasing Explanation in Interactive Visual Knowledge Discovery

Boris Kovalerchuk[(✉)] and Abdulrahman Gharawi

Central Washington University,
400 E. University Way, Ellensburg, WA 98926, USA
{BorisK,Abdulrahman.Gharawi}@cwu.edu

Abstract. Explanation and occlusion are the major problems for interactive visual knowledge discovery, machine learning and data mining in multidimensional data. This paper proposes a hybrid method that combines the visual and analytical means to deal with these problems. This method, denoted as FSP, uses visualization of n-D data in 2-D, in a set of Shifted Paired Coordinates (SPC). SPCs for n-D data consist of n/2 pairs of Cartesian coordinates, which are shifted relative to each other to avoid their overlap. Each n-D point is represented as a directed graph in SPC. It is shown that the FSP method simplifies the pattern discovery in n-D data, providing the explainable rules in a visual form with a significant decrease of the cognitive load for analysis of n-D data. The computational experiments on real data has shown its efficiency on both training and validation data.

Keywords: Visual knowledge discovery · Visual analytics
Visual data mining · Occlusion · Interactive visualization
Shifted Paired Coordinates

1 Introduction

1.1 Background and Problem

For a long time, explanation and occlusion have been among the major problems for interactive visual knowledge discovery, data mining, and machine learning in multidimensional data. This paper proposes a hybrid method, which combines the visual and analytical means to deal with these problems. The proposed method, denoted as FSP, uses visualization of n-D data in 2-D in a type of General Line Coordinates (GLC) [1, 2] known as Shifted Paired Coordinates (SPC). A set of Shifted Paired Coordinates for n-D data consists of n/2 pairs of Cartesian coordinates that are shifted relative to each other without overlap. Each n-D point A is represented as a directed graph A^* in SPC, where each node of A^* is a 2-D projection of A in a respective pair of the Cartesian coordinates.

The proposed FSP method significantly decreases the cognitive load for the analysis of n-D data and simplifies the discovery of *explainable patterns* in the n-D data. At the upper level, the steps of the FSP are: (1) *Filtering* out less efficient visualizations

© Springer International Publishing AG, part of Springer Nature 2018
S. Yamamoto and H. Mori (Eds.): HIMI 2018, LNCS 10904, pp. 505–526, 2018.
https://doi.org/10.1007/978-3-319-92043-6_42

from multiple SPC visualizations, (2) *Searching* for sequences of paired coordinates that are more efficient, and (3) *Presenting* the SPC visualizations only with better sequences to the analyst. FSP includes the randomized search for pairs of coordinates and explainable "rectangular" classification rules with maximized accuracy on training/validation data.

The computational experiments with the 9-D Wisconsin Breast Cancer data, the 33-D Ionosphere data, and the 8-D Abalone data, from UCI Machine Learning repository, show the efficiency of the FSP method, on the training and validation data. The visualization process in SPC is reversible, i.e., all n-D information is visualized and can be restored from visualization for each n-D case. This hybrid visual analytics method allows classifying data in a way that can be communicated to the domain experts such as medical doctors in the explainable/understandable and visual form.

1.2 Shifted Paired Coordinates: Challenge and Opportunity to Better Visualization

The **Shifted Paired Coordinates** (SPC) visualization of the n-D data requires the splitting of n coordinates $X_1 - X_n$ into pairs producing the $n/2$ non-overlapping pairs (X_i, X_j), such as $(X_1, X_2), (X_3, X_4), (X_5, X_6), \ldots, (X_{n-1}, X_n)$ [1, 2]. In SPC, a pair (X_i, X_j) is represented as a separate orthogonal Cartesian Coordinates (X, Y), where X_i is X and X_j is Y.

In SPC visualization design each coordinate pair (X_i, X_j) is *shifted* relative to other pairs to avoid their overlap. This creates $n/2$ scatter plots. Next in SPC, for each n-D point $\mathbf{x} = (x_1, x_2, \ldots, x_n)$, the point (x_1, x_2) in (X_1, X_2) is connected to the point (x_3, x_4) in (X_3, X_4) and so on until point (x_{n-2}, x_{n-1}) in (X_{n-2}, X_{n-1}) is connected to the point (x_{n-1}, x_n) in (X_{n-1}, X_n) to form a directed graph $\mathbf{x}^*$. Figure 1 shows the same data visualized in SPC in two different ways due to different pairing of coordinates.

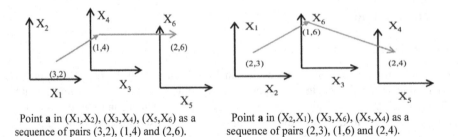

Point **a** in (X_1,X_2), (X_3,X_4), (X_5,X_6) as a   Point **a** in (X_2,X_1), (X_3,X_6), (X_5,X_4) as a
sequence of pairs (3,2), (1,4) and (2,6).   sequence of pairs (2,3), (1,6) and (2,4).

Fig. 1. 6-D point **a** = (3, 2, 1, 4, 2, 6) in Shifted Paired Coordinates.

In general, there are multiple combinatorial ways to form the pairs of coordinates for the SPCs, and to sequence the pairs. The SPC visualization graphs $\mathbf{x}_k^*$ of each given n-D point $\mathbf{x}$ differ for different sequences S_k of pairs of coordinates. Figure 1a illustrates it for a 6-D point $\mathbf{a}$ = (3, 2, 1, 4, 2, 6) visualized in pairs (X_1, X_2), (X_3, X_4), (X_5, X_6), and Fig. 1b shows this point visualized in pairs (X_2, X_1), (X_3, X_6), (X_5, X_4).

The SPC allows visualizing each individual n-D point losslessly, but together graphs of multiple n-D points occlude each other. See Fig. 2. This creates a difficulty for discovering patterns to classify cases of opposing classes in SPC. In Fig. 2, some of the areas are visibly dominated by the cases of the specific color. However, it is not sufficient to build discrimination rules to classify cases visually. It required an addition analytical process. Such process is proposed in this paper. SPC visualizations with some sequences S_k can reveal classification patterns of n-D data better than with using other sequences. The *dependence* of the visualization on the different pairing of coordinates creates a *challenge* and an *opportunity* to find the better pairs and their sequences.

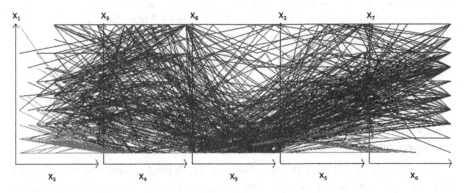

Fig. 2. A set of 688 Wisconsin Breast Cancer (WBC) data visualized in SPC as 2-D graphs of 10-D points with benign cases in red and malignant cases in blue. (Color figure online)

The challenge is that it is impractical to conduct the interactive search of the efficient sequences of pairs of coordinates for the large number of sequences. The total number of pairs of n coordinates is the number of combinations $C(n, 2) = n!/((n-2)! \cdot 2!)$, e.g., 45 for $n = 10$. Next, there are multiple different sequences of the same set of n/2 pairs, e.g., $(X_1, X_2), (X_3, X_4), (X_5, X_6), \ldots, (X_{n-1}, X_n)$ and $(X_5, X_6), (X_3, X_4), (X_1, X_2), \ldots, (X_{n-1}, X_n)$. The number of these sequences (orders) for the given n/2 pairs is (n/2)!, e.g., (10/2)! = 120 for $n = 10$. Thus, the total number of sequences of n/2 pairs of n coordinates is $(n/2)! \cdot C(n, 2) = (n/2)! \cdot n!/((n-2)! \cdot 2!)$ and for $n = 10$ it is 45·120 = 5400 assuming that any of $C(n,2)$ pairs can be used to form sequences. The analyst cannot observe all of them, to find the best one for visual separation of classes. The FSP algorithm resolves this issue by the automatic search for the best sequences, and presenting only a few of the best visualizations to the analyst.

2 Algorithms

2.1 FSP Algorithm

The FSP algorithm:

(a) *Filters* out less efficient rules and visualizations for supervised classification learning,
(b) *Searches* for sequences of pairs of coordinates and respective rules that are more efficient for supervised classification learning, and
(c) *Presents* the SPC visualizations only with better sequences to the analyst.

The main characteristic of the FSP is avoiding the interactive exploration of the exponential number of alternative sequences with the following *major steps*:

Step 1: *Random generation* of sequences of pairs of coordinates S;
Step 2: *SPC representation* of n-D data in sequences S from Step 1;
Step 3: *Machine learning* process for learning "rectangular" classification rules with high accuracy, precision and recall in sequences S from Step 2;
Step 4: *Full* automated visualization process: SPC representation of best n-D rules in the best *full sequences* S of pairs of coordinates S discovered in Step 3;
Step 5: *Simplified* automated visualization process: *SPC representation* of best n-D rules in the best *subsequences* S of pairs of coordinates S discovered in Step 3;
Step 6: *Interactive* visualization process: an analyst interactively controls and manages produced SPC visualizations.

The approach of this algorithm is in line with [4]. It produces efficient classification rules and 3-D visualization of n-D data. In that paper another visualization method called Collocated Tripled Coordinates from the class of General Line Coordinates (GLC) is combined with Machine Learning for predicting the investment strategy. The ideas of steps 1 and 2 already have been explained above. The step 3 uses rules and learning criteria presented below.

Rules and Learning Criteria. The filtering works on a set of rules such as the rules (1)–(8), listed below. Each rule is defined on an n-D point $\mathbf{x} = (x_1, x_2, \ldots, x_n)$ to be classified into some classes:

$$\text{If } (x_i, x_j) \in R_1 \text{ then } \mathbf{x} \in \text{class 1}, \tag{1}$$

$$\text{If } (x_i, x_j) \in R_1 \mathbin{\&} (x_k, x_m) \in R_2 \mathbin{\&} (x_s, x_t) \in R_3 \text{ then } \mathbf{x} \in \text{class 1} \tag{2}$$

$$\text{If } ((x_i, x_j) \in R_1 \vee (x_k, x_m) \in R_2) \mathbin{\&} (x_s, x_t) \in R_3 \text{ then } \mathbf{x} \in \text{class 1} \tag{3}$$

$$\text{If } ((x_i, x_j) \in R_1 \vee (x_k, x_m) \in R_2) \mathbin{\&} (x_s, x_t) \notin R_3 \text{ then } \mathbf{x} \in \text{class 1} \tag{4}$$

$$\text{If } ((x_i, x_j) \in R_1 \vee (x_k, x_m) \in R_2) \mathbin{\&} (x_s, x_t) \notin R_3 \text{ then } \mathbf{x} \in \text{class 1, else } \mathbf{x} \in \text{class 2} \tag{5}$$

$$\text{If } (x_i, x_j) \in R_1 \,\&\, (x_k, x_m) \in R_2 \,\&\, (x_s, x_t) \notin R_3 \text{ then } \mathbf{x} \in \text{class 1} \tag{6}$$

$$\text{If } (x_i, x_j) \in R_1 \,\&\, (x_i, x_j) \notin R_2 \,\&\, (x_i, x_j) \notin R_3 \text{ then } \mathbf{x} \in \text{class 1} \tag{7}$$

$$\text{If } \left((x_i, x_j) \in R_1 \lor (x_k, x_m) \notin R_2 \lor (x_s, x_t) \notin R_3 \text{ then} \right) \mathbf{x} \in \text{class 1} \tag{8}$$

where R_1, R_2 and R_3 are specific *rectangles*, in respective pairs of Cartesian coordinates in a given sequence S of pairs of coordinates S, e.g., R_1 can be in (X_1, X_2).

The filtering follows the common Data Mining/Machine Learning strategy of learning the rules on the training data, and testing on the validation data. The *quality of learning of the classification* and *the expected visualization* for the rules (1)–(4), (6)–(8) is measured by the **precision** and **recall** of the classification of the training and validation data, where the precision Pr is the fraction of the of cases, predicted correctly by the rule, over all the predicted cases by the rule:

$$Pr = |\{\text{cases predicted correctly by the rule}\}| / |\{\text{all predicted cases by the rule}\}|.$$

The precision Pr for the basic rule (1): If $(x_i, x_j) \in R_1$ then $\mathbf{x} \in$ class 1, is calculated as follows:

$$Pr = \frac{n1(R_1)}{n1(R_1) + n2(R_1)} \tag{9}$$

where $n1(R_1)$ is the number of points of class 1 in R_1 (i.e., the number of correctly classified cases), and $n2(R_1)$ is the number of points from the class 2 in R_1 (i.e., the number of misclassified cases). More generally, for any rule($\mathbf{x}$) such that

$$\text{If rule}(\mathbf{x}) \text{ then } \mathbf{x} \in \text{class 1} \tag{10}$$

the precision is

$$Pr = \frac{n1(rule)}{n1(rule) + n2(rule)} \tag{11}$$

where $n1$(rule) is the number of points of class 1 that satisfy the if part of the rule and $n2$(rule) is the number of points of class 2 that satisfy the if the part of the rule too.

The formula (11) is applicable to all the rules (1)–(8). For example, the precision Pr for the rule (4) is calculated as follows:

$$Pr = \frac{n1(R_1) + n1(R_2) - n1(R_1 \& R_3) - n1(R_2 \& R_3)}{n1(R_1) + n1(R_2) - n1(R_1 \& R_3) - n1(R_2 \& R_3) + n2(R_1) + n2(R_2) - n2(R_1 \& R_3) - n2(R_2 \& R_3)} \tag{12}$$

where

$n1(R_1)$ and $n1(R_2)$ are the number of points of class 1 in R_1, R_2, respectively,

$n1(R_1 \& R_3)$ is the number of graphs $\mathbf{x}^*$ of the n-D points of class 1 that have 2-D points in both R_1 and R_3,

$n1(R_2 \& R_3)$ is the number of graphs $\mathbf{x}^*$ of the n-D points of class 1 that have 2-D points in both R_2 and R_3,

$n2(R_1 \& R_3)$ is the number of graphs $\mathbf{x}^*$ of the n-D points of class 2 that have 2-D points in both R_1 and R_3,

$nb(R_2 \& R_3)$ is the number of graphs $\mathbf{x}^*$ of the n-D points of class 2 that have 2-D points in both R_2 and R_3.

Here

$n1(R_1) + n1(R_2) - n1(R_1 \& R_3) - n1(R_2 \& R_3)$ is the number of correctly classified n-D points by rule (4) and

$n2(R_1) + n2(R_2) - n2(R_1 \& R_3) - n2(R_2 \& R_3)$ is the number of misclassified n-D points by this rule.

We use the precision for rules (1)–(4) and (6)–(8) instead of the accuracy, because these rules predict only one class. All cases, which do not satisfy the condition of these rules, are not classified (*refused* to be classified). The computing accuracy would require predictions of the class for all cases as it is the case for rules (5). Therefore, rules in set of rules (5) we use the accuracy for filtering. Note that a high precision rule from sets of rules (1)–(4), (6)–(8) may covers only a few cases, but the precision value does not show it's low coverage. Therefore, we also use the **recall**

$$RC = |\{\text{cases predicted correctly by the rule}\}| / |\{\text{all cases}\}|$$

that is a fraction of cases correctly predicted by the rule to all cases to be predicted.

The **Random generation** in Step 1 consists of the two substeps:

(RS1) Randomly generate a set of pairs of coordinates from coordinates $X_1 - X_n$,
(RS2) Randomly generate sequence S_k for a set of pairs from (RS1).

The **Machine Learning process** in the Step 3 consists of the following substeps:

(ML1) Search for rectangles in each (X_i, X_j) that maximize precision or accuracy of a rule from (1)–(8) on training data for given S_k,
(ML2) Evaluate this rule on validation data,
(ML3) Repeat (RS1)–(RS2) and (ML1)–(ML2) in attempt to reach desired precision/accuracy and recall.
(ML3) Combine promising rules to get a stronger rule in precision, accuracy and recall.

The **Automated Visualization process** in Steps 4–5 consists of the following steps:

(AV1) Visualize in SPC most accurate classification results. This includes visualization of only best results.
(AV2) Remove data that are covered by best results in (IV1),
(AV3) Repeat (RS1–RS2) and (ML1–ML3) for remaining data in search for the best classification results.

The **Interactive Visualization process** in the Step 6 is as follows:

(IV1) Substitute the automatic search in (ML1) by the interactive search, where the analysts select the rectangles in SPC visualization using the GUI.

(IV2) The automatic system supports this interactive process by computing accuracies of rules based on selected rectangles and removing data covered by best results found before the next interactive selection of new rectangles starts.

Figure 3 shows the overall design of the FSP process.

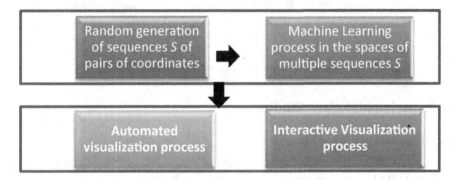

Fig. 3. Overall design of the FSP process.

2.2   Frequency Visualization Algorithm

One of the challenges of the SPC visualization for the larger data sets is a form of occlusion, caused by a limited resolution of the data on the screen. It leads to overlap of similar data including complete colocation of some lines. In addition, identical cases will collocate on any visualization at any resolution. Therefore, the task is enhancing the SPC visualization to show frequency of the lines.

The algorithm for this, denoted as the **FRE algorithm**, has the following steps:

(F1) For each consecutive pairs of coordinates (X_i, X_j), (X_k, X_m) form sets of edges $\{E_q\}$ that are collocated or nearly collocated under some threshold T,

(F2) Count the number of edges $C_q(T)$ in each of these sets,

(F3) Draw each such set of edges E_q with the adjustable width w:
   (a) *Equally* proportional to the number of edges in E_q for each node,
   (b) *Unequally* proportional to the number of edges in E_q for each node *without adjusting* to the data density focused in the given node of graph.,
   (c) *Unequally* proportional to the number of edges in E_q for each node *with adjusting to* the data density in the given node of graph.

Figure 4 illustrates the differences between the versions (a)–(c) of the algorithm FRE for the red graphs in the SPCs. The version (a) is "neutral", the analyst can use it when no specific node it set up to explore, Versions (b) and (c) are for exploring specific nodes of interest (e.g. nodes of the discovered classification rule), because the large width of edge in (a) may occlude that specific node.

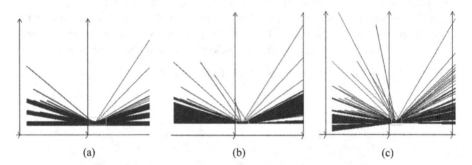

(a)   (b)   (c)

Fig. 4. Alternative frequency visualizations.

3 Experimental Case Studies

3.1 Experimental Case Study 1

The computational experiments with the 9-D Wisconsin Breast Cancer (WBC) data, from the UCI Machine Learning repository [3] presented below, show the *efficiency* of the FSP algorithm. To get the even number of coordinates and 5 pairs of coordinates, the coordinate X_5 was duplicated in X_{10} getting total 10 coordinates.

The discovered patterns were found by the search in the set of rules (1)–(8). In particular, on WBC data, the FSP algorithm found an efficient sequence of the pairs of the coordinates. This sequence of pairs is (X_5, X_1), (X_4, X_3), (X_9, X_8), (X_5, X_2), (X_6, X_7). Here X_5 is used in two pairs (X_5, X_1) and (X_5, X_2). The SPC visualization with this sequence reveals classification pattern with *precision over 90%* in all 11 random 70%:30% splits that are presented in Tables 1 and 2. The best precision on the training data is 99.3%, which is accompanied by the high precisions on the validation data (98.21%), in one of the 70%:30% splits of the given data into the training and the validation data.

The discovered rules in Tables 1 and 2 belong to the set of rules (7). The first rule in Table 1 that we denote as **WBC Rule 1** is:

$$\text{If } (x_9, x_8) \in R_1 \ \& \ (x_6, x_7) \notin R_2 \ \& \ (x_6, x_7) \notin R_3 \text{ then } \mathbf{x} \in \text{class } 1(\text{Red}, \text{Benign}) \quad (13)$$

where R_1, R_2, and R_3 are specific rectangles, in respective pairs of Cartesian coordinates (X_8, X_9) and (X_6, X_7). Table 3 shows the parameters of R_1, R_2 and R_3 in the normalized coordinates.

This rule, with a random 70%:30% data split into the training and the validation data, has the precision of 94.6% on the training data, and 96.82% on the validation data. Figures 5 and 6 show its rectangles R_1, R_2 and R_3 drawn in the SPC as magenta boxes. The difference between these figures is that Fig. 5 shows all graphs that go through rectangle R_1 (have node (x_9, x_8) in R_1), but Fig. 6 shows only graphs that in

Table 1. Number of cases that satisfy the if part of Rule 1 in 11 random 70%:30% splits of data.

70%:30% random data splits	Number of cases that satisfy if part of the Rule 1					
	Red class			Blue class		
	Training	Testing	Total	Training	Testing	Total
1	303	122	425	17	4	21
2	300	105	405	10	4	14
3	290	132	422	20	4	24
4	291	110	401	2	2	4
5	253	123	376	2	1	3
6	297	116	413	13	3	16
7	301	121	422	12	2	14
8	299	122	421	12	4	16
9	282	127	409	10	4	14
10	307	116	423	27	7	34
11	282	117	399	27	9	36
Average	291	119	411	14	4	18

Table 2. Precision, recall and coverage of Rule 1 in 11 random 70%:30% splits of data.

70%:30% random data splits	Rule precision		Rule recall (correct) coverage			Rule total coverage of cases, %
	Training, %	Testing, %	Training, %	Testing, %	Total, %	
1	94.06	96.82	44.04	17.73	61.77	64.83
2	96.77	96.33	43.6	15.26	58.86	60.9
3	93.5	97.05	42.15	19.19	61.34	64.83
4	99.3	98.21	42.3	15.99	58.29	58.87
5	98.41	99.19	36.77	17.88	54.65	55.09
6	95.8	97.47	43.17	16.86	60.03	62.35
7	96.16	98.37	43.75	17.59	61.34	63.37
8	96.14	96.82	43.46	17.73	61.19	63.52
9	96.57	96.94	40.99	18.46	59.45	61.48
10	91.91	94.3	44.62	16.86	61.48	66.42
11	91.26	92.85	40.99	17.01	58	63.23
Average	95.44	96.76	42.3	17.3	59.6	62.35

Table 3. Parameters of rectangles R_1–R_3.

Rectangle	Parameters			
	Left	Right	Bottom	Top
R_1 in (X_9, X_8)	0.0020	0.1402	0.0734	0.1028
R_2 in (X_6, X_7)	0.7214	1.001	0.3484	1.001
R_3 in (X_6, X_7)	0.0081	0.6325	0.6014	1.001

addition do not go rectangles R_2 and R_3. (do not have node (x_6, x_7) in R_2 and R_3). Thus, Fig. 6 shows only graphs that satisfy Rule 1.

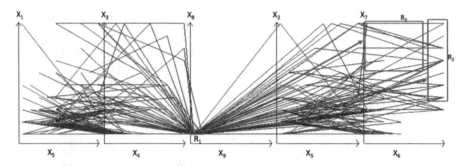

Fig. 5. WBC data in SPC as graphs representing 10-D points that go through R_1 without showing frequency of cases. Magenta boxes show rectangles R_1–R_3. (Color figure online)

In Fig. 5, the width of the lines are not adjusted their frequency, but in Fig. 6 the width of the lines is adjusted to their frequencies. The rule 1 covers 64.8% of all given 9-D points: 446 cases out of 688 cases (425 red cases and 21 blue cases) with the recall value 61.77% (425/688) as shown in Table 1. Among these 446 cases 303 Red and 17 Blue cases belong to training data and 122 Red and 4 blue cases belong to testing data.

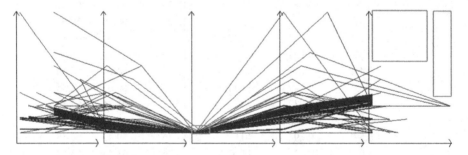

Fig. 6. WBC data in SPC as graphs representing 10-D points that satisfy Rule 1 (go through R_1 and not coming to R_2 and R_3 shown in magenta. Wide red lines show the frequency of red cases. (Color figure online)

The cases covered by rule 1 do not include only the 5.13% (23 cases) of the Red data, but include the 91.25% (219 cases) of the blue data. This rule covers only cases found in R_1 that do not come to R_2 or R_3. Rule 1 refused to predict other cases. Those other cases are dominantly blues (see Fig. 7) and must be covered by other rules. The WBC Rule 1 uses only 4 coordinates that form two pairs (X_9, X_8) and (X_6, X_7) therefore we can simplify the visualization of this rule by showing only them in SPC. It is done in Fig. 8 where each 4-D points is visualized losslessly as a single line. The advantage of this visualization is that it is easy to see and communicate to the medical

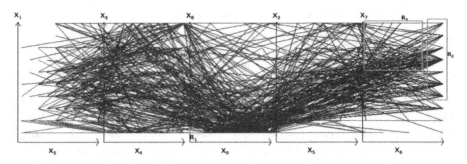

Fig. 7. Remaining WBC cases not covered by rule 1 (dominated by blue class). (Color figure online)

experts. The medical expert can easily understand this rule because it simply says that two attributes x_8 and x_9 must be in some limits identified in Table 3 and two other attributes x_6 and x_7 must not have values in some intervals that are shown visually in Fig. 8.

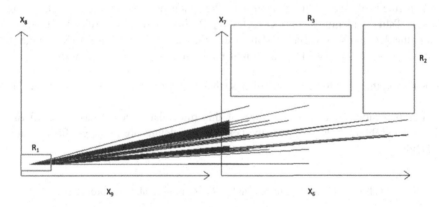

Fig. 8. WBC data in 4-D SPC as graphs in coordinates (X_9, X_8) and (X_6, X_7) that are used by Rule 1, i.e., WBC cases that go through R_1 and not go to R_2 and R_3 in these coordinates.

This allows the medical experts to analyze the consistency of this rule with the other medical domain knowledge, which is extremely difficult for the ML discrimination functions, which are "black" boxes or complex mathematical formulas.

The simple **WBC Rule 2** classifies all remaining cases (not covered by rule 1) to class 2 is

$$\text{If } (x_8, x_9) \in R_1 \ \& \ (x_6, x_7) \notin R_2 \ \& \ (x_6, x_7) \notin R_3 \tag{14}$$
$$\text{then } \mathbf{x} \in \text{class 1(Red, Benign) else } \mathbf{x} \in \text{class 2(Blue, Malignant)}$$

This rule classifies all cases that are either in R_2 or in R_3 or not in R_1 as blue. This rule has accuracy 93.60% (425 + 219)/688) on *all 688 cases* as the confusion matrix

Table 4. Confusion matrixes of WBC rule 2.

Actual class	Confusion matrix on all 688 cases			Confusion matrix on 70% of training cases (482)			Confusion matrix on 30% on testing cases (206)		
	Predicted class			Predicted class			Predicted class		
	Red	Blue	Total	Red	Blue	Total	Red	Blue	Total
Red	425	23	448	303	19	322	122	4	126
Blue	21	219	240	17	143	160	4	76	80
Total	446	242	688	320	162	482	126	89	206

shows in Table 4. Its accuracy on *training data* is 92.53% and on *validation data* it is 96.11% computed from respective confusion matrixes shown in Table 4.

We found a better **WBC Rule 3** by searching the rectangles with highest density of blue cases:

$$\text{If } ((x_1, x_6) \in R_4 \vee (x_3, x_5) \in R_5) \& (x_2, x_5) \notin R_6 \text{ then } \mathbf{x} \in \text{class 2 (Blue)} \quad (15)$$

This rule is of type of rules (4), but with the conclusion that **x** belongs to class 2, not class 1. Table 5 identifies the rectangles R_4–R_6 involved in this rule. WBC Rule 3 covers the 226 cases from class 2 (Blue), and the 12 cases from class 1 (Red) with the total precision of 94.95%. The **combined rule1 and rule 3** is as follows:

$$\text{If Rule1}(\mathbf{x}) \text{ then } \mathbf{x} \in \text{Red, else if Rule3}(\mathbf{x}) \text{ then } x \in \text{Blue, else refuse to classify } \mathbf{x} \quad (16)$$

The precision, recall and coverage of this rule relative to all cases are 95.18%, 94.62%, and 99.42% (Fig. 9). It is the performance details are in the confusion matrix in Table 6.

Table 5. Parameters of rectangles R_4–R_6 in normalized coordinates.

Rectangle	Parameters			
	Left	Right	Bottom	Top
R_4 in (X_1, X_6)	0.0010	0.9712	0.4180	1.001
R_5 in (X_3, X_5)	0.0000	0.9881	0.3154	0.7261
R_6 in (X_2, X_5)	0.0000	0.1003	0.143	0.3153

3.2   Experimental Case Study 2

The computational experiments with the 33-D Ionosphere data from the UCI Machine Learning repository [3] also show the efficiency of the FSP algorithm. To get the even number of coordinates and 17 pairs of coordinates, the algorithm in each epoch will select randomly the coordinate that will serve as coordinate X_{34}. In the following results X_{34} is a copy of X_{10}. The discovered patterns also were found by the search in the set of "rectangular" rules (1)–(8). In particular, on Ionosphere data, the FSP

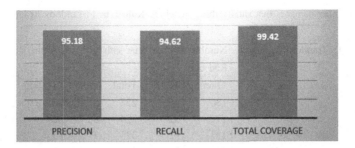

Fig. 9. The precision, recall and total coverage of combined rule 1 and rule 3

Table 6. Confusion matrix of combined rules 1 and 3 on all 688 cases.

	Predicted red class	Predicted blue class	Refusal	Total
Actual red class	425	12	2	437
Actual blue class	21	226	2	247
Total	446	238	4	688

algorithm found an efficient sequence of pairs of coordinates: (X_5, X_{26}), (X_{27}, X_{16}), (X_4, X_{11}), (X_{18}, X_{24}), (X_{28}, X_{31}), (X_{10}, X_3), (X_{23}, X_8), (X_{22}, X_{30}), (X_{21}, X_{10}), (X_{17}, X_2), (X_{15}, X_{33}), (X_{29}, X_{20}), (X_9, X_6), (X_{32}, X_{16}), (X_1, X_{25}), (X_{12}, X_{14}), (X_{19}, X_7). Here X_{10} is repeated in pairs (X_{10}, X_3) and (X_{21}, X_{10}). Similarly to the Case Study 1, the SPC visualization, with this sequence, reveals the visual classification pattern of *precision of over 90%* in all the 11 random 70%:30% splits of data into the training and the validation data (see Tables 7 and 8). The best precision on the training data is 98.36% with 100% precision on the validation data in one of these 70%:30% splits of the given data. The discovered rules in Tables 7 and 8 belong to the set of rules (4). The first rule in Table 7 that we denote as **Ionosphere Rule 1** is:

$$\text{If } [(x_4, x_{11}) \in R_1 \vee (x_5, x_{26}) \in R_8 \vee (x_{21}, x_{10}) \in R_9 \vee (x_{19}, x_7) \in R_{10}] \&$$
$$[(x_{27}, x_{16}) \notin R_3 \& (x_{28}, x_{31}) \notin R_6 \& (x_{23}, x_8) \notin R_4 \& (x_{17}, x_2) \in R_5 \& (x_{15}, x_{33}) \notin R_7 \&$$
$$(x_9, x_6) \notin R_2]$$
$$\text{then } x \in \text{class } 1(\text{Red, Good})$$

$$(17)$$

where R_1, R_2, R_3, R_4, R_5, R_6, R_7, R_8, R_9, and R_{10} are specific rectangles, in respective pairs of Cartesian coordinates (X_4, X_{11}), (X_9, X_6), (X_{27}, X_{16}), (X_{23}, X_8), (X_{17}, X_2), (X_{28}, X_{31}), (X_{15}, X_{33}), (X_5, X_{26}), (X_{21}, X_{10}), and (X_{19}, X_7). Table 9 shows the parameters of R_1–R_{10} in the normalized coordinates. This rule, with a random 70%:30% data split into the training and validation data, has the precision 98.36% on the training data and 100% on validation data. Figures 10 and 11 show its rectangles R_1–R_{10} in the SPCs as magenta boxes and cases that satisfy rule 1 in Fig. 11 and all cases in Fig. 10. Figure 12 shows the remaining cases.

Table 7. Number of cases that satisfy the Ionosphere Rule 1 in 11 random 70%:30% splits of data.

70%:30% random data splits	Number of cases that satisfy if part of the Rule 1					
	Red class			Blue class		
	Training	Testing	Total	Training	Testing	Total
1	180	45	225	3	0	3
2	172	52	224	8	2	10
3	133	92	225	7	3	10
4	129	95	224	6	2	8
5	200	24	224	9	0	3
6	160	65	225	10	2	12
7	183	42	225	11	1	13
8	158	67	225	11	2	13
9	191	34	225	13	1	14
10	184	37	221	7	0	7
11	157	66	223	12	2	14
Average	170	54	224	9	1	9

Table 8. Precision, recall and coverage of Ionosphere Rule 1 in 11 random 70%:30% splits of data.

70%:30% random data splits	Rule precision		Rule recall (correct) coverage			Rule total coverage of cases, %
	Training, %	Testing, %	Training, %	Testing, %	Total, %	
1	98.36	100	51.28	12.82	64.1	64.95
2	95.55	96.29	49	14.81	63.81	66.6
3	95	96.84	37.89	26.21	64	66.95
4	95.5	97.93	36.75	27.06	63.81	66.09
5	95.69	100	56.98	6.83	63.81	64.67
6	94.1	97.01	45.58	18.51	64.1	67.52
7	94.32	97.67	52.13	11.96	64.1	67.80
8	93.491	97.1	45.01	19.08	64.09	67.80
9	93.62	97.14	54.41	9.68	64.1	68.09
10	96.33	100	52.421	10.54	62.96	64.95
11	92.89	97.05	44.72	18.8	63.53	67.52
Average	94.99	97.93	48.43	15.38	63.81	66.38

The Rule 1 uses only 20 coordinates that form 10 pairs (X_5, X_{26}), (X_{27}, X_{16}), (X_4, X_{11}), (X_{28}, X_{31}), (X_{23}, X_8), (X_{21}, X_{10}), (X_{17}, X_2), (X_{15}, X_{33}), (X_9, X_6) and (X_{19}, X_7). Therefore, we simplify its visualization by showing only them in the SPCs (see Figs. 13, 14 and 15) with the lossless visualization of each of the 20-D points, as a single polyline. The simple **Ionosphere Rule 2** classifies all remaining cases (not covered by rule 1) to class 2 is

Table 9. Parameters of rectangles R_4–R_6 in normalized coordinates.

Rectangle	Parameters			
	Left	Right	Bottom	Top
R_1 in (X_4, X_{11})	0.23912	1.001	0.663667	0.999667
R_2 in (X_9, X_6)	0.1979	0.99869	0	0.0403333
R_3 in (X_{27}, X_{16})	0.130225	1.001	0	0.0256667
R_4 in (X_{23}, X_8)	0.29087	0.58943	1.001	0.946
R_5 in (X_{17}, X_2)	0.40433	0.9998	0	0.022
R_6 in (X_{28}, X_{31})	0.30765	1.001	0	0.0366667
R_7 in (X_{15}, X_{33})	0.25733	1.001	0.960667	0.998667
R_8 in (X_5, X_{26})	0.125675	0.575667	0.575667	0.645333
R_9 in (X_{21}, X_{10})	0.25278	0.50868	0.0476667	0.0843333
R_{10} in (X_{19}, X_7)	0.4203	0.8254	0.194333	0.432667

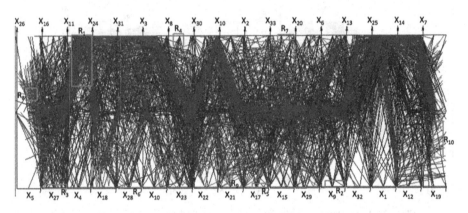

Fig. 10. 351 Ionosphere cases in the SPCs as graphs of 34-D points (good cases in red and bad cases in blue). Rectangles that are used in Ionosphere Rule 1 are in magenta. (Color figure online)

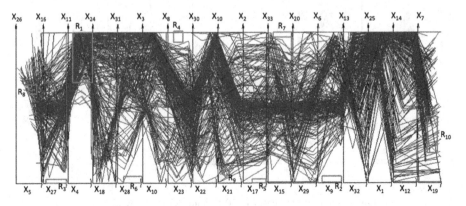

Fig. 11. 34-D Ionosphere cases covered by Ionosphere rule 1. Rectangles from rule 1 are in magenta. (Color figure online)

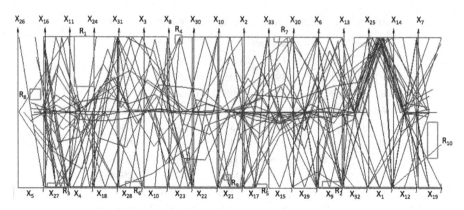

Fig. 12. Remaining Ionosphere cases (cases not covered by Ionosphere rule 1).

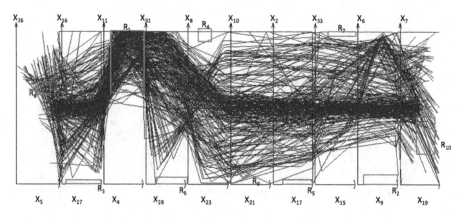

Fig. 13. Ionosphere cases in 20-D points covered by the Ionosphere rule 1.

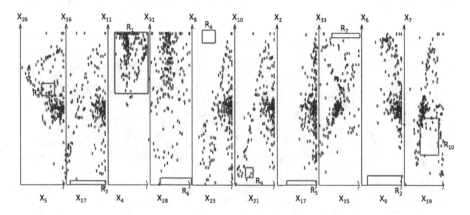

Fig. 14. Ionosphere cases in 20-D points covered by rule 1.

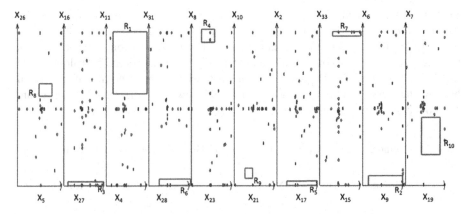

Fig. 15. Ionosphere cases in 20-D points not covered by rule 1.

$$\text{If } [(x_4, x_{11}) \in R_1 \lor (x_5, x_{26}) \in R_8 \lor (x_{21}, x_{10}) \in R_9 \lor (x_{19}, x_7) \in R_{10}] \&$$
$$[(x_{27}, x_{16}) \notin R_3 \& (x_{28}, x_{31}) \notin R_6 \& (x_{23}, x_8) \notin R_4 \& (x_{17}, x_2) \notin R_5 \& (x_{15}, x_{33}) \notin R_7 \&$$
$$(x_9, x_6) \notin R_2]$$
$$\text{then } x \in \text{class 1(Red, Good) else } x \in \text{class 2(Blue, Bad)}$$

$$(18)$$

This rule classifies all the cases rejected by rule 1 as blue with 99.14% accuracy on *all the 351 cases* (225 + 123)/351), see the confusion matrix in Table 10. Its accuracy on *training data* is 98.78% and 100% on the *validation data* based on the confusion matrixes in Table 10.

Table 10. Confusion matrixes of the Ionosphere rule 2.

Actual class	Confusion matrix on all 351 cases			Confusion matrix on training data (246, 70%)			Confusion matrix on validation data (206, 30%)		
	Predicted class			Predicted class			Predicted class		
	Red	Blue	Total	Red	Blue	Total	Red	Blue	Total
Red	225	3	228	180	3	183	45	0	45
Blue	0	123	123	0	63	63	0	60	60
Total	225	126	351	180	66	246	45	60	105

3.3   Experimental Case Study 3

The computational experiments with the 8-D Abalone data, male and infant cases, from the UCI Machine Learning repository [3] also show the efficiency of the FSP algorithm. The discovered patterns were found by the search in the set of "rectangular" rules (1)–(8). In particular, the FSP algorithm found an efficient sequence of coordinate pairs: (X_5, X_6), (X_1, X_2), (X_3, X_8), (X_4, X_7). Similarly, to Case Studies 1 and 2, the

SPC visualization, with this sequence, reveals the visual classification pattern of with the *precision of over 90%* (see Tables 11 and 12) in the 11 random 70%:30% splits of data into the training and the validation cases. The best precision is 92.06% on the training data and 96.17% on the validation data. Figure 16 shows 2870 Abalone data in the SPCs, as graphs of 8-D points, with the male cases in red, and the infant cases in blue. Rectangles used in Abalone Rule 1 defined below are in magenta. Figures 17 and 18 show cases covered this rule and Fig. 19 shows the remaining cases.

Table 11. Number of cases satisfying the if part of Abalone Rule 1 in 11 random 70%:30% splits.

70%:30% random data splits	Number of cases that satisfy if part of the Rule 1					
	Red class			Blue class		
	Training	Testing	Total	Training	Testing	Total
1	1193	304	1497	103	12	115
2	1034	396	1430	103	37	140
3	972	463	1435	86	46	132
4	1015	484	1499	105	40	145
5	1224	272	1496	126	10	136
6	1003	371	1374	98	28	126
7	989	402	1391	92	37	129
8	1077	389	1466	108	22	130
9	1146	328	1474	117	31	148
10	1201	276	1477	121	15	136
11	996	357	1353	103	35	138
Average	1078	367	1445	106	29	134

The discovered rules shown in the Tables 11 and 12 belong to the set of rules (4). The first rule in the Table 11, which we denote as the **Abalone Rule 1** is:

If $[(x_4, x_7) \in R_1 \vee (x_1, x_2) \in (R_2 \vee R_8) \vee (x_3, x_8) \in (R_4 \vee R_6)] \& [(x_5, x_6) \notin (R_3 \& R_{10} \& R_{11})$
$\& (x_1, x_2) \notin (R_5 \& R_9) \& (x_3, x_8) \in (R_4 \& R_7)]$, then $\mathbf{x} \in$ class 1 (Red, Male),

$$(19)$$

where R_1, R_2, R_3, R_4, R_5, R_6, R_7, R_8, R_9, R_{10}, and R_{11} are specific rectangles (see Table 13), in respective pairs of the Cartesian coordinates (X_4, X_7), (X_1, X_2), (X_5, X_6), (X_3, X_8), (X_1, X_2), (X_3, X_8), (X_3, X_8), (X_1, X_2), (X_1, X_2), (X_5, X_6), and (X_5, X_6). The simple **Abalone Rule 2,** which classifies all the remaining cases (not covered by rule 1) into class 2 is:

If $[(x_4, x_7) \in R_1 \vee (x_1, x_2) \in (R_2 \vee R_8) \vee (x_3, x_8) \in (R_4 \vee R_6)] \& [(x_5, x_6) \notin (R_3 \& R_{10} \& R_{11})$
$\& (x_1, x_2) \notin (R_5 \& R_9) \& (x_3, x_8) \notin (R_4 \& R_7)]$
then $\mathbf{x} \in$ class 1 (Red, Male), else $\mathbf{x} \in$ class 2 (Blue, Infant)

$$(20)$$

Table 12. Precision, recall and coverage of Rule 1 in 11 random 70%:30% splits of data.

70%:30% random data splits	Rule precision		Rule recall (correct) coverage			Rule total coverage of cases, %
	Training, %	Testing, %	Training, %	Testing, %	Total, %	
1	92.05	96.20	41.56	10.59	52.16	56.16
2	90.94	91.45	36.02	13.79	49.82	54.70
3	91.87	90.96	33.86	16.13	50	54.59
4	90.62	92.36	35.36	16.86	52.22	57.28
5	90.66	96.45	42.64	9.47	52.12	56.86
6	91.09	92.98	34.94	12.92	47.87	52.26
7	91.48	91.57	34.45	14.00	48.46	52.96
8	90.88	94.64	37.52	13.55	51.08	55.6
9	90.73	91.36	39.93	11.42	51.35	56.51
10	90.84	94.84	41.84	9.616	51.46	56.2
11	90.62	91.07	34.70	12.43	47.14	51.95
Average	91.07	93.06	37.53	12.80	50.33	55.01

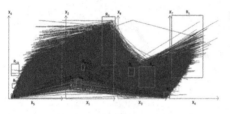

(a) Red cases visualized on the top blue cases.

(b) Blue cases visualized on the top the red cases.

Fig. 16. A set of 2870 abalone data visualized in the SPCs, as graphs of 8-D points, with the male cases in red, and the infant cases in blue. Rectangles used in Abalone Rule 1 are in magenta. (Color figure online)

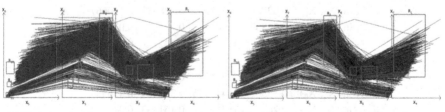

(a) Red cases on the top of the blue cases.

(b) blue cases on the top of the red cases.

Fig. 17. 8-D abalone cases covered by abalone Rule 1. (Color figure online)

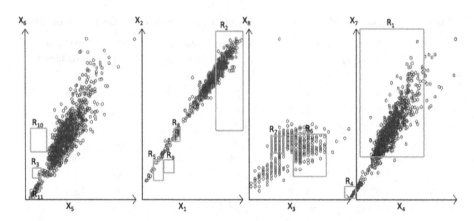

Fig. 18. 8-D Abalone cases covered by rule 1 (only nodes of graphs are shown to decrease occlusion)

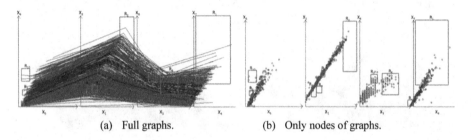

(a) Full graphs.   (b) Only nodes of graphs.

Fig. 19. Remaining Abalone cases (cases not covered by Abalone rule 1).

Table 13. Parameters of rectangles R_1–R_{11}.

Rectangle	Parameters			
	Left	Right	Bottom	Top
R_1 in (X_4, X_7)	0.09068	0.6946	0.260333	1.00
R_2 in (X_1, X_2)	0.7458	0.99955	0.421667	0.99996
R_3 in (X_5, X_6)	0.0209	0.1015	0.128333	0.190667
R_4 in (X_3, X_8)	0.94857	1.001	0	0.077
R_5 in (X_1, X_2)	0.162175	0.253525	0.242	0.113667
R_6 in (X_3, X_8)	0.46645	0.77602	0.139333	0.399667
R_7 in (X_3, X_8)	0.25837	0.38017	0.282333	0.388667
R_8 in (X_1, X_2)	0.355025	0.415925	0.355667	0.436333
R_9 in (X_1, X_2)	0.26	0.355025	0.161333	0.242
R_{10} in (X_5, X_6)	0.015075	0.167325	0.289667	0.436333
R_{11} in (X_5, X_6)	0.012	0.035675	0.0	0.0586667

with accuracy 94.91% on *all cases*, 93.33% and 98.60% on *training* and *validation data* based on the confusion matrixes Table 14.

Table 14. Confusion matrixes of Abalone Rule 2.

Actual class	Confusion matrix on all 2870 cases			Confusion matrix on training cases (2009, 70%)			Confusion matrix on validation cases (861, 30%)		
	Predicted class			Predicted class			Predicted class		
	Red	Blue	Total	Red	Blue	Total	Red	Blue	Total
Red	1497	115	1612	1193	103	1296	304	12	316
Blue	31	1227	1258	31	682	713	0	545	545
Total	1528	1342	2870	1224	785	2009	304	557	861

4 Comparison with Published Results and Conclusion

Case Study 1: the best accuracy reported for the Wisconsin Breast Cancer (WBC) dataset for the SVM in [5] is 96.995% with the 10-fold cross-validation tests. Other results are 96.84% [6] and 96.99% [7] for the SVM, and 97.28% [5] by combing SVM, C4.5 decision tree, naïve Bayesian classifier, and the k-Nearest Neighbors algorithms.

These models classify all the cases, while many of our rules refuse to classify some of the cases. Our WBC rule 2 classify all cases, but with lower accuracy 93.60%. Our better rule that combines WBC rules 1 and 3 has precision 95.18%. While in general, accuracy and precision are different, here the combination of rules 1 and 3 cover almost all cases (99.42%, only 4 cases are refused). The precision for such high coverage is almost identical to accuracy. Thus, it is quite close to the published results, but slightly lower. However, in contrast with SVM, it is *visual*, has clear *interpretation* and *explainable to a domain expert* which is very important in domains with high cost of errors where the explanation of the model is mandatory.

Case Study 2: for Ionosphere dataset, the highest accuracy reported by [8] is 98% on training and 93% on validation data using the multilayer perceptron. Other results are 94.87% by using C4.5 algorithm and 94.59% using Rule Induction RIAC algorithm [9], 97.33% by SVM with Particle Swarm Optimization and 10-fold cross-validation [10].

These models classify all the objects, while many of our rules refuse to classify some of the cases. In contrast, our Ionosphere rule 2 classify all cases. For this rule, precision is identical to accuracy that is 98.78% on training data and 100% on validation data. Thus, our results are slightly higher, than those for the published classification models.

Case Study 3: The highest accuracy reported in [11] for Abalone dataset using SVM is 99.26% with 5-fold Cross-validation for all three classes. Another result is 97.80% accuracy using a case base reasoning method [12]. Our result, that are within [93.33,

98.60]% interval, are quite close to these published results. Unlike [11], we use a more challenging for classification 70:30 split, than the 5-fold that is 80:20 split.

Our Ionosphere and Abalone rules 2 are also *visual, interpretable,* and *explainable* to a domain expert, which is critical, in many domains with mandatory model explanation. This comparison shows that the proposed FSP algorithm, with the SPC visualization, produced the results comparable with the other major machine learning algorithms, in the accuracy and precision. The FSP algorithm has the following significant advantages: it is (i) *visual* with *minimal occlusion,* (ii), *interactive,* (iii) *understandable* by the user, and (iv) *simpler* than many machine-learning algorithms. The future study is using more interpretable rules for discovery by the FSP algorithm along with the other General Line Coordinates, beyond SPC.

References

1. Kovalerchuk, B., Grishin, V.: Adjustable general line coordinates for visual knowledge discovery in n-D data. Inf. Vis. (2017). https://doi.org/10.1177/1473871617715860
2. Kovalerchuk, B.: Visual Knowledge Discovery and Machine Learning. Springer, Heidelberg (2018). https://doi.org/10.1007/978-3-319-73040-0
3. Lichman, M.: UCI Machine Learning Repository (2013). http://archive.ics.uci.edu/ml
4. Wilinski, A., Kovalerchuk, B.: Visual knowledge discovery and machine learning for investment strategy. Cogn. Syst. Res. **44**, 100–114 (2017)
5. Salama, G.I., Abdelhalim, M., Zeid, M.A.: Breast cancer diagnosis on three different datasets using multi-classifiers. Breast Cancer (WDBC) **32**, 2 (2012)
6. Aruna, S., Rajagopalan, D.S., Nandakishore, L.V.: Knowledge based analysis of various statistical tools in detecting breast cancer. Comput. Sci. Inf. Technol. **2**, 37–45 (2011)
7. Christobel, A., Sivaprakasam, Y.: An empirical comparison of data mining classification methods. Int. J. Comput. Inf. Syst. **3**, 24–28 (2011)
8. Duch, W., Kordos, M.: Multilayer perceptron with numerical gradient. In: ICANN 2003, pp. 106–109 (2003)
9. Hamilton H.J., Shan N., Cercone N., RIAC: A Rule Induction Algorithm Based on Approximate Classification. Computer Science Department, University of Regina (1996)
10. Tu, C.-J., Chuang, L.Y., Yang, C.H.: Feature selection Using PSO-SVM. IAENG Int. J. Comput. Sci. **33**(1), 111–116 (2007)
11. Sain, H., Purnami, S.W.: Combine sampling support vector machine for imbalanced data classification. Procedia Comput. Sci. **1**(72), 59–66 (2015)
12. Smiti, A., Elouedi, Z.: Maintaining case based reasoning systems based on soft competence model. In: Polycarpou, M., de Carvalho, A.C.P.L.F., Pan, J.S., Woźniak, M., Quintian, H., Corchado, E. (eds.) HAIS 2014. LNCS (LNAI), vol. 8480, pp. 666–677. Springer, Cham (2014). https://doi.org/10.1007/978-3-319-07617-1_58

Visual Guidance to Find the Right Spot in Parameter Space

Alexander Brakowski[1], Sebastian Maier[2], and Arjan Kuijper[1,2(✉)]

[1] Technische Universität Darmstadt, Darmstadt, Germany
[2] Fraunhofer IGD, Darmstadt, Germany
`arjan.kuijper@igd.fraunhofer.de`

Abstract. The last few decades brought upon a technological revolution that has been generating data by users with an ever increasing variety of digital devices, resulting in such an incredible volume of data, that we are unable to make any sense of it any more. One solution to decrease the required execution time of these algorithms would be the preprocessing of the data by sampling it before starting the exploration process. That indeed does help, but one issue remains when using the available Machine Learning and Data Mining algorithms: they all have parameters. That is a big problem for most users, because a lot of these parameters require expert knowledge to be able to tune them. Even for expert users a lot of the parameter configurations highly depend on the data. In this work we will present a system that tackles that data exploration process from the angle of parameter space exploration. Here we use the active learning approach and iteratively try to query the user for their opinion of an algorithm execution. For that an end-user only has to express a preference for algorithm results presented to them in form of a visualisations. That way the system is iteratively learning the interest of the end-user, which results in good parameters at the end of the process. A good parametrisation is obviously very subjective here and only reflects the interest of an user. This solution has the nice ancillary property of omitting the requirement of expert knowledge when trying to explore an data set with Data Mining or Machine Learning algorithms. Optimally the end-user does not even know what kind of parameters the algorithms require.

Keywords: Big data · Visualization · Parameter space · Filtering

1 Introduction and Motivation

The last few decades brought upon a technological revolution that has been generating data by users with an ever increasing variety of digital devices, resulting in such an incredible volume of data, that we are unable to make any sense of it any more. This trend will not stop and it has been confirmed, that huge amounts of data have been generated at unprecedented and ever increasing scales. In 2012 it was already estimated that 2.5 exabytes of data are generated each day and

S. Yamamoto and H. Mori (Eds.): HIMI 2018, LNCS 10904, pp. 527–544, 2018.
https://doi.org/10.1007/978-3-319-92043-6_43

the number is doubling every 40 months. Besides the sheer size of the data, almost nothing is known about the data itself. Data Mining and Machine Learning algorithms are normally good at helping us make sense out of this large volume of data. While Data Mining algorithms extract useful information out of the data, Machine Learning aims to learn from the data and make predictions. These algorithms are quite powerful learning devices, but are of high algorithmic complexity. That means, that executing them can be extremely time consuming. Because we do not know anything about the data itself, we have to use exploration tools (algorithms) to analyse the data. To make that exploration process as flexible as possible, we want to be able to efficiently apply different algorithms on the same data.

A good way to conduct that exploration process would be by applying the active learning approach [20] and embrace an iterative process, that would allow an end-user to analyse the data step-by-step and to intervene into the exploration process by steering it into different directions or to focus it to some interesting data areas. The combination of large data and the computational complexity, makes such a process almost impossible. To make the execution less time consuming, we have multiple possibilities: either improve the algorithmic complexity or reduce the data size. Reducing algorithm complexity only works to a certain degree, and even if we were able to decrease their execution time for that specific data size, this problem will persist in the future, because the data volume is probably going to continue to grow. The other option is decreasing the data size. To do so, we have to assume that every part of the data is equally important [10]. This will allow us to uniformly sample the data, and reduce it to a size our algorithm is comfortable working with. This reduction step is only done once, and the resulting data set can then be used for the Data Mining and Machine Learning algorithms.

Ideally, an end-user wants to influence the analytical process as much as possible. Since we are using Data Mining and Machine Learning algorithms, we can parametrize these to allow better control of their executions. In order to analyse the data, we are exploring the parameter space of the algorithms in the end. The goal of an end-user is to find a good parametrisation for an algorithm, so that they can get something meaningful out of the data.

Let us look at that problem from an end-user perspective: Alice has a large data set containing two features. Alice wants to know how many clusters that data set contains. Now Alice decided to use the K-Means clustering algorithm that seems to solve that issue, and only requires one parameter (named k) which is the number of clusters K-Means is trying to find in the data set. Assuming we have n entries in the data, k can be in the range of 1 to n. Now, if Alice is not applying some clever way to set the parameter, depending on what value is considered as optimal by Alice, it could take n executions of the K-Means algorithm, until Alice finds something interesting in the data.

That is obviously not an optimal way to explore a parameter space. With only a single parameter one could employ a binary search exploration, and that would decrease the required number of executions dramatically. Generally, we are not dealing with just a single parameter. In most cases the algorithms have many more parameters. Finding a parametrisation yielding interesting results in these higher dimensions is much harder, and a simple binary search can not be employed any more. The issue in higher dimensional parameter spaces is simply the sheer size of possible parameter combinations. Depending on the types of the parameters the number of combinations could be infinite. Let us assume three parameters and the discrete case where every parameter can take one-hundred different values. This would already create a parameter space with one-million different parameter combinations. In the case of continuous parameter values we are dealing with real numbers, where even between zero and one, there are an infinite amount of numbers.

Note that the very concept of a good parametrisation is vague, because good in our use case is highly subjective to the end-user's opinion. A parametrisation is good, if an end-user thinks that an end-result is meaningful. The estimation of meaningfulness of the end-result and quality of the parametrisation are mutually dependent on each other. A less meaningful end-result would make a parametrisation less good, a more meaningful end-result would make a parametrisation better. Two users exploring the same data with the same algorithm, could produce vastly different parametrisations. Taking into account that the subjectiveness of the end-result's quality estimation is a very important requirement.

This work will present a system that allows the analysis of these large data in a workflow manner by using an visual analytics server as the algorithm execution infrastructure. In that workflow each step is working with the results of the previous steps. That allows us to reduce our problem to single tasks, where our first step would be the data retrieval. The next step is the data reduction, where uniform sampling methods are used to achieve the reduction. Following that is the execution of the algorithms with some parametrizations. Within the workflow multiple steps can be executed at the same time, so that capability can be used to execute the same algorithm with different parametrizations in parallel. The parametrizations is done by an optimizations process, that takes the user preferences of end-results into account, to try other algorithm parametrizations. That means, it tries to learn the interests of the user by applying the active learning approach, i.e. iteratively querying the user for feedback. We present two different processes to achieve that. The presentation of the end-result to the end-user is done in form of a visualisations, where an end-user may express their preference. The full details of our implementation can be found in [2].

2   Related Work

This section is going to outline the related work that affect and are within the scope of this work. We will first position our work within the active learning field, because this is closely related to our work. Then we will relate our active learning component to the visualisations and data reduction. After that we will look into different approaches to visual interactive parameter space exploration, parameter tuning, and pure parameter optimization. Following that we will explore different preference gallery approaches. Then we will shorty look into the data reduction problem and an central infrastructure component for this work. We end the section by distancing ourselves from recommender systems.

2.1 Active Learning

A research area closely related to our iterative feedback process is active learning. In the literature it is often also called "query learning" or "optimal experimental design" [20]. The idea behind active learning is that the learning component can achieve greater prediction accuracy when it can choose from which data it learns. This is generally done by allowing the learning component (active learner) to iteratively pose queries to an oracle (usually a human), and learn from the feedback of that oracle [16,20]. The active learning query strategies can be organized into different categories. Most of the optimization strategies in our system would likely fall under the one called "Balance exploration and exploitation" [20], where the choice of new query points is a dilemma between choosing points near regions of high certainty and points near regions of high uncertainty in the parameter space. Since the query strategy can be changed any time in our system, this categorisation is not applicable to the whole process in our system, but only to single optimization strategies.

2.2 User-Centered Visual Analysis

The idea of user interaction loop in an visual analysis process is not a new one. The well known Card model [5,13] described this process already a long time ago. This approach provides means to transform and represent data in a form that allows human interaction. Users can therefore analyse data by exploration rather than pure reasoning. Because of the visualisations users can also immediately observe effects of their interactions with the system, resulting in easier development of an understanding for structures and connections in the data. We also want to allow an end-user to observe immediate effects of their interactions with the system, that is why we also make use of that model in some way. In later chapters we present our adoption of that model.

2.3 Interactive Parameter Space Exploration and Tuning

There are already a few approaches to achieve an interactive parameter space exploration and tuning, three of these we will discuss here. One approach to explore the parameter space was done by Pretorius et al. [17], where they provide the a tool, that allows the exploration of the parameter space in a almost real-time manner. In this approach users are actively involved in tuning the number of samples per parameter the algorithm should generate. In our work we want the algorithm to deduce on itself which parameters to tune based on the feedback given by the user. The system should also remove all the bias the user could have when going through the process, which includes hiding the concrete parameter values that generated the visualisation. Another approach to that was done by Marks et al. [14], here they choose the parameters in a way that they optimally disperse the resulting outputs. This approach has two problems: First, we are working with a black-box function of which we don't know the output values. Secondly, we do not know, how the visualizations will look like, so we can't compare them.

Torsney-Weir et al. [23] developed a tool called "Tuner", which helps with the parameter space exploration of image segmentation algorithms. This tool introduces a two-stage process, where the first stage can be seen as the pre-processing stage. Here, it first samples the complete parameter space as densely as possible within a certain time frame, and acquires all the corresponding segments. This is generally done over-night. In the second stage the user is then able to interactively explore the results of the first step.

They also employ an Gaussian Process with an acquisition function to sample the parameter space for their pre-processing [11], which is nothing else but an automatically guided Bayesian optimization. In our work the exploration of the parameter space should be guided by incorporating the feedback of the user and not constrain the exploration strategy to only Bayesian Optimization. That's also why no preprocessing steps concerning the exploration can be made. A preprocessing step by sampling the large data first would be possible.

2.4 Parameter Optimization

Grid Search is a simple methodology for finding appropriate parameter values for objective functions, which are functions of which we know nothing about. It is an exhaustive or brute-force search in an parameter space, where the points are located on a grid in the parameter space. This method can usually find good parameters in the grid, and because of it's brute force nature it does a complete search of every point in the parameter grid. To achieve that, it sets up a grid of parameter combinations and evaluates the values of the objective function at each grid point. A grid search is guided by some performance metric, typically measured by cross-validation [7]. The point with the best value is considered to be the optimum solution [9]. This method has the problem, that the objective function has to be executed an incredible amount of times, depending on the number of parameters [1]. That is a big problem in our case, since we want to reduce the number of executions necessary to a minimum.

Bayesian Optimization is a global optimization strategy for black-box functions. The goal is finding a global maximum in as less iterations as possible [4], this motivation comes from the assumption, that evaluating the objective function is very expensive. Either because of it's computational complexity, or because of its general resource requirements. It assumes a very general prior over functions, which when combined with observed parameter values and corresponding outputs describes a distribution over functions, which can easily be used to draw samples. This method works in a iterative manner, which in every iteration evaluates a objective function with parameters drawn from the distribution. These parameters are picked by using an acquisition function, which makes a trade-off between exploitation (regions where the expected value is high) and exploration (regions that are of high uncertainty).

It has been shown, that this methodology of doing parameter optimization obtains better results in fewer iterations than grid search [8, 21], the reason being the ability to reason about the quality of parameters. We will look into that optimization strategy in more detail in later sections of this work, because we will make use of it in our system.

2.5   Visualization Gallery

The system should provide the user with multiple different parameter configurations in form of visualizations, from which the user can express it's preference. This gives the system the necessary data to perform the next round of calculations. There are a lot of different ways to present these to the user, the most effective would be in form of a gallery, that just shows the visualization without the user knowing the parameter configuration of it. This removes all the pre-conceptions and influences the user might have about the parameter configurations [14]. As interesting as this approach is, it is not applicable in our case, since the system can't pre-calculate all the available input configurations, as this would require depending on the number of parameters and parameter space and incredible number of visualizations. The arrangement algorithm as described in that paper is also not applicable here, since we do not know how the visualization should look like, so there is no conclusion and comparison that could be derived from the output visualizations.

2.6   Preference Gallery for Material Design

In their work Eric et al. [6] also explored the possibilities to use the end-user feedback to find an optimal parametrisation for an algorithm. They too present the user with visualisations generated from different parametrisations and also use the Bayesian optimization method for the optimization. However they only present the user with two visualisations at every iteration and applied the method only to the Bidirectional Reflectance Distribution Function, which is a function used to render reflection on different materials. In our work depending on the strategy and iteration a diverse amount of visualisations is generated. The visualisation and the algorithm that should be optimized can also be chosen freely in our system. Further more in our system we did some improvements on the Bayesian optimization and use this optimization method merely as one of the methods that can be used for the parameter optimization. That means that our system provides a framework in which many different parameter optimization strategies can be used. So we am making a lot of use of that same method in our work, with a few improvements, and also apply it to the problem of extracting interesting or useful information from an extremely large data set by optimizing the parameters of the data analysis algorithm.

2.7   Data Reduction

The user should be provided with galleries of different parameter configurations to let the user give feedback about their preference. That means the given algorithm has to be executed many times with different parametrisations. To be able to do that within an acceptable time frame the large data set would have to be sampled first, that would allow the system to use the data for the process. Since the main part or even the focus of this work is not the data quality, a simple random sampling algorithm is sufficient for now. This does not mean, that sampling is not an important factor for assisting a user in the parameter tuning process. The quality of data can probably influence the quality of the visualizations for the gallery immensely. One random sampling algorithm would be the "Reservoir sampling", also known as "Algorithm R" by Vitter [24], which requires to iterate through all the data at least once to get a random sampling of an large data set. The same problem arises in Bernoulli and Poisson sampling, in these it is additionally necessary, that each item passes the "independent Bernoulli trial" [22], which requires even more calculation time.

2.8   Visual Analytics Server

The idea to split the work that needs to be done for data analysis into separate tasks is not a new one. The Visual Analytics Server, developed at Fraunhofer IGD, manifested the idea into a system that manages these separate tasks in form of an workflow [12,15]. We refer to this server later as our Workflow Execution Environment. This server provides the necessary infrastructure for our system to execute the data analysis algorithms by separating process the tasks into data retrieval, data reduction and algorithm execution steps in a workflow. A workflow generally consists of multiple steps, where each step has inputs, outputs and parameters. Theses steps can be connected to each others via their inputs and outputs. So depending on the connections, which also defines their dependencies, steps can be executed in parallel or sequentially. In a workflow, a step being triggered to execute will also trigger its predecessors to execute. The execution results will then be propagated down the tree branches again to the step triggering the execution. Generally the successor steps of the step triggering a run will not be triggered to execute. This allows the workflow to have a nice auxiliary property of allowing partial execution of an workflow. The re-execution of a step is only being performed, if either the parameters were changed or an explicit re-execution was requested, otherwise the step just outputs the cached result from the last run. The combination of result caching, parallel and partial execution makes this an extremely powerful and useful tool, especially for our use case.

2.9   Recommender Systems

From the first glance at the given problem of this work, one could think, that a recommender system would help to solve the problem [3]. This is not really the case, since recommender systems work on historical and community collected data [19], which are both not available here. Therefore recommender systems are also called "collaborative filtering" [18]. We are working with a black-box algorithm (or function) of which we neither know the output type nor the quality measurement. The only data we have are boolean values mapped to parameter configurations, which is the feedback data given by the user. From this and the input data we have to deduce the next round of parameter configurations. With that it should be clear, that we are not working in the same scope in this work as recommender systems do.

3   Concept

Our concept is based on a fairly simple idea: we want to help an end-user find something interesting in a data-set by providing an easy-to-use interface. That interface should allow the user insight into the data-set by providing visualizations. Because the user does not necessary know what it is looking for and is only able to provide feedback to visualizations, this interface should give an end-user the opportunity to express preferences. Helping the user find something interesting, would then require that we learn from these preferences of the user, and try to present the user with more visualizations which should theoretically be more interesting to the user. That creates an iterative feedback process, which can be stopped by the user at any point, when they found something interesting.

This learning process can either be applied to some machine learning algorithm, the algorithm that generates the visualisations or anything else that would influence the visualisations. In this work we are only going to focus on the parametrisation of the machine learning algorithms and the visualisations. In both cases we would be able to change the visualisations presented to the user by altering the parameters of these algorithms (Fig. 1).

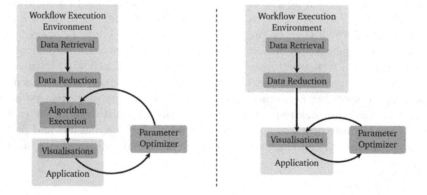

Fig. 1. The left side shows the process when applied to machine learning algorithms, on the right side the same process can be observed when applied to the algorithm generating the visualisations.

The overview of the whole system can be seen in the above figure as a schematic representation. On the left side, the process can be seen when applied to machine learning algorithms, here the parameter optimizer sets the parameters of the machine learning algorithm by incorporating the feedback from the frontend. While on the right side the same process is applied on the visualisations itself, here the machine learning algorithm execution is completely omitted. The iterative feedback process can be observed in both cases, while in the first case it starts at the algorithm execution, in the second case it starts at the frontend. The system that is managing and performing that process is divided into three different parts: the application, the parameter optimizer and the workflow execution environment.

- The Application can be seen as the controller of the whole system. It bootstraps the whole process, provides a frontend to the end-user and propagates user interactions within the system to the appropriate parts.
- The Parameter optimizer takes the process configuration and the feedback data and generates parameters for the next round of algorithm executions.
- The Workflow Execution Environment executes the parametrized algorithms in a workflow-manner, where each step is dependent on the result of the previous step and the first step is the data retrieval.

This can be done in a completely distributed fashion, because of its workflow nature (Fig. 2).

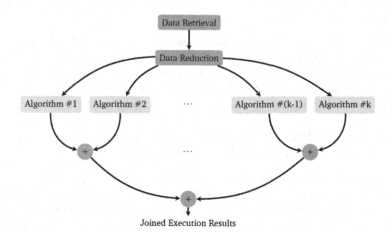

Fig. 2. A diagram of the workflow as it is being configured for the execution server. The green nodes are the algorithm execution steps, and there are a flexible number of these depending of how many visualisations should be presented to the user. Because the executions can take some time, every execution step would have to be polled for a status report periodically. To prevent that the results of the execution steps are combined together, so that at the last step at the bottom the results of all the execution steps are joined together. That would allow us to only having to poll the last step at the bottom of the workflow for a status report periodically.

3.1   Application

The application is the conductor of the whole orchestra of components that makes the system work together. This means it has to handle the user interaction in form of feedback and manage the whole process configuration, which includes the configuration of the data source, algorithm and visualisation. The end-user should be able to freely choose a data source, the algorithm and a visualisation. By incorporating all the meta information the frontend receives from the execution server, it can provide the end-user with configuration options of all the process steps. This process configuration is organized in a wizard kind of way, where the end-user is guided through the process step-by-step:

1. First the data source is chosen and depending on the data source additional parameters like table name or database have to be set
2. Next, the algorithm that the end-user wants to use for the exploration process is chosen
3. After choosing an algorithm a user can decide which parameters should be optimized and which should be fixed to a certain value
4. For the end-result display, a user can choose which visualisation it wants to use
5. Since the chosen visualisation can have their own specific data mappings, the user has to map the algorithm result features to the visualisation features

After performing the configuration, the preparations for the exploration process is done and the process can be started. The exploration process consists of the following steps:

1. Take the feedback data and generate parametrizations with the parameter optimizer (the feedback data can be empty in the first iteration)
2. Create the workflow for the execution server by incorporating the configurations set by the user and the parametrizations of step 1
3. Generate visualisations from the workflow execution and present them to the user
4. Create feedback data from the user interaction with the visualisation
5. Repeat from step 1, if user wishes to continue

This creates the iterative feedback loop, which can be stopped any time the user does not want to continue. The end-user is also able to change the process configuration at any point in the exploration. Changing the data-source in the middle of an exploration process is certainly a bad idea, because that could also change the visualisations greatly, and with that the preferences of an user would probably also change. This would cause an incorrect preference learning model for the parameter optimizer. None the less, the possibility for that exists, but has to be used with caution.

The great advantage of being able to change the configuration within the exploration process is the possibility to fixate specific parameters to a certain value at any point. That could improve the performance of the process (in terms of number of iterations), since the user may observe an optimal value for an parameter and help the optimizer by reducing the number of parameters the optimizer would otherwise have to optimize.

Switching the visualisation at any point in the process is also a very important feature. Just because it generally is not a good idea for example to use a bar-chart visualisation with the k-means algorithm, does not mean that it is not interesting to the user. Bar-charts normally visualise grouped data, which would mean in this k-means example that it would categories the data into cluster numbers and then show how many data items belong to each of the clusters. While when visualising the k-means with a scatter plot, it would show the data split into the different cluster groups.

From this example we can already deduce one very important property for the exploration process: by choosing different visualisations, the user is able to look at the data with the same algorithm from different perspectives. Perhaps the simple categorisation and counting the number of cluster items is more inter-esting and intuitive to the user. The parametrisation the user found with the bar-chart would probably also yield a good result in the scatter plot visualisation.

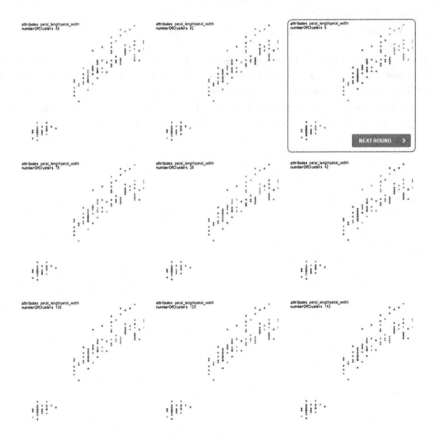

Fig. 3. First iteration

This interaction process is nearly identical to the visualisation reference model described by Card et al. [5]. The difference to the Card model is the additional possibility to indirectly influence the algorithm execution by providing visual feedback (expressing preferences) to a parameter optimizer, which parameterizes the algorithm execution.

Full details of our implementation can be found in [2].

4   Results

4.1   Process Applied to the K-Means Clustering Algorithm

First we will try to use the K-Means clustering algorithm with the iris flowers data set, because the K-Means algorithm has only one numerical parameter, and the iris data set can be visualized really well. The process is configured the following way:

Fixed parameters: attributes = petal_length, petal_width
Visualisation: Scatter Plot
Visual Mapping: x = petal_length, y = petal_width

So we are using the K-Means to cluster on the given attributes, and generate scatter plots by mapping the end-results to the visual features. The first iteration generated already a few good parameters (Fig. 3). On the top right, the third visualisation, already looks quite good, it is already possible to clearly distinguish a few groups. In the next iteration (Fig. 4) we have a really good result:

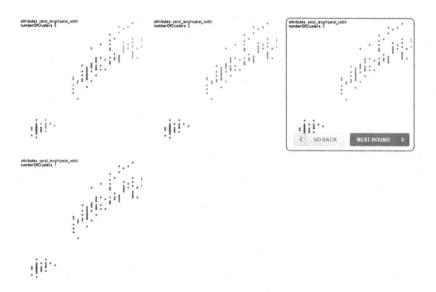

Fig. 4. Second iteration

We found a good parametrisation for the k-means; in only two iterations we were able to find a k = 3 or k = 2 for the iris flowers data set. Two iterations means, the end-user only had to express it's own preference only a single time! But more importantly our system was able to improve the visualisations from the first to the second iteration based on the feedback from the user.

4.2   Process Applied to Visualisations

Next let us try to use our system to find a good parametrisation for the visualisations itself. To keep the changes to the implementation of the system to a minimum we had to create a "pseudo"-Step that does not do any real work, but mostly just returns the given step parameters as a data output. This data can then be used as the parameters for the visualisation. As a visualisation let us use the force directed graph layout, which has seven numerical parameters, each with different bounds. The process would then be configured the following way:

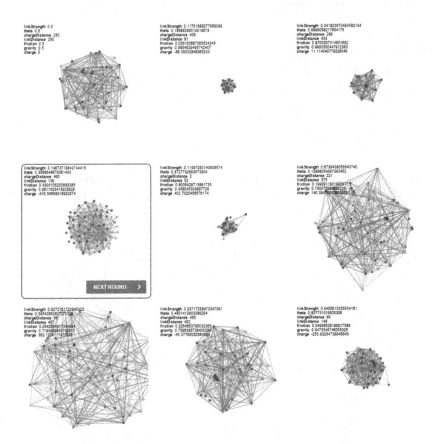

Fig. 5. The first iteration when applying the process on the visualisation itself. The preference is marked here with the grey rectangle. It was chosen because it had a nicer grouping and node distance.

Fixed parameters: None, we want to optimize all the parameters
Visualisation: Force directed graph layout
Visual Mapping: Visual mappings correspond to the step parameters

The system generates the following visualisations, depending on the iteration and the feedback given by the user (Figs. 5, 6, 7 and 8):

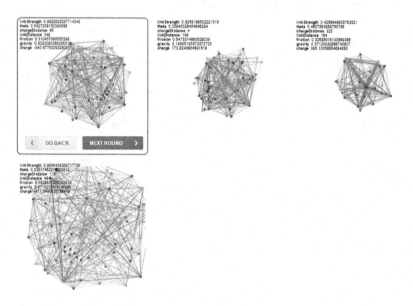

Fig. 6. The second iteration when applying the process on the visualisation itself. The marked item was chosen because some nice blue groups are forming. (Color figure online)

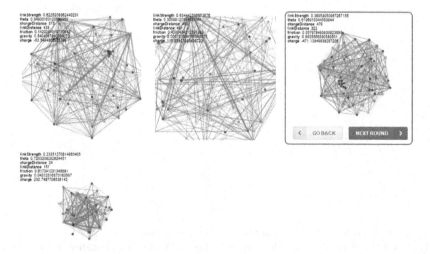

Fig. 7. The third iteration when applying the process on the visualisation itself. Again the blue group formed, and it just looks better than the others. (Color figure online)

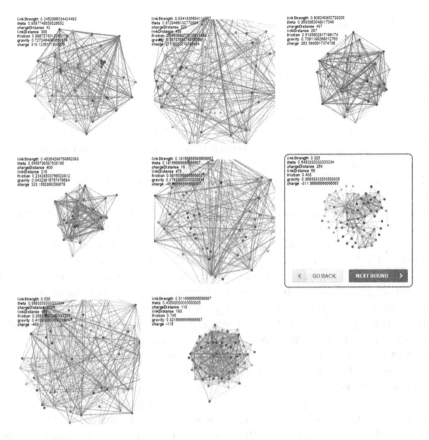

Fig. 8. The fourth iteration when applying the process on the visualisation itself. Here we can see great group formations and positioning in the selected visualisation. Another thing that happened here is that the system mixed in additional visualisations, because it detected that the parametrisations in the last view iterations were to similar to each other.

In the fourth iteration (Fig. 8) we already found something really good. Even though we were using the process to optimize seven parameters, it only took us four iterations to find the parametrisation that generated the last visualisation. Now, we obviously were quite lucky here right now, that it only took us four iterations to find something interesting, it does not always work so well and could take much longer, or not work at all. If a user does not find anything in the current process it started, it can restart the process to try it again, that often improves the exploration process immensely. Also note that in Fig. 8 we were not just simply generating parameters in the natural number space, but in the real number space!

5   Conclusions

In this work we presented a system that solved the particular problem in data analysis of extracting interesting or useful information out of an extremely large data set by using different methods for parameter space exploration. We first looked into the concept of the solution and came to the realisation that a good way to steer an exploration process is by using the feedback a user can provide to us, because we do not know anything about the data or the data analysis methods. But before applying any exploration we first had to find a method to reduce the large data set to a data size we are able to handle within our system more easily. We presented a simple uniform sampling method, that was able to reduce our data extremely fast. This reduced data could now be used for the data analysis process, by applying some parametrisation to the data analysis algorithms. These parametrisations were generated by an optimizer that was incorporating the feedback from an end-user. A very effective way to get good feedback from an end-user is by iteratively showing it visualisations and asking it to express their preference. Humans are exceptionally good at providing good feedback by expressing preferences from visual stimulations, especially because we are dealing with highly subjective quality measurements like "useful" and "interesting". This feedback provided by the user allowed us to perform a parameter space exploration, and helped us find a good parametrisation that generated a visualisation, which a user found "useful" or "interesting". We presented two different methods to find a good parametrisations based on the feedback data, one more sophisticated than the other. The first one is called Bayesian Optimization, which uses sophisticated probability models to infer the preferences from the user feedback and then make predictions based on these. The second method is simpler but still quite effective, which makes predictions by maximising a classifier. In the end we saw two use cases in how our system was able to find a good parametrisation for a given problem:

- In the first use case we took the K-Means clustering algorithm and the full iris-flower data set. We then iteratively let the system generate visualisations for different k-values. In the first iteration we already found quite good results, and in the second iteration we already found a really good k-value. Now the K-Means clustering algorithm has only one single parameter, which motivated the second use case.
- In the second use case we wanted to use our system to explore a bigger amount of parameters and apply the exploration on the visualisations itself. For that were took a smaller graph based data set and used the force directed layout algorithm for the visualisation generation. This algorithm has eight numerical parameters. Here the process took a few more iteration, but we were still able to show, that even with eight parameters, we found something interesting within 10 Iterations.

This concludes that our system is able to extract useful information out of an large data set by performing a guided parameter space exploration with the feedback provided by an end-user.

References

1. Bergstra, J., Bengio, Y.: Random search for hyper-parameter optimization. J. Mach. Learn. Res. **13**, 281–305 (2012)
2. Brakowski, A.: Visual guidance to find the right spot in the parameter space. Technical report, TU Darmstadt (2015)
3. Breyer, M., Nazemi, K., Stab, C., Burkhardt, D., Kuijper, A.: A comprehensive reference model for personalized recommender systems. In: Smith, M.J., Salvendy, G. (eds.) Human Interface 2011. LNCS, vol. 6771, pp. 528–537. Springer, Heidelberg (2011). https://doi.org/10.1007/978-3-642-21793-7_60
4. Brochu, E., Cora, V.M., de Freitas, N.: A tutorial on Bayesian optimization of expensive cost functions, with application to active user modeling and hierarchical reinforcement learning. CoRR abs/1012.2599 (2010)
5. Card, S.K., Mackinlay, J.D., Shneiderman, B. (eds.): Readings in Information Visualization: Using Vision to Think. Morgan Kaufmann Publishers Inc., San Francisco (1999)
6. Eric, B., de Freitas, N., Ghosh, A.: Active preference learning with discrete choice data. In: Advances in Neural Information Processing Systems, vol. 20, pp. 409–416. MIT Press, Cambridge (2007)
7. Hsu, C.W., Chang, C.C., Lin, C.J.: A Practical Guide to Support Vector Classification (2010)
8. Hutter, F., Hoos, H.H., Leyton-Brown, K.: Sequential model-based optimization for general algorithm configuration. In: Coello, C.A.C. (ed.) LION 2011. LNCS, vol. 6683, pp. 507–523. Springer, Heidelberg (2011). https://doi.org/10.1007/978-3-642-25566-3_40
9. Kuijper, A.: On detecting all saddle points in 2D images. Pattern Recogn. Lett. **25**(15), 1665–1672 (2004)
10. Kuijper, A.: Using catastrophe theory to derive trees from images. J. Math. Imaging Vis. **23**(3), 219–238 (2005)
11. Kuijper, A., Florack, L.: The relevance of non-generic events in scale space models. Int. J. Comput. Vis. **57**(1), 67–84 (2004)
12. von Landesberger, T., Bremm, S., Kirschner, M., Wesarg, S., Kuijper, A.: Visual analytics for model-based medical image segmentation: opportunities and challenges. Expert Syst. Appl. **40**(12), 4934–4943 (2013)
13. von Landesberger, T., Fiebig, S., Bremm, S., Kuijper, A., Fellner, D.W.: Interaction taxonomy for tracking of user actions in visual analytics applications. In: Huang, W. (ed.) Handbook of Human Centric Visualization, pp. 653–670. Springer, New York (2014). https://doi.org/10.1007/978-1-4614-7485-2_26
14. Marks, J., Andalman, B., Beardsley, P.A., Freeman, W., Gibson, S., Hodgins, J., Kang, T., Mirtich, B., Pfister, H., Ruml, W., Ryall, K., Seims, J., Shieber, S.: Design galleries: a general approach to setting parameters for computer graphics and animation. In: Proceedings of the 24th Annual Conference on Computer Graphics and Interactive Techniques (SIGGRAPH), pp. 389–400 (1997)
15. Nazemi, K., Stab, C., Kuijper, A.: A reference model for adaptive visualization systems. In: Jacko, J.A. (ed.) HCI 2011. LNCS, vol. 6761, pp. 480–489. Springer, Heidelberg (2011). https://doi.org/10.1007/978-3-642-21602-2_52
16. Osugi, T., Kim, D., Scott, S.: Balancing exploration and exploitation: a new algorithm for active machine learning. In: Fifth IEEE International Conference on Data Mining (ICDM 2005), p. 8 (2005)

17. Pretorius, A.J., Bray, M.A., Carpenter, A.E., Ruddle, R.A.: Visualization of parameter space for image analysis. IEEE Trans. Vis. Comput. Graph. **17**(12), 2402–2411 (2011)
18. Resnick, P., Iacovou, N., Suchak, M., Bergstrom, P., Riedl, J.: GroupLens: an open architecture for collaborative filtering of netnews. In: Proceedings of the 1994 ACM Conference on Computer Supported Cooperative Work, CSCW 1994, pp. 175–186. ACM, New York (1994)
19. Resnick, P., Varian, H.R.: Recommender systems. Commun. ACM **40**(3), 56–58 (1997)
20. Settles, B.: Active learning literature survey. Technical report, University of Wisconsin (2010)
21. Snoek, J., Larochelle, H., Adams, R.P.: Practical Bayesian optimization of machine learning algorithms. In: Proceedings of the 25th International Conference on Neural Information Processing Systems, NIPS 2012, vol. 2, pp. 2951–2959. Curran Associates Inc., New York (2012)
22. Tille, Y.: Sampling Algorithms. Springer, New York (2006). https://doi.org/10.1007/0-387-34240-0
23. Torsney-Weir, T., Saad, A., Moller, T., Hege, H.C., Weber, B., Verbavatz, J.M., Bergner, S.: Tuner: principled parameter finding for image segmentation algorithms using visual response surface exploration. IEEE Trans. Vis. Comput. Graph. **17**(12), 1892–1901 (2011)
24. Vitter, J.S.: Random sampling with a reservoir. ACM Trans. Math. Softw. **11**(1), 37–57 (1985)

Analyzing Reading Pattern of Simple C Source Code Consisting of Only Assignment and Arithmetic Operations Based on Data Dependency Relationship by Using Eye Movement

Shimpei Matsumoto[1]([✉]), Ryo Hanafusa[2], Yusuke Hayashi[2], and Tsukasa Hirashima[2]

[1] Faculty of Applied Information Science, Hiroshima Institute of Technology, 2-1-1 Miyake, Saeki-ku, Hiroshima 731-5193, Japan
s.matsumoto.gk@cc.it-hiroshima.ac.jp
[2] Graduate School of Engineering, Hiroshima University, 1-4-1 Kagamiyama, Higashihiroshima, Hiroshima 739-8527, Japan

Abstract. Some programming learners in the lowest achievement group do not have even a minimum skill to read a simple program correctly. Reading programs would be an essential programming learning. To efficiently support learners in the lowest group, firstly we should conduct a fundamental analysis of reading programs to unveil their features. Therefore, the authors focused on eye tracking as a method to carry out the idea. The authors have thought that utilizing eye movement helps to clarify the reasons for making programming learning difficult. Therefore, the purpose of this study is to investigate the possibility of learner's program comprehension process based on the pattern of eye movement, not the eye distribution during reading source code. In this paper, we first measure the data of eye movement during reading some source codes and propose a modeling method to represent the feature of eye movement. Then we design an experimental protocol for analyzing eye movement based on program structure. The experiment of this paper focuses on source codes based on four types of data dependency relationship that can be generated by three lines of assignment statement only. As the analysis result, we confirmed that the data dependence of each pattern appeared as the unique eye behavior of program reading.

Keywords: Programming education · Reading · Eye tracking
Data dependency relationship

1 Introduction

Though programming has been regarded as a particularly important subject in the special field of higher education institutions such as universities, every year

© Springer International Publishing AG, part of Springer Nature 2018
S. Yamamoto and H. Mori (Eds.): HIMI 2018, LNCS 10904, pp. 545–561, 2018.
https://doi.org/10.1007/978-3-319-92043-6_44

there are many students who cannot accept any concept of programming. Sagisaka and Watanabe confirmed that most of the students belonging to the lowest group of programming comprehension level could not understand the most basic terms and grammars [1]. The existence of learners who cannot accept any concept of programming will be an especially serious problem when programming is integrated into the curriculum of compulsory education. Therefore, investigating when and why beginning programming learner gives up would be important; besides, programming education needs to establish an appropriate new instructional method for them.

Previous studies suggested that some programming learners in the lowest achievement group do not have even a minimum knowledge to properly read simple programs than writing a program consisting of dozens of lines. Reading programs must be an essential programming skill. Therefore, in programming introduction education, the authors think that supporting for reading programs must play an important role. Based on this idea, in order to effectively support learners in the lowest group, firstly educational experts including us should clarify their features from basic studies on analyzing program reading. Therefore, the authors focused on eye movement as a method to promote this idea. We thought that utilizing eye movement helps to unveil the reasons for making programming learning difficult.

This paper firstly examines whether the eye tracking is useful for estimating the reading process of programs by surveying and summarizing various results of previous studies. Based on the knowledge of these previous studies, the purpose of this paper is to investigate the possibility of learner's program reading comprehension based on the pattern of eye movement, not the eye distribution. This study fucuses on programs with only assignment and arithmetic operations, simplify the structure of programs to data dependencies, and set them as analysis targets. From the data of such programs obtained by eye tracking, this study constructs a simple Markov process model expressing eye movement patterns as transition probabilities. By comparing the difference of transition probabilities between nodes and data dependencies, this study shows the possibility of analyzing the program reading process based on eye movement.

2   Design of Analysis Method

2.1   Source Code Reading

The subject of this paper is to investigate and clarify whether the program structure is detectable by the pattern of eye movement under the assumption that programmer's thinking process appears on the eye movement. In addition to variable dependencies, actual source codes contain many elements not directly related to the internal structure, such as programming language specific writing style and algorithms. These are likely to be strongly influenced by experience. There is a suggestion that skilled programmers have abundant knowledge of problem domain and they use it efficiently [2]. As the first step, this paper aims to clarify that the program structure appears on the eye movement not depending on the learner's experience. Therefore, actual source codes would be unsuitable for the experimental subject of this paper. Also, if an examiner creates the experimental tasks (source codes) manually, external factors other than the program structure may be included regardless of intention. Specifically, external factors are such as the design pattern of the program (variable name, the design of operator, and the description of assignment statements). These external factors should be eliminated as much as possible to clarify the eye pattern according to the program structure. Therefore, this paper does not generate the experimental task manually but automatically.

The source codes used as experimental tasks have no specific meaning, that is, has no purpose of processing. They are generated with arbitrary rules and depends only on the internal program structure. Also, they should be surely readable even if an examinee has just the most basic knowledge about the programming language specification and the minimum required calculation ability enough level as needed in daily life. Source codes not depending on the knowledge of problem domain will allow analysis without arbitrariness. Based on the suggestion of iterative learning of only sequential processing which is the basis of programming [3], even source codes including only a meaning of the internal program structure will be able to unveil the understanding process from the eye movement. In addition to these, to minimize the effect of memory ability, this paper designs an experimental protocol that uses source codes composed of concise instructions as the reading subject and allows examinees to read experimental subjects many times until examinees understand enough.

2.2 Flow of Analysis

The traditional analysis method of eye movement for program reading was based on time series data or the total amount of eye movement [4,5]. However, it is not easy to use the time series pattern as it is and to estimate the comprehension process of each examinee because the reading time and the reading strategy are entirely different depending on examinee. So this paper assumes that programmer's thinking behavior is similar to program slicing while reading a source code, and based on the assumption, grasps the behavior of program reading in terms of slicing. In other words, the assumption is that a programmer who understands the content of program correctly reads the source code while implicitly grasping the structure and the strength of data dependency relationship, and the comprehension process appears in the eye movement. Previous study qualitatively confirmed that the characteristic of the reading appears in eye movements depending on the program structure and the nature of programmer [6]. Therefore, this paper focuses on the fixation [7] which is a motion where gaze point stops within a range for a period and quantifies the time series fluctuation of the eye movement by fixation. Then, as a method for quantitatively analyzing the quantized data, a Markov model is employed.

The examinee's eye movement is measured during reading a source code until he/she grasps the values of all variables after the execution of the presented source code. Experiments in this paper do not limit the reading time

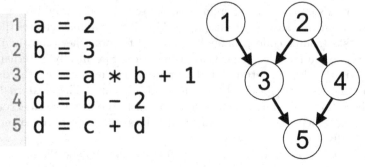

(a) A source code of C language   (b) Program dependence graph

Fig. 1. An example of a source dode and its program dependence graph

and continue to measure until the examinee satisfied. As mentioned above, the experimental task is the source code of only simple assignment statements and includes only data dependencies among program structures (not including control dependency relationship). Thus, we believe that the history of eye movement can be evaluated not strongly affected by examinee's coding skill and reading strategy. Also, Markov model is useful for evaluating the pattern of eye movement because the examinee can repeatedly read and understand the presented source code many times until they fully understand the content.

2.3 Research of Analyzing Program Comprehension Process with Eye Tracking

Studies on eye movement targeting programming learning have also been conducted since the early 1990s [8]. Eye tracking instruments have been used to acquire nonverbal features of beginning/experienced programmers to analyze skill differences. However, although jumping of eye movement is common in source code reading, it is extremely rare in reading of natural language sentences [9]. Therefore, a unique method different from the analysis of general document reading is necessary to analyze programmer's cognitive process, but its analysis method is not sufficiently established even at the present stage.

3 Analysis Technique

3.1 Program Slicing

Program slicing is a technique that focuses on the dependency relationship between sentences in a program and extracts a set of sentences having dependency with a specific sentence, and the extraction result is called as a slice [10]. A slice is a part directly or indirectly related to information a programmer wants to know. The slicing technique has mainly used to narrow the sentence including bugs in a procedural language and has achieved various developments such as static slicing and dynamic slicing [11]. Program slicing has also been reported to be important for debugging [12]. Therefore, an appropriate slicing skill will enable efficient work to quote existing sentences or to find bugs. Many learners have had a hard time finding a mistake in source codes. This work considered to be extraneous (inefficient) cognitive load. Proper slicing skill will enable programming learners to efficiently reduce non-essential work like scanning descriptive errors and to concentrate on essential learning like designing algorithm. As a means to quantify the slicing skill, the eye tracking would be effective. When reading a source code, we can assume that the learner implicitly makes a thought similar to slicing [13]. Although there are some studies focusing on slicing skill for programming education [14,15], the authors cannot find a research trying to estimate slicing skill from eye movement. If we can evaluate the slicing skill from eye movement, the analysis result will realize more appropriate instruction according to learner's skill level. Specific methods of program slicing include static slicing and dynamic slicing. The slice extraction process and its feature related to this paper are described below.

Static Program Slicing. In static program slicing, source code is analyzed statically and a dependency between statements is extracted. When the following conditions are satisfied for the statements s_1 and s_2 in the source code p, the statement s_2 depends on s_1, and the relationship is defined as the control dependency relationship.

- s_1 is the control statement.
- The result of s_1 determines whether s_2 is executed or not.

When the following conditions are satisfied, the statement s_2 depends on s_1, and the relation is called as data dependency relationship.

- In s_1 the variable v is defined.
- There is at least one route which does not redefine the variable v in s_1 to s_2.
- The variable v is referenced in s_2.

Program Dependence Graph. From the above dependency relationships, we can draw a program dependence graph (PDG) [16]. PDG is a directed graph in which edges represent dependency relationships between sentences, and nodes are statements such as control statements and assignment statements. The direction of the directed edge is expressed in the opposite direction to the direction of dependence in order to make it easy to grasp the flow of the program. An example of PDG is shown in Fig. 1. Figure 1(b) is a PDG created from slicing with the static slicing criterion (20 outputs) for the source code of Fig. 1(a). In Fig. 1(b), each node represents a row.

There are two types of dependency relationship, data dependency relationship, and control dependency relationship. In this paper, we focus only on the data dependency relationship. Programming skill to follow control syntax is considered to be highly dependent on knowledge. A learner who has not enough skill of control syntax would be difficult to follow the dependency relationship among variables involving the control syntax. Therefore, in this paper, we focus only on the data dependency relationship composed only of the most fundamental dependency relationship between variables. Figure 1 shows an example of data dependence graph in which a data dependence is expressed as a directed graph.

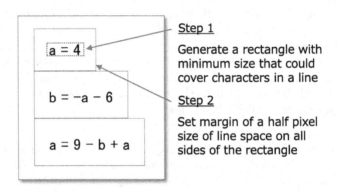

Fig. 2. Definition of the region of each line

3.2   Eye Tracking Technique

The movements of human's eyes when reading a document is roughly divided into two patterns [17]. One is a behavior called "Fixation" that gazes one point for about 100 to 500 ms, about 200 to 300 ms on average. The other is a quick movement called "Saccade" of about 30 ms that is a move from fixation to the next fixation. Saccade can be further categorized according to its characteristics such as the behavior of reading forward sentences (forward-reading) and the behavior of returning to previously read sentences (regression).

Measured raw data of eye movement contains many noises. To efficiently analyze the data of source code reading and evaluate these, only the characteristic patterns of eye movement while reading is considered to be available. Therefore, previous studies have usually observed only the history of the transition between each line a programmer gazed. Salvucci et al. proposed three methods to collate eye tracking data and process model: target tracing method, fixation tracing method, and point tracing method [18,19]. In the method of Salvucci et al., human's work is described by rule sets consisted of regular grammars and is defined as a process model. Then, by automatically matching the eye tracking data with the process model, their method estimated the executed process model. In the fixation tracing method, Hidden Markov Model (HMM) [20] is combined to map to each state, and the whole process model is expressed. By giving fixation points to the obtained HMM, the selected process model is probabilistically calculated.

Based on the knowledge of Salvucci et al., this paper focuses on the transition of fixation from the features of eye movement and processes measured data similar to the retention tracking method. In this paper, the process model corresponds to each line of the program, and we try to construct a thinking model from the transition between processes. First, focusing only on the fixation motion in eye movement, we can obtain the time series of filtered eye positions. To extract the filtered eye data, these paper employs i-vt fixation filter which is a technique to extract only the fixation [21]. To analyze programmer's comprehension process, the transition data from a line to another line is necessary. Therefore, the eye movement is converted to the transition of gazed lines, i.e., fixations are converted based on the range of lines. Figure 2 shows the method to define the row range and the details are as follows. First, in order from the top line, the minimum range of rectangle is determined for each line in which characters are written. Next, the row range is identified by adding the values of pixels corresponding to 1/2 of each line space to the place around the obtained minimum rectangles. This row range enables us to transform the raw data of eye movement into the series of the gazed line.

3.3 Markov Process

Markov process is a stochastic process in which future predicted values are determined only by current observations. In this paper, we use a simple Markov process determined by only the last state among them. By considering a part of source code including a process as state and the transition between parts as edge, we construct the state transition model and regard it as the learner's thinking model. Markov property certainly exists in the transition of eye movement [22]. The target of this paper is only the transition between the lines in source code. Therefore, it is not necessary to emphasize that there is a hidden state between the output symbol string of the observed eye movement and the internal cognitive process. Based on this reason, we adopt simple Markov model, not Hidden Markov model. Also, even in the gaze movement, we focus only on the transition of eyes from one row to another, so we do not think about the transition to the same state. For example, the eye transition according to each line Fig. 3(c) is represented by a simple Markov process as Fig. 4. The edge from node 1 to node 2 in Fig. 4 represents the probability of transition to the second line from the first line in the case when gaze line is staying on the first line.

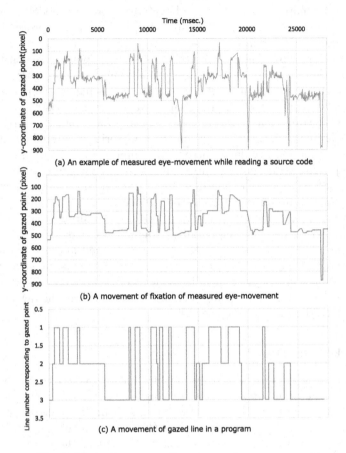

(a) An example of measured eye-movement while reading a source code

(b) A movement of fixation of measured eye-movement

(c) A movement of gazed line in a program

Fig. 3. Flow of processes for measured eye movement

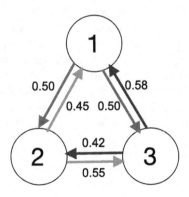

Fig. 4. Simple Markov process of eye movement

4   Experiment

4.1   Protocol

We conducted experiments with examinees who are third, fourth grade, and graduate students majoring informatics and have already learned the basics of programming, such as the foundation of C language, Java language, and the foundation of the algorithm. Regarding the achievement results related to programming, the skill levels of examinee were uniform. This paper conducted experiments in a room where the inside was invisible from the outside to make examinees relaxing. Besides, this paper gave enough consideration to make all examines concentrate on the experimental subject. The first experiment was conducted with 15 people and the second experiment was conducted with 15 people. In the second experiment, we performed on the same 14 people as the first experiment, i.e., one examinee was different from the first experiment. The reason for dividing the experiment into two was to confirm the result of the first experiment and to clarify whether it is a general trend or not by acquiring the data of different subjects. In this paper, we focused on the source code consisted of 3 lines as experiment subjects. All source code was composed only of basic assignment statements and did not include instructions such as increment/decrement and compound assignment operators. The used operators were only arithmetic and surplus operation, and it consisted only of simple statements calculatable in his/her head. In addition, we informed examinees beforehand that all types of variables are integers in advance. There are four kinds of combinations of data dependency relationship in the case of three lines of source code. All source codes were automatically generated for each data dependency relationship and used for experiments. In the first experiment, the examinees addressed 12 subjects (3 source codes for each data dependency). The experimental time was about 10–15 min. In the second experiment, the examinees addressed the 8 subjects (2 source codes for each data dependency). In order to measure the eye movement, X2-30 Eye Tracker produced by Tobi Technology Inc. was used. Examples of the source code used in the experiment is shown in Fig. 5.

Figure 6 shows the experimental situation. The examinee read the presented program and answers the values of all variables after its execution. Each source code was presented in an irregular order, and the examinee was not informed the type of data dependency. We finished gaze measurement at the time when the examinee answered the values of all variables. The timing of answer was free. For example, at the time when the value of only one variable in the presented program is found, it is possible to answer its value. The examinee continued reading until he/she had the correct value for all variables. We paid attention to examinees in advance so as not to respond at a venture. The source code was displayed at the 12.1-in. display with fullscreen. Regarding display setting of the source code, the left margin was 100 px, the upper margin was 40 px, the font size was 50 px, and the line spacing was 150 px, which were empirically determined but the comfortable settings for us. The flow of the experiment was as follows.

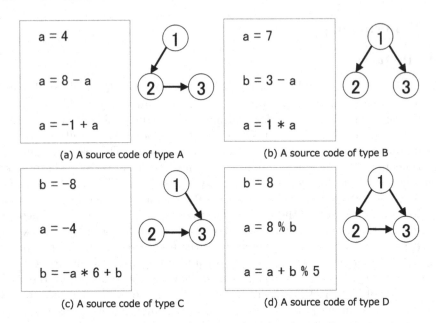

(a) A source code of type A

(b) A source code of type B

(c) A source code of type C

(d) A source code of type D

Fig. 5. Source codes used in the experiments

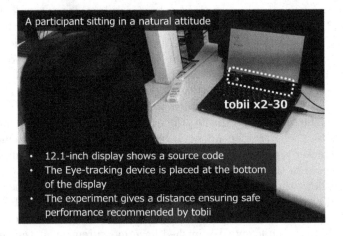

Fig. 6. A picture showing actual condition of the experiment

1. Each examinee was given to explanation about the experimental method such as the presentation method of the subject, the answer method, the experimental time, things to keep in mind when reading source code (do not close your eyes, do not move your face, and keep proper posture). After the confirmation of essential matters, calibration was performed (about one minute).
2. We explained the detail of the subject and do a preliminary exercise of the experiment with the examinee (about two minutes).

3. We measured the data of eye movement one by one. After receiving the examinee's answer, we ended the measurement (about 10–30 s per question).
4. After completing the experiment of one question, we prepared a sufficient time break (about 10–30 s). After the break, went to the next question. We presented the problem in the same way as the above procedure and finished when performing all the tasks.

4.2 Result

Markov models were generated for each reading task from the eye movement data of each examinee. As the targets of this paper were four types of data dependency relationship. This paper calculated the averages of the transition probabilities of Markov model between nodes for each Type as shown in Fig. 7. In Fig. 7, the transition probability from the node i to the node j is represented as $i \to j$, and the error bar indicates upper and lower limits of the population mean (infinity) with 99% reliability when assuming t-distribution. Also, the probability of each data structure is indicated by a solid line, and the average of the probabilities of all data structures is connected by a broken line. From the result of Fig. 7, it can be confirmed that the features corresponding to the data dependency relationship were reflected the eye movement. We can observe that there are some edges with the large difference from the average, and this pattern of difference corresponds to the data dependency relationship.

The basic reading strategy in all experimental tasks consists of the scanning operation for the whole from the top to the bottom to understand the data

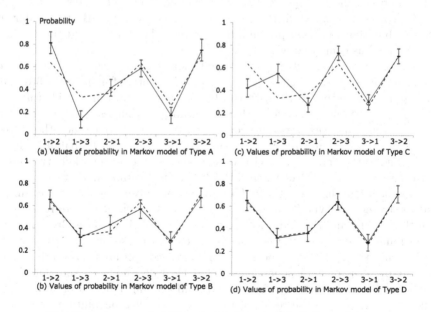

Fig. 7. Average probabilities in Markov model of all participants

dependency relationship, and the reading operation based on the structure after returning to the top. Uwano et al. showed three patterns on the programmer's eye movement: behavior to view the entire code from the top to the bottom immediately after the review started, behavior to check the declaration part of variables when each variable is first referenced, and behavior to check the previous line containing a variable when the variable appears [4,5]. While correlating Uwano's findings and time series of examinee's eye movement, we qualitatively considered the reading strategies of each structure. The basic flow of each structure depended on the order of variables to be identified, and this difference affects the eye movement.

Considering the data structure of Type A, it is supposed that this common reading strategy is based on a flow of understanding while identifying variable values in order from the top after scanning. Therefore, it can be predicted that the transition probabilities of the 1st to the 3rd line and vice versa can be lowered. In the result of Fig. 7, as expected, the transition from the 1st line to the 3rd line and the transition from the 3rd line to the first line, both of which have no data dependence, were lower than these pair transitions. The transition from the 3rd line to the 1st line was significantly lower, and it can be said that it was a tendency strongly influenced by the structure even if considering the regression (the action to return to read the sentence already read).

The common reading strategy of Type B has, firstly, the motion to understand the instructions in order from the top after scanning. Next, since there is a dependency relationship between the 3rd line and the 1st line, there is a motion to read returning to the 3rd line once confirming the 1st line. In other words, it can be predicted that there is the reading motion including the reciprocating motion between the 1st line and the 3rd line in addition to the flow of the reading of Type A. As expected, the shape of the result of Type B was similar to Type A, but the transition probabilities between the 3rd line and the 1st line was increased as compared with Type A.

The common reading strategy of Type C is as follows. After scanning, there is a reading motion to check the 3rd line, move the center of sight to the 3rd line, and repeat minute round-trip with the 1st line and the 3rd line to identify the value of the variables. Therefore, the transition probabilities from the 1st line to the 2nd line and vice versa are considered to be lower than the other three structures. In the result of Fig. 7, as expected, the transition probabilities from the 1st line to the 2nd line and vice versa were lower than the other structures.

In the case of Type D, the value of the 3rd line is determined after the value of the variable is determined in the 2nd line. Therefore, after scanning, the common reading strategy of Type D consists of the motions to repeat minute round-trip with the 1st line and the 2nd line, move the center of sight to the 3rd line, and repeat minute round-trip between the 3rd line and the 1st line and between the 3rd line and the 2nd line. Type D has the most complicated structure among the four patterns of data dependence. That is, it can be predicted that Type D follows a reading strategy that includes all reading motions of other data structures together. In the result of Fig. 7, the overall probability average and

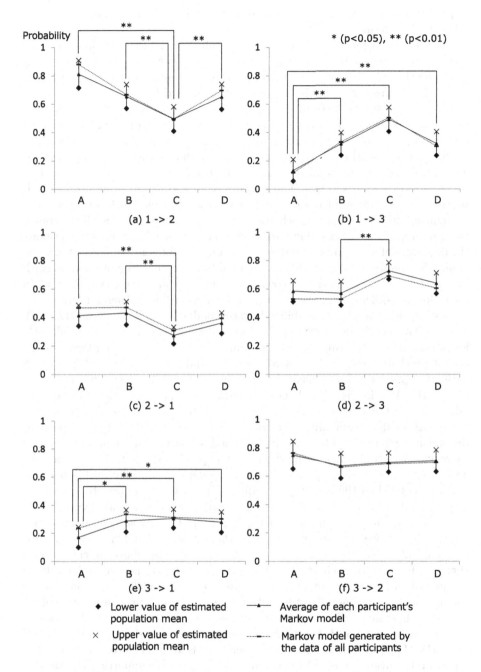

Fig. 8. Comparison of each structure's average probability

Type D probability were almost similar. Therefore, as expected, the reading of Type D was undoubtedly the result of adding together all reading motions of other data structures.

As the overall tendency, the transition probabilities of the 1st line from the 2nd line and the 1st line from the 3rd line were lower than other. The reason might be that the 1st line in many cases was a definition sentence, and it was relatively easy to compare with the content of the 3rd line.

Next, Fig. 8 shows the result of comparing the average of transition probabilities between nodes i, j for each structure. The error bar of Fig. 7 indicates upper and lower limits of the population mean (infinity) with 99% reliability when assuming t-distribution. The probability of each data structure is connected by a solid line, and the probability generated by combining all examinees' eye data is connected by a broken line. The purpose of combining all examinees' data is to examine only the influence of the data dependency relationship from the eye movement without relying on each examinee's feature. From Fig. 8, all the transition probabilities of all examinees were within confidence intervals. Furthermore, the difference of structure certainly appeared in the difference of transition probability. Therefore, we confirmed that general tendency of eye movement according to the structure could be shown by combining all examinees' data when analyzing the difference in the reading pattern according to each structure.

As for the transition probability between nodes 1 and 2 shown in Fig. 8 (a), Type C without dependency relationship between nodes 1 to 2 is significantly lower than other structures by Tukey-Kramer test ($p < 0.01$). Similarly, for the data dependency relationship between nodes 1 and 3 shown in Fig. 8(b), Type A without dependency relationship between nodes 1 and 3 was significantly lower ($p < 0.01$). Also in Fig. 8(c), Type C without dependency relationship between nodes 2 and 1 was significantly lower ($p < 0.01$) than Type A and B, but there was no significant difference with Type D. In Fig. 8(d), Type B without dependency relationship between nodes 2 and 3 had the lowest probability, but there was an only significant difference with Type C ($p < 0.01$). Between nodes 3 and 1 shown in Fig. 8(e), Type A, which has no dependency relationship between nodes 3 and 1, had the lowest probability. Type A was significantly lower than Type B and D ($p < 0.05$), and was also significantly lower than Type C ($p < 0.01$). Between nodes 3 and 2 shown in Fig. 8(f), Type B with no dependency relationship between nodes 3 and 2 had the lowest probability, but there was a significant difference from other structures. From the above, it can be said that the features of each structure appeared in the transition probability of eye movement, and the effectiveness of the proposed modeling method and the usefulness of the eye movement for estimating the reading process were clarified.

As an effort for reading source code, Orlov et al. proposed a method to convert a source code into an abstract semantic element and model an eye movement using HMM [23]. However, Orlov et al. concluded that his achievement is only one guideline because they could not obtain a result clearly supporting the validity of the model. On the other hand, this paper could obtain meaningful results from the approach similar to Orlov's one. Therefore, the findings of this paper would be valuable for further expansion and application of the Markov model-based method.

5 Conclusion

This paper investigated the possibility of estimating learner's program comprehension process during reading based on the pattern of eye movement, not the eye distribution. From the viewpoint of program slicing, the internal structure of the program was restricted to the data dependency relationship, and the features of eye transition at lines of source codes were clarified. First, we got the eye movement during programming reading process and proposed its modeling method. Then, based on the program structure, we designed an experimental protocol to analyze the eye movement during reading source code. This study constructed programs with only assignment and arithmetic operations, simplified the structure of programs to data dependency relationship, and set them as analysis targets. From the data of such programs obtained by eye tracking, this study constructed a simple Markov process model to express eye movement patterns as transition probabilities. This paper used four kinds of combinations of data dependency relationship in the case of three lines of source code at the experiment. Some source codes were automatically generated for each data dependency relationship and used for experiments. By comparing the difference of transition probabilities between nodes and data dependencies, this study showed the possibility of analyzing the program reading process based on eye movement. As a result, we could confirm the influence of each pattern's data dependency relationship as the peculiar behavior of program reading.

Although the experimental subjects in this paper were simple, we were able to show the tendency of eye movement based on data dependency relationship. Therefore, based on the result of this paper, in the future, we will identify the eye trend of learners who have a difficulty to read a source code. The analysis will be possible to discriminate learners who are out of the criteria, standard eye movement depending on data dependency relationship, and to instruct an efficient reading method. Also, the increase of the size and complexity of program will contribute to elucidating the implicit thinking process which was difficult to clarify by conventional methodology alone.

Acknowledgments. This work was partly supported by Japan Society for the Promotion of Science, KAKENHI Grant-in-Aid for Scientific Research(C), 16K01147, 17K01164.

References

1. Sagisaka, T., Watanabe, S.: Development and evaluation of a web-based diagnostic system for beginners programming course. J. Jpn. Soc. Inf. Syst. Educ. **27**(1), 29–38 (2010). (in Japanese)
2. Pennington, N.: Stimulus structures and mental representations in expert comprehension of computer programs. Cogn. Psychol. **19**, 295–341 (1987)
3. Okamoto, M., Terakawa, K., Murakami, M., Ikeda, K., Mori, M., Uehara, T., Kita, H.: Computer programming course materials for self-learning novices. In: Proceedings of World Conference on Educational Multimedia, Hypermedia and Telecommunications, vol. 2010, no. 1, pp. 2855–2861 (2010)
4. Uwano, H., Nakamura, M., Monden, A., Matsumoto, K.: Exploiting eye movements for evaluating reviewer's performance in software review. IEICE Trans. Fundam. **E90–A**(10), 317–328 (2007)
5. Uwano, H., Nakamura, M., Monden, A., Matsumoto, K.: Analyzing individual performance of source code review using reviewers' eye movement. In: Proceedings of the 2006 Symposium on Eye Tracking Research & Applications, pp. 133–140 (2006)
6. Kashima, T., Matsumoto, S., Yamagishi, S.: Knowledge acquisition with eye-tracking to teach programming appropriate for learner's programming skill. In: Proceedings of the Third Asian Conference on Information Systems, pp. 287–292 (2014)
7. Rayner, K.: Eye movements in reading and information processing: 20 years of research. Psychol. Bull. **124**(3), 372–422 (1998)
8. Ihantola, P.: Notes on eye tracking in programming education. In: Eye Movements in Programming Education, pp. 13–15 (2014)
9. Crosby, M., Stelovsky, J.: How do we read algorithms? A case study. IEEE Comput. **23**(1), 24–35 (1990)
10. Weiser, M.: Programmers use slices when debugging. Commun. ACM **25**(7), 446–452 (1982)
11. Agrawal, H., Horgan, J.: Dynamic program slicing. In: SIGPLAN Notices, vol. 25, no. 6, pp. 246–256 (1990)
12. Nishimatsu, A., Nishie, K., Kusumoto, S., Inoue, K.: An experimental evaluation of program slicing on fault localizaion process. IEICE Trans. Inf. Syst. **82**(11), 1336–1344 (1999). (in Japanese)
13. Ishio, T., Kusumoto, S., Inoue, K.: Debugging support for aspect-oriented program based on program slicing and call graph. In: Proceedings of 20th IEEE International Conference on Software Maintenance, pp. 178–187 (2004)
14. Inazumi, H., Takeuchi, S.: How to learn programming technique by using program slicing. Aoyama Inf. Sci. **29**(1), 51–78 (2001). (in Japanese)
15. Yoshida, H., Tateiwa, Y., Yamamoto, D., Takahashi, N.: A code review navigator with chunking and slicing for assembly programming exercise. IEICE technical report, vol. 109, no. 335, ET2009-81, pp. 169–174 (2009). (in Japanese)
16. Ottenstein, K., Ottenstein, L.: The program dependence graph in a software development environment. In: Proceedings of ACM SIGSOFT/SIGPLAN Software Engineering Symposium on Practical Software Development Environments, SDE 1, pp. 177–184 (1984)
17. Rayner, K.: Eye movements in reading and information processing, 20 years of research. Psychol. Bull. **124**(3), 372–422 (1998)

18. Salvucci, D., Anderson, J.: Automated eye-movement protocol analysis. Hum.-Comput. Interact. **16**(1), 39–86 (2001)
19. Ohno, T.: What can be learned from eye movement?: understanding higher cognitive processes from eye movement analysis. Jpn. J. Cogn. Sci. **9**(4), 565–579 (2002). (in Japanese)
20. Juang, B., Rabiner, L.: The segmental k-means algorithm for estimating the parameters of hidden Markov models. IEEE Trans. ASSP **38**(9), 1639–1641 (1990)
21. Olsen, A.: The Tobii I-VT Fixation Filter, Tobii Technology (2012)
22. Iwao, T., Mima, D., Kubo, H., Maejima, A., Morishima, S.: Analysis and synthesis of eye movement in face-to-face conversation based on probability model. Inst. Image Electron. Eng. Jpn. **42**(5), 661–670 (2013). (in Japanese)
23. Orlov, P.: Primary investigation of applying Hidden Markov Models for eye movements in source code reading. In: Eye Movements in Programming Education II: Analyzing the Novice's Gaze, pp. 18–20 (2015)

Development of a Pair Ski Jump System Focusing on Improvement of Experience of Video Content

Ken Minamide[1]([✉]), Satoshi Fukumori[2], Saizo Aoyagi[3],
and Michiya Yamamoto[2]

[1] Graduate School of Science and Technologies, Kwansei Gakuin University,
Sanda, Hyogo, Japan
k.m@kwansei.ac.jp
[2] School of Science and Technologies, Kwansei Gakuin University, Sanda,
Hyogo, Japan
[3] Faculty of Information Networking for Innovation and Design, Toyo
University, Kita-ku, Tokyo, Japan

Abstract. "Ski Jumping Pairs" is a video content of an imaginary sport in which two players jump using a pair of skis. The highlight of the content is incredible aerial style of players. We have already developed a VR ski jump system using HMD. In this study, we have developed a system in which users can experience Ski Jumping Pairs. First, we propose a design method by introducing an idea of composing make-believe play to enhance the VR experience of the world of video content. Then, we developed a prototype of the system. An experimental evaluation was used to demonstrate the effectiveness of the method.

Keywords: VR headset · Experience of video content · Make-believe play
Ski Jumping Pairs

1 Introduction

The VR market is entering a new phase from the sales of headsets to the spread of platforms, for example, Sony released PlayStation VR and Microsoft made Windows Mixed Reality available on Windows 10. Also, various kinds of completely new content are being developed. Specifically, a lot of content is constantly released in which users can enjoy special experiences by utilizing the high reality feeling of VR headsets. For example, content to experience live-action 3D videos such as Exploring the SHINKAI VR [1] and Skydiving [2], and interactive experiences using CG such as BIOHAZARD 7 [3] and SUMMER LESSON [4] have been developed.

We have developed a VR ski jump system [5]. This system has already been experienced by more than 600 people at the Sapporo Television Broadcasting (STV) Cup's Virtual Ski Jumping Competition in Chi Ka Ho held in the underground walking space in front of Sapporo station (Fig. 1). Participants could jump and enjoy a long-distance flight regardless of age as shown in Fig. 2.

© Springer International Publishing AG, part of Springer Nature 2018
S. Yamamoto and H. Mori (Eds.): HIMI 2018, LNCS 10904, pp. 562–571, 2018.
https://doi.org/10.1007/978-3-319-92043-6_45

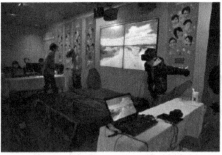

Fig. 1. A scene from the virtual ski jump tournament.

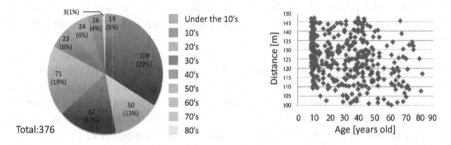

Fig. 2. Age of the participants and their scores.

In this study, we focus on improvement of experiences with HMD, and propose how to design experiences in video content. The common way was to develop a game content inspired by the video content like Super Robot Wars [6]. Conversely, we focused on the fact that we can reproduce the world of the virtual video content and perform various actions there by using HMD and VR technology. We regard such technology as a new style of make-believe play and examined how the experience of video content was improved by introducing the fun of make-believe play. For our study, we selected Ski Jumping Pairs [7] known as a media art work, and we examined how to experience the content.

2 Related Studies

2.1 Experience Ski Jumping with VR

There are many VR ski jumping systems using live-action video, and they enable users to vicariously enjoy player's experience. Various VR ski jumping systems using CG have also been developed, such as VR SKI JUMP [8] developed as the first app in Oculus. Some of these systems have been used to launch events, for example, Planica 2015 Virtual Ski Flying by Zavarovalnica Triglav [9]. On the other hand, systems with large-scale equipment developed for training are also reported such as the Staurset's system [10].

2.2   Experience Video Content

Focusing on the experience of the video content, various systems are being developed. For example, BANDAI NAMCO ENTERTAINMENT released DRAGONBALL XENOVERSE in 2015 [11], and Rare Ltd. released GoldenEye 007 in 2011 [12]. In these cases, video content is made into games. Flow theory is well-known as a method for immersing in games. However, this theory is not applicable when there is a significant difference between players' skill levels because flow theory is proposed to explain one player's balance of skill level and the difficulty of the game.

As for multi-player VR content, Mikulus Kinect Online was developed by Needle in 2013 [13]. This is a technological prototype to realize that two players can play the role of CG character Hatsune Miku in the same VR space, so to speak, it is a "cosplay" using VR. In commercial software, Sony Interactive Entertainment Inc. has released RIGS Machine Combat League for PS VR in 2016 [14]. This is a game content in which multiple players manipulate robots and fight with each other from a first-person perspective.

On the other hand, in this study, we focused on studies of make-believe play from the standpoint of improving the experience. Specifically, we focused on the elements of fun in make-believe play proposed by Yagi [15]. He classified the fun of make-believe play as five elements: "role- fun to become someone", "object- fun to create or use something", "action- fun to do something", "space- fun to consider as somewhere", and "interaction- fun to play with someone". He said that the weights of these elements varied depending on the types of make-believe play.

3   Developing Pair Ski Jump System

3.1   Targeted Video Content

Ski Jumping Pairs is video content of an imaginary sport in which two players jump using a pair of skis. The highlight of the content is incredible aerial style of players. According to Wikipedia, it became very popular by the internet streaming service and it was screened in various film festivals around the world. In addition, Ski Jumping Pairs - Reloaded - for PS2 was been released by KAMUI in 2006 [16]. This game has interesting live comments on various aerial styles and actions selected by the user, so it attracts fans of video content.

3.2   Method

We propose that we can enhance the experience of video content by deciding the details of the five elements by Yagi that players will experience, and by assigning them an appropriate weight according to the video content. In the Ski Jumping Pairs experience, we first decided the content of five elements as shown in Fig. 3. We assumed that it had two features, jump and pose as "action", and jump by two players as an "interaction". Then we changed the weight of the "action" and "interaction" elements as shown in Fig. 4. There could be many designs of the balance of weights, for example, an action-oriented mode in which each player concentrated on "action" by improving aerial styles completion, and an interaction-oriented mode in which each player concentrated on "interaction" by requiring cooperation.

"role"	· · ·	Become a virtual ski jumper
"object"	· · ·	Wearing HMD as a goggle
"action"	· · ·	Jumping, posing
"place"	· · ·	Being in virtual ski jump stadium
"interaction"	· · ·	Jumping with another player

Fig. 3. Elements and details of fun.

4 System

4.1 System Configuration

We developed a prototype of the system targeting two-player, side-by-side at game centers etc. Therefore, we configured the system with an HMD (Oculus VR, Oculus Rift DK2), an HMD head tracking camera, a motion sensing input device (Microsoft, Kinect v2), and a PC for each player (Fig. 5).

Two PCs shared data such as jump timing and character pose by UDP with P2P. HMD was used to present the virtual world. Kinect v2 was used to sense the pose of the players and to reflect it on the character IN the virtual world. As the development environment, we used Windows 8.1 for PCs OS, and we used Unity 5.3.4 and C# for development language. The SDK we used for HMD was Oculus SDK 0.8, and for Kinect v2, Kinect SDK 2.0. We used LightWave 11.5.1 for modeling CG objects.

4.2 CG Model and Jump

We made a CG model with the motif of Okurayama Ski Jump Stadium (Fig. 6). A jump stand for skiing has an altitude of 133.6 m, a total length of 368.1 m, and a slope length of 403.8 m. The altitude of the starting point is 307 m. When creating CG models, we created a jump stand for skiing to be this size. Also, in order to make it easy to understand the timing of jumping, we colored the take-off point pale pink and dark pink.

We also adjusted the virtual jump. First, the actual duration of jumping was about three seconds, but we changed it to about five seconds because there was a feedback that it was too short. Also, in order to give a feeling of flight, as shown in Fig. 7, we set the height of the jumping orbital from the slope to be high and made a direction in landing towards the ground to make it easy for players to recognize that they have landed. The range of distances was set at five levels. The longest distance was set about 4.35 s when the CG character just passed through the take-off point. Other levels were set around 4.35 s. The distance was set randomly. All distances were set over 100 m so that even small children or elderly people can enjoy the jumping experience.

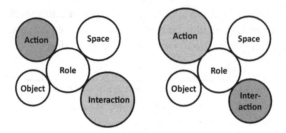

Fig. 4. Weights of elements.

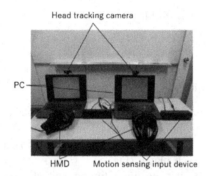

Fig. 5. System configuration.

Seen from the starting gate   Seen from the landing Bahn

Fig. 6. Scenery in CG. (Color figure online)

4.3 Developed Mode

In Ski Jumping Pairs, the score is mainly calculated by the distance and aerial styles similar to real life ski jumping competitions. In our virtual ski jumping, we scored the distance according to the timing of the jump and scored the "performance (waza)" instead of the aerial styles, and combined them as the score of the jump. We calculated "performance" based on the difference between the player's pose and an ideal pose. As shown on the left side of Fig. 8, after displaying the ideal "performance" of the video content on the display beforehand, and players jumped while taking a pose after remembering it (Fig. 8 right).

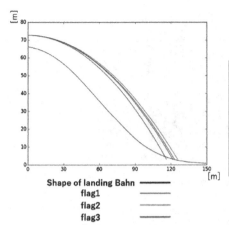

Timing of the jump[s]	Distance[m]	flag
-1.35~-0.55	110~120	3
-0.55~-0.05	120~135	2
+-0.05	135~147	1
+0.05~+0.15	120~135	2
+0.15~+0.35	110~120	3
+0.35~	100~110	4

Fig. 7. Timing and orbit in jump.

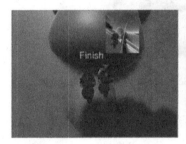

Fig. 8. Example pose and posing.

Figure 9 is an example of "performance" we developed in this system. These are reproduced representative "performance" of video content. In detail, players take different poses in sliding, in jumping, and after landing, so that they can enjoy various "performance" during one jump.

After landing, the score is displayed on the scoreboard as shown in Fig. 10. We created an action-oriented mode and an interaction-oriented mode shown in Sect. 3.2 by changing the display method of this score. In an action-oriented mode, we displayed players' individual scores on the timing of the jumps and the degree of completion of "performance". By displaying independent scores, players can concentrate on improving timing and "performance." They can raise the elements of "action", and improve the experience. In an interaction-oriented mode, we calculated the difference of the timing when two players jumped and displayed the same value to both players. The degree of completion of "performance" was the average of two players. In this mode, players enjoy communication. They can raise the elements of "interaction", and improve the experience.

Fig. 9. Type of poses.

Fig. 10. Scoreboard.

5   Evaluation

5.1   Purpose and Method

We performed an experiment to evaluate in which mode the experience of this video content is more enjoyable for users.

We showed participants the video of Ski Jumping Pairs to introduce it. Then we told participants that the prototype system is designed to experience the world of Ski Jumping Pairs. We instructed them how to play and explained that there are two scoring methods. When we discussed the scoring methods, we avoided using the names of the modes. In the experiment, we asked participants to wear HMDs. Then, we asked participants to experience each mode five times. In the action-oriented mode only, we asked them not to tell their own score to each other. This was because we wanted them concentrate on their play and prevent competition. When they experienced each mode, we asked players not to remove the HMD unless complications arose, such as VR sickness or fatigue. The order of the modes was randomly assigned. After the experience, we asked participants to answer the questionnaire of five items shown in Fig. 11 on a seven-point scale. Item (1) is a question about the experience of the video content. Item (2) is a question about the evaluation of the content. Items (3) to (5) are questions about elements of fun. After that, changing modes, we asked participants to experience another mode in the same procedure. During the experience, we recorded scores. Participants are twenty people aged 19 to 22 years (eighteen males and two females).

Fig. 11. Experimental scene.

5.2 Results

Figure 12 shows the results of the questionnaire. The results show that we could develop a fun and enjoyable platform regardless of the mode because the average value of item (2) was very high, 6.4, in each mode. A Wilcoxon signed-rank test was conducted in order to reveal differences between the action-oriented mode and the interaction-oriented mode, and these results are also shown in Fig. 12. Item (5) shows significant differences at a significance level of 1%. Items (1) and (4) are even more different, at a significance level of 5%. The interaction-oriented mode was evaluated higher for each of these items.

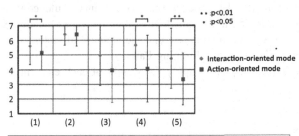

(1) I could experience the world of "Ski Jumping Pairs".
(2) I could enjoyed the experience.
(3) I was able to cooperate with another player.
(4) I communicated with another player for cooperation.
(5) I communicated with another player triggered by the score.

Fig. 12. Result of evaluation.

In addition, the difference between the jumping timing of the two participants is shown in Fig. 13. A T-test was conducted in order to reveal differences between the action-oriented mode and the interaction-oriented mode, and these results are also shown in Fig. 13. The second jump and the fourth jump of experience show significant differences at a significance level of 1%. The average of all trials shows differences at a significance level of 5%. It shows that the difference between the jumping timing of the two players was smaller in the action-oriented mode than in the interaction-oriented mode.

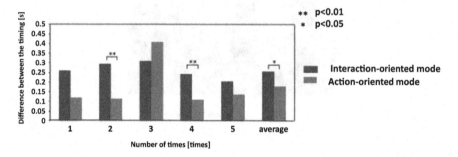

Fig. 13. Transition in score.

5.3 Discussion

There was a significant difference at item (1). This means that the experiences of the content changed according to the weight of each element. As shown in the results of items (4) and (5), participants felt that they communicated in the interaction-oriented mode, that is, they increased the element of "interaction". Also, as a result of the score, the timing of the jumps in the action-oriented mode was better than that in the interaction-oriented mode. This means that participants could concentrate on their jumps during the action-oriented mode, that is, they increased the element of "action".

From these results, we can conclude that the element of "interaction" is important for the content experience of Ski Jumping Pairs.

In the future work, we plan to apply the weight of each element to various video content.

6 Summary

In this study, we developed a system in which users can experience Ski Jumping Pairs. We used this system to evaluate how to experience the world of video content better. According to the results, the experience of the world of video content which two players cooperate was improved by increasing the rate of the element of "interaction".

Acknowledgement. This research was partially supported by JSPS KAKENHI 16H03225, etc.

References

1. SHINKAI VR. https://www.youtube.com/watch?v=XZv-KkWVBr0
2. SkyDive in 360° Virtual Reality via GoPro. https://www.youtube.com/watch?v=S5XXs RuMPIU
3. BIOHAZARD 7 resident evil official web site. http://www.capcom.co.jp/biohazard7/
4. SUMMER LESSON official web site. http://summer-lesson.bn-ent.net/
5. Hashimoto, E., Yamamoto, M., Shigeno, Y., Aoyagi, S.: Development of the pair ski jumping simulator using a wide field of view head-mounted display. In: Proceedings of the 78th National Convention of IPSJ, pp. 4-595–4-596 (2016). (in Japanese)
6. Super Robot Wars official web site. http://www.suparobo.jp/
7. Ski Jumping Pairs official web site. http://www.jump-pair.com/
8. VR SKI JUMP. http://ouka.s108.coreserver.jp/html/vr_skijump.html
9. Planica 2015 Virtual Ski Flying. https://www.youtube.com/watch?v=_yRKmXhVOOk
10. Staurset, E.M.: Creating an immersive virtual reality application for ski jumping: Norwegian University of Science and Technology, Master's thesis (2015)
11. DRAGONBALL XENOVERSE official web site. http://dbx.bn-ent.net/01/
12. GoldenEye 007 official web site. https://www.nintendo.co.jp/wii/sjbj/
13. How to create VR online game: Utilization of Photon Cloud in Mikulus Kinect Online. http://www.heistak.com/2013/12/04/mikulus-kinect-online-photon-cloud/
14. RIGS Machine Combat League official web site. http://www.jp.playstation.com/software/ title/pcjs50017.html
15. Yagi, K.: Gokkoasobi no tankyū - seikatsu hoiku no sōzō o mezashite; Shindokusyosya (1998). (in Japanese)
16. Ski Jumping Pairs –Reloaded. http://www.jp.playstation.com/software/title/slps25616.html

Risk Reduction in Texting While Walking with an Umbrella-Typed Device for Smartphone

Sohichiro Mori[1($\boxtimes$)] and Makoto Oka[2]

[1] Meihokan High School, 6-7-22, Kita-Shinagawa, Shinagawa,
Tokyo 141-0001, Japan
07o2lclasssuzu@gmail.com
[2] Tokyo City University, 1-28-1, Tamazutsumi, Setagaya,
Tokyo 158-8557, Japan

Abstract. It is widely known that texting while walking is a dangerous behavior. To reduce the risks, we proposed an umbrella-typed device, called ii-kasa to manipulate smartphones. In this paper, we investigated whether ii-kasa reduce the risks in texting while walking. In the experiment, we recorded the gaze patterns using the eye-mark recorder. As the results, the average time and its variance of the eye fixations with ii-kasa were smaller than the ones of the smartphone. The results also showed that the participants in the ii-kasa condition watched the broader areas of their circumstances and paid more attention to their circumstances than in the smartphone condition. These results indicate that ii-kasa makes the risks of texting while walking and human cognitive loads reduce.

Keywords: Texting while walking · Umbrella · Smartphone

1 Introduction

Many accidents involving walking while texting with smartphones such as collusions between a person and a person, a person and a car, and person and a bicycle, have happened and it have become social issues.

Though walking while texting is, of course, a dangerous behavior, people must never cease it, because there are many situations where they need to operate their smartphones while walking. For example, we sometimes go to the destination using navigations system, and we sometimes try to meet our friends exchanging our current places through SNS.

When it is rain, the situations become worse. We have to hold the umbrella in one hand and hold the bag and the smartphone in the other hand. Consequently, we pay more attention to our hands not to drop them and pay less attention to the circumstances.

In texting while walking, our gazes distribute mainly in the low space and our visual fields become narrow. To aiming to make texting while walking safer, we proposed ii-kasa [1] which is the system where the screen of the smartphone is

© Springer International Publishing AG, part of Springer Nature 2018
S. Yamamoto and H. Mori (Eds.): HIMI 2018, LNCS 10904, pp. 572–581, 2018.
https://doi.org/10.1007/978-3-319-92043-6_46

displayed on the canopy of the umbrella and we can manipulate it by the movement of the umbrella by the acceleration sensors. With this system, we can keep our gazes in front and aims to be able to pay more attentions to the circumstan ces. However, we did not still investigate the safety of the system.

In this paper, we investigate the gaze patterns with our umbrella system and with smartphones while walking and also investigate the safety of our system.

2 Related Works

The risks of texting while walking are widely well-known. For example, National Geographic conducted a social experiment [2]. They partition the sidewalk into two lanes, one is lane only for texting while walking, on the other lane texting while walking is prohibited. As the results, only few people who were texting while walking changed the lanes and most of them concentrated on the smartphones and even did not notice the existence of the lanes.

To overcome the risks of texting while walking, some research conducted so far. Wang et al. [3] investigated that crossing the street while texting is more dangerous than doing without the smartphones and proposed WalkSafe. In WalkSafe, the circumstance situations are monitored by the camera on the smartphone, and if a car is approaching, the warning sound and vibration come up. The detections of the car are performed by machine learning with the images obtained by the camera.

Kodama et al. [4] described that people cannot stop texting while walking though people recognize its risks. He also proposed the system that support texting while walking using image processing sensor. In this system, the image obtained by the camera are overlaid on the display of the smartphone. When something are approaching, the center of the upper part of the display become red.

Ito et al. [5] dealt with using a smartphone while riding motor bikes. They mounted the acceleration sensor on the gloves and manipulate the smartphone by the gestures. The output from the smartphone are displayed by the sounds. They confirmed that the vibrations caused by the motor bike do not affect the acceleration sensor and the users can control the smartphone with this system.

3 ii-Kasa

As we mentioned earlier, we proposed the system called ii-kasa [1] aiming to reduce the risks of texting while walking. In ii-kasa, the screen of the smartphone is projected on the canopy of the umbrella (Fig. 1) to keep the gaze point front and not to make the visual field narrow. We can manipulate the smartphone by swinging and moving up and down the umbrella. The acceleration sensor is attached on the frame and detect the actions of the umbrella (Fig. 2). For example, in the map application, the map is zoomed in in moving it upward and it is zoomed out in moving it downward. Scrolling the map is also available by swinging the direction to want to watch.

Fig. 1. Screen of ii-kasa

Fig. 2. Sensor part of ii-kasa

4 Experiment

To investigate whether ii-kasa reduce the risks of texting while walking, we conducted an experiment.

4.1   Method

First, the participants were asked to read the sentences aloud on the screen of ii-kasa while walking on the designated root, and, after that, they performed the same task with a smartphone. Here, the root was about one hundred meters long and the participants had to turn the corners four times in one trial. As this root is in the university campus and a public space, some people sometimes walked. They wore the eye-mark recorder to investigate where they were watching during the experiment. The used eye-mark recorder was NAC EMR-8.

4.2   Participants

Five participants were involved in this experiment. All of them were university students and are 22years old. Two of them were wearing the glasses but all of them are visually healthy.

4.3   Results

First of all, as the results of F-test of the fixation time between ii-kasa condition and the smartphone condition, the variance of the ii-kasa condition is significantly smaller than the one of the smartphone condition ($p = 0.00000152$). We, therefore, also conducted the Welch's t-test of the fixation time, the average time of one fixation in ii-kasa condition is significantly shorted than the one in the smartphone condition ($p = 0.0032$) (Fig. 3).

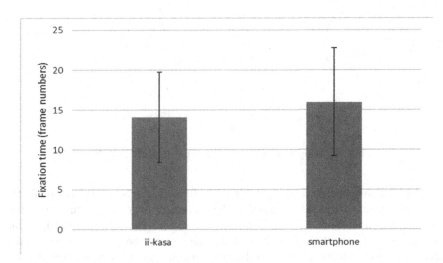

Fig. 3.   Average fixation time

This means the participants in ii-kasa condition watched the screen shortly at one time and averagely paid attention to the screen and the circumstances. On the other hand, in the smartphone conditions, being larger of the variance and the average means that very long fixations are happened. As people are hard to notice the change of the circumstances when the fixation time on the screen is long. Actually, in the experiment, despite one participant almost caused the collision with one pedestrian and the pedestrian got out of his way, he did not notice it.

Above results, however, only show the average of all fixation and does not tell whether they watch the screen or their circumstances. Here, the eye-mark data recorded by the eye-mark recorder are pairs of the points of x-axis and y-axis on each video frames. They are on the relative coordinate and we do not know what the participants were watching because they asked to walk and their heads were not locked. So, we divided the pictures by 5° of the gaze and put the fixation point (Fig. 4) on them. From these data, we calculated the fixation time on each cell. Figures 5 and 6 is the heat-map of the ratios of the fixation time to the trial times in ii-kasa conditions and in the smartphone condition respectively. In these figures, the frames on the maps indicate the position of the screen in each condition.

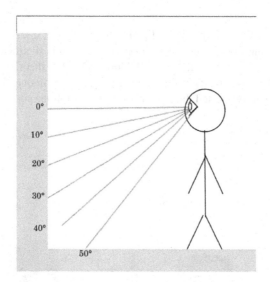

Fig. 4. Mapping to the abstract coordinate of the outside

Figure 5 shows that, in ii-kasa condition, the fixation points are widely distributed and the participants frequently watched the below area of the canopy. This means that they paid attentions both to their front and the screen. On the other hand, in the smartphone condition, (Fig. 6) their fixation points gather on the screen and hardly watch their circumstances. This means that they concentrated to watch the screen of the smartphone and paid less attentions to their circumstances.

Comparing the data in the ii-kasa condition and the smartphone condition within each participant, these trends are shown more clearly, because the set positions of the smartphone were different by the participants. The heat-maps of the fixation time of two participants are from Figs. 7, 8, 9 and 10. While, in Fig. 7, Participant 1 almost

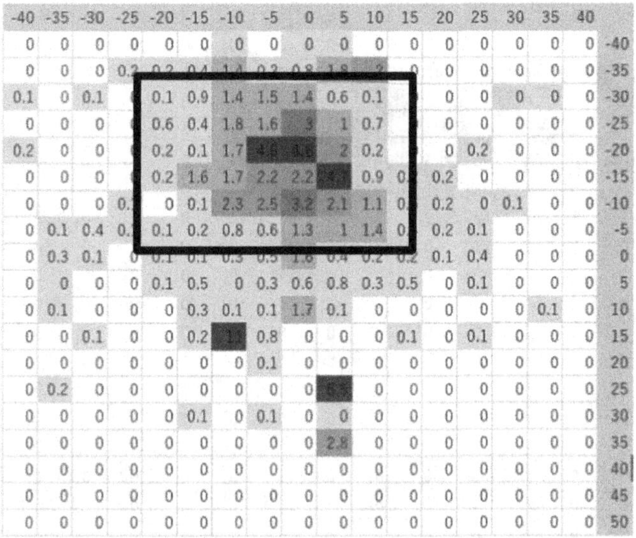

-40	-35	-30	-25	-20	-15	-10	-5	0	5	10	15	20	25	30	35	40	
0	0	0	0	0	0	0	0	0	0	0	0	0	0	0	0	0	-40
0	0	0	0.3	0.2	0.4	1.1	0.2	0.8	1.8	2	0	0	0	0	0	0	-35
0.1	0	0.1		0.1	0.9	1.4	1.5	1.4	0.6	0.1		0	0	0	0	0	-30
0	0	0		0.6	0.4	1.8	1.6	3	1	0.7		0	0	0	0	0	-25
0.2	0	0		0.2	0.1	1.7	4.6	3.6	2	0.2		0	0.2	0	0	0	-20
0	0	0		0.2	1.6	1.7	2.2	2.2	4.1	0.9	0	0.2	0	0	0	0	-15
0	0	0	0.	0	0.1	2.3	2.5	3.2	2.1	1.1	0.	0.2	0	0.1	0	0	-10
0	0.1	0.4	0.	0.1	0.2	0.8	0.6	1.3	1	1.4	0	0.2	0.1	0	0	0	-5
0	0.3	0.1	0	0.1	0.1	0.3	0.5	1.6	0.4	0.2	0.2	0.1	0.4	0	0	0	0
0	0	0	0	0.1	0.5	0	0.3	0.6	0.8	0.3	0.5	0	0.1	0	0	0	5
0	0.1	0	0	0	0.3	0.1	0.1	1.7	0.1	0	0	0	0	0	0.1	0	10
0	0	0.1	0	0	0.2	1.1	0.8	0	0	0	0.1	0	0.1	0	0	0	15
0	0	0	0	0	0	0	0.1	0	0	0	0	0	0	0	0	0	20
0	0.2	0	0	0	0	0	0	0	6.5	0	0	0	0	0	0	0	25
0	0	0	0	0	0.1	0	0.1	0	0	0	0	0	0	0	0	0	30
0	0	0	0	0	0	0	0	0	2.8	0	0	0	0	0	0	0	35
0	0	0	0	0	0	0	0	0	0	0	0	0	0	0	0	0	40
0	0	0	0	0	0	0	0	0	0	0	0	0	0	0	0	0	45
0	0	0	0	0	0	0	0	0	0	0	0	0	0	0	0	0	50

Fig. 5. Heat-map of the average gaze fixation rates in ii-kasa condition

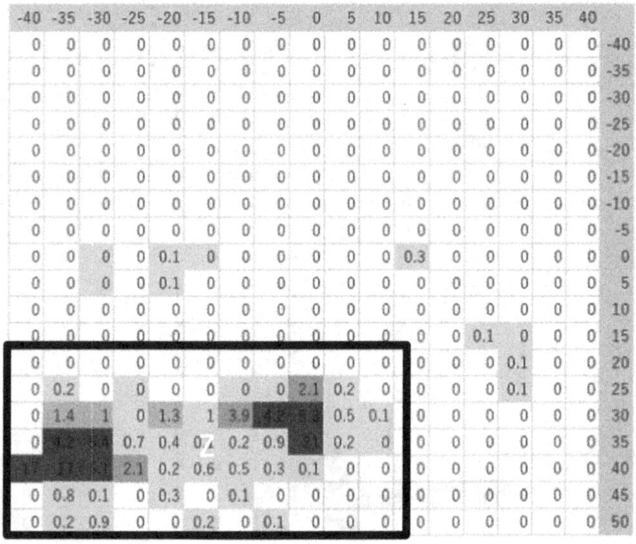

-40	-35	-30	-25	-20	-15	-10	-5	0	5	10	15	20	25	30	35	40	
0	0	0	0	0	0	0	0	0	0	0	0	0	0	0	0	0	-40
0	0	0	0	0	0	0	0	0	0	0	0	0	0	0	0	0	-35
0	0	0	0	0	0	0	0	0	0	0	0	0	0	0	0	0	-30
0	0	0	0	0	0	0	0	0	0	0	0	0	0	0	0	0	-25
0	0	0	0	0	0	0	0	0	0	0	0	0	0	0	0	0	-20
0	0	0	0	0	0	0	0	0	0	0	0	0	0	0	0	0	-15
0	0	0	0	0	0	0	0	0	0	0	0	0	0	0	0	0	-10
0	0	0	0	0	0	0	0	0	0	0	0	0	0	0	0	0	-5
0	0	0	0	0.1	0	0	0	0	0	0	0.3	0	0	0	0	0	0
0	0	0	0	0.1	0	0	0	0	0	0	0	0	0	0	0	0	5
0	0	0	0	0	0	0	0	0	0	0	0	0	0	0	0	0	10
0	0	0	0	0	0	0	0	0	0	0	0	0	0.1	0	0	0	15
0	0	0	0	0	0	0	0	0	0	0	0	0	0	0.1	0	0	20
0	0.2	0	0	0	0	0	0	2.1	0.2	0	0	0	0	0.1	0	0	25
0	1.4	1	0	1.3	1	3.9	4.2	5.5	0.5	0.1	0	0	0	0	0	0	30
0	4.2	4.4	0.7	0.4	0.7	0.2	0.9	3.1	0.2	0	0	0	0	0	0	0	35
3.1	1.9	5	2.1	0.2	0.6	0.5	0.3	0.1	0	0	0	0	0	0	0	0	40
0	0.8	0.1	0	0.3	0	0.1	0	0	0	0	0	0	0	0	0	0	45
0	0.2	0.9	0	0	0.2	0	0.1	0	0	0	0	0	0	0	0	0	50

Fig. 6. Heat-map of the average gaze fixation rates in the smartphone condition

watched only the screen, he did not watch only the screen but also the wide area outside of the screen (Fig. 8). Though Participant 2 watched a little bit wider area than Participant 1 in the smartphone condition (Fig. 10), he always watched down the low area and could not pay attention to his circumstances.

-40	-35	-30	-25	-20	-15	-10	-5	0	5	10	15	20	25	30	35	40	
0	0	0	0	0	0	0.17	0	0	0	0	0	0	0	0	0	0	-40
0	0	0	1.18	0.84	0.59	0.42	0.51	1.1	0	0	0	0	0	0	0	0	-35
0.68	0	0.59	0	0	4.22	2.87	6.16	1.94	1.18	0.59	0	0	0	0	0	0	-30
0	0	0	0	2.87	1.52	4.64	5.57	10	2.78	3.38	0	0	0	0	0	0	-25
0.76	0	0	0	1.1	0	2.45	2.53	7.43	3.54	0.84	0	0	0	0	0	0	-20
0	0	0	0	0	0.08	3.29	1.69	6.24	6.41	2.87	0	0	0	0	0	0	-15
0	0	0	0	0	0.42	2.36	0	0.68	0	0.34	0	0	0	0.08	0		-10
0	0	0	0	0	0	0.68	0.08	0.17	0	0	0	0	0	0			-5
0	0.17	0	0	0	0	0	0	0	0.84	0	0	0	1.1	0	0		0
0	0	0	0	0	0	0	0	0	0	0	0	0	0	0	0	0	5
0	0	0	0	0	0	0	0	0	0	0	0	0	0	0	0	0	10
0	0	0	0	0	0	0	0	0	0	0	0	0	0	0	0	0	15
0	0	0	0	0	0	0	0	0	0	0	0	0	0	0	0	0	20
0	0	0	0	0	0	0	0	0	0	0	0	0	0	0	0	0	25
0	0	0	0	0	0	0	0	0	0	0	0	0	0	0	0	0	30
0	0	0	0	0	0	0	0	0	0	0	0	0	0	0	0	0	35
0	0	0	0	0	0	0	0	0	0	0	0	0	0	0	0	0	40
0	0	0	0	0	0	0	0	0	0	0	0	0	0	0	0	0	45
0	0	0	0	0	0	0	0	0	0	0	0	0	0	0	0	0	50

Fig. 7. Heat-map of participant 1's fixation in ii-kasa condition

-40	-35	-30	-25	-20	-15	-10	-5	0	5	10	15	20	25	30	35	40	
0	0	0	0	0	0	0	0	0	0	0	0	0	0	0	0	0	-40
0	0	0	0	0	0	0	0	0	0	0	0	0	0	0	0	0	-35
0	0	0	0	0	0	0	0	0	0	0	0	0	0	0	0	0	-30
0	0	0	0	0	0	0	0	0	0	0	0	0	0	0	0	0	-25
0	0	0	0	0	0	0	0	0	0	0	0	0	0	0	0	0	-20
0	0	0	0	0	0	0	0	0	0	0	0	0	0	0	0	0	-15
0	0	0	0	0	0	0	0	0	0	0	0	0	0	0	0	0	-10
0	0	0	0	0	0	0	0	0	0	0	0	0	0	0	0	0	-5
0	0	0	0	0	0	0	0	0	0	0	0	0	0	0	0	0	0
0	0	0	0	0	0	0	0	0	0	0	0	0	0	0	0	0	5
0	0	0	0	0	0	0	0	0	0	0	0	0	0	0	0	0	10
0	0	0	0	0	0	0	0	0	0	0	0	0	0	0	0	0	15
0	0	0	0	0	0	0	0	0	0	0	0	0	0	0	0	0	20
0	0	0	0	0	0	0	0.2	9.9	0.3	0	0	0	0	0	0	0	25
0	0	0	0	6.2	4.8	19	21	25	2.5	0.5	0	0	0	0	0	0	30
0	0	0	0	0	0	0	0.2	9.9	0.3	0	0	0	0	0	0	0	35
0	0	0	0	0	0	0	0	0	0	0	0	0	0	0	0	0	40
0	0	0	0	0	0	0	0	0	0	0	0	0	0	0	0	0	45
0	0	0	0	0	0	0	0	0	0	0	0	0	0	0	0	0	50

Fig. 8. Heat-map of participant 1's fixation in the smartphone condition

Here, Fig. 10 shows that his gazes are distributed in the left parts. This was caused he hold the smartphone in the left hand and relatively set it at the left side from him.

-30	-25	-20	-15	-10	-5	0	5	10	15	20	25	30	35	40	45	50	
0	0	0	0	0	0	0	0	0	0	0	0	0	0	0	0	0	-40
0	0	0	0	0	6.42	0	0.75	8.3	10.2	0	0	0	0	0	0	0	-35
0	0	0	0	0	0	0	0	0	0	0	0	0	0	0	0	0	-30
0	0	0	0	0	0	0	0	0	0	0	0	0	0	0	0	0	-25
0	0	0	0	0	1.89	14.7	7.92	1.89	0	0	0	0	0	0	0	0	-20
0	0	0	0	0	7.55	2.64	3.02	0.75	10.6	0	0	0	0	0	0	0	-15
0	0	0	0	0	0	1.13	4.53	5.66	2.64	0	0	0	0	0	0	0	-10
0	0	0	0	0	0	0	0	0	0	0	0	0	0	0	0	0	-5
0	0	0	0	0	0	0	0	0	0	0	0	0	0	0	0	0	0
0	0	0	0	0	0	0	0	2.64	0	0	2.26	0	0	0	0	0	5
0	0	0	0	0	0	0	0	0.38	0	0	0	0	0	0	0	0	10
0	0	0	0	0	0	4.15	0	0	0	0	0	0	0	0	0	0	15
0	0	0	0	0	0	0	0	0	0	0	0	0	0	0	0	0	20
0	0	0	0	0	0	0	0	0	0	0	0	0	0	0	0	0	25
0	0	0	0	0	0	0	0	0	0	0	0	0	0	0	0	0	30
0	0	0	0	0	0	0	0	0	0	0	0	0	0	0	0	0	35
0	0	0	0	0	0	0	0	0	0	0	0	0	0	0	0	0	40
0	0	0	0	0	0	0	0	0	0	0	0	0	0	0	0	0	45
0	0	0	0	0	0	0	0	0	0	0	0	0	0	0	0	0	50

Fig. 9. Heat-map of participant 2's fixation in ii-kasa condition

-40	-35	-30	-25	-20	-15	-10	-5	0	5	10	15	20	25	30	35	40	
0	0	0	0	0	0	0	0	0	0	0	0	0	0	0	0	0	-40
0	0	0	0	0	0	0	0	0	0	0	0	0	0	0	0	0	-35
0	0	0	0	0	0	0	0	0	0	0	0	0	0	0	0	0	-30
0	0	0	0	0	0	0	0	0	0	0	0	0	0	0	0	0	-25
0	0	0	0	0	0	0	0	0	0	0	0	0	0	0	0	0	-20
0	0	0	0	0	0	0	0	0	0	0	0	0	0	0	0	0	-15
0	0	0	0	0	0	0	0	0	0	0	0	0	0	0	0	0	-10
0	0	0	0	0	0	0	0	0	0	0	0	0	0	0	0	0	-5
0	0	0	0	0	0	0	0	0	0	0	0	0	0	0	0	0	0
0	0	0	0	0	0	0	0	0	0	0	0	0	0	0	0	0	5
0	0	0	0	0	0	0	0	0	0	0	0	0	0	0	0	0	10
0	0	0	0	0	0	0	0	0	0	0	0	0	0	0	0	0	15
0	0	0	0	0	0	0	0	0	0	0	0	0	0	0	0	0	20
0	0	0	0.2	0	0	0	0.7	0	0	0	0	0	0	0	0	0	25
0	4	3	0.2	0.3	0	0	0.3	0.3	0	0	0	0	0	0	0	0	30
0	21	27	3.5	1.8	0.3	0.9	2	0	0	0	0	0	0	0	0	0	35
0.3	4.4	12	2	0.2	0.2	1	1.3	0.3	0	0	0	0	0	0	0	0	40
0	4	0.5	0	1.4	0	0.4	0	0	0	0	0	0	0	0	0	0	45
0	1.1	4.3	0	0	0.9	0	0.3	0	0	0	0	0	0	0	0	0	50

Fig. 10. Heat-map of participant 2's fixation in the smartphone condition

This means he would not pay attentions not only to the front to him but also the right side of him.

Bartmann et al. [6] said that the depth of the visual processing and the width of the visual field have the relationship of tradeoff and, if the visual processing are high, much mental load is required and, consequently, the visual field to be able to pay attention to become narrow. Despite the tasks in this experiment were easy and all participants could complete all tasks accurately, the visual field in using smartphone is narrower than the one in ii-kasa. This means people in using ii-kasa require less cognitive load than in using the smartphone

To conclude, these results show ii-kasa reduce not only the risks of texting while walking but also the cognitive load and contributes for people to be able to pay attentions to the circumstances even in using the mobile services while walking.

5   Considerations

We consider that ii-kasa benefits from three aspects, safe, function, and health. From the functional aspects, it is easier to manipulate the smartphone by one hand and ii-kasa prevents the malfunction and the bungles caused by dropping rain droplets on the touch screen. From the healthy aspects, it relieves the loads on the neck and shoulders caused by keeping the head downward.

In this paper, we investigated the safe aspects. In using ii-kasa, we can keep our gazes in front and pay attention to the wider area in our circumstances than the smartphone. In the research for car navigation system [7], it is known that the driver's gazes frequently move from the outside and the navigation system shortly in amounting it at the high position and is safer than in mounting it at the low position. These results and our results show that we will frequently repeat to watch our circumstances and the screen to watch in a short time and it makes us to keep paying attentions to our circumstances when itis located the close area to our head-on the gazes and we can keep our gazes in front. We consider ii-kasa makes us available this behavior and, in this reason, ii-kasa contributes to reduce the risks in texting while walking.

6   Conclusion

In this paper, by analyzing the gaze patterns, we investigate whether ii-kasa reduce the risks in texting while walking. As the results, the average fixation time once in ii-kasa condition is significantly shorter than the one in smartpone condition and its variance is significantly smaller than the one in smartphone condition. This means that people evenly watch the screen and the circumstances shortly at one time, while they gaze the screen deeply in a long time.

The results also showed the gazes were widely distributed not only on the screen but also outside of the screen in ii-kasa condition, while the gaze converged in the narrow area on the screen in the smartphone condition.

It can be said, therefore, that ii-kasa contribute to keep our gaze in front and to pay more attentions to our circumstances than the smartphone and reduce not only the risks of texting while walking but also human cognitive load.

In this research, however, the experiment was conducted in the paths where there is little traffic and which the participants are well known. We also need to investigate the safety in more crowded and the participants' unknown streets.

In this paper, we only investigate the benefits from the safe aspects. In the future works, we need to investigate the benefits from the functional and healthy aspects. From the functional aspect, as we mentioned earlier, we can operate the smartphone by swinging and moving ii-kasa up and down using acceleration sensor. We have already been able to identify the acceleration caused by user's operation from the vibration caused by waling. We, however, we do not investigate its usability. We need to investigate it in the near future.

When we designed ii-kasa aiming to make texting while walking safer, using see-through typed head-mounted display was another candidate. The reason why we did not adopt it is that we do not want the users to immerse the virtual world and the visual distance from eyes to the screen is very short. If it is short, we always need to change our focus and it is hard to recognize both of the information on the screen and the circumstances once. In the near future, we will compare ii-kasa to the see-through typed head-mounted display and hope to indicate the advantages of ii-kasa.

References

1. Mori, S., Oka, M.: Proposal of interaction used umbrella for smartphone. In: Yamamoto, S. (ed.) HIMI 2017. LNCS, vol. 10273, pp. 579–588. Springer, Cham (2017). https://doi.org/10.1007/978-3-319-58521-5_45
2. Cellphone Talkers Get Their Own Sidewalk Lane in D.C. https://www.yahoo.com/tech/cellphone-talkers-get-their-own-sidewalk-lane-in-d-c-92080566744.html
3. Wang, T., Cardone, G., Corradi, A., Torresani, L., Campbell, A.T.: WalkSafe: a pedestrian safety app for mobile phone users who walk and talk while crossing roads. In: HotMobile 2012 (2012). https://doi.org/10.1145/2162081.2162089
4. Kodama, S., Enokibori, Y., Mase, K.: Examination of safe-walking support system for "texting while walking" using time-of-flight range image sensors. In: UbiComp/ISWC 2016, pp. 129–132 (2016). https://doi.org/10.1145/2968219.2971407
5. Ito, A., Yamabe, T., Kiyohara, R.: Study of smartphone operation UI for motorcycles using acceleration sensor. In: DPSWS 2014, pp. 109–114 (2014). (in Japanese)
6. Bartmann, A., Spijkers, W., Hess, M.: Street environment, driving speed and field of vision; Vison in Vehicles-III, pp. 381–389 (1991)
7. Zheng, R., Nakano, K., Ishiko, H., Hagita, K., Kihira, M., Yokozeki, T.: Eye-gaze tracking analysis of driver behavior while interacting with navigation systems in an urban area. IEEE Trans. Hum.-Mach. Syst. **PP-99**, 1–11 (2015). https://doi.org/10.1109/thms.2015.2504083

Evaluation of Discomfort Degree Estimation System with Pupil Variation in Partial 3D Images

Shoya Murakami[✉], Kentaro Kotani, Satoshi Suzuki,
and Takafumi Asao

Department of Mechanical Engineering, Kansai University, Osaka, Japan
{k202465,kotani,ssuzuki,asao}@kansai-u.ac.jp

Abstract. The purpose of this paper was to examine whether the changes in pupil diameter can refrect on the degree of discomfort by various levels of partial 3D images as well as other validated characteristics. Moreover, we discuss the effectiveness of the systems while guiding visual attention by partial 3D images. Images chosen from IAPS (International Affective Picture System) were used to make 3D images. Power spectrum ratio of the pupil variation, called S/C value, generated by stimuli images to those by control images was calculated. The relationship between VAS scores for the impression regarding projected images and the S/C values was set to the major concern for this study. As a result, the average S/C values in 2D neutral images ranged 0.634 to 1.318, whereas the average S/C values in partial 3D neutral images ranged 0.412 to 1.552. VAS scores in 2D neutral images ranged 3.6 to 8.5 and that in partial 3D neutral images ranged 1.2 to 7.4. Moreover, correlation coefficients between VAS scores and S/C values in 2D neutral images was 0.116 and those in partial 3D neutral images was -0.114. In partial 3D images, this negative correlation coefficient was found in consistent with the previous study, whereas the correlation coefficients for both images were relatively low. It was suggested that S/C value was likely to use as a candidate for discomfort measure with a modification of collection technique for VAS scores during the experiment.

Keywords: Partial 3D · Pupil variation · Visual attention guidance

1 Introduction

In recent years, numerous visual information is sent to us through the medium of outdoor advertising and web pages, however, we cannot process all the information due to limited capacity for visual and memory functions, resulting in ineffective collection of useful information [1]. Providers for such advertising information were eager to optimize the technology to send such information effectively and thus, visual attention guidance has been receiving attention among providers as well as researchers.

The technology of visual attention guidance can make users focusing on an intended part of visual images for a longer period of time than usual and leave a significant impression on users [2]. Therefore, both of users and publishers can get the benefits by using the visual attention guidance [1].

© Springer International Publishing AG, part of Springer Nature 2018
S. Yamamoto and H. Mori (Eds.): HIMI 2018, LNCS 10904, pp. 582–593, 2018.
https://doi.org/10.1007/978-3-319-92043-6_47

However, it was reported that people felt the guidance to be factitive when they found themselves being forced to guide their visual attentions, and that spoiled the impression of the image of information turning into negative impressions [3]. This led to the study about visual attention guidance without users' awareness. Hagiwara et al. [4] reported a study on the visual attention guidance by applying change of colors. However, the disadvantages of this method were that the quality of information at a guidance destination were changed and users may be confused by finding the color of the part changing if they know its original color [1]. Hata et al. [1] also reported a study on visual attention guidance using image resolution control. The disadvantages of this method were that the information was forced to be changed by lowering resolutions except the area of guidance destination. Both of these methods apparently change the quality of information.

In our study, use of partial 3D images was used for a possible solution to maintain the quality of information in applying visual attention guidance. Partial 3D image refers to an image converted to 3D by augmenting cross parallax to a specific area of the 2D image [5]. To eliminate any undesired parallax at the other areas, a fixed level of non-cross parallax is augmented [5]. It was reported that the area augmented parallax was being focused on [6] and such cases were happened regardless of whether the viewers were aware of 3D at the area [7]. Therefore, use of partial 3D images was a potential alternative to make users guide their visual attentions without giving a feeling like "forcing to be guided". By using this method, the modification of information at a guidance-destination was minimum and the information shown other than the area of a guidance destination stayed unchanged.

However, how much negative feeling can be surpressed by using the partial 3D images should be evaluated. This study focused on the effect of use of partial 3D images for avoiding unpleasant feelings about visual attention guidance.

The final purpose of our study is to investigate usefulness of the unaware visual attention guidance by partial 3D images by comparing the degree of discomfort between the aware and unaware visual attention guidances. However, the subjective evaluation may suffer from lower reliability of unpleasant feelings because people have different standards for positive or negative feelings between aware and unaware visual attention guidance. Thus, it is necessary to construct the objective evaluation to estimate the degree of discomfort.

In this study, we used the pupil diameter variation to estimate the emotion. The pupil is controlled by the automatic nervous systems and reflects emotional variation [8]. Kawai et al. [9] reported that the pupil variation was changed by whether positive or negative feelings induced by images visually given to the participants.

To estimate the degree of discomfort, based on the previous study [9], we built the systems to estimate the degree of discomfort with the power spectrum of the pupil variation.

The purpose of this paper is to examine whether or not the evaluation systems for the degree of discomfort is valid at partial 3D images. We further discuss the effectiveness of the systems while guiding visual attention by using partial 3D images.

2 Pupillary Response

When the light with constant brightness is given to the pupil, a pupil contraction is generated [8]. If the light level becomes low, the pupil size is back to the normal size [8]. The higher the light level is, the larger the amount of the pupil contraction is and the longer the pupil stays in its contraction [8]. The pupil size gets smaller when the light level gets high [8]. In the pupillary response to light, in case the light level is gradually increasing, the pupil size does not vary [8]. However, in case of the low light level, the pupil size varies when the light level suddenly increases [8]. In case the light stimulus keeps being given to the pupil with a duration of one second, the pupil generally starts contracting after the latency for 0.2–0.3 s [10]. Moreover, the pupillary contraction reaches to the maximum a second after the onset of the light capture, then the pupil dilates and it is back to normal [10].

Kawai et al. [9] calculated the power spectrum ratio of the pupil diameter variation while presenting control images (non-stimuli) and stimuli (positive, neutral, negative) images. As a result, they found that the power spectrum ratio in each stimulus was different below the frequency components of 0.4 Hz [9]. Moreover, Kawai et al. [11] reported that the pupil diameter was contracted after presenting positive images and dilated after presenting negative images. However, Hess [12] reported that the pupil was dilated not only when we were being given the workload but also the arousal was increased, and the pupil was contracted when the arousal was decreased. Murakawa et al. [13] reported that the pupil diameter was dilated when the visual stimuli were of particular interests. Thus, it is important to give participants the workload and the stimuli, which gives minimum interest to focus on positive/negative impressions to participants.

3 Methods

3.1 Participants

Five undergraduate students (M_{age} = 21.8 years; 5 males) participated in this experiment. All participants had corrected-to-normal vision.

3.2 Experimental Setup

Figure 1 shows the schematic view of experimental device. Pupillary responses, specifically pupil area, were recorded by using an eye-tracker (EyeLink II, SR-Research) at 250 Hz of sampling rate. EyeLink II was a head-mounted device, as shown in Fig. 2 and detected the pupil by infrared ray radiation. When infrared rays reached the iris, the part except for the pupil reflected the rays [14]. The eye tracker detected the non-reflection part as the pupil and generated output the area as pixel data. Camera units provided an accurate measure of pupil size across variations in eye shapes and camera angles.

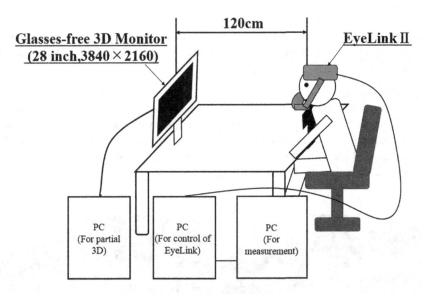

Fig. 1. Schematic view of experimental device

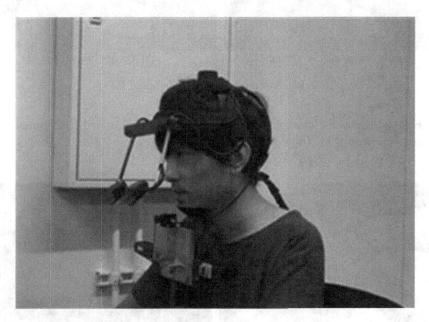

Fig. 2. Posture during experiment using eye tracker

A total of four neutral images were used in this experiment. The images were chosen from the IAPS (International Affective Picture System) [15]. They were expressed by three criteria (valence, arousal, and dominance), each of which were evaluated with nine grades. Based on the previous study [16], using these three criteria, the neutral images were chosen as valence of 5.0, arousal of 3.0 and dominance of 6.0 in this study. The size of visual stimulus was set to 1024 × 768 pixels. The distance from participants to the display was 120 cm. 3D images were displayed with a 4 K glasses-free 3D monitor (Let's Corporation).

Image processing

Control and partial 3D images: The pupil diameter varied with the luminance change of the presented image. Therefore, the control image of each stimulus image was created such that there were brightness differences between the control and stimulus images. Figure 3 shows the example of control images. Control images were generated by using the procedure by kawai et al. [9].

(a) Stimulus image selected from the IAPS   (b) Control image

Fig. 3. Conversion to control image

Figure 4 shows the example of partial 3D images. In this study, the partial 3D images were created by Tridef 3D Photo Transformer.

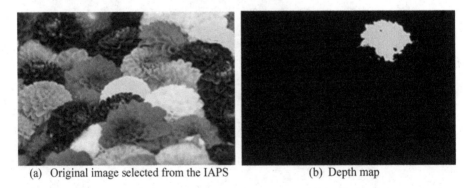

(a) Original image selected from the IAPS   (b) Depth map

Fig. 4. Sample image of partial 3D conversion, the area for 3D conversion is shown in grey on depth map (right)

3.3   Experimental Procedure

Before the experiment, we showed participants several partial 3D images to minimise their emotional changes caused by the initial interest in partial 3D. Figure 5 shows the time chart of the experimental session including the time sequence for presenting images. In the experiment, this session was performed for each image. A control image, two stimuli images (2D, partial 3D) and three masking images were presented in each session. Participants were allowed to blink while the monitor was displaying the masking images, but we instructed them not to blink while the monitor was displaying the control images and stimuli images in order to obtain secure data for pupil area. Participants were asked to watch the images while this session. After finishing all the sessions, Visual Analogue Scale (VAS) was given to participants for collecting the degree of positive/negative feelings for the images.

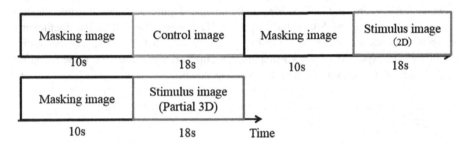

Fig. 5. Time chart of the experimental session

3.4   Data Analysis

The pupil area partially lost by blink were linearly interpolated between 100 ms around the time of blink based on the previous study [17] and a 5-points moving average was applied to the time course of the pupil areas for smoothing [9]. Fast Fourier Transformation for 2048 units in time-series data of the pupil area (8.192 s) was proceeded. Latency for the changes in pupil diameter was reported as 0.2–0.3 s by Kondo et al. [18]. We, however, set the latency for two seconds after presenting the images since the pupil variation was affected by the emotional changes generated by bright changes when images were changed. Based on the previous study [9], the power spectrum ratio of the pupil variation in stimuli images to that in control image was calculated from Eq. (1).

$$S/C = \frac{\sum_{f=0.3Hz}^{1.6Hz} P(t)}{\sum_{f=0.3Hz}^{1.6Hz} P(c)} \tag{1}$$

P(t):   Power spectrum of the pupil variation caused by presenting stimulus image
P(c):   Power spectrum of the pupil variation caused by presenting control image

Generally, it was suggested that the peak frequency range for most pupil responses were lower than 1.6 Hz [14]. Takahashi and colleague also reported that pupil variation contained 0.05–0.3 Hz of frequency components was considered as pupillary noise [14]. Thus, the frequency range of the pupil variation was defined between 0.3 and 1.6 Hz in this study. In the previous study [9], when positive or neutral stimuli were presented, S/C value showed below 1.0 and when negative stimulus was presented, it was reported that S/C value was above 1.0. In this study, the relationship between VAS score and S/C value was quantitatively evaluated.

4   Results

Figure 6 shows temporal changes in Participant A's pupil area. Figure 7 shows the power spectrum of Participant A's pupil area variation. Figure 8 shows the relationship between VAS score and S/C value in 2D image. Figure 9 shows the relationship between VAS score and S/C value in partial 3D image. The colleration coefficients between VAS score and S/C value in Figs. 8 and 9 were 0.116, −0.114, respectively, which were unexpectedly low. Participant E's pupil area was not able to be recorded while Image d was presented, due to excessive noises.

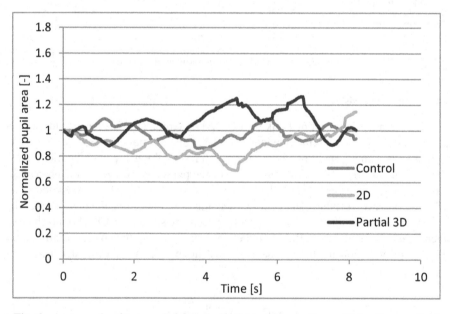

Fig. 6. An example of temporal change in Participant A's pupil area (Color figure online)

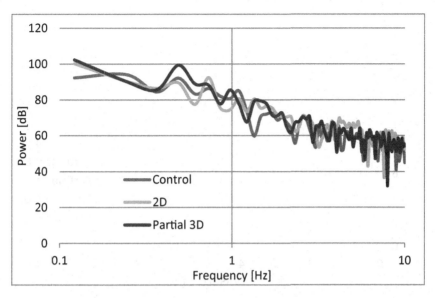

Fig. 7. Power spectrum of (Participant A) pupil area variation (Color figure online)

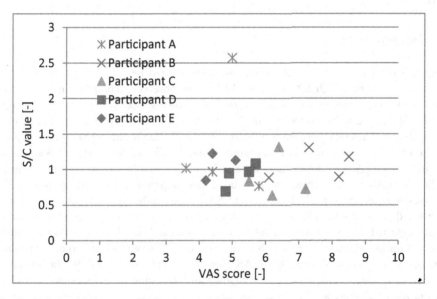

Fig. 8. Relationship between VAS scores and S/C values for each participant when participants monitored in 2D images

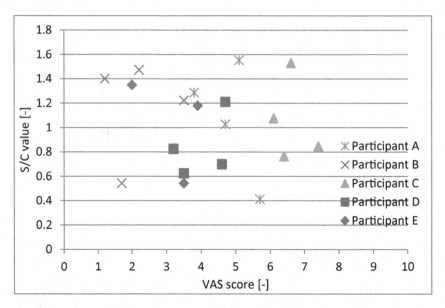

Fig. 9. Relationship between VAS scores and S/C values for each participant in partial 3D images

5   Discussion

Figure 8 revealed that most S/C values were distributed near 1.0 except for the participant A, where a high S/C value (= 2.573) was observed. The result can be interpreted that the emotion was hardly varied by showing neutral images. It was also observed that S/C values were not correlated with VAS scores when 2D neutral images were shown. These unexpected results were possively due to the potential polarization caused by glasses-free 3D monitor. Figures 8 and 9 indicated that some VAS scores in partial 3D images were lower than those in 2D images and S/C values did not seem to relate to VAS scores. Thus, we concluded that impression induced by partial 3D images may affect both VAS scores and S/C values. When partial 3D images were presented, relatively high S/C values were obtained at low VAS score conditions and it was suggested that S/C value was likely to use as a candidate for discomfort measure. Reason for independency between S/C values and VAS scores when partial 3D neutral images were presented was that VAS scores did not correspond to the intended responses because VAS scores were obtained after finishing all the sessions, resulting in unfocused responses.

Figure 10 shows the relationship between VAS scores and S/C values (except for a high value outliers) of each image when 2D images were given. Figure 11 shows the relationship between VAS scores and S/C values of each image when partial 3D images were given. Comparing Figs. 10 and 11, S/C values in Image *a* and Image *c* tended to be above 1.0 and the values in Image *a* and Image *c* tended to be below 1.0 in Fig. 11. Thus, it was concluded that the compartments given by partial 3D conversion may affect their feelings.

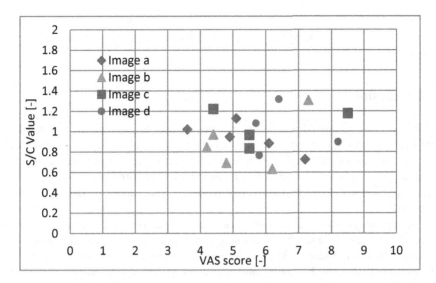

Fig. 10. Relationship between VAS scores and S/C values for each 2D images

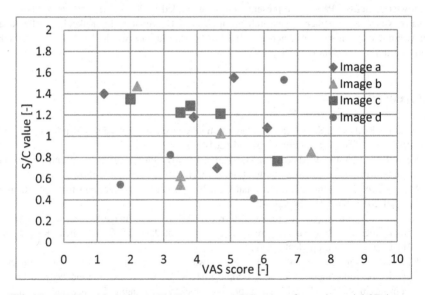

Fig. 11. Relationship between VAS scores and S/C values for each partial 3D images

In this study, whether or not participants made a specific impression, irrespective of comfort or discomfort, in neutral images was not clarified. It was possible that participants happened to gaze at the partial 3D image, which made such specific impression although selected images were designed to be free from impression of comfort or discomfort for all participants. Thus, it is possible that the responses obtained by VAS was not matched with the intended responses associated with changes in pupil sizes.

6 Conclusion

In this study, we examined whether or not the systems to estimate the degree of dis comfort based on the previous study [9] was valid at partial 3D image presentations. Moreover, we discussed the effectiveness of the systems while guiding visual attention by partial 3D images. As a result, VAS scores did not have correlation with S/C values. Therefore, it was not clear whether or not the systems were valid at partial 3D images at this point.

The goal of our study was to investigate usefulness of the visual attention guidance by partial 3D image. The results cannot judge whether the negative emotion was due to the visual attention guidance. Thus, we need to evaluate the effectiveness of the systems to the partial 3D with modification of strategy to obtain VAS scores more clearly to show their impressions for each image.

Acknowledgements. Part of the present study was funded by Environmental control based on human environment interaction research group Kansai University, and Kakenhi of the Japan Society for the Promotion of Science (17H01782). The authors would like to thank Keitaka Akamatsu during the data collection.

References

1. Hata, H., Koike, H., Sato, Y.: Visual attention guidance using image resolution control. 18th Inf. Process. Soc. Japan Interact. 57–64 (2014)
2. Veas, E.E., Mendez, E., Feiner, S.K., Schmalstieg, D.: Directing attention and influencing memory with visual saliency modulation. In: CHI 2011, pp. 1471–1480 (2011)
3. McCay-Peet, L., Lalmas, M., Navalpakkam, V.: On saliency, affect and focused attention. In: CHI 2012, pp. 541–550 (2012)
4. Hagiwara, A., Sugimoto, A., Kawamoto, K.: Saliency-Based Image Processing for Guiding Visual Attention. Information Processing Society of Japan SIG Technical Report, vol. 2011-CVIM-177(5), pp. 1–8 (2011)
5. Kim, S., Takahashi, M., Watanabe, K., Kawai, T.: The effects of functional binocular disparity on route memory in stereoscopic images. In: IS&T International Symposium on Electronic Imaging, vol. 2016, no. 5, pp. 1–6. Society for Imaging Science and Technology (2016)
6. Koido, Y., Kawai, T.: Cognitive characteristics of partial stereoscopic image conversion. Papers of Japan Ergonomics Society Proceeding In: The 52th Conference of Japan Ergonomics Society, pp. 252–253. Japan Ergonomic Society (2011)
7. Kandachi, H., Koido, Y., Kim, S., Morikawa, H., Mitsuya, R., Kawai, T., Watanabe, K.: Partially converted stereoscopic images and the effects on attention and memory. Ergon. Jpn. Ergon. Soc. 49(Supplement), 34–35 (2013)
8. Matsunaga, K.: Psychology of the Pupil Movement. pp. 7–30. Nakanishiya publishing (1990)

9. Kawai, S., Takano, H., Nakamura, K.: Development of Emotional Estimation System by Pupil Diameter Variation. The Institute of Electronics, Information and Communication Engineers Technical Report MBE, ME and Bio Cybernetics, 112(3), pp. 7–11 (2012)
10. Lowenstein, O., Loewenfeld, I.E.: The pupil. In: Davson, H. (ed.) The Eye, vol. 3, pp. 255–337. Academic Press, New York (1969)
11. Kawai, S., Takano, H., Nakamura, K.: Emotional Estimation by Pupil Diameter Variation. 11th Forum Inf. Technol. **11**(3), 421–422 (2012)
12. Hess, E.H.: Attitude and Pupil Size. Sci. Amerian **212**(4), 46–54 (1965)
13. Murakawa, S., Nishina, D., Ueki, M., Yokota, M.: Relationships between the image features on a river landscape. Trans. AIJ J. Archit. Plan. Environ. Eng. **64**(524), 53–60 (1999)
14. Takahashi, K., Tsukahara, M., Toyama, K., Hisada, M., Tamura, H.: Neural Networks and Biological Control, pp. 224–233. Asakura shoten, Tokyo (1976)
15. Lang, P.J., Bradley, M.M., Cuthbert, B.N.: International Affective Picture System (IAPS): Affective ratings of pictures and instruction manual. Technical Report A-8, University of Florida (2008)
16. Yamakawa, Y., Takano, H., Nakamura, K.: Estimation of trait anxiety using pupil diameter variation during visual stimuli. Proc. 13th Forum Inf. Technol. **13**(2), 373–374 (2014)
17. Nakamori, S., Mizutani, N., Yamanaka, T.: Changes in pupil size to visually liking for pictures of faces. Int. J. Affect. Eng. **10**(3), 321–326 (2011)
18. Kondo, Y., Nishimura, Y., Ishii, H., Shimoda, H., Yoshikawa, H.: A study on an objective examination method of eye strain by using eye-sensing display. Hum. Interface Symp. **2**, 643–648 (2006)

Can I Talk to a Squid? The Origin of Visual Communication Through the Behavioral Ecology of Cephalopod

Ryuta Nakajima[✉]

University of Minnesota Duluth, Duluth, USA
lazymonk22@gmail.com

Abstract. The quest of modernity has come to its final phase in the form of post modernism. Many past attempts to define "individualism" and "self" encountered the wall of linguistics structure and categorization, the governing principals of human consciousness. Postmodernism tends to recycle the façade of preexisting methods and theories, thereby creating fragmentation and dislocation. Simultaneously, computer technology is rapidly reshaping our visual culture by offering more streamlined production and distribution possibilities. Considering this environment, it is essential to investigate its effect and implication on the visual culture, by asking existential questions such as: Why do we make images? Where do they come from and what is their primary function? In order to pursue these rather difficult questions, my work focuses on the adaptive coloration of cephalopods' (squid, octopus and cuttlefish) as comparative models that can code and re-map visual information such as paintings, photographs, and videos. The genetically and evolutionarily pure empirical data of the squid and cuttlefish not only uncover certain key information needed to understand the origin of visual communication, but also function as a catalyst that can redirect our culture away from the over-stimulated hyper reality. This, in turn, can create a valuable interdisciplinary platform to discuss the current trends in both art and science.

Keywords: Cephalopod · Contemporary art · Communication
Zombies · Cuttlefish and body pattern

1 Introduction

Body patterns play an important role in predator/prey interactions, such as crypsis (Endler 1978) and disruption (Cott 1940) for many animals. They are also used for inter- and intraspecific communication such as agonistic or mating display. While most animals have unchangeable to slightly changeable body patterns (Cott 1940), coleoid cephalopods such as the octopus, squid and cuttlefish can rapidly change their body colors and textures, exhibiting unrivaled speed, complexity and variety of appearances. These appearances, for camouflage and communication, consist of a combination of chromatic, textural, postural and locomotor components (Hanlon and Messenger 1988; 1996; Moynihan 1985; Packard 1972; Packard and Hochberg 1977). Chromatic components are produced by neurally controlled ink-filled organs called

© Springer International Publishing AG, part of Springer Nature 2018
S. Yamamoto and H. Mori (Eds.): HIMI 2018, LNCS 10904, pp. 594–606, 2018.
https://doi.org/10.1007/978-3-319-92043-6_48

chromatophore (Messenger 2001), which are connected to a set of radial muscles that expand and contract the pigmented sac, changing its affective surface area (Hanlon 1982). Neurally controlled light reflective cells, iridophores, produce blue, green and pink colors while leucophores produce white color (Messenger 2001; Wardill et al. 2012). In the case of octopuses and cuttlefishes, chromatic components are also enhanced by altering physical skin texture from smooth to three dimensional by controlling dermal muscles called Papillae (Holmes 1940; Packard and Hochberg 1977). Postural components are defined by positional orientation of flexible muscular arms, tentacles, mantle, head and fins (Packard and Sanders 1971). Locomotor components are expression and movement of the entire body of the individual (Roper and Hochberg 1988). Each of these components can appear for seconds (acute) or for hours (chronic) and can be displayed in wide varieties of combinations to create the total appearance of the animal (Packard and Hochberg 1977; Hanlon and Messenger 1996).

My discovery of the remarkable ability of the cephalopods to change body pattern helped me realize uncanny similarities between the process of layer painting, which fuses multiple independent layers of information into a singular and comprehensive pictorial plane and the structural and cognitive process of cephalopods' camouflage, which is also composed of multiple layers of chromatophores creating a singular whole by filtering information provided by its environment. In short, cephalopods' camouflage parallels the process of painting and other image-making practices. Although the objectives for creating images are very different from each other—survival and reproduction for cephalopods and aesthetics and metaphysics for artists—the fundamental triple-step structure (exterior information, individual interpretation and visual output) remains similar between them.

In the past 9 years, I have conducted many types of body pattern experiments with pharaoh cuttlefish (Sepia pharaonis) as a study subject at the National Resource Center for Cephalopods in Galveston, Texas and at the University of the Ryukyus in Okinawa. In these experiments, I replaced the sediment found in the natural habitat of the pharaoh cuttlefish, such as rocks, sand and seaweed, with 20th century paintings, photographic documentations of 20th century events, and short videos, in order to solicit its camouflage behavior. The cuttlefish responded to visual information from each image by interpreting visual attributes of the image into artificially triggered camouflage patterns, which were photographed and video recorded for further analysis. Furthermore, the data gathered from the analysis was used as the fundamental visual structure informing my creative works.

Through this project, I seek to revisit our cultural past through the eyes of a cuttlefish. Our raison d'être and vision of the future rely heavily on reevaluation of the past, which relies, in turn, on the idea of the linear progress of time. However, as Aby Warburg, art historian and founder of Warburg Institute, suggested at the turn of the 20th century, the cultural past of humanity may have followed a much more complex, nonlinear path that intertwines diverse cultural differences with time. The behavioral ecology of cephalopods as an interpretive cultural model not only brings together two different academic fields as a hybrid model, but also may present new and exciting insights into our cultural past. In these ways, this project attempts to present an alternative linguistic structure of the visual language that opens up future possibilities in art and humanity.

1.1   Model of Representation

Umwelt, self-world (Von Uexküll and Mackinnon 1926), is a cognitive reconstruction of the information or signs gathered from our sensory organs. Each biological sensor (visual, audio, tactile, temperature, olfactory, etc.) is specifically tuned to select important exterior and environmental information which is vital to our own survival. This information is, then, sent to the brain for processing and codifying for simple rapid output response to the exterior world (Okutani 2013). Artistic creation, like painting, according to Paul Klee, is a mere reflection of the environment that surrounds the artist (Klee 1924). Hence, it is the representation of the *Umwelt* of the artist and the artist's function is in the translation rather than the creation. Similarly, cephalopods have their own *Umwelt* that is described by their sensory system, their brain and their output to the world in visually traceable body pattern responses. In these ways, both artists and cephalopods are producing visual representation of their *Umwelt*. If this hypothesis is true, then, using a comparative study of these two different visual response bio-systems, it is probable to detect the fundamental visual schematic of the total reality, which may otherwise be unattainable due to the limitation of both sensory systems and cognitions (Fig. 1).

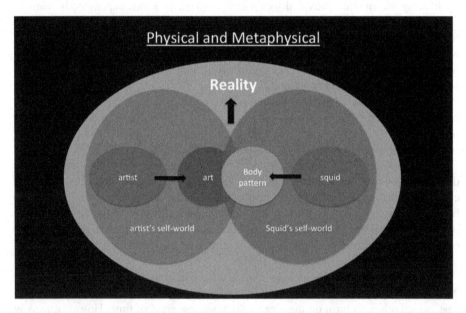

Fig. 1. The diagram is showing the relationship between the self-world of artist and squid and how the overlapping area may direct towards the larger reality. (Color figure online)

1.2   Can a Cuttlefish Match Paintings?

Upon conducting the experiment of replacing a cuttlefish's natural substrate (sand, rocks, corals, etc.) with varieties of 20th century paintings from the geometric abstraction of Piet Mondrian to the Japanese anime-influenced "super flat" paintings of

Takashi Murakami, it is clear that the cuttlefish is able to match its' body appearance to the given images. This experiment, which was conducted from 2009 through 2011, tested over 100 individual animals with 80 stylistically unique paintings and yielded hours of video data. A careful review of still images taken from the videos showed certain specific trends. Due to their physiological limitation of chromatophores and iridophores, the cuttlefish could not produce a highly saturated color combination (Fig. 2).

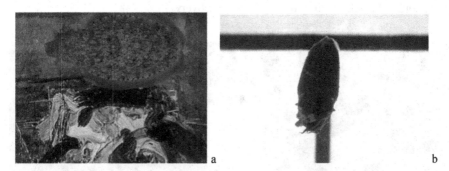

Fig. 2. These images show *Sepia pharaonis* matching its body pattern to Frank Auerbach (a) and Piet Mondrian (b)'s painting. (Photographed by Ryuta Nakajima.)

In addition, their responses to dangerous predators depicted in some of the paintings did not include defensive postures that the cuttlefish would exhibit in nature. On the other hand, cuttlefish performed very well to both gestural and geometric abstraction in paintings, matching their patterns more accurately to the paintings. Such paintings are constructed mainly with formal measures, qualities and weights, such as line, tone, colors, shapes, hues, and salutations, without, in most cases, any figure or object representation (Klee 1924). Ironically, this lack of specific objects and figures produces images that resemble natural landscapes or cityscapes and contain organic elements. It is precisely within these natural elements, organized according to a certain aesthetic and geometric principle, that the cuttlefish exhibited an ease in matching its body patterns, which are a simplified and geometrically organized representation of Nature. It is also important to note that there seems to be a certain visual threshold where the cuttlefish differentiates between animate and inanimate. For instance, the painting of Frank Auerbach is a gestural, figurative abstraction in which most viewers see a figure; the reaction of a group of cuttlefish indicated more of a landscape with no sense of a figure.

1.3 Cephalopod Art

The data collected are used to produce various types of artwork for exhibition. In the early stage, paintings and drawings were used to track the body pattern change of the cuttlefish in relationship with the substrate images. After gaining access to laboratory animals, the visualization methods changed to predominantly photograph, video and sculpture installation. One of the most comprehensive solo exhibitions was produced in collaboration with the Minneapolis Institute of Art [MIA] and the University of the Ryukyu Cephalopod Research Laboratory. In this exhibition, there were six large-scale photographs (Fig. 3) of the cuttlefish disguises formed in response to selected art works from the MIA collection, 51 cuttlefish sculptures with images taken from the videos painted on them as their body patterns. This exhibition created a sense of synergy between art and science, nature and culture, animate and inanimate. In this way, it displayed the interconnection between nature and culture by using a cephalopod as a metaphorical vehicle inviting curiosity and discovery.

Fig. 3. Large-scale photograph produce for an exhibition at Minneapolis Institute of Art.

1.4   Creating a Catalog of Body Patterns of the Pharaoh Cuttlefish

Over 100 species of cuttlefish have been described (Jereb et al. 2005). The pharaoh cuttlefish, *Sepia pharaonis* (Ehrenberg 1831 1999), is distributed in tropical coastal waters in the Indo-Pacific region (Norman and Reid 2000; Nabhitabhata and Nilaphat 1999). Although there were previous studies done of the species, many focused on the camouflage and its visual cues with no extensive catalogue of body patterns that could be used for more accurate analysis of the responses that further experiments required. In light of this, I constructed a catalogue of the chromatic, postural and locomotor behaviors for use as a tool in behavioral monitoring, quantitative analysis, and species identification. I consolidated the data from 2010, 2011, 2012 and 2014 for a total of 325 HD videos and 9,799 still images obtained and analyzed. Through this process, 53 chromatic, 3 textual, 11 postural and 9 locomotor components were identified and described (Nakajima and Ikeda 2017). This study was conducted in an artificial environment and the complete repertoire of *S. pharaonis* body pattern may be larger than what the study describes. However, compared with studies of other related cuttlefish species and their ethogram, the number of components represented are compatible and reasonable (Fig. 4).

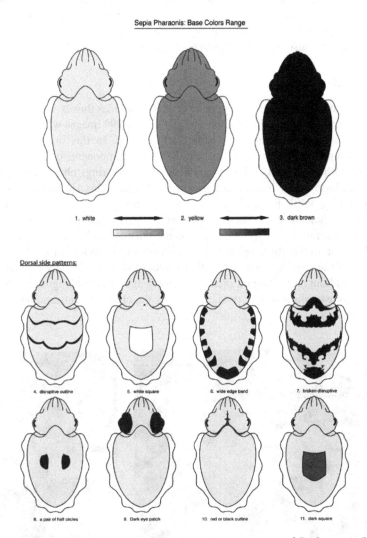

Fig. 4. An example of illustration describing chromatic components of *S. pharaonis* (Nakajima and Ikeda 2017)

2 Communication

The catalogue clearly represents the remarkable variety of body pattern expression that *S. pharaonis* can produce. However, isolating body pattern, that used for camouflage can be categorized into three major forms: uniform, mottled and disruptive (Hanlon and

Messenger 1988; 1996) with variations within each category (Zylinski et al. 2009), which may not require the entire collection of chromatic components and possible body patterns. The natural progression of question would be to ask What are all these components for and what are the cues that trigger the appearance of them? and ultimately asking such questions as Are they using body pattern for communication and is there an overlap with our communication system? While these questions are certainly important, they are also extremely problematic and abstract.

2.1 Agonistic Behavior

Agonistic behavior is a display of aggression which includes threats, retreats, placation and conciliation (Scoot and Fredericson 1951). Of the 9,799 images which were used to create the catalog, one image was particularly unique. In this image (Fig. 5.) the chromatic components of the cuttlefish and the visual components of the Ngady mask showed incredible similarity. Both contain the dark eye ring, blue mantle-like pink lines on each arm and a mottled pattern on the mantle. This similarity between the cuttlefish and the mask is clearly not manifested from the necessity to camouflage; rather both are functioning as an embodiment of aggression. The chromatic components and the overall body pattern expression are intended to induce certain fear and a startling reaction to the object upon which the pattern is projected. In short, the pattern functions as a trans-species communication symbol that is genetically programed and can be read without depending on rhetorical knowledge of visual language. Hence, this particular agonistic display may possibly possess the clue to understanding the shared visual schematics.

Fig. 5. *S. pharaonis* is displaying fully articulated agonistic body pattern with dark eye ring, blue mantle-like, pink lines on each arm, flat and expanded body shape etc., reacting to Ngady mask show below. There is an uncanny similarity between the chromatic components of the cuttlefish and the mask. (Color figure online)

2.2   Zombies' Chromatic Components Analysis

In order to further understand the relationship between agonistic display found in both cuttlefish body pattern and Ngady mask, zombies were used as a comparative model. Zombies are an imaginary construct that is designed to induce an interspecific emotional response of fear. Zombie folklore originated in Haiti in the 17th century when African slaves were brought to Haiti to work on the sugar cane plantations. Although there are many versions of cultural fascination with reanimated human corpses, for this study, 100 zombie samples are taken from the Post George Romero's film *Night of the Living Dead,* which later became an American pop culture icon. The sampled Zombies are from commercial films, Halloween costumes, makeup tutorials, etc., from an Internet image search. Every image was deconstructed and analyzed for chromatic components (Fig. 6).

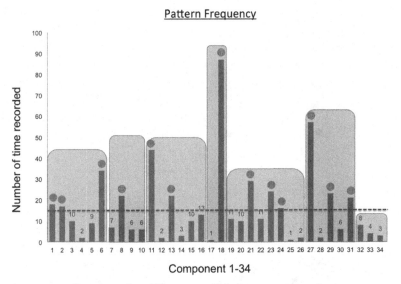

Sample number: 100, average: 15.9, 6 components per image

Fig. 6. Zombie chromatic components frequency chart

34 chromatic components (Fig. 7.) were detected which were divided into 7 facial zones, where each component is mutually exclusive within the zone. The average number of occurrences is 15.9 times with an average number of chromatic components per image of 6 components. The component with the highest number of occurrences was #18, dark eye patches, at 87 times. 12 components (1, 2, 6, 8, 13, 18, 21, 23, 24, 27, 29 and 31) were above the mean of which 6 components had significantly higher occurrences within the zone. Based on this simple observation and analysis, it appears that there are four dominant components (pale complexion, dark eyes, white pupils and bloody mouth) that are absolutely necessary to create a zombie and four more components (missing lips, dark lips, cuts, and decomposition) that are subdominant and

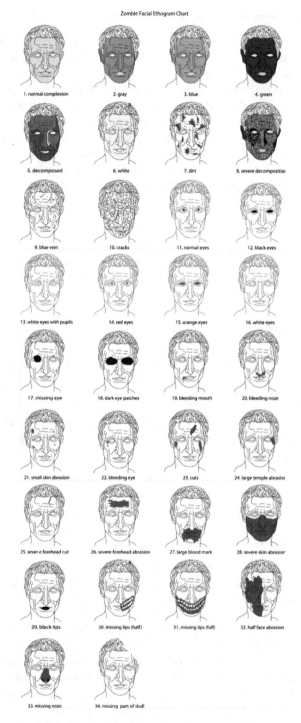

Fig. 7. Zombie chromatic components chart

enhance overall appearances. The four dominant components can be combined together to create a total appearance that might be considered the archetype of zombie, which is the visualization of human interspecific fear. Combining the archetype zombie with four subdominant components produces 12 different overall appearances that cover most of the visualization of zombies produced by the media industries (Fig. 8).

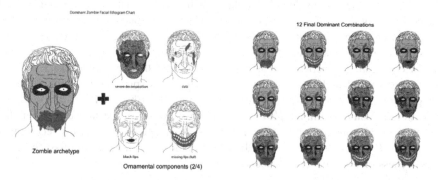

Fig. 8. 4 dominant Zombie chromatic components are combined to produce the zombie visual archetype. Four ornamental components are added to produce 12 possible total appearances.

Comparing the chromatic components of cuttlefish, mask and zombie archetype, there is a considerable similarity between them. Although this is a pseudo-scientific experiment and comparison between these three seemingly dissimilar visual elements, it seems to me that there is a definite collation between chromatic components that induce fear responses across species. These components function as signs or visual cues that communicate potential danger in engaging with the displayer. In this way, it reduces the chance of predation and physical engagement with the receiver of the information. Hence, the body pattern has to be trans-specifically effective as a basic level of inter- and intraspecific communication (Fig. 9).

Fig. 9. Chromatic components are very similar between cuttlefish, zombie and Ngady mask

2.3   Cuttlefish Cosplay

I further investigated the notion shared visual components to induce a sense of fear, I asked Myu Amatsuka, a professional Japanese cosplayer, to recreate 3 selected body patterns of cuttlefish. The objective of this experiment was not communicated to her during this experiment. 3 body patterns were produced and photographed. The result (Fig. 10.) also shows considerable similarity with zombie make-ups. Although, there are evidences of Amatsuka's interpretation has been influenced by *Ika musume* (squid girl), a famous Japanese anime character, three manifestations expressed the gradual transformation of Amatsuka into something resembling zombie like character rather than a cuttlefish. Furthermore, it is important to note that the Amatsuka's postural presentation has also changed as the make up changed expressing a potential shift in her psychological state. Here, the chromatic components have effected not only the viewers reception of them, but also effected the psychology of the subject applying the makeup. It is also interesting that the final outcome (image on the right) is very similar to the Ngady mask as well. From these comparative studies, the trans-specifically effectiveness of body pattern is reconfirmed.

Fig. 10. Cosplayer's interpretation of three distinctive body patterns of cuttlefish.

3   Conclusion

From art to science, my work tries to find a new and more direct communication method. I believe that artwork is not a product of self expression, rather it is a product of biological necessity that helps us connect with our environment. The human sensory system has been fine-tuned in the course of evolution to detect and isolate specific information which is important for spatial navigation, predator prey interaction, selecting mating partner and so on. In order to reduce the margin of error, the system is designed to omit and simplify millions of variables and details that constantly surround us. Similarly, cephalopods, in their self-world, are doing the same. By comparing the human and the cephalopod visual communication systems, I hope to isolate important visual cues that induce certain emotional reactions without depending on linguistic categorization or logical understanding. Further more, if one is able to create a visual communication model that is based on this minimal yet effective visual communication system, that increases the level of empathy and understanding with others that helps to cultivate a new consciousness.

References

Cott, H.B.: Adaptive Coloration in Animals. Methuen, London (1940)

Endler, J.A.: A predator's view of animal color patterns. In: Hecht, M.K., Steere, W.C., Wallace, B. (eds.) Evolutionary Biology. EBIO, vol. 11, pp. 319–364. Springer, Boston (1978). https://doi.org/10.1007/978-1-4615-6956-5_5

Nabhitabhata, J., Nilaphat, P.: Life cycle of cultured pharaoh cuttlefish Sepia pharaonis Ehrenberg, 1831. Phuket Mar. Biol. Center Spec. Publ. **19**, 25–40 (1999)

Hanlon, R.T.: The functional organization of chromatophores and iridescent cells in the body patterning of Loligo plei (Cephalopoda: Myopsida). Malacologia **23**, 89–119 (1982)

Hanlon, R.T., Messenger, J.B.: Adaptive coloration in young cuttlefish (Sepia officinalis L.): the morphology and development of body patterns and their relation to behaviour. Phil. Trans. R. Soc. Lond. B **320**(1200), 437–487 (1988)

Hanlon, R.T., Messenger, J.B.: Cephalopod Behaviour. Cambridge University Press, Cambridge (1996)

Holmes, W.: The colour changes and colour patterns of Sepia officinalis L. J. Zool. **110**(1-2), 17–35 (1940)

Jereb, P., Roper, C.F.E., Vecchione, M.: FAO species catalogue for fishery purposes. In: Cephalopods of the World. Food and Agriculture Organization of the United Nations, Rome, pp. 1–19 (2005)

Klee, P.: On modern art. In: Herbert, R.L. (ed.) (1924)

Messenger, J.B.: Cephalopod chromatophores: neurobiology and natural history. Biol. Rev. **76**(4), 473–528 (2001)

Moynihan, M.: Communication and Noncommunication by Cephalopods. Indiana University Press, Bloomington (1985)

Nakajima, R., Ikeda, Y.: A catalog of the chromatic, postural, and locomotor behaviors of the pharaoh cuttlefish (Sepia pharaonis) from Okinawa Island, Japan. Mar. Biodiv. **47**(3), 735–753 (2017)

Norman, M., Reid, A.: A Guide to Squid, Cuttlefish, and Octopuses of Australasia. CSIRO Publishing, Collingwood (2000)

Okutani, T.: 『日本のタコ学』. Tokai University Press, Hadano-shi (2013)

Packard, A., Sanders, G.D.: Body patterns of Octopus vulgaris and maturation of the response to disturbance. Anim. Behav. **19**(4), 780–790 (1971)

Packard, A., Hochberg, F.G.: Skin patterning in octopus and other genera. In: Symposium of the Zoological Society, London, vol. 38, pp. 191–231 (1977)

Packard, A.: Cephalopod and fish: the limit of convergence. Biol. Rev. **47**, 241–307 (1972)

Roper, C.F.E., Hochberg, F.G.: Behavior and systematics of cephalopods from Lizard Island, Australia, based on color and body patterns. Malacologia **29**(1), 153–193 (1988)

Scott, J.P., Fredericson, E.: The causes of fighting in mice and rats. Physiol. Zool. **24**(4), 273–309 (1951)

Von Uexküll, J., Mackinnon, D.L.: Theoretical biology (1926)

Wardill, T.J., Gonzalez-Bellido, P.T., Crook, R.J., Hanlon, R.T.: Neural control of tuneable skin iridescence in squid. Proc. Roy. Soc. Lond. B: Biol. Sci. **279**(1745), 4243–4252 (2012)

Zylinski, S., Osorio, D., Shohet, A.J.: Cuttlefish camouflage: context-dependent body pattern use during motion. Proc. Roy. Soc. Lond. B: Biol. Sci. **276**, 3963–3969 (2009)

Text and Data Mining and Analytics

Discovering Significant Co-Occurrences to Characterize Network Behaviors

Kristine Arthur-Durett[1], Thomas E. Carroll[1(✉)], and Satish Chikkagoudar[2]

[1] Pacific Northwest National Laboratory, Richland, WA, USA
`{Kristine.Arthur-Durett,Thomas.Carroll}@pnnl.gov`
[2] U.S. Naval Research Laboratory, Washington, DC, USA
`satish.chikkagoudar@nrl.navy.mil`

Abstract. A key aspect of computer network defense and operations is the characterization of network behaviors. Several of these behaviors are a result of indirect interactions between various networked entities and are temporal in nature. Modeling them requires non-trivial and scalable approaches. We introduce a novel approach for characterizing network behaviors using significant co-occurrence discovery. A significant co-occurrence is a robust concurrence or coincidence of events or activities observed over a period of time. We formulate a network problem in the context of co-occurrence detection and propose an approach to detect co-occurrences in network flow information. The problem is a generalization of problems that are encountered in the areas of dependency discovery and related activity identification. Moreover, we define a set of metrics to determine robust characteristics of these co-occurrences. We demonstrate the approach, exercising it first on a simulated network trace, and second on a publicly-available anonymized network trace from CAIDA. We show that co-occurrences can identify interesting relationships and that the proposed algorithm can be an effective tool in network flow analysis.

Keywords: Cyber situation awareness
Significant co-occurrence detection · Temporal relationship discovery
Robust correlation

1 Introduction

Characterizing network behavior can help inform the operation and state of a networked computing environment. Characterization is a process of describing attributes and activities. There are a number of existing characterizations that illuminate a system's relationship to a network [13], role [17], and activities [4,18]. Characterizations improve situation awareness, helping network operators and cyber practitioners understand purpose and intent of the component, along with its relation to the network. In this paper we introduce a novel network behavior characterization method based on discovering significant co-occurrence

© Springer International Publishing AG, part of Springer Nature 2018
S. Yamamoto and H. Mori (Eds.): HIMI 2018, LNCS 10904, pp. 609–623, 2018.
https://doi.org/10.1007/978-3-319-92043-6_49

within network flow information. A *significant co-occurrence* is a robust concurrence or coincidence of events or activities observed over a period of time. As the definition suggests, the analysis occurs in the temporal plane, examining for correlation of events or activities.

The method discussed in this paper came from our research into dependency discovery. Advanced network defenses such as application reconfiguration and network address hopping often need to understand dependencies to maintain network and service function. Dependencies between components can be a surprisingly complex web of relationships. Direct dependencies (e.g., web service requires a database to store and retrieve data) were documented by architects and owners, but indirect relationships (e.g., the database would admit connections slowly if it could not reverse map the network address to a qualified hostname) were unnoticed or overlooked. Automated methods to discover these relationships are needed to overcome this gap in knowledge. Dependencies exhibit both topological and temporal characteristics and we found it difficult to meaningfully combine both aspects into existing graphical representations. We modified our approach to first extract temporal structure, before performing any type of topological analysis. We tried an approach based on Self-Organizing Maps (SOMs), an unsupervised, two-dimensional artificial neural network [14], to extract temporal relationships from sets of time-encoded event vectors. The result was that the SOM would overgeneralize relationships, learning aspects of the data not relatable to the problem. We believe this is a natural consequence of the low information content of network flow. Each record is a 13-tuple, and while a single record has informational merit, a set of records is highly redundant. This results in component analysis focusing on elements that have high entropy but little interest, such as linearly incrementing or randomly chosen port numbers or unmeaningful differences in network addresses. We finally fell to an approach that discovers significant co-occurrences between aggregates of data.

Later research identified that significant co-occurrences are useful for detecting related activity, such as coordinated interactions over remote shell sessions. This is useful because: (1) identified behaviors are coordinated both in time and topology, and (2) it permits identification of relationships in the time domain that cannot be inferred from graph representations.

This paper is organized as follows. In Sect. 2, we describe a characterization problem and formulate it as significant co-occurrence discovery. We introduce a robust algorithm to discover significant co-occurrences in Sect. 3 and we then evaluate our approach in Sect. 4. In Sect. 5, we discuss the outcomes of the evaluation, additional measures that we investigated, and work to enrich the models. We then conclude in Sect. 6.

Related Work. In earlier work, we examined flow information to discover network service and application dependencies. We introduced the idea of encoding network flow information into time series [6]. We then used cross-correlation to identify dependencies as a function of the lag between classes of time series. This showed an improvement in sensitivity and specificity when compared to earlier approaches. We then refined this approach by contextualizing flow behavior [7].

This is done by aggregating flow series based on a set of frequently occurring patterns of elements. We did not guarantee long-term stationarity (that is, mean and variance does not change with time) in the time series, which is an underlying assumption of the cross-correlation methodology. The approach proposed in this paper employs a cross-correlation analog that is not sensitive to stationarity.

Upon further investigation, we generalized our approach as a method to detect significant co-occurrence relationships. Work by Jalali and Jain [12] considers a similar problem formulation, but uses counts and conditional probabilities to extract relationships. NSDMiner [16] and Sherlock [1] use a similar count-based approach to extract dependencies. In an evaluation comparing NSDMiner and our approach [6], we demonstrated that cross-correlation was less sensitive to flow volumes than count-based approaches. CloudScout [21] uses cross correlation to detect dependencies, but did not consider that the cause and effect did not occur concurrently. Tor correlation attacks take advantage of correlating traffic patterns at different locations [15]. The Tor injection and correlation attack described in [8], which deanonymizes users, is performed over flow information and can be thought of as an exercise of a co-occurrence detection of the reference signal (created by the injection).

2   Problem Formulation

The completeness of network flow information is dependent on the positions of the flow probes. It is a simple fact that a probe will only report flows that it observes in its observation domain (not withstanding sampling policies, capture

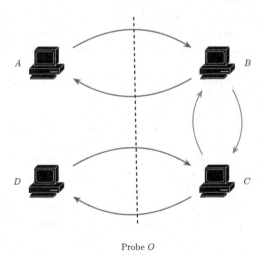

Probe O

Fig. 1. The communications between systems A, B, C, and D are coordinated. Probe O can observe communications that the dashed line cuts, which are A-B and C-D interactions. Due to probe O's position in the network, it cannot observe $B \leftrightarrow D$ communications.

speed, resource limitations, etc.). Due to the incompleteness of data, constructed graph topologies may be missing significant relationships, complicating the process of computing graph walk statistics, centrality measures, graph distances, and other graph theoretic measures.

To illustrate the problem, consider Fig. 1 where four systems, A, B, C, and D are coordinating communications. Probe O can observe communications that are cut by the vertical dashed line. In this example, communications between $A \leftrightarrow B$ and $C \leftrightarrow D$ are observed, but the probe does not observe $B \leftrightarrow D$. While the relationship between B and D cannot be observed, there may exist temporal structure between $A \leftrightarrow B$ and $C \leftrightarrow D$ to infer a hidden relationship. Specifically, if $A \leftrightarrow B$ are recurrently co-occurring with $C \leftrightarrow D$, we may infer a relationship between $A \leftrightarrow B$ and $C \leftrightarrow D$. Since there are four communication channels between $A \leftrightarrow B$ and $C \leftrightarrow D$ (traffic destined to A, B, C, and D), there are four opportunities for co-occurrences (in practice this often limited as a probe sees on direction only, either due to configuration or asymmetric routing).

The problem as formulated is the correlation in time of observable activities in network flow information and/or packet traces. The conceptual *mechanism* is that if components exhibit some form of functional relationship, that we should observe that network activity at one component will cause an incidence of co-occurring network traffic at another component. In our parlance, we map network activity to a stream of timestamped *events*, and then discover significant co-occurrences between the events. A *co-occurrence* is a concurrence or coincidence of events, discovery of which entails observation for these types of recurrent relationships over a time period. *Significant co-occurrences* are then, given some metric, robust co-occurrences. Significant co-occurrences are of interest in multiple domains as causal relationships often emit co-occurrences and significant co-occurrences are frequently determined on the path to causal identification. In the context of our problem, co-occurrences do not strictly mean causations. Information systems and architectures have not conventionally been built for experimentation, that is, to allow an invention that may reduce network redundancy and may impair system function. Moreover, passive approaches are preferable and feasible to more data sources.

The problem as formulated has evolved with our understanding and work on several related problems that have common threads of time structure and practical causation/correlation (i.e., without intervention) discovery. Returning to our prior work on dependency discovery, our insight was that significant co-occurrences are established among aggregates of systems and components. For instance, a dependency exists between set of clients accessing a web application running on a cluster (a set of servers). The set of clients and set of servers are their own *event class*. The dependency exists between the classes and not necessarily among the individual elements contained therein. We encoded the problem as the counts of connection initializations and terminations, analyzing classes for significant co-occurrences of these events.

The problem introduced at the start has distinguishing attributes that require modified aggregation and encoding. Moreover, aspects of collection influence the design. In terms of collection, connection initializations and terminations are not reliably observed or inferred. Moreover, to reduce resource requirements, high-speed collectors don't maintain full connection state. Instead, they export flow information based on resource thresholds and timers, and not necessarily at the proper end of flow. Furthermore, we don't expect to observe many connection initializations (or terminations) when looking for activity relations. To provide better quality encoding for the related activity, we aggregate flow information by unique source and destination network address and transport attributes. We then encode the number of bytes transferred during each period. The significant co-occurrences are identified, which correlates variations in the time series.

3   Approach

Our proposed approach is founded on co-occurrence detection, followed by identifying significant co-occurrence relationships. A *co-occurrence* is a repeated observation of (concurrent or delayed) "first this, then that." A *co-occurrence relationship* is revealed through time and is a primitive aspect of causal relationships.

Returning to Fig. 1, our approach is tooled to detect the co-occurrence of *A-B* communications with *C-D* communications. More specifically, we are looking for co-occurrences of changes in datarate, that is, the changes in *A-B* datarate co-occur with changes in *C-D* datarate. We conceive of a *mechanism* where the systems are organized such that *A-B* interactions result in effects along a series of systems, which eventually transits *C-D*. As we observe relative changes in *A-B* datarate, we expect to observe relative changes in *C-D* datarate. This mechanism fits where information is being passed from one node to the next, such as what occurs in Tor and other low-latency virtual/overlay networks and the chaining of remote terminals.

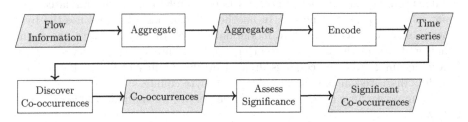

Fig. 2. Diagram of the analytic workflow. A rectangle represents a process, a parallelogram shows an input or output.

We break our approach to detect significant co-occurrence relations in network flow information into four stages:

1. Aggregate flow information,
2. Encode aggregates as time series,
3. Detect co-occurrences within the time series, and
4. Identify *significant co-occurrence relationships* over the time dimension.

The relationships of the stages and the corresponding inputs and outputs are diagrammed in Fig. 2. The combined effect of State 1 and 2 is to represent the flow information as time series. This input is then consumed by Stage 3 to analyze for co-occurrences. Lastly, the significance of the co-occurrences are assessed. We now describe the stages in full.

3.1   Aggregation

The analysis workflow begins by dividing the set of flows into flow aggregates. The flows in an *aggregate* share common features, such as all are sourced from or destined to the same endpoint. The exact details of the procedure are dependent on the causal mechanism and its effects. Once understood, rules can be defined to categorize flows into aggregates. In our previous work [6,7] we investigated dependencies in enterprise computing networks. A *dependency* exists between component a (say, a web server) and b (say, a database), if a's function is impaired when b becomes impaired (i.e., fails, faults, or is degraded). For this case, flows that have destination address, protocol, and destination port in common are merged into a single aggregate. For the question introduced in this paper, the dynamics are not expressed *en masse* but are expressed *singularly* between pairs of endpoints. Consequently, we create aggregates based on common source and destination addresses, protocol, and destination port.

3.2   Encode

Next, we encode aggregates as time series. A feature of interest from the flows in the aggregate is accumulated along an equally-spaced time dimension. We choose equally-spaced time series over other time series representations as it promotes efficient computation of correlation to be performed in the next step. In one encoding, the number of flow starts or completions are accumulated as a count in the respective time bin. In another possible encoding, we use the byte (flow volume) or packet count to model data rates. In this encoding, the count is accumulated in the bins representing the time interval the flow was active. Before accumulating, the count is divided by the number of bins. The primary encoding discussed in this paper is based on byte counts.

Assuming we have b_t bytes in the t time period, we compute the log ratio as follows:

$$r_t = \log \left(\frac{b_t + 1}{b_{t-1} + 1} \right), \tag{1}$$

which gives the normalized datarate changes. Computing the log compresses the dynamic range of the ratio, preventing false positive results that result from high dynamic range.

3.3   Co-Occurrence Detection

We correlate the time series to detect co-occurrence relationships between flow aggregates. To determine the correlation between each pair of time series, we used an alternative coefficient presented by Erdem et al. [9]. This coefficient effectively captures the cross-dependence of the two time series over time, but does not have the requirement of stationarity of the time series, in contrast to the more commonly applied Pearson's product moment correlation coefficient (see [2] for more information about Pearson's). We quickly appreciated the alternative coefficient as we could not guarantee stationarity over the length of the time series in prior work.

Given two time series, X and Y, the Erdem coefficient ρ_O is defined as:

$$\rho_O = \frac{\alpha_{xy}}{\sqrt{\alpha_{xx}}\sqrt{\alpha_{yy}}},$$

where $\alpha_{xy} = \mathbf{E}\left[(X_t - X_{t-1})(Y_t - Y_{t-1})\right]$, $\alpha_{xx} = \mathbf{E}\left[(X_t - X_{t-1})^2\right]$, and $\alpha_{yy} = \mathbf{E}\left[(Y_t - Y_{t-1})^2\right]$. Expanding the definition, we observe that

$$\alpha_{xy} = \frac{1}{T-1}\sum_{t=2}^{T}(X_t - X_{t-1})(Y_t - Y_{t-1}),$$

is the first-order autocorrelation of X_t and Y_t [3]; the definitions of α_{xx} and α_{yy} directly follow. The definition as specified is for the case when the cause and effect simultaneously occur at the same time t. We can generalize the definition for a delay model, for which the effect *lags* ℓ time periods behind the cause. For $\ell = 0, \pm 1, \pm 2, \ldots$, the lagged variant of the coefficient is $\rho_O[\ell] = \alpha_{xy}[\ell]/\sqrt{\alpha_{xx}}\sqrt{\alpha_{xy}}$, where:

$$\alpha_{xy}[\ell] = \begin{cases} \frac{1}{T-1}\sum_{t=2}^{T-\ell}(X_t - X_{t-1})(Y_{t+\ell} - Y_{t+\ell-1}) & \ell = 0, 1, 2, \ldots \\ \frac{1}{T-1}\sum_{t=2}^{T+\ell}(Y_t - Y_{t-1})(X_{t-\ell} - X_{t-\ell-1}) & \ell = 0, -1, -2, \ldots \end{cases} \tag{2}$$

Using the log ratio defined in (1), we can rewrite (2) to

$$\alpha_{xy}[\ell] = \begin{cases} \frac{1}{T-1}\sum_{t=2}^{T-\ell} r_{xt}r_{yt+\ell} & \ell = 0, 1, 2, \ldots \\ \frac{1}{T-1}\sum_{t=2}^{T+\ell} r_{yt}r_{xt-\ell} & \ell = 0, -1, -2, \ldots \end{cases} \tag{3}$$

Using the fast Fourier transform (FFT) to compute the discrete frequency transform (DFT) of the time series, convolution in the frequency domain can be used to compute the coefficients $\rho_O[\ell]$ for all lags $\ell = 0, \pm 1, \pm 2, \ldots, n$ providing significant computational speedup and improved precision over the naïve approach computed in the time domain. Prior to performing the FFT, the time series is doubled in length by padding with zeros. Convolution allows us the opportunity to smooth the results. We continue to have success using Lanczos' σ factors [10, "Lanczos' σ factors" and "The σ Factors in the General Case"] in increasing the magnitude of the coefficient of true correlations:

$$\sigma_i = \begin{cases} 1 & i = 0, \\ \frac{\sin(2\pi i/2(T-1))}{2\pi i/2(T-1)} & i = 1, 2, \ldots, 2(T-1) \end{cases} \tag{4}$$

3.4 Significant Co-Occurrence

A significant co-occurrence is a robust pattern found in the observations. In order to identify these significant co-occurrences, we developed several criteria. Fundamentally, we associate significance with values of ρ_O. As ρ_O approaches one, we assume a greater significance. Unfortunately, our time series encoding combined with the character of the problem produces a sparse representation, where many elements of the time series are zero. We developed additional criteria to measure the amount of information in an encoded time series, which supports the interpretation of ρ_O and significance conclusions. The intuition for the criteria is as follows. Assume that two co-occurrences are measured to have equal ρ_O, we consider the time series with more non-zero elements to be of greater significance. Additionally, we have found the zeroes between the first and last non-zero elements to also support the claim of significance. Given equal quantity of non-zero elements, the longer this length, more significant the co-occurrence. We developed two measures of this criteria: span and occupancy. For an encoded time series, we define *span* as the length of the run of the first to last non-zero elements, while *occupancy* is the quantity of non-zero elements. Formally, we define span as:

$$s = j - i,$$

where i and j are the indices of the first and last non-zero elements, respectively. Occupancy is defined as:

$$o = count\{x_i | x_i \neq 0\}. \tag{5}$$

To speed the computation and increase scalability, encoded time series must satisfy minimum span and occupancy conditions. The time series that satisfy these conditions are then submitted for correlation computation. The distribution of ρ_O is then examined and a threshold determined. Finally, we were interested in novel pairings. To find these, we sorted the IP/port pair labels and counted the number of occurrences. We then kept records that had a count of 2 or less. The last step was necessary as certain time series were characterized by low variance. The correlation coefficient can be considered a measure of covariance between two time series; this has the effect of driving the value of ρ_O higher if one or both time series has a low variance. Therefore, time series with a low variance will tend to falsely present as significant co-occurrences with multiple pairings.

4 Evaluation

We evaluated our proposed approach using two methods. For the first method, we build a simple mechanism in a testbed to coordinate remote shell sessions. The goal is to prove out the analysis for unambiguous, optimal series of coordinated interactions. For the second, we apply the approach to a publicly-available dataset to identity significant co-occurrences. Unlike the first evaluation, there isn't additional available information to support conclusions.

Fig. 3. A diagram of the first evaluation's setup. The setup comprised twelve virtual machines, coordinating a VNC session and ten SSH remote shell sessions in a series circuit.

Table 1. Significant co-occurrence coefficients for the first evaluation. The table has been compressed to save space. Observe that closer the sessions are in the sequence, higher the correlation coefficient.

	VNC	SSH1	SSH2	SSH3	···	SSH8	SSH9	SSH10
VNC	—	0.977	0.977	0.976	···	0.976	0.976	0.977
SSH1	0.977	—	1.000	1.000	···	1.000	0.999	0.999
SSH2	0.977	1.000	—	1.000	···	1.000	0.999	0.999
SSH3	0.976	1.000	1.000	—	···	1.000	0.999	0.999
···	···	···	···	···	···	···	···	···
SSH8	0.976	1.000	1.000	1.000	···	—	1.000	1.000
SSH9	0.976	0.999	0.999	0.999	···	1.000	—	1.000
SSH10	0.977	0.999	0.999	0.999	···	1.000	1.000	—

The purpose of the first synthetic test is to evaluate the analysis on unambiguous, optimal series of coordinated remote shell session interactions. A diagram of the setup is shown in Fig. 3. In a testbed environment, twelve Linux virtual machines were created. A harness was constructed that would begin with a VNC sessions and then login via SSH into a series of ten remote hosts (the first host executes the VNC viewer, connecting to the VNC server). On the downriver host, a command was executed that would mirror input to output. A script would arbitrarily generate input that would traverse the series of hosts and then return to be displayed in the VNC viewer. We captured the packets on the network, recording network traffic for a duration of one hour. The YAF tool suite [11] was used to transform the trace into unidirectional flow information. The active timeout (or hard timeout, where the flow entry is exported once expired) and the idle timeout (the flow entry is exported if no traffic is received before timeout) were each set to 10 s. Flow was aggregated and encoded into 30 s interval time series by payload length. Time series representing inbound and outbound traffic were then merged into a single time series. A summary of results is given in Table 1. All coefficients were near one as, again, this was evaluated under optimal conditions.

For the second evaluation, we applied our approach to the CAIDA Anonymized Internet Traces 2016 Dataset [5]. This publicly-available dataset is an hour-long packet trace that was captured on January 21st, 2016 at the Chicago Equinix Internet Exchange. The trace, comprising network and

Table 2. CAIDA Anonymized Internet Traces 2016 Datasets summary information.

First timestamp	January 21, 2016 12:59:11.415 UTC
Last timestamp	January 21, 2016 14:02:51.405 UTC
Capture length	20 B to 80 B
Unique IPv4 addresses	17,722,741
IPv4 flows	165,476,840
IPv4 byte/flow	34,629 B flow^{-1}± 296 B flow^{-1}
IPv4 datarate/1 min average	410.5 Gbit s^{-1}± 3.36 Gbit s^{-1}
Unique IPv6 addresses	476,500
IPv6 flows	3,323,221
IPv6 byte/flow	68,332.0 B flow^{-1}± 806 B flow^{-1}1
IPv6 datarate/1 min average	21.0 Gbit s^{-1}± 0.2 Gbit s^{-1}

Table 3. Notable significant co-occurrences that were identified.

Pair 1	Pair 2	Coefficient
194.84.75.170 → 66.216.158.246:ssh	194.84.75.170 → 66.216.152.59:ssh	1.
209.169.193.11 → 138.38.133.180:telnet	209.169.208.163 → 138.38.133.180:telnet	1.

transport headers of incoming and outgoing packets, was collected with the assistance of a high-speed monitor. Dataset documentation does not mention lost or dropped packets. Summary information about the dataset is presented in Table 2. Our objective was to identify plausibly related remote shell sessions.

We began the analysis by transforming the packet trace to flow information. The packet trace arrives in a series of files, demarcated by direction and minute timestamp. We merged the files by direction in increasing order of timestamp, followed by merging the directions into a common trace. The YAF tool suite was used to transform the trace into unidirectional network flow information. The active timeout (or hard timeout, where the flow entry is exported once expired) and the idle timeout (the flow entry is exported if no traffic is received before timeout) were each set to 10 s. We next filtered the flow information by application, keeping records characteristic of remote shell sessions, and discarding the rest. We filtered on the combination of protocol and well-known or common ports, preserving flows sourced or destined to TCP ports 22 (ssh), 23 (telnet), 512–514 (BSD UNIX "r" commands), 3389 (remote desktop protocol), 5900–5963 (VNC), 5938 (TeamViewer), and 6000–6063 (X11); along with UDP 3389 (remote desktop protocol), 5938 (TeamViewer), and 60000–61000 (Mosh). Flow was then encoded and aggregated into 30 s intervals, producing time series of 128 elements in size.

Analysis was performed and 131,161 co-occurrences were identified with a correlation coefficient greater than or equal to 0.80. Upon examination, we noted that 122,001 results involved TeamViewer. Investigating the packet traces, we observed that the flows exhibited 1 min-periodic traffic pattern. A TeamViewer protocol description [19, 20] supported our suspicion that the purpose of this communication was to maintain connectivity (keep a firewall from considering the connection dead, among other reasons) and inform status. While these co-occurrences were significant, they were not notable in the context of our objective. We further winnowed the results by offset (preferring alignments of −1, 0, or 1), then by uniqueness of network and transport pairs. The treatment identified 43 co-occurrences that were deemed significant *and* notable. While all 43 co-occurrences were plausible, two co-occurrences, which are identified in Table 3, comprise two nodes connecting to a service on a third. Time of use and the structure further supports the conclusion that these sessions were possibly coordinating their interactions.

Once again analyzing the CAIDA dataset, we investigated for significant co-occurrences of HTTP communications. HTTP operates using TCP on well-known ports 80 and 443, and alternate ports 4433, 8000, 8008, 8080, 8443, and 8888. A significant co-occurrence opens up the possibility that the channel is used for something other than strictly document transfer, such as anonymizing network traffic. Using the aggregation and encoding approach as described in the treatment of the remote shell session, we identified 1172 significant co-occurrences among the 27,789,076 flow records. For this problem we defined significance as correlations with $\rho_O \geq 0.8$, an offset of zero, occupancy greater than or equal to ten, and equal occupancy and span of the aggregate time series. After reviewing the significant co-occurrences, we argue that they are plausible. Additional measures need to be defined to assess notability, that is, the interest of an analyst to the problem.

The evaluation demonstrates approach finding significant co-occurrences that could be interesting to cyber practitioners and analytics. The datasets themselves have limitations that prohibit further validation. As the packet trace does not provide payloads and the data itself was passively collected from an anonymous population, it is difficult to improve our confidence in the results. Significance is built with repeated observations: given more data observed over a longer duration we would improve our confidence in the results. We are further evaluating the approach using internal datasets, where it continues to show promise.

5   Discussion

We utilized several measures and metrics to identify significant correlations. As is evident from the evaluation and prior work, specific aggregations, encodings and measures are required to model aspects of the temporal behaviors exhibited by complex systems. The span and occupancy measures were found to be useful for filtering the time series during the encoding phase. Additional measures and metrics were used during the significant correlation detection stage. As discussed below, some of those metrics explored did not prove to be useful to the context of the problem in hand and further work is required to explore additional metrics. Additionally, the relationship calls being made do not have any directionality or order associated to them. Further exploration to address the directionality and "*causal*" nature of relationships is underway. Additional attributes can be added to the generated models to make them more useful to analysts.

5.1 Measures and Metrics

During initial attempts at significant correlation detection, we investigated several metrics to identify "interesting" characteristics of co-occurring time series. We considered a time series to contain more significant information if it has more non-zero entries or a wide span of non-zero entries.

Attempts to identify significant correlations involved scaling the cross correlation peaks using span and occupancy. Peaks between time series with a high occupancy and/or span were weighted more heavily than sparsely populated time series. Two different weighting factors were developed from the ratio of the occupancy and the span of a time series to its total length. The peak cross correlation value is then scaled using the weighting factors of both contributing time series. A pair of time series with full span and occupancy have a weighting of one; thus, the weighting does not reduce the peak. Significant correlations were defined to be outliers based on the *Interquartile Range* (IQR) of all peak values.

However, these weighting metrics significantly compressed the dynamic range of all peak values. This led to difficulty in identifying outliers. Significant co-occurrence relationships were occasionally compressed to the point where they were unidentifiable. This led us to conclude that these measures were not useful in this context. However, the basic concepts of span and occupancy proved useful in initial filtering of time series. Encoded time series are not submitted to correlation calculations unless they meet a minimum threshold of both span and occupancy.

5.2 Model Enrichment

Additional attributes and statistics may be calculated to enrich the generated models and improve their significance and understandability to end users. For example, identification of recurrent patterns in either the encoded time series or the occurrence of significant relationships can provide experts with information on the typical behavior of a system. *Recurrent* patterns are behaviors that have a periodicity to their occurrence; these events may take place hourly, daily, or weekly.

Recurrent Events. Recurrent events are event classes that occur with some periodicity. For example, connections to a time card certification system may peak on the day that certifications are due. Identification of recurrent events informs experts on when they can expect a given event class to typically occur. In turn, this allows them to develop rules for other applications or detect unusual behaviors.

Identification of recurrent events follows naturally from the calculation of cross-correlation coefficients described in Sect. 3.3. Given flow records, we may encode time series based on occurrence count, byte or packet counts. The bin sizes are chosen based on the periodicity being explored. Once the time series is encoded, we may slightly modify the Erdem coefficient ρ_0 to calculate the autocorrelation.

Substituting a time series X into (3) for both the X and Y inputs gives:

$$\alpha_{xx}[\ell] = \frac{1}{T-1} \sum_{t=2}^{T-\ell} r_{xt} r_{xt+\ell} \qquad \ell \in (0, \infty) \tag{6}$$

Analogously to the original cross-correlation approach, this coefficient describe the covariation of a time series with itself at various lag and is contained within the range $[-1, 1]$. A coefficient closer to 1 indicates a strong positive correlation, or that the time series tends to change in a similar manner over time. By its definition, the autocorrelation coefficient will always equal 1 at the zero lag because a time series is always correlated with itself.

When autocorrelation is performed on a time series, a resulting peak indicates the period at which the event class tends to occur. For example, if data is binned by the minute and there is a peak in the autocorrelation at lag 10, this event tends to reoccur every 10 min. The lack of a peak in the autocorrelation implies that the event class does not display any recurrent behavior at the current level of granularity. However, care should be taken to verify that periodicity cannot be identified at a different granularity.

Recurrent Correlations. It may be of further interest to explore *recurrent correlation* present within the dataset. A recurrent correlation is a significant co-occurrence that will occur at regular time intervals. For example, an individual may initiate a remote session connection once a day. Identification of these patterns allow experts to determine typical joint behaviors of event classes.

Determination of a recurrent correlation follows similar steps as those for recurrent events. The difference exists in the encoded time series that is the input to the autocorrelation function. For this case, the time series are encoded based on the presence of significant relationships. The time series bins contain counts of significant co-occurrences during that time interval. A peak in the resulting autocorrelation indicates that there is periodicity in this relationship.

5.3   Further Application

The work presented shows how Erdem's coefficient can be expanded to find lagged correlations between time series. This approach furthers experts' understanding of relationships within a network. An understanding of significant correlations can aid in related domains, such as mission assurance. When contemplating the risks involved in completing an organization's mission, it is necessary to understand the underlying dependencies in the networks that support the organization. Unexpected relationships may lead to additional considerations when evaluating the criticality of mission support systems.

6 Conclusion

This paper presents a general method to characterize network behaviors that manifest in the temporal domain. The characterized behaviors describe interactions between networked components. By analyzing the data in the temporal domain, indirect relationships (an indirect relationship is one in which no graph path exists between components) may be inferred. The proposed method employs significant co-occurrences detection which is a generalization of our prior work on dependency discovery. Formulations of different mechanisms governing interactions, influence the design of aggregation and encoding stages. We showed the method can be applied on flow information to related problems such as coupled remote shell sessions. Further analysis can be applied to determine the notability of the co-occurrence to network defenders and operators.

Acknowledgment. Portions of the research were funded by PNNL's Asymmetric Resilient Cybersecurity (ARC) Laboratory Research & Development Initiative. This work was performed while Satish Chikkagoudar was at Pacific Northwest National Laboratory. The views expressed in this paper are the opinions of the Authors and do not represent official positions of the Pacific Northwest National Laboratory, the Department of the Navy, or the Department of Energy.

References

1. Bahl, P., Chandra, R., Greenberg, A., Kandula, S., Maltz, D.A., Zhang, M.: Towards highly reliable enterprise network services via inference of multi-level dependences. In: Proceedings of the ACM SIGCOMM Conference on Data Communications (SIGCOMM), pp. 13–24 (2007)
2. Box, G.G.P., Jenkins, G.M., Reinsel, G.C.: Time Series Analysis: Forecasting and Control, 4th edn. Wiley, Hoboken (2008)
3. Brockwell, P.J., Davis, R.A.: Time Series: Theory and Methods, 2nd edn. Springer, Heidelberg (1991). https://doi.org/10.1007/978-1-4899-0004-3
4. Bruillard, P., Nowak, K., Purvine, E.: Anomaly detection using persistent homology. In: Proceedings of the Cybersecurity Symposium (CYBERSEC). IEEE (2016)
5. CAIDA: The CAIDA Anonymized Internet Traces 2016 Dataset (2016). http://data.caida.org/datasets/passive-2016/README-2016.txt. Accessed 29 Mar 2016
6. Carroll, T.E., Chikkagoudar, S., Arthur-Durett, K.: Impact of network activity levels on the performance of passive network service dependency discovery. In: Proceedings of the Military Communications Conference (MILCOM), pp. 1341–1347. IEEE (2015)
7. Carroll, T.E., Chikkagoudar, S., Arthur-Durett, K.M., Thomas, D.G.: Automating network node behavior characterization by mining communication patterns. In: 2017 IEEE International Symposium on Technologies for Homeland Security (HST), pp. 1–7. IEEE (2017)

8. Chakravarty, S., Barbera, M.V., Portokalidis, G., Polychronakis, M., Keromytis, A.D.: On the effectiveness of traffic analysis against anonymity networks using flow records. In: Faloutsos, M., Kuzmanovic, A. (eds.) PAM 2014. LNCS, vol. 8362, pp. 247–257. Springer, Cham (2014). https://doi.org/10.1007/978-3-319-04918-2_24

9. Erdem, O., Ceyhan, E., Varli, Y.: A new correlation coefficient for bivariate time-series data. Phys. A: Stat. Mech. Appl. **414**, 274–284 (2014)

10. Hamming, R.W.: Numerical Methods for Scientist and Engineers, 2nd edn. Dover, New York (1986)

11. Inacio, C.M., Trammell, B.: YAF: yet another flowmeter. In: Proceedings of the 24th Large Installation System Administration Conference (LISA 2010). USENIX (2010)

12. Jalali, L., Jain, R.: A framework for event co-occurrence detection in event streams. CoRR abs/1603.09012 (2016)

13. Joslyn, C., Cowley, W., Hogan, E., Olsen, B.: Discrete mathematical approaches to graph-based traffic analysis. In: Proceedings of the International Workshop on Engineering Cyber Security and Resilience (ECSaR) (2014)

14. Kohonen, T.: Self-Organization and Associative Memory, 2nd edn. Springer, Heidelberg (1987). https://doi.org/10.1007/978-3-642-88163-3

15. Murdoch, S.J., Danezis, G.: Low-cost traffic analysis for Tor. In: Proceedings of the 2005 IEEE Symposium on Security and Privacy (S&P 2005), pp. 183–195. IEEE (2005)

16. Natarajan, A., Ning, P., Liu, Y., Jajodia, S., Hutchinson, S.E.: NSDMiner: automated discovery of network service dependencies. In: Proceedings of the 31st IEEE International Conference on Computer Communications (INFOCOMM 2012). IEEE (2012)

17. Oler, K., Choudhury, S.: Graph based role mining techniques for cyber security. In: Proceedings of the FloCon. CERT (2015)

18. Sayegh, N., Elhajj, I.H., Kayssi, A., Chehab, A.: SCADA intrusion detection system based on temporal behavior of frequent patterns. In: Proceedings of the 17th IEEE Mediterranean Electrotechnical Conference (MELECON). IEEE (2014)

19. Thomas, B.: Teamviewer authentication protocol (part 1 of 3), 31 January 2013. https://www.optiv.com/blog/teamviewer-authentication-protocol-part-1-of-3. Accessed 1 Jan 2018

20. Thomas, B.: Teamviewer authentication protocol (part 2 of 3), 31 January 2013. https://www.optiv.com/blog/teamviewer-authentication-protocol-part-2-of-3. Accessed 1 Jan 2018

21. Yin, J., Zhao, X., Tang, Y., Zhi, C., Chen, Z., Wu, Z.: CloudScout: a non-intrusive approach to service dependency discovery. IEEE Trans. Parallel Distrib. Syst. **28**(5), 1271–1284 (2017)

Exploring the Cognitive, Affective, and Behavioral Responses of Korean Consumers Toward Mobile Payment Services: A Text Mining Approach

Minji Jung, Yu Lim Lee[(✉)], Chae Min Yoo, Ji Won Kim,
and Jae-Eun Chung

Department of Consumer and Family Sciences, Sungkyunkwan University,
25-2 Sungkyunkwan-ro, Jongno-gu, Seoul, Korea
ylee168@naver.com

Abstract. The purpose of this study was to examine the cognitive, affective, and behavioral responses of Korean consumers toward mobile payment services based on the tri-component model by using a text-mining technique. Samsung Pay was chosen it is used in both online and offline transactions. We targeted social media data posted during the period between 1 July 2016 and 31 December 2016, which was about one year after Samsung Pay was launched. We conducted word frequency analysis, clustering analysis, and association rules using R programming. The results were the following. First, the 50 most frequently used words referenced the brand names of the mobile devices, payment methods, and the procedures and unique functions of Samsung Pay compared to other types of payment methods. Second, we classified the terms into 24 categories (11 categories of cognitive responses, 10 categories of affective responses, and 3 categories of behavioral responses) based on the tri-component model. The results of the clustering analysis based on the 24 categories showed a clear split between positive and negative responses at the macro level. The positive responses were further clustered into four groups, while the negative responses were fused into two groups at the micro level. Third, the association rules produced 65 rules, and we found that economic benefits played a great role in the positive feelings and continuous use of mobile payment services. This study offers valuable implications that may help mobile payment marketers with delivering services that correspond to consumer values and expectations, thus increasing consumer utility and satisfaction.

Keywords: Mobile payment service · Tri-component model · Text mining

1 Introduction

Since the emergence of online-to-offline (O2O) e-commerce, the mobile payment market has undergone rapid growth [44]. Mobile payment allows consumers to easily pay expenses anytime and anywhere via smartphone devices. This leads to reduction of transaction costs, allowing for more efficient payments [28] that, in turn, may change consumer behavior.

© Springer International Publishing AG, part of Springer Nature 2018
S. Yamamoto and H. Mori (Eds.): HIMI 2018, LNCS 10904, pp. 624–642, 2018.
https://doi.org/10.1007/978-3-319-92043-6_50

Because mobile payment comprises high technology that lay consumers might not fully understand, they may encounter unexpected difficulty when using the service. Through research on consumption experiences with smart technology, Choi [7] suggested the existence of a discrepancy between the utility that consumers pursue through smart technology and the utility that the technology actually provides. Due to this discrepancy, the utility and satisfaction derived from smart technology product use often fail to be maximized [7]. It is therefore necessary to examine diverse consumer responses to such services so that their needs and wants may be better understood.

Analysis of social media data can be useful for understanding consumer behavior because users tend to freely share about their consumption experiences in these media [20]. Thus, to examine the diverse and genuine responses of consumers toward mobile payment services, it would be useful to analyze remarks that are shared in social media. Most previous research on the subject, however, has failed to fully capture the dynamic responses of mobile payment service users because the findings were based on limited and structured data generated by surveys [2].

According to the tri-component model, consumer attitudes toward products are formed based on three dimensions: (1) cognitive; (2) affective; and (3) behavioral components [43]. Based on this model, the purpose of this study was to examine the cognitive, affective, and behavioral responses of Korean consumers toward mobile payment services by analyzing the massive data shared through social media using a text-mining technique. Samsung Pay was chosen because it is used in both online and offline transactions, thus offering a more dynamic consumption experience than that of services that provide only one option. This study contributes to the existent literature by providing in-depth understanding of consumers' true and lively opinions about Samsung Pay and offering valuable practical implications that may help marketers build services that correspond more closely to the values and expectations of their users, thus increasing consumer utility and satisfaction.

2 Literature Review

2.1 Mobile Payment Service

Mobile payment services are defined as any mobile commerce service that involves payment transactions via a mobile device [42]. Consumers are required to register their financial information on a mobile payment application only once, after which they may quickly complete payments and transactions using simple identification. These services, therefore, provide a fast and convenient means of completing financial transactions anytime and anywhere [27].

The mobile payment market in South Korea has expanded alongside the rapid growth of smart device (smartphones, tablet PCs, etc.) usage. In 2016, there were more than 30 mobile payment services available in Korea, each with its distinguishable features. Samsung Pay, launched by the smart device manufacturer Samsung, supports both near field communication (NFC) and magnetic secure transmission (MST) solutions, allowing users to pay in "brick-and-mortar" stores as well as through online payment. It also offers a management system in which consumers can digitally manage their membership cards in one application, a transportation card service, and links to various services of alliance companies. Samsung Pay reached 8 billion dollars of accumulated transmission amount in 2017, and it has held first-place among Korean mobile payment services since its launch based on accumulated transaction amount [10]. Its major competitors are Naver Pay—launched by Naver, the largest internet portal site in Korea—and KakaoPay, which is managed by KakaoTalk, the largest message application company in Korea.

2.2 Social Media and Text Mining

Social media is defined as a collective category of Internet-based applications that allow the creation and exchange of user-generated content; these applications include Facebook, Twitter, Instagram, blogs, and other online communities [20]. Consumers tend to trust product or service information generated by social media more than that from mainstream media sources because the former tends to be produced and exchanged by actual users [16]. Thus, information spread through social media can affect the attitudes or behavior of those who use the products that are being discussed [31]. Thus, it would be desirable to analyze data from social media to obtain insights about consumption trends.

Recently, text-mining techniques have been used to analyze large amounts of social media datasets. It involves using a natural language processing (NLP) application to extract and process useful information from a large set of informal or semi-structured data to identify hidden patterns [32]. Although many researchers have analyzed textual data from social media using text-mining techniques in various consumer-behavior fields [38], few have examined consumer responses to mobile payment services [2]. Therefore, additional research on this issue is required to fill this gap.

3 Conceptual Framework

3.1 Tri-Component Model

During consumption, there exists an interaction between consumer and product that is manifested through purchase and usage [9]. The consumption experience encompasses the subjective mental events surrounding these acts [15]. These experiences induce a consumer response, including a change of attitude or a behavioral response toward the product or service, i.e., the internal processing of an organism in accordance with an external stimulus [4].

In an attempt to understand the relationship between attitude and behavior, the tri-component model proposes that attitude comprises three dimensions: (1) cognitive; (2) affective; and (3) conative or behavioral components [46]. These three components are inter-related. For example, the cognitive component—which includes knowledge, opinions, beliefs, and thoughts—can influence behavioral responses [22, 29] and affection [47].

Previous studies on mobile payment services primarily addressed the cognitive determinants of intention to use the service and overlooked the affective factors. For example, perceived compatibility, individual mobility [39], perceived usefulness, and perceived ease of use [21] were found to increase intention to use, whereas perceived cost and perceived risk were associated with a decrease [29]. Because of this lack of research examining the consumer affect derived from mobile payment services [22], there is a need for new research to address both the cognitive and affective perspectives. Thus, the present study will explore the following research questions:

(RQ 1) What are the cognitive, affective, and behavioral responses of Korean consumers toward Samsung Pay?

(RQ 2) How are these three responses related?

4   Research Methods

In the present study, we employed the following process of extracting and analyzing data to examine the above research questions: (1) crawling of social media data, (2) cleaning and preprocessing of the data, (3) text-mining analysis, and (4) visualization. The following is a brief description of these processes:

(1) *Crawling social media data*

Using Trend-up[1], a bigdata analysis platform, we crawled documents (postings) that included the terms "Samsung Pay" or "Sam Pay" (an abbreviation of Samsung Pay in Korean) from Facebook, Twitter, Instagram, and other online community sites. These types of social media were selected because the short length of their content allowed for high concentrations of relevant information, thus increasing the accuracy of the analysis [26]. We excluded retweets because they merely duplicated the original information, but we included Instagram hashtags (#) in the analysis. We targeted documents posted during the period between 1 July 2016 and 31 December 2016, which was about one year after the Samsung Pay service was launched. We collected a total of 19,529 documents posted by consumers along with relevant information including user ID, posted date, and the type of site.

(2) *Cleaning and preprocessing data*

Since social media data tend to have frequent word variants and misspellings as well as incomplete or incorrect content, data cleaning and preprocessing were necessary to improve the quality of the data [36]. Data cleaning refers to the process of detecting or removing noise or irrelevant documents such as advertisements, content containing no opinions or experiences (e.g., questions and answers), and content consisting of less than three terms, which may not provide sufficient data to infer responses regarding Samsung Pay [25]. Thus, these irrelevant documents were manually removed from the collected documents. We also performed normalization by modifying term variants, misspellings, and abbreviations. For example, all occurrences of "Samsung smartphone" were replaced with "Galaxy" because "Samsung

---

[1] The platform was established by a bigdata analysis company in Korea known as Tapacross.

smartphone" and "galaxy" referred to identical devices. After data cleaning, the number of documents was reduced to 1,609.

Next, we performed a preprocessing procedure known as term extraction through *stop-terms elimination, part-of-speech tagging,* and *stemming* using a Korean morphological analyzer and a KoNLP package with R programming. First, we removed stop words from the documents by using a list of stop words to reduce the noisy information that did not carry relevant content. Over 500 stop words were chosen and added to the list, including pronouns, prepositions, conjunctions, etc. In addition, all punctuation (e.g., # and @) and emoticons were removed from the data set. Second, we performed *part-of-speech tagging,* which classifies each term as a noun, verb, adjective, or other part of speech. Following this stage, *word stemming* (Cao et al. [7]) was used to reduce inflected or derived terms to their stem, base, or root forms (e.g., "buying," "bought," and "buys"). These terms were consolidated into single terms. After these term-extracting procedures, 2,466 terms were identified from 1,609 documents, resulting in a 2,466 × 1,609 matrix.

(3) *Text mining*

Text mining is defined as the discovery of new, interesting, and previously unknown information by automatically extracting information from written resources [12]. We employed the following three methods in text mining to explore the research questions of the study using R programming. To explore RQ 1—which addressed the identification of the cognitive, affective, and behavioral responses of Korean consumers toward Samsung Pay—word frequency analysis and clustering using an *agglomerative* hierarchical method was performed. To examine RQ 2—which addressed the relationships among the three types of responses—we utilized association analysis to extract association rules.

(4) *Visualization*

Visualization is the process of depicting or mapping information. In this study, we employed a dendrogram graph to represent the hierarchy of the groups obtained as a result of clustering as well as a network graph to visualize the extracted association rules. The results of text mining and visualization are discussed in detail in Sect. 5.

5   Results

5.1   Word Frequency Analysis

First, we performed frequency analysis to identify trends in the varying responses toward mobile payments; the top 50 terms are presented in Table 1. In the table, positive adjectives such as "convenient" (n = 473), "novel" (n = 77), and "best" (n = 43) as well as negative adjectives such as "inconvenient" (n = 63) and "troublesome" (n = 33) were found, indicating Korean consumer opinions about Samsung Pay. Information about the unique functions of Samsung Pay compared to other types of payment services (e.g., fingerprinting (n = 37) and iris scanning (n = 31)) were also found. Interestingly, "wallet" (n = 261) and "card" (n = 230) were the 5th and 6th most frequently mentioned terms. These results may suggest the replacement of traditional wallets or credit cards with Samsung Pay. In addition, consumer emotions such as "like" (n = 209), "irritate" (n = 41), and "dislike" (n = 27) as well as behaviors such as "use" (n = 617), "buy" (n = 106), and "change" (n = 46) were found in the table.

5.2 Clustering

Clustering is a process in which terms that belong to a similar topic are clustered together, allowing each cluster to represent a unique theme detected from the data set. The present study employed agglomerative hierarchical clustering, which groups small clusters into larger ones [11]. Before conducting *agglomerative* hierarchical clustering, we conducted the following two preliminary steps to enhance the efficiency and accuracy of clustering and to clearly interpret the results.

First, the results of the above step still yielded too many terms with lower levels of frequency, inhibiting meaningful results from subsequent clustering analyses. Inclusion of these rare terms would have caused excessive noise to enter the clustering process [18]. Thus, based on the results of frequency analysis, we initially selected terms that appeared more than three times, resulting in about 700 terms in the data set. Furthermore, we manually excluded terms that were too general to infer consumer cognitive, affective, and behavioral responses toward Samsung Pay, reducing the number of target terms to 300.

Second, we classified the remaining terms into three groups: cognitive, affective, and behavioral responses. To identify affective responses, we used Laros and Steenkamp [24]'s list of emotions as a reference. After this classification, each response was further divided into positive or negative responses according to valence, resulting in six dimensions (i.e., 3 (cognitive; affective; and behavioral responses) × 2 (positive and negative valences)). In each dimension, we classified together the terms that shared a common meaning or theme and then labeled each category. To ascertain the exact meaning of certain terms (e.g., wallet, battery, and coin), we referred to the original documents in which the terms were mentioned. In order to increase the reliability of these procedures, three researchers cross-checked the results of these categorizations. Whenever discrepancies emerged, discussions were held until all researchers reached a consensus. After these procedures, we obtained 24 categories (11 categories in cognitive responses; 10 categories in affective responses; 3 categories in behavioral responses) that were later used as data for *agglomerative* hierarchical clustering. Tables 2, 3 and 4 show the results of the three consumer responses (i.e., cognitive, affective, and behavioral) regarding valence, category, and other relevant terms. Due to the page limit, we selected only some of the representative terms for each category. The numbers in the brackets indicate the total number of times that all terms in the same category were mentioned in the sample documents.

As shown in Table 2, the positive categories for the cognitive responses included *Usefulness, Innovativeness, Ubiquity, Speed, Ease of use, Economic benefits* and *Security,* reflecting the specific benefits provided by Samsung Pay. The first category, *Usefulness,* comprised terms such as "convenience," "integrated," and "wallet." "Integrated" meant that most credit cards could be integrated into mobile devices, reducing the number of cards carried in a wallet. "Wallet" implied that Samsung Pay

Table 1. The result of term frequency analysis

No.	Terms	N	No.	Terms	N
1	use	617	26	registration	46
2	convenient	473	27	best	43
3	Galaxy	426	28	today	41
4	smart phone	277	29	irritate	41
5	wallet	261	30	gear	41
6	card	230	31	reason	40
7	payment	230	32	calculate	40
8	like	209	33	available	38
9	iPhone	193	34	a whole new world	38
10	SAMSUNG	110	35	fingerprinting	37
11	buy	106	36	deposit and withdrawal	36
12	first	85	37	instant	35
13	money	81	38	fast	34
14	novel	77	39	troublesome	33
15	think	76	40	support	31
16	need	67	41	iris scanning	31
17	function	66	42	come	31
18	transportation card	65	43	mind	29
19	inconvenient	63	44	Korea	28
20	go out	62	45	dislike	27
21	different	57	46	bank	26
22	people	56	47	feel	26
23	online	50	48	T-money	26
24	take out	46	49	know	25
25	change	46	50	innovation	25

could replace a wallet. The terms in this category were the most frequently mentioned in the sample documents (n = 668). The second-most-frequently mentioned category, *Innovativeness* (n = 404), included terms such as "novel," "a whole new world," "perfect," and "innovative," etc. It represented how Korean consumers recognized the innovativeness and novelty of Samsung Pay.

Ubiquity (n = 148) primarily contained terms that represented places where consumers could easily use Samsung Pay, such as "taxi," "bus," "bank," "vending machine," etc. It can be used in almost any existing POS machine that swipes cards because of MST technology and provides consumers with vast purchase possibilities, timely access to financial assets, and an alternative to cash payments for services such as transportation, bank services, and vending machines. The fourth category consisted of terms (e.g., "fast," "immediate," and "in seconds") that tautologically declared *Speed* (70), which measured consumer perceptions of the quick operation of Samsung Pay.

Ease of use (n = 43) contained terms such as "easy," "simple," and "fingerprinting." This category indicated the extent to which Korean consumers perceived

Table 2. Classification of cognitive responses

Valence	Category	Terms (translated in English)
Positive	Usefulness (668)	convenient, integrated, wallet
	Innovativeness (404)	novel, a whole new world, revolution, perfect, smart, new, innovative, technology
	Ubiquity (148)	taxis, bus, bank, vending machine, transportation card, deposits and withdrawals, anywhere, gas station, compatibility, MST (magnetic secure transmission)
	Speed (70)	fast, immediate, in seconds
	Ease of use (43)	easy, simple, fingerprinting, iris scanning
	Economic benefits (14)	discount, event, point, membership service
	Security (10)	secure, safe
Negative	Inconvenience (105)	difficult to refund, update problem, ID (identification) card, coin
	Risk (41)	fall out of hand, battery, damage, failure, system error, internet
	Onerousness (41)	handing over, difficult, heavy, instruction, cumbersome, UI (inconvenient User Interface design)
	Non-ubiquity (24)	affiliates of Shinsegae group, E-mart, CGV, Starbucks, place where it does not work, IC (Integrated Circuit) card exclusive, small restaurant

Table 3. Classification of affective responses

Valence	Category	Terms (translated in English)
Positive	Like (224)	like, nice
	Passionate (92)	best, awesome, jackpot, obsessed, killer app, irreplaceable, God-like Samsung pay, love
	Attractive (42)	attractive, fun, desirable, envious
	Grateful (17)	thankful, relievable
	Satisfied (16)	satisfied, recognized
	Sophisticated (4)	cool, sophisticated
Negative	Irritated (47)	irritated, anger
	Dislike (31)	dislike, disapproving
	Embarrassed (13)	embarrassed, ashamed
	Distrustful (10)	distrustful, afraid, fear

Table 4. Classification of behavioral responses

Valence	Category	Terms (translated in English)
Positive	Intention for continued use (15)	cannot switch, stick to
	Becoming accustomed to (12)	habit, becoming accustomed to
Negative	Impulsive purchasing (15)	impulsive purchasing, excessive purchasing, going over budget, spending more money

Samsung Pay as a service that is easy and simple to use; users only need to select their card in the Samsung Pay app, input their fingerprint, and hold it close to the POS machine to complete the payment procedure. *Economic benefits* (n = 14) included terms such as "discount," "event," "point," and "membership service." Consumers can earn reward points for every purchase made through Samsung Pay and redeem these points for gift cards or other products. *Security* (n = 10) included terms such as "secure" and "safe." The relevant sample documents showed that Korean consumers perceive Samsung Pay as a safe service to use because it has the option to require a fingerprint to authorize every payment.

On the other hand, the negative categories for the cognitive responses included *Inconvenience, Risk, Onerousness,* and *Non-ubiquity. Inconvenience* (n = 105) included terms such as "difficult to refund," "update problem," "ID card," and "coin." The relevant sample documents described several inconvenient situations caused by Samsung Pay such as difficulty with obtaining refunds, unwanted automatic updates of the application, the irreplaceability of ID cards, and the need to carry coins due to unavailability for small transactions. The *Risk* category (n = 41) (e.g., "fall out of hand," "damage," "system error," and "failure") reflected Korean consumer concerns about the payment procedures and system errors of Samsung Pay. For example, some consumers recognized the risk of damaging their mobile devices when handing them to cashiers for offline payment transactions. Others worried that the phone battery might run out or that the network connection could fail in the middle of a payment transaction.

Onerousness (n = 41) (e.g., "handing over," "difficult," "heavy," "instruction," "cumbersome," and "UI design") reflected cumbersome situations such as handing one's phone to the cashier or explaining to cashiers how to conduct payments using the app. In addition, due to the small screens and ambiguous layout designs, some consumers expressed that the user interface could be complicated and that the instructions for making payments were difficult to find. Finally, *Non-ubiquity* (n = 24) addressed the offline stores where Samsung Pay was not available (e.g., "affiliates of Shinsegae group," "E-mart," "CGV," and "Starbucks").

Next, the classification of affective responses is represented in Table 3. The positive categories for the affective responses included *Like, Passionate, Attractive, Grateful, Satisfied,* and *Sophisticated. Like* (n = 224), the most frequently mentioned category of affective responses, included terms such as "like" and "nice." The second-most-frequently mentioned category, *Passionate* (n = 92), contained terms such as "best," "irreplaceable," "god-like," and "love," indicating the strong affection of Korean consumers toward Samsung Pay. The *attractive* category (n = 42) included terms such as "attractive," "fun," "desirable," and "envious." The category of *Grateful* (n = 17) mainly contained terms such as "thankful" and "relievable." The *Satisfied* category (n = 16) indicated the extent to which consumers felt "satisfied" with Samsung Pay or "recognized" by others. The *Sophisticated* category (n = 4) included terms such as "cool" and "sophisticated." The negative categories of affection consisted of *Irritated* (n = 47), *Dislike* (n = 31), *Embarrassed* (n = 13), and *Distrustful* (n = 10). Overall, the terms in these negative categories were mentioned much less frequently than those related to positive affection.

Classification of the behavioral responses is shown in Table 4. We found only three categories of behavioral responses: *Intention for continued use* and *Becoming accustomed to* (positive) and *Impulsive purchasing* (negative). We classified impulsive purchasing as a negative response because it may result in potentially negative consequences such as regret, guilt, or financial problems [37]. Compared to the frequencies of the cognitive and affective responses, the frequency of the behavioral responses was much lower, as the highest level of frequency was only 15. *Intention for continued use* (n = 15) reflected the continued use Samsung Pay over other services, as indicated by terms such as "cannot switch" and "stick to." *Becoming accustomed to* (n = 12) accounted for when consumers grew comfortable with using Samsung Pay and developed it as a habit. *Impulsive purchasing* (n = 15) showed that some consumers tended to make rash purchase decisions because of the convenience of mobile payment.

Finally, the results of the agglomerative hierarchical clustering are displayed using a dendrogram, allowing estimation of the number of clusters based on the distances of the fused clusters [17] and illustrating the fusions made at each stage of the analysis. The nodes of the dendrogram represent *clusters* and the *heights* represent the distances at which the clusters are joined [11]. The higher the *height* of the fusion, the less similar the clusters. *Ward*'s method has been widely used in hierarchical methods; it produces groups that minimize the increase in the total within-cluster error sum of squares [41, 48]. In the present study, the distances between words were calculated with *dist()*, and the terms were clustered with *hclust()* using R programming. The dendrogram was cut at a seven-height threshold. *Ward's* method was used as the agglomerative method in this study.

The dendrogram of the clusters appears in Fig. 1. The words were split clearly into positive and negative responses at the macro level. The positive responses were further clustered into the following four groups: {*Like, Satisfied*}, {*Attractive, Passionate, Intention for continued use, Security, Becoming accustomed to, Grateful, Economic benefits*}, {*Ease of use, Impulsive purchasing*}, {*Speed, Ubiquity, Sophisticated*}. The negative responses were further clustered into two groups: {*Dislike, Onerousness, Risk, Distrustful*} and {*Irritated, Inconvenience, Non-ubiquity, Embarrassed*}.

While some of these groups (clusters) appeared to be imperfectly formed and ostensibly contrary to intuition, the relationships patterns that appeared frequently in these clusters were (1) cognitive-affective responses (e.g., *Risk, Distrustful*; *Non-ubiquity, Embarrassed*), (2) cognitive-behavioral responses (e.g., *Ease of use, Impulsive purchasing*), and (3) affective-behavioral responses (e.g., *Passionate, Intention for continued use*).

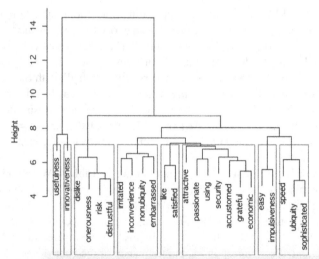

Note: using = intention for continued use; accustomed = becoming accustomed to; economic = economic benefits; impulsiveness = impulsive purchasing

Fig. 1. Dendrogram of clustering

The remaining clusters, *Usefulness* and *Innovativeness*, were fused rather arbitrarily at a higher height. As the most frequently mentioned categories in the sample documents, they served as the two dominant themes associated with Samsung Pay in social media.

5.3   Association Analysis

An association analysis is typically performed for the purpose of discovering relationships among a set of responses in a data set, and association rules are applied for this purpose [43]. In the present study, we used the *Apriori* algorithm to generate association rules [1]. The parameters for our *Apriori* algorithm were *support, confidence*, and *lift*.

We followed Mahgoub et al. [30]'s definition of association rules. Given a set of terms $A = \{w_1, w_2, \cdots, w_n\}$ and a collection of indexed documents $D = \{d_1, d_2, \cdots, d_m\}$, where each document d_i is a set of terms such that $d_i \subseteq A$, let W_i be a set of terms. A document d_i is said to contain W_i if and only if $W_i \subseteq d_i$. An association rule is an implication of the form $W_i \Rightarrow W_j$, where $W_i \subset A$, $W_j \subset A$, and $W_i \cap W_j = \varphi$.

The rule $W_i \Rightarrow W_j$ has *support* (s) in the collection of documents D if s % of documents in D contains $W_i \cup W_j$. *Support* measures how frequently an association rule occurs in the entire set of transactions and is calculated by the following formula:

$$Support(W_iW_j) = \frac{Support\ count\ of\ W_iW_j}{Total\ number\ of\ documents\ D}$$

The rule $W_i \Rightarrow W_j$ has *confidence* (c) in the collection of documents D if among those documents that contain W_i, c % of them also contains W_j together. Confidence measures the strength of the association rules and is calculated by the following formula:

$$Confidence(W_iW_j) = \frac{Support(W_iW_j)}{Support(W_i)}$$

Lift is another important measure for association rules [3]. To rank the most interesting rules, we used the lift index, which measures the (symmetric) correlation between W_i and W_j. *Lift* is the ratio of *confidence* to the probability that a consequent W_j occurs in the data set (Park et al. 2017). If lift is larger than 1, the rule is potentially useful for predicting consequents in future data sets [34].

$$Lift(W_iW_j) = \frac{confidence(W_iW_j)}{Support(W_j)}$$

Rules that involve many terms are hard to interpret and can potentially generate a high number of rules [33]. Thus, the association rule needs to be constrained by both minimum *support* and *confidence* thresholds to find interesting rules. In this study, due to the relatively small size of the data set, a minimum *support* threshold of 5% was used to ensure that as many rules were generated as possible [14]. A minimum *confidence* threshold of 50% was used to ensure the degree of interest of these rules. Furthermore, the rules were constrained by *lift* value, which was greater than 1 and sorted by its value. As a result, the association rules produced a total of 385 rules in this study.

In addition, based on our domain knowledge, we decided which sets of antecedents and consequents should be selected to extract useful information in this study. As mentioned above, the results of clustering analysis indicated the following relationship patterns: (1) cognitive-affective responses, (2) cognitive-behavioral responses, and (3) affective-behavioral responses. Because consumer cognitive responses can influence affect and behavioral responses [22], we constrained cognitive responses to appear only in the antecedent, while affective and behavioral responses appeared only in the consequent of association rules. We excluded the relationship between affective and behavioral responses and removed the rules that were counter-intuitive to mobile payment service responses. These procedures further reduced the number of association rules to 65. Because the significance of a rule is determined by using the lift value, the top 15 association rules for responses are listed in Table 5 based on lift values.

For the association between *Economic benefits* and *Grateful*, the confidence value was 1.00, i.e., the maximum level of association strength, suggesting that 100% of the documents addressing *Economic benefits* also addressed *Grateful*. The lift value (3.43) implies that *Economic benefits* was 3.4 times more likely to involve *Grateful* than other responses. In addition, *Economic benefits* tended to lead to *Satisfaction* (lift = 2.00; confidence = 0.67), *Intention for continued use* (lift= 2.40; confidence = 0.50), and *Becoming accustomed to* (lift = 1.71; confidence = 0.50). *Security* tended to occur

Table 5. Top 15 association rules of responses

	Antecedent		Consequent	Support	Confidence	Lift
1	{economic benefits}	=>	{grateful}	0.25	1.00	3.43
2	{economic benefits}	=>	{Intention for continued use}	0.13	0.50	2.40
3	{security}	=>	{satisfied}	0.21	0.71	2.14
4	{economic benefits}	=>	{satisfied}	0.17	0.67	2.00
5	{security}	=>	{grateful}	0.17	0.57	1.96
6	{security}	=>	{attractive}	0.21	0.71	1.91
7	{ease-of-use}	=>	{impulsive purchasing}	0.29	0.70	1.87
8	{ease-of-use}	=>	{attractive}	0.29	0.70	1.87
9	{ease-of-use}	=>	{satisfied}	0.25	0.60	1.80
10	{economic benefits}	=>	{becoming accustomed to}	0.13	0.50	1.71
11	{non-ubiquity}	=>	{embarrassed}	0.33	0.67	1.78
12	{ease-of-use}	=>	{grateful}	0.21	0.50	1.71
13	{risk}	=>	{dislike}	0.29	0.50	1.71
14	{onerousness}	=>	{embarrassed}	0.38	0.60	1.60
15	{ease-of-use}	=>	{passionate}	0.25	1.00	1.60

alongside positive feelings such as *Satisfied* (lift = 2.14; confidence = 0.71) and *Grateful* (lift = 1.96; confidence = 0.71).

Interestingly, *Ease of use*, a cognitive response, was associated with not only positive but also negative responses. It tended to lead to certain positive feelings such as *Attractive* (lift = 1.87; confidence = 0.70) and *Passionate* (lift = 1.60; confidence = 1.00) and occurred with *Impulsive purchasing* (lift= 1.87; confidence = 0.70).

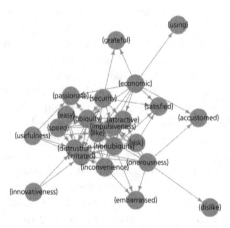

Note: using = intention for continued use; accustomed = becoming accustomed to
economic = economic benefits; impulsiveness = impulsive purchasing

Fig. 2. Networks of association rules of responses.

In addition, *Non-ubiquity* (lift = 1.78; confidence = 0.67) and *Onerousness* (lift = 1.60; confidence = 0.60) tended to occur alongside *Embarrassed*.

Figure 2 visualizes all of the extracted association rules. The nodes characterize the antecedent or the consequent of the association rules. The most central terms were affective responses, such as *Like* and *Attractive*, which were related to various cognitive responses. On the other hand, *Intention for continued use* was only connected to *Economic benefits* (Fig. 2).

6 Discussions, Implications, and Limitations

This study explored the cognitive, affective, and behavioral responses of Korean consumers toward Samsung Pay, a mobile payment service in Korea, and the relationships among these responses based the tri-component model. Data were collected from social media and analyzed using a text-mining technique. A discussion, implications, and limitations are provided as follows.

The results of the manual classifications of the preprocessed data indicated 24 categories (11 categories in cognitive responses, 10 categories in affective responses, and 3 categories in behavioral responses). The positive categories of the cognitive responses were *Usefulness, Innovativeness, Ubiquity, Speed, Ease of use, Economic benefits,* and *Security. Usefulness* was the most frequently mentioned category of all. The results indicated that Korean consumers perceive Samsung Pay as a useful service because it can replace "wallets" and "integrate" multiple credit cards. These useful features have been overlooked in previous studies that employed survey measurements using rather abstract questions [6, 28]. *Innovativeness*, the second-most-frequently mentioned category, referred to the novelty of Samsung Pay as reflected by the term "a whole new world." *Ubiquity* referred to the availability of products/services for consumers to use in a variety of transactions anytime and anyplace. In this study, Korean consumers seemed to perceive ubiquity as one of Samsung Pay's premier features because it allows usage "anywhere" such as "taxis" and "vending machines." This theme has been neglected in previous studies, which examined similar issues by only using surveys [19, 40]. We infer that these perceptions of *Ubiquity* are possible because of Samsung Pay's MST solution, which allows the service to be compatible with existing card readers.

The negative categories for the cognitive responses included *Inconvenience, Risk, Onerousness,* and *Non-ubiquity. Inconvenience* referred to situations such as difficulty with obtaining refunds, unwanted automatic updates of the application, and the irreplaceability of ID cards. It would be desirable for Samsung Pay to improve the system by allowing consumers to receive quicker refunds, to mount ID cards, and to eliminate updating problems. *Risk* reflected some concerns related to technological malfunctioning and the possibility of damaging mobile devices during offline transactions such as handing a device to a cashier. The latter is an issue only relevant to Samsung Pay because of its use of MST solutions. Samsung Pay is advised to resolve these problems to reduce perceived risk. Interestingly, the issue of personal information extrusion was not extracted in this study, although it has served as an important determinant for adopting mobile payment services in previous research [35, 45]. Furthermore, *Onerousness*, the counterpart to perceived *Ease of use,* reflected situations in which consumers had to give their devices to a cashier to make offline transactions or to instruct a cashier about how to conduct transactions using Samsung Pay. Therefore, Samsung Pay is advised to encourage affiliated retailers to educate their clerks about payment methods as well as to enable consumers to directly connect their mobile devices to a payment terminal when making a payment. Also, it would be desirable for Samsung Pay to present consumers with a more well-designed interface with a clearer layout and wider screens. Finally, despite the importance of *Ubiquity, Non-ubiquity* also emerged from the study, particularly in offline purchasing situations in which Korean consumers are refused to pay using Samsung Pay because of the unavailability of the service in the store. It appears that these negative responses tend to be more prevalent among offline consumers than online consumers.

For affective responses, we discovered various types of positive and negative affects that have been overlooked in previous studies [22]. The positive categories included *Like, Passionate, Attractive, Grateful, Satisfied,* and *Sophisticated,* and the negative categories consisted of *Irritated, Dislike, Embarrassed,* and *Distrustful.* Overall, the negative terms were mentioned less frequently than the positive terms. Three behavioral categories were extracted: *Intention for continued use, Impulsive purchasing,* and *Becoming accustomed to.* The first two had already been examined in previous studies [13], but *Becoming accustomed to* was newly revealed in the present study and may help consumers build a switching barrier for Samsung Pay by decreasing the costs usage. Previous studies have focused only on adoption and the usage of mobile payment services [8, 29].

The results of the agglomerative hierarchical clustering based on the 24 categories of responses toward Samsung Pay illustrate that the responses were clearly divided into two groups: positive vs. negative responses at the macro level. Furthermore, we found three patterns of clustering at the micro level: cognitive-affective responses, cognitive-behavioral responses, and affective-behavioral responses. These findings confirm that cognitive, affective, and behavioral responses are interrelated, as suggested by previous studies [22, 29, 47].

Based on the results of association analysis, this study uncovered interesting relationships among the three types of responses. First, *Economic benefits* led to *Grateful* and *Satisfied.* Kim and Rha [22] argued that economic benefits were the key determinant of positive affects toward mobile payment services, but they could not identify the specific types of affection because they were limited by their survey method. This study revealed that Korean consumers tend to feel satisfaction and gratitude upon receiving economic benefits from promotions such as discounts and reward points. Furthermore, *Economic benefits* was found to occur alongside behavioral responses such as *Becoming accustomed to* and *Intention for continued use,* implying that *Economic benefits* can serve as an incentive for new users to become comfortable with Samsung Pay and eventually decrease the possibility of switching to other payment services. Therefore, it is important for mobile payment service firms to establish effective promotion programs that offer economic benefits to consumers to increase their satisfaction, gratitude, and ultimately, loyalty.

Next, *Security* was found to lead to *Satisfied,* demonstrating that Korean consumers were satisfied with the security of their payment transactions thanks to the enhanced security solutions of Samsung Pay. Several researchers mentioned the security of mobile payment services as an important prerequisite for popular use of mobile payment, arguing that risks related to security and privacy may reduce positive feelings [23, 35]. Mobile payment service firms are thus advised to provide reliable and secure services in order to enhance consumer satisfaction.

Furthermore, we found that *Ease of use* occurred alongside *Attractive* and *Satisfied*, suggesting that Korean consumers tended to be attracted to and satisfied with Samsung Pay if its payment procedures were simple for them to use. This result is consistent with those of previous studies, which found that perceived ease of use is a key factor leading to an individual's attitude toward mobile payment services [28, 35]. On the other hand, *Ease of use* also led *Impulsive purchasing*, implying that Korean consumers may need the self-discipline to guard against rash purchase decisions when utilizing easy-to-use services to make payments.

Finally, *Non-ubiquity* and *Onerousness* were found to lead Korean consumers to feel *Embarrassed*. Considering that issues related to non-ubiquity and onerousness were prevalent in offline stores, it would be beneficial for Samsung Pay to educate retailers about usage of the service and to extend its affiliations with more retailers, banks, and card providers to prevent consumers from experiencing embarrassment when using the service.

This study contributes to the consumer-research discipline and to practitioners in the mobile payment service industry by providing an in-depth investigation of consumer responses toward mobile payment services. However, this research has several limitations, which may provide avenues for future study. First, because we only focused on the use of Samsung Pay among Korean consumers, caution should be taken when generalizing these findings. To enhance external validity, future studies need to examine these issues while using more diverse sample sets of consumers and types of mobile payment services. Second, we performed clustering manually based on a theory before conducting agglomerative hierarchical clustering, which might have limited the size of the sample documents deemed appropriate for analysis. Therefore, future research is advised to employ more advanced analytic tools such as machine-learning techniques to extract further insight from bigdata. Finally, the present study analyzed only two types of associations (i.e., cognition-affection & cognition-behavior). Therefore, future studies are advised to examine more diverse associations such as affection-behavior.

References

1. Agrawal, R., Srikant, R.: Fast algorithms for mining association rules. In: Proceedings of 20th International Conference on Very Large Data Bases. VLDB, vol. 1215, pp. 487–499, September 1994
2. An, J., Lee, S., An, E., Kim, H.: Fintech trends and mobile payment service analysis in Korea: application of text mining techniques. Inf. Policy **23**(3), 26–42 (2016)
3. Angeline, D.M.D.: Association rule generation for student performance analysis using apriori algorithm. SIJ Trans. Comput. Sci. Eng. Appl. (CSEA) **1**(1), 12–16 (2013)

4. Bagozzi, R.P.: Attitude formation under the theory of reasoned action and a purposeful behaviour reformulation. Br. J. Soc. Psychol. **25**(2), 95–107 (1986)
5. Cao, Q., Duan, W., Gan, Q.: Exploring determinants of voting for the "helpfulness" of online user reviews: a text mining approach. Decis. Support Syst. **50**(2), 511–521 (2011)
6. Cheong, J.H., Park, M.C., Hwang, J.H.: Mobile payment adoption in Korea: switching from credit card. In: Paper Presented at the 15th Biennial Conference (2004)
7. Choi, A.: A study on utility & disutility in consumers' smart-tech experience. Master's Degree thesis, Seoul National University (2014)
8. Dai, H., Hu, T., Zhang, X.: Continued use of mobile technology mediated services: a value perspective. J. Comput. Inf. Syst. **54**(2), 99–109 (2014)
9. Desmet, P., Hekkert, P.: Framework of product experience. Int. J. Des. **1**(1), 57–66 (2007)
10. DMC Media: 2016 the status and prospects of mobile payment service market, Issue & Trend. Digital Media Convergence Company (2016)
11. Everitt, B.S., Landau, S., Leese, M., Stahl, D.: Cluster Analysis. Wiley series in probability and statistics. Wiley, Hoboken (2014)
12. Fan, W., Wallace, L., Rich, S., Zhang, Z.: Tapping the power of text mining. Commun. ACM **49**(9), 76–82 (2006)
13. Go, C., Han, E.K.: Mobile simple payment attributes and intention to use: focused on non-consumers. J. Korean Soc. Hazard Mitig. **16**(2), 191–196 (2016)
14. Goh, D.H., Ang, R.P.: An introduction to association rule mining: an application in counseling and help-seeking behavior of adolescents. Behav. Res. Methods **39**(2), 259–266 (2007)
15. Holbrook, M.B., Hirschman, E.C.: The experiential aspects of consumption: consumer fantasies, feelings, and fun. J. Consum. Res. **9**(2), 132–140 (1982)
16. Hong, J., Oh, I.: Image difference of before and after an incident using social big data analysis. Int. J. Tour. Hosp. Res. **30**(6), 119–133 (2016)
17. Hotho, A., Nürnberger, A., Paaß, G.: A brief survey of text mining. LDV Forum **20**(1), 19–62 (2005)
18. Huang, A.: Similarity measures for text document clustering. In: Proceedings of The Sixth New Zealand Computer Science Research Student Conference, Christchurch, New Zealand, pp. 49–56 (2008)
19. Johnson, V.L., Kiser, A., Washington, R., Torres, R.: Limitations to the rapid adoption of M-payment services: understanding the impact of privacy risk on M-payment services. Comput. Hum. Behav. **79**, 111–122 (2018)
20. Kaplan, A.M., Haenlein, M.: Consumer of the world, unite! the challenges and opportunities of social media. Bus. Horiz. **53**(1), 59–68 (2010)
21. Kim, C., Mirusmonov, M., Lee, I.: An empirical examination of factors influencing the intention to use mobile payment. Comput. Hum. Behav. **26**(3), 310–322 (2010)
22. Kim, H.J., Rha, J.Y.: Application of the stimulus-organism-response model on consumer's continued intention to use mobile payment services. J. Korean Home Manag. Assoc. **34**(4), 139–156 (2016)
23. Kleijnen, M., Wetzels, M., De Ruyter, K.: Consumer acceptance of wireless finance. J. Finan. Serv. Mark. **8**(3), 206–217 (2004)
24. Laros, F.J.M., Skeenkamp, J.B.E.M.: Emotions in consumer behavior: a hierarchical approach. J. Bus. Res. **58**(10), 1437–1445 (2005)
25. Li, N., Wu, D.D.: Using text mining and sentiment analysis for online forums hotspot detection and forecast. Decis. Support Syst. **48**(2), 354–368 (2010)
26. Liu, B.: Sentiment analysis and opinion mining. Synth. Lect. Hum. Lang. Technol. **5**(1), 1–167 (2012)

27. Liébana-Cabanillas, F., Muñoz-Leiva, F., Sánchez-Fernández, J.: A global approach to the analysis of user behavior in mobile payment systems in the new electronic environment. Serv. Bus. **12**(1) 25–64 (2018)

28. Liébana-Cabanillas, F., Sánchez-Fernández, J., Muñoz-Leiva, F.: Antecedents of the adoption of the new mobile payment systems: the moderating effect of age. Comput. Hum. Behav. **35**, 464–478 (2014)

29. Lu, Y., Yang, S., Chau, P.Y., Cao, Y.: Dynamics between the trust transfer process and intention to use mobile payment services: a cross-environment perspective. Inf. Manag. **48** (8), 393–403 (2011)

30. Mahgoub, H., Rösner, D., Ismail, N., Torkey, F.: A text mining technique using association rules extraction. Int. J. Comput. Intell. **4**(1), 21–28 (2008)

31. Maynard, D., Bontcheva, K., Rout, D.: Challenges in developing opinion mining tools for social media. In: Proceedings of the @NLP can u tag# consumergeneratedcontent, pp. 15–22 (2012)

32. Mostafa, M.M.: More than words: social networks' text mining for consumer brand sentiments. Expert Syst. Appl. **40**(10), 4241–4251 (2013)

33. Ordonez, C., Omiecinski, E., De Braal, L., Santana, C.A., Ezquerra, N., Taboada, J.A., Garcia, E.V.: Mining constrained association rules to predict heart disease. In: Proceedings IEEE International Conference on Data Mining. ICDM 2001, pp. 433–440. IEEE (2001)

34. Park, S.H., Synn, J., Kwon, O.H., Sung, Y.: Apriori-based text mining method for the advancement of the transportation management plan in expressway work zones. J. Super comput. **74**(3), 1283–1298 (2018)

35. Rahki, T., Mala, S.: Adoption readiness, personal innovativeness, perceived risk and usage intention across customer groups for mobile payment services in India. Internet Res. **24**(3), 369–392 (2014)

36. Rahm, E., Do, H.H.: Data cleaning: problems and current approaches. IEEE Data Eng. Bull. **23**(4), 3–13 (2000)

37. Rook, D.W.: The buying impulse. J. Consum. Res. **14**(2), 189–199 (1987)

38. Salehan, M., Kim, D.J.: Predicting the performance of online consumer reviews: a sentiment mining approach to big data analytics. Decis. Support Syst. **81**, 30–40 (2016)

39. Schierz, P.G., Schilke, O., Wirtz, B.W.: Understanding consumer acceptance of mobile payment services: an empirical analysis. Electron. Commer. Res. Appl. **9**(3), 209–216 (2010)

40. Shin, D.H.: Towards an understanding of the consumer acceptance of mobile wallet. Comput. Hum. Behav. **25**(6), 1343–1354 (2009)

41. Ward, J.H.: Hierarchical groupings to optimize an objective function. J. Am. Stat. Assoc. **58**, 234–244 (1963)

42. Weber, R.H., Darbellay, A.: Legal issues in mobile banking. J. Bank. Regul. **11**, 129–145 (2010)

43. White, C.J., Scandale, S.: The role of emotions in destination visitation intentions: a cross-cultural perspective. J. Hosp. Tour. Manag. **12**(2), 168–178 (2005)

44. Xiao, S., Dong, M.: Hidden semi-markov model-based reputation management system for online to offline (O2O) e-commerce markets. Decis. Support Syst. **77**, 87–99 (2015)

45. Yang, S., Lu, Y., Gupta, S., Cao, Y., Zhang, R.: Mobile payment services adoption across time: an empirical study of the effects of behavioral beliefs, social influences, and personal traits. Comput. Hum. Behav. **28**(1), 129–142 (2012)

46. Yuan, J.J., Morrison, A.M., Cai, L.A., Linton, S.: A model of wine tourist behavior: a festival approach. Int. J. Tour. Res. **10**(3), 207–219 (2008)

47. Zajonc, R.B., Markus, H.: Affective and cognitive factors in preferences. J. Consum. Res. **9** (2), 123–131 (1982)
48. Zhao, Y.: R and Data Mining: Examples and Case Studies. Academic Press, Cambridge (2012)

An Exploration of Crowdwork, Machine Learning and Experts for Extracting Information from Data

Fabion Kauker[(⊠)], Kayan Hau, and John Iannello

Biarri Networks Pty Ltd., Melbourne VIC 3000, Australia
fabion.kauker@biarri.com

Abstract. The growing use of data to derive insights and information presents many challenges and opportunities. Further, the increased awareness of the potential of crowdworking and machine learning technologies has created a need to understand the benefits and caveats of these approaches. By reviewing current research and then comparing a novice based crowdworking approach against experts and machine learning benchmarks we seek to assess the trade-offs. The task specifically requires users to interpret satellite imagery and determine the location of residences or businesses. We are able to demonstrate that a novice approach can provide value where the data collected meets an accuracy tolerance that closely matches the expert users. Further, the potential for equivalent results is shown to be possible based on potential improvements to the system and user familiarity with the task.

Keywords: Business integration · Data · Machine learning · Crowdwork
AI · GIS · Mapping · Address geolocation

1 Introduction

We undertake a literature review of research and applications of computer systems for data gathering/analysis. We focus on the use of human computation to derive information [17, 43, 59]. Examples include the research of user experience in these systems, which can be termed an "auto-ethnography" [37], exploring the perspectives of use cases which involve data collection [6]. The data science community relies on the availability and/or creation of labelled data to create supervised learning models [60]. This data must be structured in such a way that there are abundant examples of an input and output, other data sets are being linked [23]. Understanding the cost, methods, quality and assumptions for this data is paramount. Some examples of data gathering include image/video creation/analysis, object identification/localization, social interaction, market research, sentiment analysis, audio transcription, language translation and data cleaning. These labelling tasks can be completed by a number of types of participants, including domain experts, paid participants who are trained and even crowdworkers. Moreover the framing of these tasks for crowdwork can be termed a Human Intelligence Task (HIT) [73]. We aim to determine which method/s perform best at extracting useful information from data.

© Springer International Publishing AG, part of Springer Nature 2018
S. Yamamoto and H. Mori (Eds.): HIMI 2018, LNCS 10904, pp. 643–657, 2018.
https://doi.org/10.1007/978-3-319-92043-6_51

The three categories of approach examined are novice, expert and automated. For the novice-based approach we utilize the Amazon MTurk system and explore the trade-offs encountered when gathering derived data from satellite imagery. Examples of such systems include the identification of objects like boats or roof tops. The research conducted aims to gather and improve the quality of address locations. Some of the uses of this data include emergency services dispatch and infrastructure planning. Whilst data sets are available and techniques like geocoding can be used to enhance/augment the data, it can be challenging finding the ground truth. This data is constantly evolving as cities develop, and data capture is typically latent if it exists [11]. Typically, the more accurate and descriptive a data set the higher price it commands. The approach seeks to test the financial viability of a low-cost alternative to current commercial products. By utilizing currently open data platforms the research is able to be rapidly explored. We also seek to compare the effort and assumptions of the variations in these types of systems [22].

The specific task termed "address geolocation" requires the model or participant to analyze the spatial context of a satellite image and identify the living residence or business by creating a marker at that location. There has been previous automated work that examines the use of street view data to estimate demographics [13]. Our task however, requires the user in the manual case to interpret an image, move the mouse to where the point is to be created and click once. This is captured, validated and stored as shown in Fig. 1.

Fig. 1. Before and after – address points created on rooftops/residences.

An automated approach is compared with a crowd of novices and the expert group. The results examine the trade-offs between timeliness, quality and quantity. The specifics of the approach and workflow differ for each and demonstrate that it is important to take into account a number of factors when selecting the approach. However, our interaction with the process remains the same. In all cases the baseline data is sourced from OpenStreetMap (OSM) and openaddress.io [39, 65]. The base satellite imagery is delivered via the Google Maps API [62]. The effort, means, rate and quality of the task varies for each method.

The automated approach follows a typical machine learning methodology. Using supervised learning the existing data is sourced from the SpaceNet challenge and is overlaid on satellite imagery. From this, samples are taken showing just the imagery and then with the desired rooftop polygons which are exported as images. This input is then used to train the models. Upon training completion, the model is used to predict where the rooftops should be located. Setting up and running this process takes time, custom software development, specialized hardware, as well as the utilization of various open source packages.

From the SpaceNet challenge there are a number of competitive submissions including the work of XD_XD whose model is able to get a score of 0.885 on the Las Vegas, NV data and an overall average score of .693 across the four areas [65]. This highlights the variance of result across geographies. One of the challenges of maintaining a modeling approach is to ensure that the same level of accuracy can be achieved consistently. The robustness of the model is subject to the techniques used in the model as well as the parameters set during training.

The novice system utilizes a software driven approach where users interact with the tasks through their device. Images are created at a resolution that enables users to easily distinguish rooftops in residential areas. However, it is challenging to determine which rooftops are residences and which are garages or storage facilities. This requires a user to decide and apply a human level of understanding. It may be possible to describe the heuristic and interaction steps; however, this is not part of the scope of this research. In order to validate the data collection, a set of verification examples is used to validate the results.

Lastly, the expert approach places less demands on software development and is able to use off the shelf open source software, including a Geographic Information System (GIS) called QGIS [42]. The expert group is defined as users that have at least three months experience using a GIS package and can easily create, interpolate and manipulate data.

In industry it is also typical to combine these approaches to ensure that they yield the ideal results. Typically, on larger scale applications a crowdsourced approach can be used to train a machine learning system [73]. This can even be done recursively. This system can in turn be used to further validate the crowd work and retrain the model/s, with exceptions being handled by experts. The increased accuracy of machine learning systems to near expert levels raises the importance of assessing the requirements for such systems and the trade-offs involved [30, 44, 55]. By comparing the three types of solutions we are able to assess the suitability for this use case as well as develop a deeper understanding of what is required to create, deploy, maintain and asses these types of systems.

2 Literature Review

When assessing what solution approach may be the best fit, it is important to consider the implications and defensibility of the solution. Since the inception of computation, practitioners have sought to create information systems that are able to meet or exceed human level performance; whether this be through taking advantage of structured data, calculation or by using machine learning techniques to test potential solutions to NP-hard problems. The success of such implementations has relied on the knowledge of the practitioners, as well as the current state of technology. The evolution of Artificial Intelligence (AI)/Machine Learning (ML) systems is challenged by the trade-offs of black box v. glass box implementations. This is further extended to the concept of Explainable AI (XAI) [12]. Understanding the performance, applicability and capability of a model to a specific task is explored by Holzinger et al. [19]. Presenting an extensive historical perspective and a demonstration of how a human in the loop may present improvements to the Traveling Salesman Problem (TSP). The combination of ML and humans is termed interactive Machine Learning (iML) [45]. The research conducted is motivated by potential applications in protein folding. A domain problem that has previously been explored through crowdsourcing, this is demonstrated by FoldIt [9] a community of over 57,000 participants who outperformed traditional algorithmic approaches. These approaches are able to be successful in applications where the solution space is complex and of a high dimension.

However, when using systems like Amazon's Mechanical Turk (MTurk) to cheaply and rapidly gather or fill gaps in data, the quality of the resulting data/labels must be assed [3, 5–7, 10, 16, 24, 26, 27, 50, 51]. The matching of workers to tasks has also been explored [61]. MTurk has also been successfully used to label images. ImageNet has a large number of labelled images which are used as a performance benchmark [48, 63]. A potential for bias or variation in results may be created by assumptions made in the methods/structure/content of tasks. There is potential to mitigate this and process the data directly into a machine learning algorithm and the approach has been successfully tested [46]. Snow et al. [51] explore the gathering of data annotations and show that it is possible to yield results at a comparable level of quality for a reduced cost in less time, which can then be used to develop ML models. There also exist novel art experiments that seek to explore the potential of the technologies [28, 35, 74].

There also exist fully automated, automatic Machine Learning (aML) approaches that can meet or exceed human expert level of competence. One approach involves the use of games and video games as a test bed to achieve the goal of human level performance in a rich controlled environment. Examples of specific applications exist where practitioners are able to successfully develop simulations to model complex real-world scenarios. However, these come at a high cost and require extensive domain knowledge (Laird et al. 2001). More recently, human level performance and/or mastery has been demonstrated in language translation, sentiment analysis, driving and some types of medical imaging [21, 30, 41, 44, 56]. More recently games such as Go, chess and poker have been tested, as well as the computer games, Starcraft and Dota 2 [34, 36, 38, 40, 58]. These approaches do not bake in the rules or strategies humans have used explicitly. Rather, the behavior is learned from sample data as well the utilization of a technique called reinforcement learning.

The use of information systems and computation to augment human performance in tasks is also explored by Best-Rowden et al. [3] and demonstrates that it is possible to enhance performance by combining human and computer systems [52]. Whilst such systems may be successful there are examples where the converse is true. There are also applications that require a human, the CAPTCH system for example [8]. Farid and Dressel (2018) present an example where alternative models and humans are able to nearly match and/or out perform a current model "COMPAS". This presents a number of challenges regarding the deployment and evolution of models that directly impact individuals and society. This reinforces the importance of considering the modeling assumptions when deploying systems for production. The impact of false positives, false negatives, data representation, overfitting, model retraining/replacement need to be explored. However, with techniques like deep learning it is not always evident why systems present certain behaviors. The effectiveness of AI and the progress towards general intelligence is discussed further by Katz [25]. The philosophical exploration of the impact of AI is important but not within scope of this research. Whilst the definition of what is "human level" performance is grey, we limit it to a specific outcomes-based evaluation for a specific task.

Given the previous examples it is not clear that one type of system is suitable for all tasks nor which is best suited for a specific task. Results are given for specific examples and it remains challenging to extrapolate from this. Therefore, we aim to test the applicability of novices, experts and automated approaches. The methods assume the use of computing for data gathering, it is important to be cognizant of the different human computer interactions, whether they be the practitioner programming the system or the participant completing the task [2, 4, 18, 31, 57, 70].

3   Motivation

When solving data gathering problems through the use of computers there are many options. It is often easier to take the approach that is most familiar or has the least amount of effort required. Rather than examine the trade-offs and select the most effective. The availability of cheap easy to access digital labour is key to Amazon MTurks offering. There has been some work done to estimate the demographics, size of the work force and their availability [13, 47]. However, it is hard to get a measure of the suitability of these workers for specific tasks. Both in terms of the timeliness and quality of work that is completed. Is it worth engaging in a 58MTurk process? How does the resulting work compare to the alternatives of experts and automated modeling approaches?

There are many incomplete data sets and many physical mediums that require digitization. This data can be used to shed light on new insights, make storage and retrieval easier and can even be used to automate processes. The curation of data sets is a very involved process. When seeking to determine the quality of a data set benchmarks regarding quality and quantity are useful. However, often it is not until an automated modeling process ingests the data that all facets of a data set can be reconciled. Mislabelling and categorization are key issues. Therefore, these techniques can also be used as a validation mechanism to improve the state of data sets.

4 Experiment Design

Data gathering for home and business location is key to a number of domains. Everything from the postal service through to infrastructure planning. There is no global body that oversees a database that links an address to a point on earth [49]. We seek to collect and validate the address geolocation points for houses and businesses. In order to draw valid comparisons, the same task is assigned to the novice and expert user groups. The results are then compared against a known data set sourced from openaddress.io. Whilst this data is not perfect it does provide good measures quantitatively and qualitatively against a known set of data that is used today. This is then compared against the current results from the SpaceNet challenge, where polygons are extracted from satellite imagery.

However, this must only be used as a proxy as the derived information regarding whether the building is a living residence or business is not given. For simplicity it is assumed that given a reasonable amount of time a model could be developed that takes the raw imagery and/or the footprints and predicts whether a polygon fits these categorizations. Labelled data is currently available at OSM buildings as well as from machine learning projects [64, 66, 69, 71]. The performance of such a system could be able to match the current level of accuracy for rooftop extraction.

An area approximately 2 km squared in Las Vegas is selected to conduct the experiment. This area is then used in both the expert and novice process. The ground truth is established from the data available on openaddress.io. Each process follows the same steps, but the actions vary for each. The steps are: select, distribute, complete and finally review. It may be necessary to run this process multiple times to ensure that the data is more accurate. To make the data usable in downstream tasks a data cleaning process may be undertaken.

4.1 Novice

We have created a framework which enables the requestor to select an area to conduct data gathering. This consists of a web-based application that interacts with a database. Upon initiating an area, it is cut up into discrete areas at a zoom level with high detail such that the user can clearly distinguish houses in the satellite imagery. The imagery is presented via the Google Maps API. The tasks are then created on Amazon MTurk. Instructions are given through an animated gif that shows an example task being completed. A set of text instructions is also given:

> *"Locate all of the locations on the screen. Just move the mouse to the location and click. Once you have found all the locations click "Submit" When you have completed your sequence you will be given a unique payment code."*

Upon accepting a task, the user is taken to the first location to examine. Once complete, the application automatically moves the map to the next location. Users are unable to move the map themselves to ensure that the data is only captured for areas of interest. The user is asked to move the mouse to the desired locations and click on the points of interest. These are then submitted via the click of a button. The map then moves to the next location. This minimizes the need for the user to scroll through the map and adjust the zoom level. Upon completing five locations the user is given a code to receive payment in the MTurk platform. Therefore, a collection of five subtasks is one HIT. The results are then verified by querying the MTurk API. This shows which workers have submitted the code. This is cross referenced against the data inputted into the system and only submissions of good quality are accepted (Fig. 2).

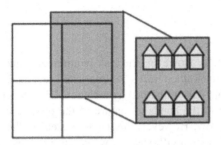

Fig. 2. Left: example work area grid, the whole region to be covered; Right: a single sub task area as seen by the user.

4.2 Expert

Users are given a polygon area which matches the area covered by the Novice group. The experts utilize free open source software QGIS to create the data. A layer is created and then they are able to click on the locations where they deem a house or business to be. From a HCI point of view the users have a very rich set of features to move, zoom and add other data views. This is contrasted against the very constrained environment created for novices. It can be challenging to master GIS software as it is very complex [54].

In order to create the data, the experts must create a new file, add a satellite image, make the file editable and then zoom to a level that is interpretable. Once this is done the user can then begin to enter points, this is done by clicking the mouse on the desired location and then moving to the next location. Users must pan and zoom themselves. Typically, this can be done using the keyboard. Users must also save their data and transact with files in a desktop computing environment. Once complete the new data is then sent/shared to the requester through a file sharing mechanism. During this process the whole area is only ever assigned in its entirety. This differs from the novice implementation which spreads the work between multiple users.

5   Analysis of Approaches

We first examine the effort required to develop each system. There are also trade-offs in terms of ongoing effort both in communication and software development. This is shown in Table 1.

We examine each system based on the categories in Table 1. The expert system requires less effort on day one due to using off the shelf software and users who are

Table 1. The effort required for process, communication and software development

Approach	Day 1 effort	Prod. effort	Scalability	Extensibility	Prod. rate
Expert (fully manual)	Med.	Med.	Low	High	Low
Novice (MTurk)	High	Med.	High	Med.	Med.
Automated (machine learning)	High	Low	High	Low	High

versed in the GIS domain. However, the task requirements and expectations must be communicated. Data must then also be transacted manually. Further, very little transparency is available during the task completion. This is captured by the medium operational effort. Due to the hands-on nature of the process and the limited pool of users the scalability is low. However, this is contrasted by the high level of extensibility. If it can be done with a GIS system, we can use this user base. For example, if we want to gather swimming pools instead, or we wanted more in-depth analysis, we could leverage this group's skill set.

The novice system requires approximately one-man month of effort to create the initial framework. However, once complete it takes seconds to create new tasks in new areas. In an ideal scenario the operational effort is low to medium. However, due to the variance in worker interpretation of the task and the quality of work effort must be spent on developing the users so that results can be improved. One method of doing this is to create a set of training tasks where users can obtain a qualification to undertake task that require it. By engaging with the users, results can be improved both in quantity and quality. There is a high level of perceived scalability, however this does not take into account on-boarding and the feedback process. The process is extensible in that it could be modified to gather polygons rather than points and could also be used for other objects (e.g. utility poles or even roads). This may affect the production rate of the data.

When examining the automated system, we draw information from the write up of the winning entry of the SpaceNet challenge by XD_XD [72] as well as the progress of image analysis [15, 29, 33]. Specifically, the detailed documentation and modeling completed for the Las Vegas, NV data set. The initial effort is high requiring a large volume of labelled satellite input data (24 GB). This is then combined with specific high-end GPU hardware, either physical or cloud and machine learning techniques which take hours to train the model. Once a model is trained the production of new predictions is relatively straightforward. But it is assumed that a data pipeline exists that is able to take raw input, pass it to the model and then transform it back into the desired format [20]. The production rate and scalability can be increased by creating more instances of the model on more computing resources. The challenge with such an automated system is that it will be tuned to produce a very specific set of results. It may require another model with new data and even modeling approach to gather results for a different task. Due to these limitations this specific model is rated as low in extensibility.

6 Results

6.1 Novice

The novice system created 100 tasks which were accepted by 27 unique workers. On average a worker completed 3.7 tasks, which translates to 20 screen views of a map. The workers who performed the most tasks and inputted the most points included one worker who completed 18 tasks inputting 175 points and another who completed 14 tasks and inputted 868 points. The most prolific worker in terms of points per task completed 5 tasks with 384 points. Some areas of the map are empty, and some contain houses and business. These are not pre-filtered to give an indication of the productivity even when verifying that there are no points of interest. In total 3365 points are created over a 24-h period. This translated to an average of 28 points per task and 12.2 points per subtask. Below in Fig. 3 we plot the number of points created against the number of tasks completed. The users sometimes found the instruction unclear and entered the data in the incorrect location. This added time to the data processing and meant some results were discarded. Many of the tasks were completed concurrently, therefore we are interested in the number of clicks per second.

We analyze this data based on the clicks per second and the number of clicks for a given subtask. This gives us a measure of the "fatigue" per subtask to ensure that there are not too many points that need to be created per subtask. We can see that the time taken to create points is consistent even when entering a large number of points. Examining the quantitative and qualitative aspects of the data we find 2346 unique points using a radius of seven meters to eliminate duplicates. The base data set contains 2449 unique points. This is a difference of 4.2% which is acceptable given that the base data set is not expected to be 100% accurate. We also examine the distance from the collected points to the baseline data.

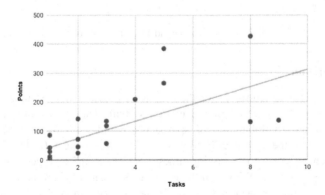

Fig. 3. Number of points vs. tasks completed

The distance from the baseline has a median of 6.1 m. Further 90% of data points gathered are within 11 m. This is good result and informs how the resulting data can be used. It also demonstrates that the instructions and usability could be improved to ensure the gathered data more closely matches the desired location.

6.2   Expert

Four experts were selected two from within the organization and two from UpWork (a crowdsourcing contract work platform). The work was conducted over a 24-h period, from initiation to completion. By using two user groups a broader sample could be taken. The number of points collected varied for each of the submissions. The values were 2388/2351/2376/2443 all of which are within 4% of the baseline data set. Whilst this is equivalent to the novice data the majority of the submissions are better than this. With regard to the accuracy the median distance from an expert point to its base line point is 3.02 m. The time taken per point varies from 0.19 to 1.1 points per second. This large range suggests that the internal resources are more familiar with the task as they are approximately at least twice as fast (Fig. 4).

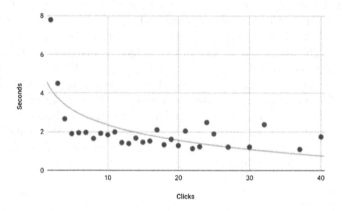

Fig. 4. Clicks per second by number of clicks

6.3   Comparison

The fact that experts perform better than novices, provides a good target to aim for when examining the potential of the novice system. The novice process could be improved. Further a combination of approaches could yield further improvements. With regard to cost the experts utilized were paid the following $10/$15/$50/$50 USD per hour. Each of which took varying times 3.5/1.1/0.62/0.5 h respectively. Whereas the MTurk rate was $10 USD for all the tasks. This equated to approximately $5 USD per hour. Based on the accuracy from the novice system it does present a valid alternative for some use-cases. Further it was easier to complete the area with the distributed and divided tasks and interaction with the MTurk API. With large areas which contain inaccurate and incomplete data the novice approach could provide value. Given that the automated approach can be of value (Figs. 5 and 6).

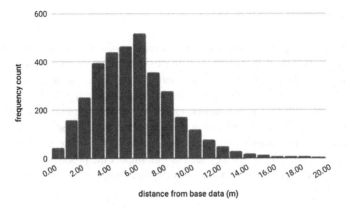

Fig. 5. Novice distance from baseline distribution

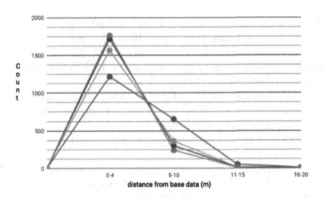

Fig. 6. Experts distance from baseline distribution

7 Future Work

Having examined the effectiveness of the approaches, a larger test upon different geographies and other data collection items would demonstrate a richer sample set with which to draw comparisons. The novice system could be improved to yield better accuracy, this could partially be done by adding a requirement for qualifications to tasks. It may also be beneficial to recruit from the current Turk pool and offer private Turk tasks. A limitation of the current assessment is the assumptions made regarding the automated system. To remedy these a machine learning approach should be developed either as a fully independent process or integrated as a human in the loop process that utilizes the data gathered. With these extensions further analysis about the suitability of novice, expert and automated approaches can be conducted.

8   Conclusion

There are many examples of the use of each of these types of systems, their benefits and performance. However, it is still challenging to determine which approach is the best fit. Currently for data collection of residential and business locations the expert approach is superior, though the scalability of the novice approach is desirable for applications that seek to cover a larger survey area. The automated approaches also demonstrate potential but requires some customization for this use case. The requester and worker computer interactions and interfaces of the systems used should be explored further to determine the key factors that impact the productivity and quality of results. This will enable the improvement of the novice system to approach expert level performance [1].

References

1. Adelson, B.: When novices surpass experts: the difficulty of a task may increase with expertise. J. Exp. Psychol. Learn. Mem. Cogn. **10**(3), 483–495 (1984)
2. Bannon, L.J.: From human factors to human actors: the role of psychology and human-computer interaction studies in system design. In: Readings in Human–Computer Interaction, pp. 205–214. Elsevier (1995)
3. Best-Rowden, L., Bisht, S., Klontz, J.C., Jain, A.K.: Unconstrained face recognition: establishing baseline human performance via crowdsourcing. In: IEEE International Joint Conference on Biometrics, pp. 1–8 (2014)
4. Carroll, J.M.: HCI Models, Theories, and Frameworks: Toward a Multidisciplinary Science (2003)
5. Casey, L.S., Chandler, J., Levine, A.S., Proctor, A., Strolovitch, D.Z.: Intertemporal differences among MTurk workers: time-based sample variations and implications for online data collection. SAGE Open **7**(2), 215824401771277 (2017)
6. Chen, J.J., Menezes, N.J., Bradley, A.D.: Opportunities for Crowdsourcing Research on Amazon Mechanical Turk
7. Chen, X., Golovinskiy, A., Funkhouser, T., Chen, X., Golovinskiy, A., Funkhouser, T.: A benchmark for 3D mesh segmentation. In: ACM SIGGRAPH 2009 Papers on - SIGGRAPH 2009, vol. 28, no. 3, p. 1 (2009)
8. Chew, M., Tygar, J.D.: Image recognition CAPTCHAs. In: Zhang, K., Zheng, Y. (eds.) ISC 2004. LNCS, vol. 3225, pp. 268–279. Springer, Heidelberg (2004). https://doi.org/10.1007/978-3-540-30144-8_23
9. Cooper, S., et al.: Predicting protein structures with a multiplayer online game. Nature **466** (7307), 756–760 (2010)
10. Corney, J.R., Torres-Sánchez, C., Jagadeesan, A.P., Yan, X.T., Regli, W.C., Medellin, H.: Putting the crowd to work in a knowledge-based factory. Adv. Eng. Inform. **24**(3), 243–250 (2010)
11. Daly, T.M., Nataraajan, R.: Swapping bricks for clicks: crowdsourcing longitudinal data on Amazon Turk. J. Bus. Res. **68**, 2603–2609 (2015)
12. DARPA-BAA-16-53: Explainable Artificial Intelligence (XAI). Defense Advanced Research Projects Agency, p. 1 (2016)

13. Difallah, D., Filatova, E., Ipeirotis, P.: Demographics and dynamics of mechanical turk workers. In: Proceedings of the Eleventh ACM International Conference on Web Search and Data Mining - WSDM 2018, pp. 135–143 (2018)
14. Gebru, T., et al.: Using Deep Learning and Google Street View to Estimate the Demographic Makeup of the US, February 2017
15. Girshick, R., Donahue, J., Darrell, T., Malik, J.: Rich feature hierarchies for accurate object detection and semantic segmentation
16. Goodman, J.K., Cryder, C.E., Cheema, A.: Data collection in a flat world: the strengths and weaknesses of mechanical turk samples. J. Behav. Decis. Mak. **26**(3), 213–224 (2013)
17. Grier, D.A.: Human computation and divided labor. In: Michelucci, P. (ed.) Handbook of Human Computation, pp. 13–23. Springer, New York (2013). https://doi.org/10.1007/978-1-4614-8806-4_3
18. Hancock, P.A., Jagacinski, R.J., Parasuraman, R., Wickens, C.D., Wilson, G.F., Kaber, D. B.: Human-automation interaction research: past, present, and future (2013)
19. Holzinger, A., Plass, M., Holzinger, K., Crian, G.C., Pintea, C.-M., Palade, V.: Glass-box interactive machine learning with the human-in-the-loop a glass-box interactive machine learning approach for solving np-hard problems with the human-in-the-loop (2017)
20. Sturrock, H.: Predicting sprayable structures using machine learning, June 2017. http://www.disarm.io/progress_updates/2017-06-29-predicting-sprayable-structures-using-machine-learning/. Accessed 19 Feb 2018
21. Huval, B., et al.: An empirical evaluation of deep learning on highway driving, April 2015
22. Ipeirotis, P.G.: Analyzing the Amazon Mechanical Turk marketplace. ACM Mag. Students **17**, 16–21 (2010)
23. Jacob, B.: The (near) future of data is linked – distinct values: data.world. https://blog.data.world/the-near-future-of-data-is-linked-75f4c011f9cf. Accessed 27 Oct 2017
24. Kalantari, M., Rajabifard, A.: To crowdsource or not to crowdsource?: Crowdsourcing. GIM Int. **26**, 31–35 (2012)
25. Katz, Y.: Manufacturing an artificial intelligence revolution. SSRN Electron. J. (2017)
26. Kittur, A., Chi, E.H., Suh, B.: Crowdsourcing user studies with Mechanical Turk. In: Proceeding of the Twenty-Sixth Annual CHI Conference on Human Factors in Computing Systems - CHI 2008, p. 453 (2008)
27. Kittur, A., et al.: The future of crowd work. In: Proceedings of the 2013 Conference on Computer Supported Cooperative Work - CSCW 2013, p. 1301 (2013)
28. Koblin, A.: The Sheep Market. http://www.aaronkoblin.com/work/thesheepmarket/. Accessed 27 Oct 2017
29. Levin, G., Newbury, D., McDonald, K., Alvarado, I., Tiwari, A., Zaheer, M.: Terrapattern: Open-Ended, Visual Query-By-Example for Satellite Imagery using Deep Learning. http://www.terrapattern.com/faq. Accessed 27 Oct 2017
30. Litjens, G., et al.: A survey on deep learning. Med. Image Anal. **42**, 60–88 (2017)
31. Gillies, J.M., Fiebrink, R., Tanaka, A., Caramiaux, B., Mackey, W., Garcia, J., Bevilacqua, F., Heloir, A., Nunnari, F., Amershi, S.: Human-centered machine learning. In: Proceedings of the 2016 CHI Conference-Extended Abstract on Human Factors Computer Systems, pp. 3558–3565 (2016)
32. Mark, D.M., Freundschuh, S.M.: Spatial concepts and cognitive models for geographic information use. In: Nyerges, T.L., Mark, D.M., Laurini, R., Egenhofer, M.J. (eds.) Cognitive Aspects of Human-Computer Interaction for Geographic Information Systems. NATO ASI Series (Series D: Behavioural and Social Sciences), vol. 83, pp. 21–28. Springer, Dordrecht (1995). https://doi.org/10.1007/978-94-011-0103-5_3
33. Mattern, S.: Mapping's intelligent agents. Places J. (2017)

34. McAfee, A., Brynjolfsson, E.: Machine, Platform, Crowd: Harnessing Our Digital Future (2017)
35. McCarthy, L.: jwz: Social Turkers. https://www.jwz.org/blog/2014/01/social-turkers/. Accessed 27 Oct 2017
36. Moravčík, M., et al.: DeepStack: expert-level artificial intelligence in heads-up no-limit poker. Science **356**(6337), 508–513 (2017)
37. Moss Motors: Crowdwork for Machine Learning: An Autoethnography. http://blog. fastforwardlabs.com/2017/09/26/crowdwork-for-ml.html. Accessed 27 Oct 2017
38. OpenAI, "Dota 2". https://blog.openai.com/dota-2/. Accessed 15 Feb 2018
39. OpenStreetMap, "OpenStreetMap," Open Database License (ODbL) (2016)
40. Pan, Y.: Heading toward artificial intelligence 2.0. Engineering **2**(4), 409–413 (2016)
41. Pang, B., Lee, L., Vaithyanathan, S.: Thumbs up? In: Proceedings of the ACL-02 Conference on Empirical Methods in Natural Language Processing - EMNLP 2002, vol. 10, pp. 79–86 (2002)
42. QGIS Development Team, "Welcome to the QGIS project!," QGIS (2016)
43. Quinn, A.J., Bederson, B.B.: Human computation. In: Proceedings of the 2011 Annual Conference on Human Factors in Computing Systems - CHI 2011, p. 1403 (2011)
44. Le, Q.V., Schuster, M.: Research Blog: A Neural Network for Machine Translation, at Production Scale. https://research.googleblog.com/2016/09/a-neural-network-for-machine. html. Accessed 27 Oct 2017
45. Robert, S., Büttner, S., Röcker, C., Holzinger, A.: Reasoning under uncertainty: towards collaborative interactive machine learning. In: Holzinger, A. (ed.) Machine Learning for Health Informatics. LNCS (LNAI), vol. 9605, pp. 357–376. Springer, Cham (2016). https:// doi.org/10.1007/978-3-319-50478-0_18
46. Rodrigues, F., Pereira, F.: Deep learning from crowds, September 2017
47. Ross, J., Irani, L., Silberman, M.S., Zaldivar, A., Tomlinson, B.: Who are the crowdworkers?: Shifting demographics in mechanical turk. In : Proceedings of the 28th of the International Conference Extended Abstracts on Human Factors in Computing Systems - CHI EA 2010, p. 2863 (2010)
48. Russakovsky, O., et al.: ImageNet large scale visual recognition challenge. Int. J. Comput. Vis. **115**(3), 211–252 (2015)
49. Sebake, M.D., Coetzee, S.M.: Address data sharing: organizational motivators and barriers and their implications for the South African spatial data infrastructure. Int. J. Spat. Data Infrast. Res. **8**, 1–20 (2012)
50. Sheehan, K.B.: Crowdsourcing research: data collection with Amazon's Mechanical Turk. Commun. Monogr. **85**(1), 140–156 (2018)
51. Snow, R., O'Connor, B., Jurafsky, D., Ng, A.Y.: Cheap and fast—but is it good?: Evaluating non-expert annotations for natural language tasks. In: Proceedings of the Conference on Empirical Methods in Natural Language Processing. Association for Computational Linguistics, pp. 254–263 (2008)
52. Takagi, H.: Interactive Evolutionary Computation: Fusion of the Capabilities of EC Optimization and Human Evaluation
53. Tiecke, T.: Open population datasets and open challenges | Engineering Blog | Facebook Code. https://code.facebook.com/posts/596471193873876/open-population-datasets-and-open-challenges/. Accessed 27 Oct 2017
54. Turk, A.G.: An overview of HCI for GIS. In: Nyerges, T.L., Mark, D.M., Laurini, R., Egenhofer, M.J. (eds.) Cognitive Aspects of Human-Computer Interaction for Geographic Information Systems. NATO ASI Series (Series D: Behavioural and Social Sciences), vol. 83, pp. 9–17. Springer, Dordrecht (1995). https://doi.org/10.1007/978-94-011-0103-5_2

55. Uszkoreit, J.: Research Blog: Transformer: A Novel Neural Network Architecture for Language Understanding. https://research.googleblog.com/2017/08/transformer-novel-neural-network.html. Accessed 27 Oct 2017
56. Van Etten, A.: You Only Look Twice (Part II)—Vehicle and Infrastructure Detection in Satellite Imagery. https://medium.com/the-downlinq/you-only-look-twice-multi-scale-object-detection-in-satellite-imagery-with-convolutional-neural-34f72f659588. Accessed 27 Oct 2017
57. Kostakos, V., Musolesi, M.: Avoiding pitfalls when using machine learning in HCI studies. Interactions **24**, 34–37 (2017)
58. Vinyals, O., et al.: StarCraft II: a new challenge for reinforcement learning, August 2017
59. von Ahn, L.: Human computation. In: 2008 IEEE 24th International Conference on Data Engineering, pp. 1–2 (2008)
60. Witten, I.H., Frank, E., Hall, M.A., Pal, C.J.: Data Mining: Practical Machine Learning Tools and Techniques
61. Yuen, M.-C., King, I., Leung, K.-S.: Task matching in crowdsourcing. In: 2011 International Conference on Internet of Things and 4th International Conference on Cyber, Physical and Social Computing, pp. 409–412 (2011)
62. Boats – GBDX Stories – Solving the hardest geospatial problems at scale. http://gbdxstories.digitalglobe.com/boats/. Accessed 27 Oct 2017
63. Topcoder - The SpaceNet Challenge - Crowdsourcing Geospatial Vision Algorithms. http://crowdsourcing.topcoder.com/spacenet. Accessed 27 Oct 2017
64. Large Scale Parsing. http://buildingparser.stanford.edu/dataset.html. Accessed 27 Oct 2017
65. Google Maps APIs | Google Developers. https://developers.google.com/maps/. Accessed 27 Oct 2017
66. ImageNet. http://www.image-net.org/. Accessed 27 Oct 2017
67. Staggers, N., Norcio, A.F.: Mental models: concepts for human-computer interaction research. Int. J. Man Mach. Stud. **38**(4), 587–605 (1993)
68. Dstl Satellite Imagery Competition, 1st Place Winner's Interview: Kyle Lee | No Free Hunch. http://blog.kaggle.com/2017/04/26/dstl-satellite-imagery-competition-1st-place-winners-interview-kyle-lee/. Accessed 27 Oct 2017
69. Project Sunroof. https://www.google.com/get/sunroof#p=0. Accessed 27 Oct 2017
70. Mapillary. https://www.mapillary.com/. Accessed 27 Oct 2017
71. OpenAddresses. https://openaddresses.io/. Accessed 27 Oct 2017
72. Google Books Library Project – Google Books. https://www.google.com/googlebooks/library/. Accessed 27 Oct 2017
73. Amazon Mechanical Turk. https://www.mturk.com/mturk/welcome. https://requester.mturk.com/help/faq#what_is_amazon_mechanical_turk. Accessed 27 Oct 2017
74. AI Experiments. https://experiments.withgoogle.com/ai. Accessed 27 Oct 2017

Correcting Wrongly Determined Opinions of Agents in Opinion Sharing Model

Eiki Kitajima[✉], Caili Zhang, Haruyuki Ishii, Fumito Uwano,
and Keiki Takadama

The University of Electro-Communications,
1-5-1 Chofugaoka, Chofu, Tokyo 182-8585, Japan
k1010048@edu.cc.uec.ac.jp

Abstract. This paper aims at achieving a stable high accuracy of opinion sharing in a distributed network with the agents which have initial opinions. Specifically, the network is composed of multi-agents, and most agents form their opinions according to the neighbors opinions which may be incorrect while a few agents only can receive outside information which is expected to be correct but may be incorrect with noise. To order for the agents to form the correct opinions, we employ Autonomous Adaptive Tuning algorithm (AAT) which can improve the rate of correct opinion shared among the agents where incorrect opinions are filtered out during the opinion sharing process. However, AAT is hard to promote agents to form the correct opinions when all agents have their initial opinions. To tackle this problem, we proposed Autonomous Adaptive Tuning Dynamic (AATD) for the network where initial opinions of all agents are unknown. The intensive experiments have revealed, the following implications: (1) the accuracy rate of the agents with AATD is stably 70%–80% regardless initial opinion state in small network, while the accuracy rate with AAT varies from 0% to 100% depending on the state of the initial opinion; and (2) AATD is robust to different complex network topology in comparison with AAT.

Keywords: Community computing · Distributed network
Emergent behaviour · Self-organisation

1 Introduction

In our network society, wrong information is sometimes unfortunately distributed. For example, when a witness tells the information with individuals close to himself after obtaining new information, they form their own opinions and share the information forward. Repeating this process, all individuals in the network shared with the information based on mutual relationship between them. However, such an information is not always correct due to a lack of the mechanism which can filter conflicted information to share the correct opinion. In order to solve such a problem, Opinion Sharing Model (OSM) [3] was proposed

© Springer International Publishing AG, part of Springer Nature 2018
S. Yamamoto and H. Mori (Eds.): HIMI 2018, LNCS 10904, pp. 658–676, 2018.
https://doi.org/10.1007/978-3-319-92043-6_52

for analyzing sharing opinion in a distributed network composed of multi-agents [2]. In OSM, most agents form their opinions according to the neighbors opinions which may be incorrect while a few agents only can receive outside information which is expected to be correct but may be incorrect with noise. To increase the accuracy rate of sharing correct opinion, Autonomous Adaptive Tuning (AAT) [6] was proposed, but AAT only works well under the condition where all agents have not yet formed their opinions, which is highly impossible in real society, *i.e.*, all individuals has their own initial opinion more or less towards a topic. To tackle this problem, this paper proposes Autonomous Adaptive Tuning Dynamic (AATD) in the network where initial opinions of all agents are determined.

The paper is structured as follows. Section 2 starts by describing Opinion Sharing Model, and Sect. 3 explains AAT. Section 4 describes the proposed AATD and its modifications. Section 5 shows the experimental results and Sect. 6 discusses them. Finally, Sect. 7 concludes this work.

2 Opinion Sharing Model

2.1 Sensor and Normal Agents

For capturing dynamics of sharing opinion in distributed network system, Glinton proposed Opinion Sharing Model (OSM) composed of multi-agents. The agents constitute their complex network and communicate their opinion with each other. Concretely, the agents receive the opinions from the directly connected agents (*i.e.*, neighbor agents), form their opinion referring to the received opinions, and send their formed opinions to the neighbor agents. By repeating this process, the agents eventually share the same opinion among them. What should be noted here is that, a few agents (called the *sensor agent*) who have the sensor function can observe the information from the environment in addition to the opinions from the neighbor agents, while most agents (called the *normal agent*) who do not have the sensor function can only receive the opinions from the neighbor agents. Since the observation of the sensor agents are not accurate, they may form the incorrect opinion, which are spread in their network.

2.2 Description of Opinion Sharing Model

In OSM, the network $G(A, E)$ consists of a large set of agents A and a set of their connection E as described in Eqs. (1) and (2).

$$A = \{i^1...i^N\}, N = |A| \tag{1}$$

$$E = \{(i, j) : i, j \in A\} \tag{2}$$

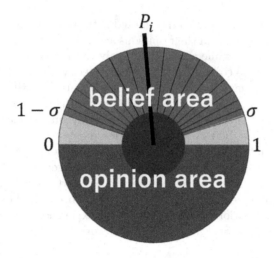

Fig. 1. Visualization of the agent

In the above equations, $i(\in A)$ indicates the agent (*i.e.*, the normal or sensor agent), i^1 indicates the 1 *th* agent. $N(=|A|)$ indicates the number of the agents, and (i, j) indicates the connection between the agents i and j. In OSM, $(i, j) = (j, i)$ because the connection is undirected. $D_i = \{j : \exists(i, j) \in E\}$ indicates the number of neighbor agents of the agent i. A set of sensor agents are represented by $S \subset A|S \ll N$. For simplicity, the sensor agents can only observe environment binary information B($= \{white, black\}$). Note that such a simplification satisfies the purpose of capturing the dynamics of opinion sharing in complex networks [7].

When the agent i receives opinions from its neighbor agents, the agent updates its belief $P_i(b = white)$ or $P_i(b = black) = 1 - P_i(b = white)$, which are the possibilities of believing *white/black* as the correct opinion. After updating the belief, the agent forms its own opinion o_i if its belief exceeds the threshold σ or $1 - \sigma$. For clear understanding, we visualize the agent with belief and opinion in Fig. 1. In this Fig. 1, the upper semicircle represents the belief area while the lower semicircle represents the opinion area. In the belief area, the thick black bar (which length is larger than the radios of the circle) represents the current belief while the thin black lines with the fixed intervals indicate the degree (location) of the belief which can be changed through its update. In the opinion area, on the other hand, the color of the lower semicircle changes when the agent forms its opinion. For example, if the agent forms $o = white$, the color of the opinion area becomes white.

The belief of the agent is updated from its previous value according to Bayes' theorem as shown in Eq. (3). In this equation, P_i^k and P_i^{k-1} respectively indicate the current (k *th*) and previous ($k - 1$ *th*) belief of the agent i when the current step is represented by k. When the sensor agents observe the information from the environment, its belief is updated according to the accuracy $r(0.5 < r \ll 1)$.

When the normal agent receives the information from its neighbors, on the other hand, its belief is updated according to the importance levels $t(0.5 < t \ll 1)$ instead of r in Eq. (3), which indicates the influence of neighbor agents on the normal agent. For example, the belief is not updated in the case of $t = 0.5$, meaning that the agent ignores the received opinions, while the belief turns $P_i^k = 1(or\ P_i^{k-1})$ in spite of value of P_i^{k-1} in the case of $t = 1(or\ 0)$.

$$P_i^k = \frac{C_{upd}P_i^{k-1}}{(1 - C_{upd})(1 - P_i^{k-1}) + C_{upd}P_i^{k-1}} \tag{3}$$

$$where \begin{cases} C_{upd} = r & if\ s_i = white \\ C_{upd} = 1 - r & if\ s_i = black \end{cases}$$

$$O_i^k = \begin{cases} undeter, initial, & if & k = 0 \\ white, & if & P_i^k \geq \sigma \\ black, & if & P_i^k \leq 1 - \sigma \\ o_i^{k-1} & otherwise \end{cases} \tag{4}$$

Figure 2 shows the changes of the belief until the agent forms its opinion. The figure shows that the agent updates belief P when receiving opinions from the neighbor agents and forms the white opinion when $P > \sigma$. In detail, the thresholds σ and $1 - \sigma$ are the confidence bounds, which range is $0.5 < \sigma < 1$ where the agents have confidence that the agent can form its opinion.

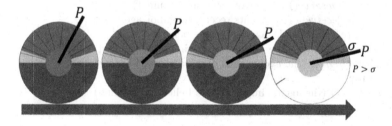

Fig. 2. The agent's belief P is updated when receiving opinions from the neighbor agents. The agent forms the white opinion when $P > \sigma$, while, the black opinion when $P < 1 - \sigma$

Figure 2 shows the sharp hysteresis loop of the function of updating opinion proposed by Pryymak et al. [6]. The belief P_i^k of the agent i should be less than $1 - \sigma$ (larger than σ) to change the opinion to black (white) after it forms the opinion white (black) (Fig. 3).

2.3 Performance Metrics of a Model

The model is simulated until dissemination rounds $M = \{m_l : l \in 1...|M|\}$ end. Every round update belief step $k \in K$. At the end of each round m_l, the

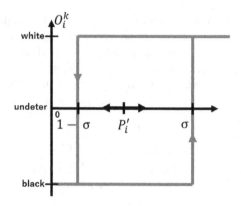

Fig. 3. Update rule of opinion

conclusive opinions are observed. Each round is limited by the enough step which the agents converge to their own opinion. When each round finishes, the current state expires. After the new round start, the agents reset their opinion and belief. Algorithm 1 indicates opinion sharing by the agents in OSM.

Algorithm 1. Opinion Sharing

1: **for all** $Round : m \in M$ **do**
2:     **for all** $Step : k \in K$ **do**
3:         $SensorObserve()$ (sensor observe environment)
4:         $AgentReceive()$ (the agents receive neighbor agents)
5:         $UpdateBelief()$ (the agents update belief)
6:         $FormOpinion()$ (the agents form opinion by updated belief)
7:         $AgentSend()$ (the agents send opinion to neighbor agents)
8:     **end for**
9:     $Initialize()$ (the agents initializes it's belief and opinion)
10: **end for**

In order to measure the average accuracy of the opinions of the agents at the end of each round, Glinton proposed the accuracy metric which is the proportion of the agent numbers that form correct opinion in the community [4].

$$R = \frac{1}{N|M|} \sum_{i \in A} |m \in M : o_i^m = b^m| \cdot 100\% \tag{5}$$

Furthermore, Pryymak proposed performance index for the single agent [6]. When its opinion is formed correctly, the agent cannot perceive. Therefore, the agents should be conscious of how often their own opinion is formed correctly. Pryymak denote it as the awareness rate of the agent h_i.

$$h_i = \frac{|m \in M : o_i \neq undeter.|}{|M|} \tag{6}$$

This myopic metric can be calculated locally by each agent and it is important metric for AAT algorithm that is described in Sect. 3.

3   Autonomous Adaptive Tuning (AAT) Algorithm

In this section, we explain Autonomous Adaptive Tuning (AAT) algorithm. The algorithm is designed for improving the accuracy R by communicating the agent's opinions with each other in the various complex networks. In this algorithm, the agents automatically update their belief only relying on the local information. This algorithm is as follows. The accuracy R increases when the dynamics of the opinion sharing is in the phase change between the stable state (when the opinions are not shared out in the community $\forall i \in A : h_i \ll 1$) and the unstable one (when the opinions are propagated on a large scale $h_i = 1$). Accordingly, it is necessary that the agents share each opinion in smaller groups before large cascade occurs without reacting to the incorrect opinions in surplus. In order to set optimum parameter of the issue, this algorithm regulates importance levels of the agents severally.

3.1   Description of AAT

This algorithm has three stages for tuning importance levels. Also Algorithm 2 shows that how AAT runs in OSM.

- The each agent running AAT build candidates of the importance levels to reduce the search space for the following stages. This step runs only one at the first time of the experiment. (BuildCandidate())
- After each dissemination round, the agent estimates the awareness rates of the candidate levels that are described in Sect. 2.3. (EstimateAwarenessRate())
- The agents select the importance level by estimated the awareness rates of the candidate levels for next round. Then the agents consider how close it is to the target awareness rates. It is necessary that the importance levels are tuned gradually while considering an influence of own neighbors. (SelectImportanceLevel())

In the following sections, we describe three stages of AAT algorithm in detail.

3.2   Candidate Importance Levels

In this section, we describe how the agent running AAT estimates the candidates of importance levels T_i in Eq. (7). By estimating the set of candidate importance levels, the agent reduces the continuous problem of selecting the importance level to use, t_i, from the range $[0.5, 1]$ to a discrete problem. Through the number of the sensor agents is much smaller than the total number of the agents, we focus on the normal agents that update their belief using opinions of only neighbor agents. Figure 4 describe the sample dynamics of the agents belief, where the agent i

---

Algorithm 2. Opinion Sharing Model with AAT

1: $BuildCandidate()$ (Sec.3.2)
2: **for all** $Round : m \in M$ **do**
3:     **for all** $step : k \in K$ **do**
4:         $Opinion\ Sharing\ Model$ (see Algorithm.1)
5:     **end for**
6:     $EstimateAwarenessRate()$ (Sec.3.3)
7:     $SelectImportanceLevel()$ (Sec.3.4)
8:     $Initialize()$ (the agents initializes it's belief and opinion)
9: **end for**

---

receives black opinions $o(b = black)$ and more white opinions $o(b = white)$. Starting from its prior P_i, the agent receives 6 black opinions from its neighbor agents, update its belief until it exceeds $1 - \sigma$ and form the black opinion. After that, the agents receive 11 white opinions, update belief until it exceeds σ and form the white opinion. In this dynamics, the most important moment is $k = 6$ and $k = 21$ because it is only time the agent sends new opinion to its neighbor agent. Consequently, we focus on how many times the agent updates its belief until changing the own opinion. According to the opinion update rule in Sect. 2, we consider the case when the belief of the agent match one of the confidence bounds $P_i^k \in \{\sigma, 1 - \sigma\}$. If we consider that the maximum number of opinions that the agent can receive is limited to the number of its neighbors, $|D_i|$, we can decrease the number of candidates of importance levels. The agent should find the importance levels as its belief coincides with one of the confidence bound $P_i^l \in \sigma, 1 - \sigma inl \in 1...|D_i|$ updates (see Eq. (3)). After solving this problem, the agent can get set of the candidate of importance levels that lead to opinion formation by receiving $1...|D_i|$ opinions.

$$T_i = \{t_i^l : P_i^l(t_i^l) = \sigma, l \in 1...|D_i|\} \cup t_i^l : P_i^l(t_i^l) = 1 - \sigma, l \in 1...|D_i| \qquad (7)$$

Consequently, the set of candidate importance levels is limited to twice the number of neighbors, $|T_i| = 2|D_i|$. This is the necessary and sufficient set of candidate importance levels in which the agent forms the opinion after different update steps and it should be initialized only once. After this stage, the agent has to estimate the most optimal importance level from its set of candidate importance levels.

3.3   Estimation of the Agents Awareness Rates

In this section, we describe how the agent selects the importance level that the network achieves high R. As mentioned above, AAT is based on the observation as follows, the accuracy R of the network improved when the opinion sharing dynamics is in the transition phase between stable state and unstable one. If all agents select minimal importance levels of candidates, only few agents closer to the sensor agents can form the opinion and most of the rest can't. In the

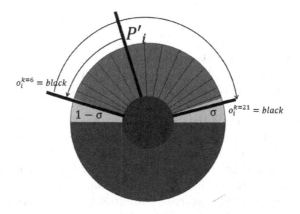

Fig. 4. The agent i with initial belief P'_i firstly receive 6 black opinions and secondly 15 white opinions.

other hand, if all agents select the maximum importance levels of candidates, the incorrect observation of the sensor agent may be shared in the network. In order to estimate the optimal parameters about the importance levels, the agents have to select the minimal importance levels to form their opinion. In the OSM, there are two things below, such that in order to maximize the accuracy R.

- Each agent has to form its opinion. Consequently, each agent should reach a high level of its awareness rate h_i, because the agents without determined opinions drop in the accuracy of the community.
- Each agent has to form the opinion as late as possible with only local information, after the agent gathers the maximum number of the opinions of the neighbors.

To satisfy these terms, the agent has to select the minimal importance level $t_i^l \in T_i$ from the candidates, such that it can form its opinion ($h_i = 1$).

However, since the sensor agents observe by random noise, the dynamics of opinion sharing like phase transition behaves stochastically. The agents cannot form their opinion until the opinions are shared on the large scale and their awareness rate are suffered. The agents should select the requisite minimum importance levels t_i^l from the candidates T_i to make sharing correct opinion successful. Then the awareness rate of the importance levels approaches the target awareness rate h_{trg}. The target awareness rate is slightly lower than maximum, $h_i = 1$. Each agent solves the following optimization problem:

$$t_i = \arg\min_{t_i^l \in T_i} |h_i(t_i^l) - h_{trg}| \tag{8}$$

We need to calculate $h(t_i^l)$ using $t_i^l \in T_i$ in Eq. (8). But we can only know the awareness rate with the importance level that the agent selects in the round. So Pryymak proposed means of estimation about other awareness rate $h(t_i^l)$ from analyzing the agent's belief update in opinion sharing in two cases. Following are them.

- Case 1 (the agent formed his opinion)
  If the agent form opinion $o_i^m \neq undeter$ with t_i in round m, all the importance levels $t_i^l \geq t_i$ higher than the selected importance level will be easier to form opinion $|P_i(t_i^l)| > |P_i(t_i)|$.
- Case 2 (the agent received the opinions)
  If the agent can't form the opinion, we can compare the number of updates it has observed and the number required for the candidate level, t_i^l, to form the opinion. Specifically, the minimal number of belief updates required to form the opinion with the candidate level, t_i^l, can be calculated by recursively updating the agent's belief starting its initial belief P_i' until it exceeds one of the confidence bounds: $[1 - \sigma, \sigma]$ for updates with t_i^l. Pryymak denote this function as $u(t_i^l, P_i', \sigma)$. At the same time, during the round the agent can observe the maximum number of updates starting P_i' that denoted this value as u_i^m.

In Fig. 4, it is received on the last belief update step, $u_i^m = |21 - 12| = 9$. If the number required for t_i^l is smaller u_i^m, these t_i^l will lead the opinion formation.

Combining these cases, Pryymak proposed the boolean function *OpinionFormed*. The function returns *True* if the agent might have formed the opinion or received the opinions from the neighbor in the current round, m using the importance level t_i^l with the actual importance level t_i:

$$OpinionFormed(t_i^l, t_i, m) = \left(o_i^m \neq undeter \wedge t_i^l \geq t_i\right)$$
$$\vee \left(u_i^m \geq u(t_i^l, P_i', \sigma)\right) \tag{9}$$

To estimate the awareness rates for the candidate levels, Pryymak proposed the estimate awareness rate $\hat{h}_i$ using Eq. (6). The measure shows the proportion that the agent will form the opinion with t_i in all the round.

$$\hat{h}_i = \frac{|m \in M : OpinionFormed(t_i^l, t_i, m) = True|}{|M|} \tag{10}$$

Algorithm 3 describes the means of the estimation of awareness rate. If the agent can't receive the opinion in the round, the agent can't form the opinion with any importance levels (Line 1). if the agent receives any opinions, it update each awareness rate in the candidates of importance levels according to Eq. (9) (Line 1). When $OpinionFormed(t_i^l, t_i, m) = True$ or $False$, the agent update the estimate awareness (Line 4, 5).

3.4   Stratagem of Select Importance Levels

The agent affects the dynamics and the awareness rates of all agents with the interdependence of the opinion of the agents and the neighbors. If the

Algorithm 3. Estimate Awareness Rate

1: **if** $OpinionRecieved : u_i^m \neq 0$ **then**
2:    **for all** $CandidateLevels : t_i^l \in T_i$ **do**
3:       **if** $OpinionFormed(t_i^l, t_i, m) = True$ **then**
4:          $\hat{h}_i = UpdateAverageAwareness(\hat{h}_i^l, 1)$
5:       **else**
6:          $\hat{h}_i = UpdateAverageAwareness(\hat{h}_i^l, 0)$
7:       **end if**
8:    **end for**
9: **end if**

---

agent greedily selects the optimal importance level following the definition of its optimization problem (Eq. (8)), it may extremely change the local dynamics of the network. The agent has to select a strategy without dramatic changes in its dynamics, in order to estimate awareness rates of the network accurately and solve faster. To select such the strategy, the agent has to focus on the fact as follows. The awareness rate of the agents for its importance levels increase monotonously. Because the minimum importance level t_i^{min} requires many updates against the maximum importance level t_i^{max}, if the importance levels are sorted in ascending order. In this fact, the agent employs a hill-climbing strategy. If the awareness rate of the current importance level $t_i = t_{i_l}$ is lower than the target $\hat{h}_{i_l} < h_{trg}$, the agent employing the hill-climbing strategy increases the importance level to closest lager one (it i.e., $l = l+1$). If the awareness rate of the close importance level is lower than the target $\hat{h}_i^{l-1} > h_{trg}$, the agent uses this importance level in the next round (*i.e.*, $l = l - 1$). The agents employed the hill-climbing strategy deliver the higher accuracy than the greedy strategy [6].

Algorithm 4. Select importance levels

1: **if** $h^t \leq h_{trg}$ **then**
2:    **if** $undeter\ counts \geq determined\ counts$ **then**
3:       $t = t_{post}$
4:    **end if**
5: **end if**
6: **if** $h^{t_{prior}} > h_{trg}$ **then**
7:    **if** $undeter\ counts < determined\ counts$ **then**
8:       $t = t_{prior}$
9:    **end if**
10: **end if**

---

4   Autonomous Adaptive Tuning Dynamic (AATD) Algorithm

AAT can not learn the importance levels in OSM with the initial opinions. AATD is an algorithm that realizes accurate opinion sharing regardless of whether it has initial opinions by adding changes to AAT. AATD is modified on estimation of the awareness rates of AAT.

4.1   First Modification of Estimation of the Agents Awareness Rates

In the network where all agents already formed opinions, the agents with AAT always judge they are forming opinions and update $\hat{h}_i^l$ from Eq. (10). At the result, the awareness rates of all importance levels in candidates become 1 and is fixed. That is because AAT calculates awareness rates from whether the agents form opinions or not. So, the agents with AAT can't learn the appropriate importance level. From this reason, as a first improvement, AATD calculates the awareness rate from whether the agent change his opinion or not. With this improvement, the problem that awareness rates is fixed to 1 is solved. Figure 5 shows that AAT updates the awareness rate only when the agent determine the opinion from undetermined state (a) but AATD updates the awareness rate when the agent change their opinion state (a), (b) and (c). The red line is the range of AAT can update the awareness rate. The green line is the range of AATD can update the awareness rate.

4.2   Second Modification of Estimation of the Agents Awareness Rates

However, there is still a problem that the agents can't learn the appropriate importance level. When updating awareness rates based on the number of times of received opinions, there is a possibility of updating the awareness rates of the importance levels with that the agent can't form any opinions. And the agent may selects the inappropriate importance level. To solve the problem, AATD calculates the awareness rate based on the importance levels with that agents can form opinions updating the number of received opinions. Figure 6 indicates that AAT updates the awareness rates of the importance levels with that the agent can't form the opinion and AATD updates only awareness rates of the importance levels with that the agent can form the opinion, updating the number of 5 received opinions ($u_i^m = 5$). Left table means the candidates of importance levels in the agent i. $u(t_i^l, P_i', \sigma)$ is the number of updating belief P_i from initial belief P_i' to reach σ or $1 - \sigma$ with candidate t_i^l. For example, 0.973 of Black Opinion Formed and $u(t_i^l, P_i', \sigma) = 1$ means the agent form the black opinion when received 1 black opinion. On the other hand, 0.700 of White Opinion Formed and $u(t_i^l, P_i', \sigma) = 1$ means the agent form the white opinion when received 1 white opinion. The area surrounded by the red line is the range of update in AAT. The area surrounded by the green line is the range of update

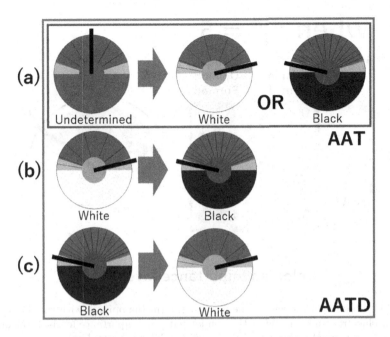

Fig. 5. The range of updating awareness rate in AAT and AATD (Color figure online)

in AATD. Blue area is the range that the agent can form the opinion and Orange area is the range that the agent can't form the opinion. Right illustration of the agent means the agent select $t = 0.671$.

4.3 AATD Estimation of the Agents Awareness Rates

Below are the criteria used for estimating the awareness rate of AATD.

– Case 1 (the agent formed his new opinion)
  If the agent form new opinion $o_i^m \neq init$ with t_i in round m, the all higher importance levels $t_i^l \geq t_i$ than the selected importance level will be easier to form the new opinion $|P_i(t_i^l)| > |P_i(t_i)|$.
– Case 2 (the agent received the opinions)
  Suppose the agent receives the opinion but can not form the new opinion. In that case, the agent will be able to form the new opinion if it uses the importance levels higher than the importance levels to form the new opinion by updating the number of received times. Let $t(u_i^m)$ be the importance levels to form the opinion by updating the number of received times u_i^m.
  If the agent can't form the opinion, we can refer to comparing the number of updates it has observed and the number required for the candidate level, t_i^l, to form the opinion. Specifically, the minimal number of belief updates required to form the opinion with the candidate level, t_i^l, can be calculated by recursively updating the agent's belief starting its initial belief P_i' until it

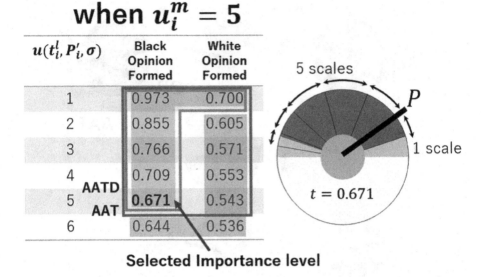

Fig. 6. Candidates of the agent (the agent can form the opinion with the importance levels in blue area and can't form the opinion with the importance levels in the orange area) and the view of the agent with $t = 0.671$. (Color figure online)

exceeds one of the confidence bounds: $[1 - \sigma, \sigma]$ for updates with t_i^l. Pryymak denote this function as $u(t_i^l, P_i', \sigma)$. At the same time, during the round the agent can observe the maximum number of updates starting P_i' that denoted this value as u_i^m. Candidates $t_i^l > t(u_i^m)$ is likely to form the opinion.

We redefine *OpinionFormed* as *AATD OpinionFormed* based on the criteria for estimating the above new awareness rate.

$$AATD\ OpinionFormed(t_i^l, t_i, m) = \left(o_i^m \neq \underline{init} \wedge t_i^l \geq t_i \right)$$
$$\vee \left(\underline{t(u_i^m) \leq t_i^l} \right) \tag{11}$$

5   Experiments

To evaluate an effectiveness of AATD, we conduct the experiments. In the experiments, we evaluate the accuracy R of different network topology with $N \in 100...2000$ which expected degree $d = 6$. We consider the following network typologies widely used in this paper: (1) a small-world network with $P_{rewire} = 0.12$ of randomized connections [5]. (2) a scale-free network [1]. New opinions are introduced through a small number of sensors ($|S| = 0.05N$ with accuracy $r = 0.55$) that are randomly distributed across the network. To simulate a gradual introduction of the new opinions, only 10% of sensors make the new observations after the preceding opinion cascade has stopped. Finally, all agents are initialized with the same confidence bound $\sigma = 0.9$, initial opinion $o_i^1 \in undeter, white, black$, and individually assigned priors P_i' that are drawn from a normal distribution $N(\mu = 0.5, s = 0.1)$ within the range of the confidence bounds $(1 - \sigma, \sigma)$ with $o_i^1 = undeter$. With $o_i^1 = [white, black]$, P_i' is biased by o_i^1. In every round m, Each round stops after 1000 sensors' observations. After this number of observations, the opinions of the sensor agents converge to the true state and the sharing process stops. In the end of each round agents reset their beliefs and opinions to the initial values. AATD and AAT tune the importance levels in the 200 rounds, then the metrics are measured in every rounds.

5.1 Selection of Target Awareness Rate

We experiment AAT with the different h_{trg} to decide the optimal value for sharing correct opinion. The result is shown Fig. 7. The vertical axis is accuracy R and the horizontal axis is the target awareness rate h_{trg}. The blue line is R in scale-free and the orange line is R in small-world with AAT. The result indicates that R is highest accuracy rate in $h_{trg} = 0.85$ in small-world, $h_{trg} = 0.9$ in scale-free. We used 0.85 as h_{trg} in the following experiments because we want to focus small-world property in small network.

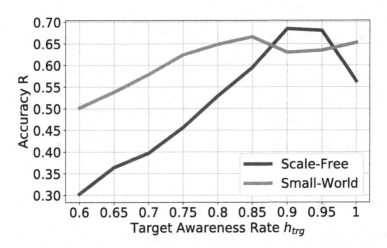

Fig. 7. The accuracy depend on the target awareness rate h_{trg} (10 instances of each topology with $N = 1000$ and $d = 6$) (Color figure online)

5.2 Accuracy of Opinions with Network Size

Figure 8 shows accuracy R of AAT and AATD when all agents have no opinions or initial opinions with $h_{trg} = 0.85$. The vertical axis is accuracy R and the horizontal axis is network size. The blue, the orange and the green line is result when all agents have the white opinions (correct initial state), the black opinions (incorrect initial state) and no opinions (undetermined initial state) respectively. With undetermined initial state, the accuracy R of AATD is almost same as AAT with respect to all network sizes (the green line in (a), (b), (c) and (d)), whether small world or scale free and the accuracy R is around 0.7. When the network size is 50...100, R is around 0.7 regardless of initial opinion state in AATD ((b) and (d)). In AAT, there is a large difference between R with determined initial opinions ((a) and (c)). Therefore, it can be said that it is stable at around 0.7 in AATD while unstable in AAT in a network with small size. However, when the network size is higher 200, the accuracy R of AATD with white initial opinion decreases greatly.

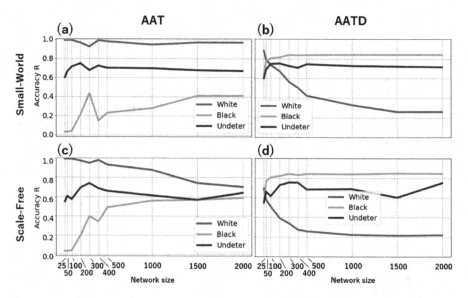

Fig. 8. The accuracy depend on the initial opinion (*white, black, undeter*) and network model (Color figure online)

6   Discussion

6.1   Effectiveness of AATD in Small Network

The agents with AATD can share the correct opinion in small network, unlike AAT, especially when the agents of network have the initial opinions. If all agents with AAT have the initial opinions, Eq. (9) returns true because of $\left(o_i^m \neq \right.$ $\left. \underline{undeter} \wedge t_i^l \geq t_i \right)$ in the equation. In this situation, the agents with AAT cannot learn the importance levels in order to change their own initial opinions. Figure 9 shows all importance levels which the agent with AAT has when all agent's initial opinions are white. The figure shows candidates of importance levels of the agent in Round 1 and 200. There are the importance levels in the blue area and the awareness rates of the importance levels in the orange area, the importance level surrounded by the red line is one that the agent selects. Figure 9 show that two candidates of the importance levels of the agent in round 1 and 200. A place surrounded by the red line with the black arrow pointing in the two candidates is the pairs of the selected importance level t and the awareness rate of t. In Fig. 9, the agents always select $t = 0.504$ as its selected importance level and in round 200, the awareness rates of all importance levels become 1 because the agents with AAT have formed white opinion from round 1 from above reason. From this result, the agents with AAT cannot learn the importance levels in this situation. Figures 10 and 11 shows the rate of the opinions which the agents with AAT and AATD formed for each round in small network (the network size is 100). The left, center and right side of these figures show the rates of the opinions when all agents have no opinions (undetermined initial state), white opinions (correct initial state) and black opinions (incorrect initial state) respectively in the first round. The vertical axis is the rate of the correct opinions (Accuracy R) and the horizontal axis is the round number. The blue, the orange and the green areas indicate the rate of the correct opinions (white opinions), incorrect opinions (black opinions) and undetermined situation, respectively. Note that the white opinion is the correct opinion in this situation. From these figures, most of agents with AAT have been shared the correct opinions with correct initial state at the beginning of the rounds, while they cannot share the correct opinions with incorrect initial state. On the other hand, the agents with AATD can form the correct opinions in both situations. In addition, the agents with AAT and AATD form the correct opinions to same extent with undetermined initial state which situation originally assumed by AAT. From the results, the agents with AAT can not learn the importance level along to each situation, while the agents with AATD can learn and share the correct opinions in small network.

6.2   Ineffectiveness of AATD in Big Network

Generally, the agents with AATD search the appropriate importance level in order from the small value of the candidates of importance levels, and select the importance level whose awareness rate is nearest $h_{trg} = 0.85$, but less than that value. Since all agents can select the large value of the importance level to share the opinion by this mechanism, AATD can make the agents be able to form the correct opinion. However, when the network size is large, the agents far from the sensor agents can't receive opinions in early rounds, and can't update the awareness rates until the neighbor agents form opinions. Since the denominator of Eq. (10) is the number of the rounds, the agents update awareness rates decreasing in this situation. At the result, the agents can't have the awareness rate nearest $h_{trg} = 0.85$, and always become selecting the maximum importance level. That is, the agents become easily influenced from received opinions, and the agent network become very weak for wrong observation from some sensor agents. Figure 8 indicates that with correct initial state (white), as the network size increases, accuracy R suddenly drops in AATD.

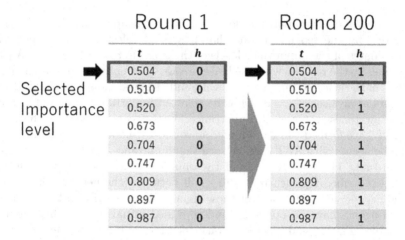

Fig. 9. Candidates of importance levels of the agent in Round 0 and 200 in AAT

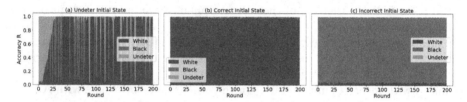

Fig. 10. AAT learning results with different initial settings (network size $= 100 h_{trg} = 0.85$) (Color figure online)

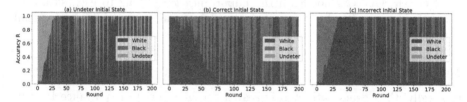

Fig. 11. AATD learning results with different initial settings (network size $= 100 h_{trg} = 0.85$) (Color figure online)

6.3   Relationship Between Parameters of Network Model and Effectiveness of AAT and AATD

The network structure of small-world model varies with the value of rewire probability p_{rewire} and average degree d. We verified the robustness of the Accuracy R for these parameters with small network size ($N = 100$) provided by AAT and AATD, changing these values. Figure 12 indicates the relationship between d and R. The vertical axis is accuracy R and the horizontal axis is average degree d. The white, the orange and the green line is the result with correct initial state, incorrect initial state and undetermined initial state. From the figure, a change of R occurs at the boundary of $d = 6$ with incorrect initial state in AAT, while regardless of the value of d, R is stable in AATD.

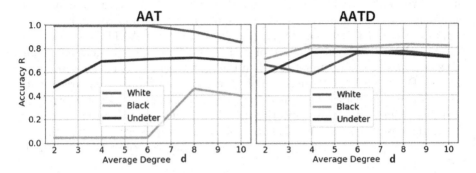

Fig. 12. Average degree d and Accuracy R (network size $= 100 h_{trg} = 0.85$) (Color figure online)

Figure 13 indicates the relationship between p_{rewire} and R in AAT and AATD. The vertical axis is accuracy R and the horizontal axis is p_{rewire}. The white, the orange and the green line is the result with correct initial state, incorrect initial state and undetermined initial state. The figure show that R is not affected by p_{rewire} both in AAT and AATD.

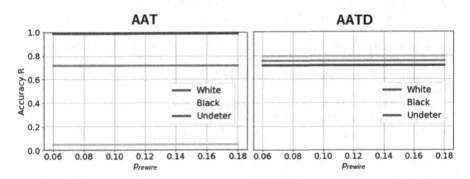

Fig. 13. p_{rewire} and Accuracy R (network size $= 100 h_{trg} = 0.85$) (Color figure online)

7   Conclusion

In this paper, we developed AATD which is extended Autonomous Adaptive Tuning Algorithm to share correct opinions in a network which all agents already form the opinions. AATD realizes opinion sharing with high accuracy regardless of initial opinion state especially when network size is small. Specifically, AATD improves the equation of *OpinionFormed* to return *True* when the agents changes their opinions from initial opinions, unlike when they determines their opinions in AAT. Also, AATD improves the equation to return *True* as to importance levels smaller than one to form the opinion with update number of the received opinions, unlike as to importance levels to form the opinion with update number less than the received opinions in AAT. From the experimental results, we can derive those things: (1) the agents with AATD can share the correct opinions without influence of each situation where all agents have initial some opinions or no opinions, when the network size is small (50–200); and (2) AATD is robust to different complex network topology in comparison with AAT. Our future work is to extend the assumption of small size of network for adopting to more complex network like distributed network system in reality. To tackle this issue, we are going to propose a new algorithm that estimates awareness rate without influence by not being able to receive the opinions in initial rounds.

References

1. Barabsi, A.L., Albert, R.: Emergence of scaling in random networks. Science **286**(5439), 509–512 (1999)
2. Eck, A., Soh, L.K.: Dynamic facts in large team information sharing. In: Proceedings of the 2013 International Conference on Autonomous Agents and Multi-agent Systems, AAMAS 2013, pp. 1217–1218. International Foundation for Autonomous Agents and Multiagent Systems, Richland (2013)
3. Glinton, R.T., Scerri, P., Sycara, K.: Towards the understanding of information dynamics in large scale networked systems. In: 2009 12th International Conference on Information Fusion, pp. 794–801, July 2009
4. Glinton, R.T., Scerri, P., Sycara, K.: An investigation of the vulnerabilities of scale invariant dynamics in large teams. In: The 10th International Conference on Autonomous Agents and Multiagent Systems, AAMAS 2011, vol. 2, pp. 677–684. International Foundation for Autonomous Agents and Multiagent Systems, Richland (2011)
5. Newman, M.E.J., Watts, D.J.: Renormalization group analysis of the small-world network model. Phys. Lett. A **263**(4), 341–346 (1999)
6. Pryymak, O., Rogers, A., Jennings, N.R.: Efficient opinion sharing in large decentralised teams. In: The 11th International Conference on Autonomous Agents and Multiagent Systems, AAMAS 2012, pp. 543–550 (2012)
7. Watts, D.J., Dodds, P.S.: Influentials, networks, and public opinion formation. J. Consum. Res. **34**(4), 441–458 (2007)

Prediction of Standing Ovation of TED Technology Talks

Shohei Maeno and Tetsuya Maeshiro[(⊠)]

Graduate School of Library, Information and Media Studies, University
of Tsukuba, 1 – 2 Kasuga, Tsukuba, Ibaraki 305-0821, Japan
s1411572@u.tsukuba.ac.jp, maeshiro@slis.tsukuba.ac.jp

Abstract. This research aims at the prediction of whether speeches of TED talk can cause audience standing ovation after the end of the talk. The phenomenon of audience standing ovation that we can see in TED talk is one of the objective evidence of the effect that speeches give to audience. We gathered TED talk data that we used as data to experiment the prediction. The methods of this present research consist of quantitative analysis according to speech content and machine learning technique by convolutional neural network. As a result, we achieved 77.11% accuracy and 0.63 F-measure from the prediction using TED talks of Technology topic. Our method used in this study is useful to predict occurrences of standing ovations, although improvement is necessary. Compared to other studies, our contribution, on the one hand, is that we focused on speech content as the effect of standing ovation. On the other hand, we incorporated quantitative analysis especially in terms of what features are effective to standing ovation and eventually apply those features to machine learning technique.

Keywords: Standing ovation · Machine learning · TED talk

1 Introduction

This research experimented the prediction of whether speeches in TED talk can cause audience standing ovation after the end of the talk by quantitative analysis focusing on speech content and machine learning technique using convolutional neural network. TED talk is one of the most famous speech conference spoken in English around the world. TED talk has been held not only in Vancouver, Canada but also in other places and other types of TED talk has emerged resulted from the popularity, named TED Global, TED Women, TEDx and so on. And TED talk has come to be famous in Japan because finally it was held at the University of Tsukuba. One of the reasons for that popularity and the trend is that there are a lot of speeches that are interesting and impressive to audience and visitor to its website. That reason can also be supported from the evidence that TED talk generates not only applause and laugh in the middle of the talk but also audience dramatic reaction called standing ovation after the end of the talk. It is suggested that these phenomena seen in TED talk are resulted from the emotional changes of audience. And at a glance, it might be useful to do questionnaire and qualitative analysis gained from subjects after listening to speeches to evaluate

© Springer International Publishing AG, part of Springer Nature 2018
S. Yamamoto and H. Mori (Eds.): HIMI 2018, LNCS 10904, pp. 677–684, 2018.
https://doi.org/10.1007/978-3-319-92043-6_53

speeches because the emotional change is psychological term. But those attempts will be a problem of lack in objectivity and must be redefined. This is because the evaluation toward speeches is probably varying from person to person which is guaranteed by the rate scored by viewers on TED website (Fig. 1). That is why we had to focus on audience listening to speeches on-site in order for evaluation significantly objectively. And audience standing ovation seen after the end of the talk is the just objective evidence of the effects that speeches generate and it is one of the most powerful expressions from audience [1]. Consequently, in this study to perform more objective analysis, we attempted to do quantitative analysis of elements effective to standing ovation by focusing on speech content. And we extracted the elements from that analysis as features and applied them into machine learning technique, that is, we can predict whether standing ovation will occur or not after the talk is over.

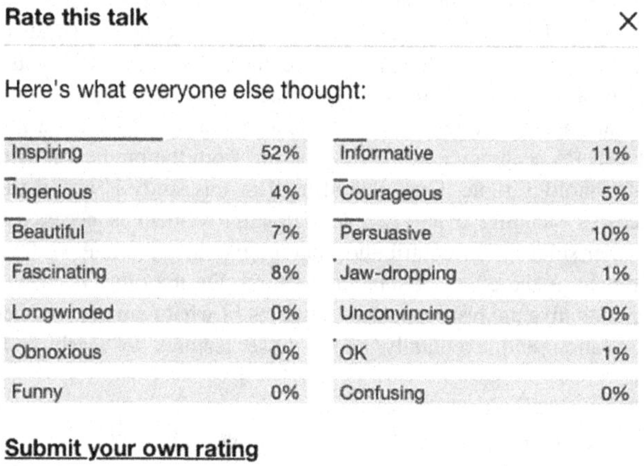

Rate this talk   ✕

Here's what everyone else thought:

Inspiring	52%	Informative	11%
Ingenious	4%	Courageous	5%
Beautiful	7%	Persuasive	10%
Fascinating	8%	Jaw-dropping	1%
Longwinded	0%	Unconvincing	0%
Obnoxious	0%	OK	1%
Funny	0%	Confusing	0%

Submit your own rating

Fig. 1. The example of impressions the listener had.

In previous studies, there are researches analyzing the factors of standing ovation with the exception of TED talk. They are focusing on the characteristics of audience and the geographically distribution of audience. Szilagyi [2] picked up five items as audience characteristics and described that those characteristics can make a difference in occurrence of standing ovation in each person. In addition, he explained that whether each person stands or not is interacted by other people around him/her, which means that he/she is likely to decide to stand if people around him/her are standing. As with Szilagyi, Miller [3] demonstrated that the decision making of whether standing or not is not always dependent on the impression that people have, and that even if he/she has a negative impression, there is always possibilities that the interaction between audience makes him/her be willing to stand with exception of their impression. It can be certainly said that in TED talk there is audience who are standing because of other audience who are standing around them. That is, audience standing ovation that we can see in TED talk can be happened by just the interaction between each person in audience. But in every speech that we gathered as data, audience standing ovation cannot be seen in all audience, which means there is audience who do not stand and who are not affected by other audience. That is why the statements of them are not necessary applicable to audience in TED talk. And the characteristics of audience can be the factor of standing ovation in TED talk, too. It is because that TED talk is held world widely and the nationality of audience is varying, so that the reaction can be changeable according to the culture. Although considering these characteristics to deepen this analysis is important attitudes, it is demanding task to clarify its characteristics in each speech. That is why we found it difficult to take the characteristics of audience into account. By contrast to researches demonstrated above, Strapparava [4] experimented the prediction of the reaction in audience by making use of political speeches. But its attempt did not focus on the factors causing its reaction according to speech content. That is why, this present research can make a difference from previous researches with respect to the perspective to speech content causing standing ovation and the method of quantitative analysis.

We can apply this study to the factor of laugh and applause seen in the middle of the talk. And the contribution is to suggest the perspective and insight to not only speech in English but also other language, and to allow for building objective evidence of effect seen in human-human interaction.

Section 2 explains methods in this research and Sect. 3 refers results and discussion. Finally, this paper concludes in Sect. 4.

2 Method

2.1 Data

As of August 2017, there were 472 speeches of the topic Technology in TED talk. The topic is assigned by TED talk organizers. We watched all of them and got target speeches from them that we used as data in this research. There are some speeches in TED talk that we cannot see the situation of audience after the end of the talk. Although those speeches may cause standing ovation or may not, we can never use them as data because of lack in objective evidence. From speeches that have objective evidence of situation of audience, we gathered 173 speeches as speech data that cause audience standing ovation, and 299 speeches as speech data that did not cause. In addition, we removed speeches even though it caused standing ovation in the case that it can be seen in only a few audience. The criterion of standing ovation is defined subjectively that if there seem to be over 10% audience indicating standing ovation, it will be used as speech data of standing ovation.

2.2 Feature Extraction

The factor of effect that speeches give to the audience consists of verbal and non-verbal element. Verbal element is mainly speech content and non-verbal element is, for example, facial expression and body-language. We attempted to extract features by doing quantitative analysis focusing on speech content from two reasons. First, we can gain every speech content from TED website. It can be possible to gain non-verbal element like body-language but it is difficult to get those element equally from every speech because of the state of TED movie, which means that there are not always speeches that we can see speakers constantly. Second, the slogan of TED talk "Ideas spreading worth" implies that the most important thing that speaker has to focus is speech content itself. That is why audience is possibly influenced the most by what speaker talks. All in all, we decided to start doing analysis according to content. Features that we extract from quantitative analysis are following.

Verbs, Adjective, Adverbs. Verb, adjective, adverb, noun are main elements of English sentences. Those elements are necessary to understand the characteristic of speech content and differ from speakers and the genre of speech. We used verb, adjective, adverb as features and did not consider noun because it can be thought that noun is specifically dependent on the genre of speech and there are significantly rare nouns. In addition, if all of verbs, adjectives, adverbs are regarded as features, the number of them will be over 4000 words. And including too much features as input parameters will be a problem of over-fitting in machine learning. That is why we had to deal with the fact that there are a great number of words. Two solutions are come up with. First, it is to make word groups according to synonyms and to reduce the number of words. Second, it is to set the criteria of word frequency in speech and delete words that emerge less times of the criteria. Former method seems to be difficult to attempt because each word has several definitions and it changes meaning according to surrounding words and context. On the other hand, latter method is realistic in this case because setting word frequency criterion is independent of context and word definition and it can be dealt with automatically. And it is also desirable to the problem of over-fitting because setting the criterion removes words that emerge less in speech and is unique. We finally set a criterion of word frequency and we attempted the prediction with refined learning data made by its setting. That criterion is 150 times, which means for example, if word 'think' is emerged 160 times of the total number emerged in all speeches included in topic, 'think' is included in the word list of 150. The type of input parameters is the number of emergence of a target word divided by the sum number of words in each speech.

11 Words. 11 words include 'any', 'because', 'but', 'if', 'I', 'never', 'not', 'than', 'without', 'world', 'you'. The reason of including 'any' is that it is possible to make a difference in an impact that sentences generate. For example, in case of the following sentence, "With this in mind, we developed a completely new robotics system to support the rat in any direction of space.", it may be thought that the impact of the sentence gets weaker if 'any' is 'some'. With respect to 'because', if what speakers want to insist is not easy to understand for audience in speech, it may be difficult to make an emotional change in audience because it will be lacking in persuasiveness. 'but' has the way to suggest the opposite thing and to stress in sentences. With 'but' in sentences, it can make a difference in terms of emphasis. For example, following sentence "Yes, they are a humanitarian responsibility, but they're human beings with skills, talents, aspirations, with the ability to make contributions if we let them.", can make an emphasis on content followed by 'but'. And although 'but' has other usage except for suggesting opposite, it is difficult to capture its difference of definition because it requires us to realize context in sentences, that is why we did not consider the definition of 'but'. And with respect to 'if', it can be thought that the usage of 'if' includes the supposition and it can give the imagination to audience. And compared to the speech that does not include 'if', it can make audience understand easily what a speaker explains. In speech, it is important to express the originality of speakers. And we can catch its situation as the way of expression with 'I'. There are other ways to express his/her insistence but we focused on the way to express. The usage of extreme expression like 'never' can make a great impact in sentence compared with other sentences. 'not' is weaker than 'never' in expression but it surely makes a difference in giving an impact to audience because it uses negation with 'not'. 'than' can cause audience's attention because it suggests comparison of more than two things. Using 'without' make a stress before 'without'. It is because the sentence with 'without' means that it is enough and valid without the things after 'without'. And TED talk has speakers from around the world so that thye sometimes deliver an ambitious thing in speech. It can be thought that sometimes 'world' implies how large its scale is. With respect to 'you', using it generates difference impact to audience because speech contains of persuasiveness in expressing. According to these 11 words, there are two input parameters, first is word frequency defined as the number of emergence of word divided the sum number of words in speech. Second is the figure indicating the number of words constituting in the middle of each words of feature. 11 words are emitted from speakers unconsciously but its distribution of words make a different impact to audience. For example, the less number of words between words, the more it gets impact. The following sentence can be example here. "A goal I had there was to draw more people in to work on those problem, because I think there are some very important problems that don't get worked on naturally." It includes 'I' in above sentence and the number of words between first 'I' and second 'I' is the figure referred the second input parameter. And its criterion has 0–9, 10–19, 20–29, 30–39, 40–49, over 50. The reason of setting those numbers is that the shorter count can make greater impact and if the space between words expands widely its impact get less so that the count over 50 is grouped as the same category. The type of parameter is defined as involved count for criteria divided the sum number of words in speech.

Not ~ But Syntax. In speech, making stresses is an important element because audience is likely to pay more attention to speech compared to the speech delivering with the same pace. And there seems to have an impact in the sentence with 'not ~ but' syntax. In its sentence, the insistence of a speaker is located in content followed by 'but'. Putting the negation before what a speaker wants to say is useful to stress it. And seemingly the more words a sentence includes, the less impact it makes. But including 'not ~ but' is likely to increase the total number of words compared to the case that it includes only the words followed by 'but'. That is why this syntax is regarded as feature that make an impact in speech. As input parameters, the existence of 'not ~ but' and the frequency of the sentence including is defined. Former one is expressed by binary and latter one is the number of sentences involving 'not ~ but' divided by the total number of sentences in speech.

4 Phrases. 4 phrases include 'I think', 'I believe', 'want to', 'need to'. From consideration of TED slogan 'Ideas worth spreading', delivering speaker's statement is regarded as one of the most important element TED talk emphasizes on. But it is difficult to count how many times speakers express their statements, that is why the way to deliver is focused as phrases 'I think', 'I believe'. In speech, the desire and hope that speakers hold is sometimes expressed as the usage of, for example 'want to', 'need to'. In particular, the situation of explaining the events happened widely in the world is paid attention with its phrase. As the type of input parameters, the existence of phrases is expressed by binary and the frequency of phrase occurrence is defined as the number of sentences including phrase divided by the total number of sentences.

2.3   Experiment

In this research, we attempted to experiment the prediction focusing on Technology topic because that topic is one of the critical topic in TED talk and includes the most speeches among all topics. Gathered data from speech classification is divided into both 75% training data and 25% test data and trained by convolutional neural network. The training data made in terms of prepared criteria in 2.3.2 section is named over150. The architecture of convolutional neural network consists of two convolutional layers with max pooling and three full-connected layers. Batch learning with whole training data is adopted and the learning count is fifty times. Its process is repeated 500 times and we took results.

3   Results and Discussions

Table 1 explains the results of the prediction focusing on Technology topic. Training data has 354 speech data and test data has 118 data. The figures followed by over in Table 1 are the criteria of the word emergence. And the evaluation of the prediction is performed by the figure of accuracy, F-measure, Precision (Positive), Recall, Specificity, Precision (Negative). F-measure is defined as the Eq. (1).

$$\text{F-measure} = \frac{2 \times \text{Precision (Positive)} \times \text{Recall}}{\text{Precision (Positive)} + \text{Recall}} \tag{1}$$

And other figures are following.

$$\text{Precision (Positive)} = \frac{TP}{TP + FP} \tag{2}$$

$$\text{Recall} = \frac{TP}{TP + FN} \tag{3}$$

$$\text{Specificity} = \frac{TN}{TN + FP} \tag{4}$$

$$\text{Precision (Negative)} = \frac{TN}{TN + FP} \tag{5}$$

Here, with respect to TP, FP, TN, FN, T (True) and F (False) is suggesting whether the prediction is correct or not. When the prediction is speech causing standing ovation is P (Positive), on the other hand, if speech not causing standing ovation is selected, it is N (Negative).

Looking at Tabel 1, we achieved 77.11% accuracy and 0.63 f-measure from the prediction in Technology topic. The value of accuracy explains it is possible to predict, but the predictive model constructed needs to be improved according to f-measure. On the contrary, specificity and precision (Negative) demonstrats that the prediction for the speeches of not causing standing ovation achieved better than the speeches of causing standing ovation. Thus, it might be said that the value of specificity and precision (Negative) is because of the number of training data. There are 1.73 times data of not causing standing ovation than causing standing ovation, meaning that the number of data makes an effect to the result and much data leads to better understanding of the chunk of data. The issue of the data scarcity is difficult to achieve soon but it is possible to achieve better prediction to improve the feature extraction and to figure out the characteristic of speeches.

Table 1. Results of prediction focusing on technology topic and the number of inputs.

	Over150
Accuracy	77.11%
F-measure	0.63
Precision (Positive)	0.60
Recall	0.67
Specificity	0.82
Precision (Negative)	0.83

4 Conclusions

In this study, we experimented the prediction of whether the speech can cause audience standing ovation or not by quantitative analysis using speech content and machine learning with convolutional neural network. We experimented the prediction using speeches of the topic Technology and achieved 77.11% accuracy, 0.63 F-measure. All in all, our own selection of features is possible to predict. But it does not mean we can predict this task with other topics. There are a lot of topics in TED talk like Science, Global issues, Communication and so forth. That is why we are going to attempt to try other topics. In addition, we are also exploring not only the prediction for topics but also the prediction with all speech without considering topics.

References

1. Erickson, T.: 'Social' systems: designing systems that support social intelligence. AI Soc. **23** (2), 147–166 (2009)
2. Szilagyi, M.N., Jallo, M.D.: Standing ovation: an attempt to simulate human personalities. Syst. Res. Behav. Sci. **23**, 825–838 (2006)
3. Miller, J.H., Page, S.E.: The standing ovation problem. Complexity **9**(5), 8–16 (2004)
4. Strapparava, C., Guerini, M., Stock, O.: Prediction persuasiveness in political discourses. In: Proceedings of the 7th International Conference on Language Resources and Evaluation, pp. 1342–1345 (2010)

Interacting with Data to Create Journalistic Stories: A Systematic Review

Daniele R. de Souza[1]([✉]), Lorenzo P. Leuck[2], Caroline Q. Santos[1,3],
Milene S. Silveira[1], Isabel H. Manssour[1], and Roberto Tietzmann[2]

[1] School of Technology, PUCRS, Porto Alegre, Brazil
danieleramosdesouza@gmail.com
[2] School of Communications, Arts and Design, PUCRS, Porto Alegre, Brazil
[3] Department of Computer Science, UFVJM, Diamantina, Brazil

Abstract. With the increasing amount of data available in digital media, new professional practices emerged in journalism to gather, analyze and compute quantitative data that aims to yield potential pieces of information relevant to news reporting. The constant evolution of the field motivated us to perform a systematic review of the literature on data-driven journalism to investigate the state-of-art of the field, concerning the process, expressed by the "inverted pyramid of data journalism". We aim to understand what are the techniques and tools that are currently being used to collect, clean, analyze, and visualize data. Also, we want to know what are the data sources that are presently being used in data journalism projects. We searched databases that include publications from both fields of computing and communication, and the results are presented and discussed through data visualizations. We identified the years with the highest number of publications, the publications' authors and the fields of study. Then, we classified these works according to the changes in quantitative practices in journalism, and to the contributions in different categories. Finally, we address the challenges and potential research topics in the data journalism field. We believe the information gathered can be helpful to researchers, developers, and designers that are interested in data journalism.

Keywords: Data journalism · Computational journalism
Computer-assisted reporting · Systematic review

1 Introduction

With the increasing amount of data generated by different sources, and made available online, the journalism industry has sought changes in search of relevance. The demands for information by the online audiences have been continuously redefined as communication and information technologies have evolved, and this has given rise to a new term in this field: digital journalism. According to Kawamoto [7], the definition of digital journalism is changing along with the change of technologies and the new ways of the area. The author conceptualizes this term as the "use of digital technologies to research, produce, and deliver news and information to an increasingly computer-literate audience".

Data journalism, on the other hand, differs from the traditional journalism possibly "by the new possibilities that open up when it combines the nose for news with the ability to tell a compelling story, with the sheer scale and range of digital information now available" [6]. And those possibilities can come at any stage of the journalists work.

Beyond the power of technologies to support data journalism, researchers consider them has also influenced the news producing and the news-consuming process. Thus, digital journalism can also be seen as a combination of computing and computational thinking applied to the news production activities: data gathering, organization, sense-making and data dissemination [5]. Therefore, nowadays, journalists are faced with the need to acquire technological skills and learn how to use tools as Google Sheets, MS Excel, Open Refine, Tableau, among others.

In this study, we take a closer look at how researchers have discussed the relationship between access, manipulation, and presentation of large-scale data and journalistic stories. We want to understand what characterizes the data-driven journalism process and what elements or factors should be considered in this field. The main question this study attempts to answer is *"how are media professionals interacting with data to create journalistic stories and what is the current state of art of research in this field?"*. More specifically, our main contributions include:

- main techniques/tools that are being used to collect, clean, analyze, and visualize data;
- primary data sources that are being used in data journalism projects and research; and
- gaps identified in this field.

The remainder of this paper is organized as follows. In Sect. 2 we present the background and related work. The proposed methodology used for the systematic literature review is described in Sect. 3. The obtained results are presented in Sect. 4, and our final considerations and suggestions for future work are presented in Sect. 6.

2   Background

With the increasing amount of personal and public information available in digital spaces and networks, new professional practices emerged in journalism during the last decades to gather, analyze and compute quantitative data that aims to yield relevant information to reporting [3]. Journalism and computer science combined efforts as programmers approached newsrooms and journalists started acquiring programming skills. The constant evolution of the field motivated us to perform a systematic review of data journalism, following the guidelines proposed by Kitchenham and Charters [9].

The quantitative practices of journalism can be defined as Computer-Assisted Reporting (CAR), Data Journalism (DJ) and Computational Journalism (CJ) [3]. CAR has its roots in Philip Meyers precision journalism, which was modeled using empirical methods for data gathering and statistical analysis to answer questions posed by reporting. It introduced the computational thinking to newsrooms and was considered an innovative form of investigative journalism up to the early 2000s. It was superseded by DJ as it goes beyond the idea of investigative reporting, focusing on data analysis, its presentation and the production of data-driven stories. CJ is also a descendant of CAR and differentiates itself from DJ because it is built around abstraction and automation, producing computable models, algorithms that can prioritize, classify, and filter information.

The term "inverted pyramid often defines news writing", a writing architecture that proposes the presentation of the most relevant information at the beginning of the text, followed by hierarchically decreasing contents regarding interest [1]. This architecture is useful for news outlets because readers can quit reading at any time and still get the most important parts of the story [10]. The "inverted pyramid" become even more critical on the web and other digital media, since users spend 80% of their time looking at information above the page fold and, although users do sometimes scroll, they allocate only 20% of their attention to elements below the fold [11].

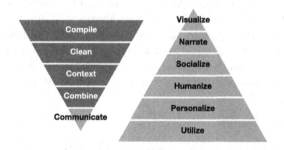

Fig. 1. The inverted pyramid of data journalism [1].

Conversely, Bradshaw [1] proposed the "inverted pyramid of data journalism" (Fig. 1) to explain the data journalism process to support those working with such content as journalists, developers, or designers. He presented as an inverted pyramid because it begins with a large amount of data that becomes increasingly focused to the point of communicating the results. It is composed of five stages: compile (gathering of data sources), clean (data preparation and error cleanup), context (inquire the sources, its biases, and purposes), combine (link data reporting with news story writing) and communicate (visualize, narrate, socialize, humanize, personalize and use) the results.

3   Methodology

According to Kitchenham [8], a systematic literature review (SLR) is a method for evaluation and interpretation of topics that are relevant to a research question, subject or event of interest. To conduct this study, we followed the guidelines of Kitchenham and Charters [9] to structure and organize our research. We defined our research goal and questions, and we established our research protocol. The main goal of this study is to investigate the state of art of data journalism research regarding its process as stated in the "inverted pyramid of data journalism". Therefore, we designed the research questions as follows:

- **RQ1:** What are the techniques/tools that are used to collect data?
- **RQ2:** What are the techniques/tools that are used to clean data?
- **RQ3:** What are the techniques/tools that are used to analyze data?
- **RQ4:** What are the techniques/tools that are used to visualize data?
- **RQ5:** What are the data sources that are used in data journalism projects?

3.1   Search Strategy

This systematic review is focused on data journalism research conducted by researchers from computing or communications field of study or even by researchers from both areas working together. For this reason, we searched[1] databases that include publications from both fields of study (ACM Digital Library, IEEE Xplore, Elsevier ScienceDirect, and Scopus). Scopus contains Google Scholars top 10 academic journals in the communication area. Our search string was adapted according to the database, always including the four keywords we selected, which contemplate the three quantitative professional practices of journalism [3]: "Computer-Assisted Reporting", "Data Journalism", "Data-Driven Journalism", and "Computational Journalism". It is possible to see in Table 1 the list of consulted databases and their respective search strings.

3.2   Selection Strategy

Initially, we analyzed each publication retrieved from the initial search to remove duplicates and to include or exclude studies according to document type and language. The inclusion criteria were applied in the first filter: *(i)* English only; and *(ii)* conference and journal papers.

Publications with at least one of the following exclusion criteria were removed: duplicated, other languages, abstract only, book, and magazine. Subsequently, we conducted a **title, keywords, and abstract review**. In this phase, our inclusion criteria were:

- Fits into one of the three types of quantitative journalism (computer-assisted reporting, data journalism, or computational journalism) [3];

---

[1] Last search update: February 2nd 2018.

Table 1. Search string according to database

Database	Search string
ACM Digital Library	*("computer-assisted reporting" OR "data journalism" OR "data-driven journalism" OR "computational journalism")* - Refined by: Journal and Proceedings.
IEEE Xplore	*(((("computer-assisted reporting") OR "data journalism") OR "data-driven journalism") OR "computational journalism")* - Refined by: Conferences Journals and Magazines.
Science Direct	*TITLE-ABSTR-KEY("computer-assisted reporting" OR "data journalism" OR "data-driven journalism" OR "computational journalism")[All Sources(Computer Science,Social Sciences)]*
Scopus	*TITLE-ABS-KEY("computer-assisted reporting" OR "data journalism" OR "data-driven journalism" OR "computational journalism") AND (LIMIT-TO(DOCTYPE, "ar") OR LIMIT-TO(DOCTYPE, "cp")) AND (LIMIT-TO(SUBJAREA, "SOCI") OR LIMIT-TO(SUBJAREA, "COMP")) AND (LIMIT-TO(LANGUAGE, "English")) AND (LIMIT-TO(SRCTYPE, "j") OR LIMIT-TO(SRCTYPE, "p"))*

- Contributes to research on data journalism in the communications or computer science field of study.

Studies that did not fulfill one of these criteria were removed. Finally, we conducted a **full-text review**. In this phase, we performed a more detailed analysis of the papers' content.

3.3 Data Extraction Strategy

We extracted for each study retrieved from the initial search the following data: year, title, keywords, abstract, authors, authors' country, authors' affiliation, publication name, and source database. In the full-text review, we extracted for each paper the following data, to answer our research questions and to classify all studies:

- Data collection tool/technique;
- Data cleaning tool/technique;
- Data analysis tool/technique;
- Data visualization tool/technique;

– Data source;
– Paper category;
– Type of quantitative journalism.

We analyzed the selected studies focusing on answering our research questions according to the data journalism process (data collection, cleaning, analysis, and visualization), as well as the data sources used in journalism projects. The collected dataset allowed us to identify which years had the highest number of publications and classified these works according to the changes in quantitative practices in journalism. We determined the publications' authors and their respective affiliations and countries of origin, as well as data concerning the publications' fields of study. We also did classify publication contributions in different categories, such as data journalism concepts, tools for journalists, application of data analysis/visualization techniques, and case studies. The results are then presented and discussed through visualizations.

4   Results

We applied the strategies of search, selection, and data extraction described in Sect. 3. Figure 2 presents the number of remaining papers according to each phase of the process. We obtained 273 papers from the initial search in the selected databases. First, we removed duplicates and applied a filter according to language and document type, resting 230 papers. After that, we executed a title keywords and abstract review, which left us with 111 papers that fulfilled our inclusion criteria. Finally, we performed a full-text review, remaining with 101 papers (Appendix A).

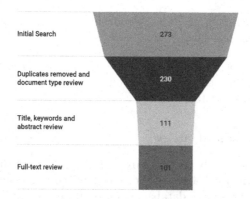

Initial Search	273
Duplicates removed and document type review	230
Title, keywords and abstract review	111
Full-text review	101

Fig. 2. Number of papers in each phase of the systematic review process.

Initially, we classified papers into six categories: data journalism concepts, case studies, new techniques, tools for journalists, application of existing techniques, and data journalism education. The treemap presented in Fig. 3 shows the final number of papers for each category.

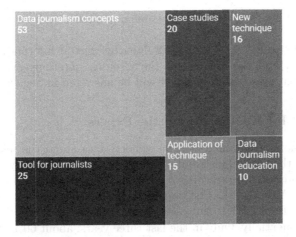

Fig. 3. Selected papers classified by category.

Most of the papers discussed data journalism concepts, showing that this field is a new subject, still under development. In several cases, the research work employed in those studies serve as a link between communication professionals and academia, or as a way to understand significant cultural shifts and disruptive technologies. For this purpose, the primary method usually is literature reviews, as we can see in S35, S40, and S85.

The definition of the "case studies" category is very much like the one of the "data journalism concepts" category. However, in the former, the papers are related somewhat more to the data journalism practice than the theory itself. A closer look is held in the newsrooms routines, journalists workloads, communication demands, etc. S97, S99, and S101 show us a regular practice of this kind of study, the interview.

The "data journalism education" category is almost a meta-category, since it concerns not only the didactic and courses of data journalism, but also the way it is announced and understood. New modes of knowledge production in the area and experiences of postgraduate classes can be found in S62 and S80.

From the categories "tool for journalists", "new techniques" and "application of existing techniques", we could classify a series of systems, scripts, and interfaces used in data journalism. The difference among them is small, although utterly significant to our systematic review.

In the "tools for journalists" category, the authors present prototypes or final versions of systems developed by them to support specific stages or even the entire data journalism process [S37, S40]. On the other hand, the "new techniques" category encompasses scripts, methods or improvements to support any of the data journalism stages, as well as the entire data process [S22, S38]. The "application of existing techniques" category refers to the use of existing programs, not necessarily created by the authors. This is the most interdisciplinary category since it is at the threshold of communication with information technology [S88, S90].

The following Subsects. 4.1 and 4.2 describe general information about the selected papers, and the answers of the research questions presented in the Sect. 3. Our primary research question will be answered in Sect. 5.

4.1 General Information About the Papers

In this subsection, we summarize some general data about the selected papers. Figure 4 shows the paper's distribution according to publication year. As we can see, scientific production regarding data journalism began to grow significantly in recent years. Although the first papers were published in 1996, we can observe a linear growth from 2011 to 2014. From 2015, the number of published papers has grown exponentially. Only in the last three years, about 60% of the selected papers have been published. It is important to mention that, by the time we performed and updated the initial search, it is possible that not all the papers published in 2017 have been uploaded to the databases.

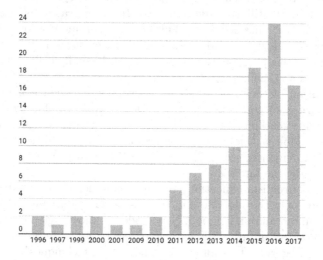

Fig. 4. Papers publication by year.

We found 195 distinct authors among the selected papers we analyzed, and 24 of them authored more than one article. From these, 7 published three or more papers, which we consider the top authors. Table 2 presents the list of authors who have more publications and the corresponding reference for each paper.

Analyzing only the selected papers authored by the top authors, we identified four networks of authors who recurrently publish together, which are shown in Fig. 5. A total of 18 papers generated the four networks, which comprise 35 authors with 89 connections between them (network density = 0,138, where 1

Table 2. Top authors of the selected papers.

Author	Papers	References
Diakopoulos N.	10	S9, S10, S11, S20, S42, S48, S58, S75, S81, S87
Li C.	8	S13, S22, S35, S37, S38, S52, S89, S100
Yang J.	6	S13, S22, S35, S37, S38, S89
Yu C.	5	S13, S22, S35, S38, S89
Wu Y.	4	S22, S35, S37, S89
Hassan N.	4	S35, S38, S52, S100
Agarwal P. K.	3	S22, S37, S89
Broussard M.	3	S43, S56, S62
Lewis S. C.	3	S28, S34, S45

represents a fully connected network). The thicker an edge of the graph, the more papers the connected authors published together.

The graphs showed in Fig. 5 represents how the top authors collaborated in their papers. As we can see, Diakopoulos has the most significant number of papers among our selection. However, his network has 22 connections and 14 co-authors (none of them in the list of top authors). Li C's, on the other hand, has 15 co-authors and 57 connections, making it the biggest and most connected network in the graph. Lewis SC's and Lewis J's are equal in size, but Lewis J's has three connections, one more than Lewis SC's.

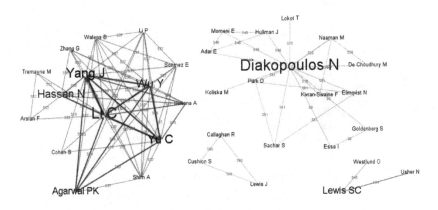

Fig. 5. Authors network.

Figure 6 shows an alternative visualization for the graph containing the authors network. In this case, we present the connections between authors from different fields of study by using different colors. The red edges starts from the

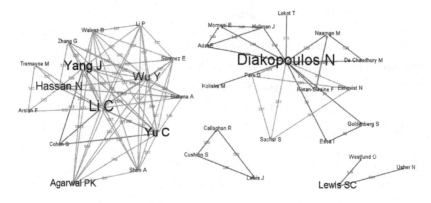

Fig. 6. Authors network according to field of study. (Color figure online)

nodes of the graph referring to the authors from the computing field of study and the green ones relate to the authors from the communications area. As we can see, the largest authors network comprises mainly papers published by authors from the computing area, except by Cohen S. The authors network generated by connections of Diakopoulos N. is mostly from the communications field of study.

We also analyzed the selected papers according to the country of authors' institutions (Fig. 7). There are 51 papers published by authors from the United States (49.03%), followed by 9 from the United Kingdom (8.65%) and 5 from Austria (4.8%). Despite the low percentage of publications from countries other than United States, papers from our sample indicate that the diversity in data journalism's research and practice is increasing.

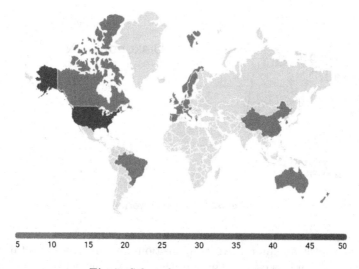

Fig. 7. Selected papers by country.

4.2   Answers to Research Questions

In this subsection, we describe the results of our SLR that answer the five research questions previous presented. The answers to the research questions were obtained by extracting and combining data from the 101 selected papers.

Several papers from our selection proposed new tools, both under development or ready to use, for supporting all the stages of the data journalism process. Some examples are Vox Civitas, SRSR, FactMinder, FactWatcher, Icheck, TATOOINE, News Context Project, TweetTalk, Readdit, The Openprocurement.mk, and YDS.

Data Collection

We identified 22 papers that helped us to answer our RQ1 (*"What are the techniques/tools that are used to collect data?"*). In S2, authors investigated the tools that daily newspapers were using in computer-assisted reporting from the application of an online survey questionnaire between December 1993 and March 1994. They mentioned that computers were becoming more and more valuable to reporters, not only for news writing but also for news gathering. For the authors, graphical user interfaces products, such as Windows and OS/2, that goes beyond DOS software, arrived to facilitate this task.

Regarding the scientific production of recent years, there are some papers presenting tools to support data collection (S17, S19, S20, S71, S76). Some of them were explicitly proposed for helping journalists in their practice (S20, S71). There are also some papers that describe new techniques or tools for the extracting information from different data sources (S10, S13, S26, S35, S37, S54, S59, S88, S91, S93). Some of them have the specific goal of supporting journalists in the discovery of news events (S10, S35, S91) or checking of facts (S26, S37). We also found one study proposing a database integration tool (S54).

We found only few papers in which authors report using existing tools and APIs for data collection. In S30 and S40, Twitter API was accessed directly or indirectly. In the latter case, a command line scraping tool was used to communicate with de API. S31, S41, and S58 mention the use of APIs from specific news sites, such as The Guardian Open Platform and The Times Community API. In S50, data was collected directly from open data government sites.

Data Cleaning

Our RQ2 (*"What are the techniques/tools that are used to clean data?"*) can be answered by 24 selected papers. Data cleaning methods and algorithms in S10, S20, S19, S26, S35, S37, S43, S54, S59, S71, S91, and S93 are used as part of closed systems, that is, tools developed for specific purposes that do not foresee changes or appropriation of functions.

In other instances, data cleaning occurs through independent steps, generated from well-established platforms and frameworks. Among these, S12 and S50 bring us examples of tabular data manipulations with Excel and Google Refine. In S2, we could find that researchers were using tools like XyWrite, WordPerfect, and Word. This tools can be considered Excel and Google Refine predecessors. An R script can be found in S40 and relational database management cases in

S14 and S89 with PostgreSQL and SQL. There are also examples in which the authors took an algorithmic approach: in-depth accounts (S15), classification and clustering engine (S7); text pre-processing (S30); clustering framework (S31); SpeakerRecognitionAPI, the Custom Recognition Intelligent Service (CRIS) and the Speech API (S72).

Data Analysis

There are 33 papers in our selection that address our RQ3 (*"What are the techniques/tools that are used to analyze data?"*). From these, several existing tools and techniques are used to support data analysis in a general way (S14, S15, S17, S19, S54, S71, S79 S93). Some papers deal with the analysis of specific types of content, such as text analysis (S10, S81, S88), audio analysis (S12, S72), and video analysis (S9). We found some papers that take advantage of users collaboration to analyze data (S72). Other papers take advantage of algorithms to automate the data analysis process. Among these, different analysis techniques are being used, such as natural language processing (S52, S55, S88) and clustering (S30, S76). Some papers are focused on sentiment analysis (S28, S30, S52).

We identified that some papers have the main goal of automatically discovering newsworthy themes in databases (S22, S35, S43, S52, S59, S61, S91), such as interesting facts (S22) and significant events (S61). Some papers aim to analyze data in order check facts (S26, S37, S89, S98). We also found some papers focused on the analysis of data retrieved from social media (S20, S30). Finally, there are only few papers in our selection that reported to use existing tools, from the most classic ones (S2) to some more modern and robust (S12, S40, S50).

Data Visualization

To answer RQ4 (*"What are the techniques/tools that are used to visualize data?"*), we analyzed all papers that referred to use visualizations techniques and/or tools to present data. In the other cases, the authors did not provide neither the technique's name nor the tool used, and it was not possible to identify only by reading these papers. Among 101 papers, 31 mentioned the usage of some visualization technique or tool, which allowed us to extract what most used techniques in the data-driven journalism research (S2, S9, S10, S11, S12, S14, S17, S19, S20, S26, S30, S31, S35, S37, S40, S41, S43, S46, S49, S50, S54, S55, S59, S66, S71, S72, S76, S77, S81, S89, and S93). Most quoted visualization techniques are tables, graphs, charts, maps, and the tools quoted more than once time are D3.js and Tableau.

The papers ranged from theoretical discourse to the development and use of tools, besides discussing visualization techniques more used by them. It is interesting to observe that, in the midst of so many free and online tools that are available to support data journalists, we perceived that they are not widely used.

Data Sources

Our answer to RQ3 (*"What are the data sources that are used in data journalism projects?"*) is based on the sources referred in 43 papers (S9, S10, S11, S14, S15, S19, S20, S21, S22, S23, S26, S28, S30, S31, S35, S37, S38, S40, S41, S43, S46, S48, S49, S50, S52, S54, S55, S59, S61, S66, S71, S72, S76, S77, S79, S81, S88, S89, S90, S91, S93, S98, and S100). The others 58 did not mention anything about data sources.

Most of these papers reported different data sources used in data-driven journalism, ranging from media outlets, as BBC and The Guardian, to social media, like Twitter, Youtube, and Facebook. Besides that, government sources, political datasets, Wikipedia, NBA dataset, among others were also mentioned.

5   Discussion

Our goal with this study was to plot a state of art landscape on data-driven journalism research. In this section, we analyze the primary results that we obtained by conducting the systematic review in comparison to the main theories that support our work: the inverted pyramid of data journalism and its process' stages [1], as well as the types of quantitative journalism [3] that evolved across time. We also discuss the implications for research in this field. Finally, we address the challenges and potential research topics in the data-driven journalism field.

This study seeks to contribute to the domain of digital journalism, or data-driven journalism, which has increased due to the advance of information and communications technologies. Although we were able to find some contributions on this field, we perceived this field as new and so there are still ongoing discussions on it.

Since the beginning of computers' use in social and human sciences research/work, professionals and researchers have been benefited from facilities provided by technologies. The same happened to journalism. It started with computer-assisted reporting (CAR), in which technology facilitated the news producing and its workflow [12]. CAR was used as support to create investigative reporting, involving data collection, analysis, presentation, and archive [4]. Figure 8 shows the 101 papers (discussed in this study) by type of quantitative journalism across time.

The data we collected to answer the systematic review's research questions converge with the chronological classification proposed by Coddington [3] for the different types of quantitative journalism. As we can see in the graphic presented in Fig. 8, Computer-assisted reporting was the forerunner of data journalism, using basic computing and statistical methods as an extension of reporters' skills. In this period, graphic interfaces and computer programs such as Word and Excel helped journalism professionals from news gathering to news reporting. There are only a few papers about this type of quantitative journalism that were published from 1996 to 2001.

Data journalism came up with the possibility of analyzing large amounts of data in a way that was not possible since no human being would be able to

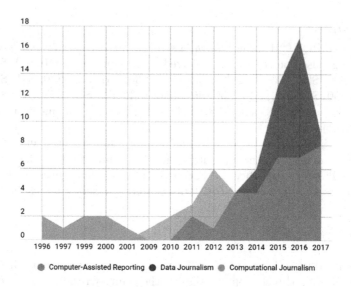

Fig. 8. Papers by type of quantitative journalism across time.

perform such task without machine help. Besides that, data journalism is characterized by its focus on data analysis and presentation, providing the readers with a new experience of interaction with the news stories, as well as with the opportunity for the public to collaborate with journalistic investigations in the data gathering stage (a process named as crowdsourcing). The rise of such activity poised discussions on data transparency from public and private players and to have access to government data portals. Scientific production regarding data journalism began to grow after 2007 and remains on the rise.

Computational journalism, in turn, arose from the direct influence of computer science and concerns the use of algorithms for the automation of data journalism processes. It is related to discussions about its implementation in the newsrooms and the resistance of some journalists who feel threatened, primarily by the news bots. As Fig. 8 shows, papers regarding computational journalism are growing exponentially in recent years. We believe there still plenty of room for conducting studies in this area considering the benefits they can bring to the practice of journalism today.

In the journalism field, graphics have been used for a long time to present statistical and non-statistical data and to allow audiences to view the information guided by the author [2], a visual solution called infographic. Until not long ago, some researchers compared infographics with data visualization, claiming that this latest emphasizes the interaction, which allows audiences to conduct a customized analysis of the data while the first could be just an editorial decision. However, there is a trend of journalists and researchers, as Cairo [2], to understand infographics and data visualizations as an organic continuum field of study, and, nowadays, there is little difference between them.

In this systematic review, we perceived that there is not much research that focuses on discussing visualization techniques or ideal tools for journalism, and not even research offering guidance to applying visualization in data-driven journalism. Thus, we believe that there is a gap in this field about the ways media outlets could incorporate visualization in the digital news. Likewise, media professionals, journalists, and researchers should improve their studies in a documentation and description of techniques and tools used by them, enabling future professionals to continue their work.

To answer our primary research question (*"how are media professionals interacting with data to create journalistic stories and what is the current state of art of research in this field?"*), we found that the journalistic skills to deal with large volumes of data are improving over time. Also, nowadays, they have a range of tools that support their work activities, as we identified in this SLR.

Considering these changes in the journalist's profession, there is a need to redefine the curricula of journalism courses, including mathematics and programming disciplines on it to better prepare the new generation of journalism professionals for the challenges of the data age. It is also interesting to unite communications and computing areas even further, creating repositories with tools to support the data journalism process in its different stages (compile, clean, context, combine, and communicate) [1]. Cohen et al. [S12] describe this idea:

> *"Journalists need to partner with computer scientists, application developers, and hardware engineers. For decades, the computing community has empowered individuals to seek information, improving their lives in the process. Few fields have done more to give citizens the tools they need to govern themselves. Few fields today need computer scientists more than public interest journalism."*

6   Final Considerations

This study presented a systematic review of the state-of-the-art research on data-driven journalism. The selected papers encompass studies conducted by researchers from computer science and communication fields or even by researchers from both areas working together. From the 273 papers retrieved from the automated search, we selected 101 that we considered relevant for investigation. These papers address several topics concerning data-driven journalism.

We analyzed the collected data in order to answer our primary research question (*"how are media professionals interacting with data to create journalistic stories and what is the current state of art of research in this field?"*). Besides that, we answered some systematic review's research questions according to the process' stages (data collection, cleaning, analysis, and visualization) and sources used in data journalism projects.

Our findings showed that the relationship between journalists and data to create news stories is still seen as a recent practice, albeit a growing one. Media outlets increasingly apply the data collection, cleaning, analysis, and visualization process as an efficient way to tell a persuasive news story. Therefore, computational tools have been incorporated into the news producing routine.

Our contributions were: *(i)* presentation of an overview of existing research in data-driven journalism; *(ii)* identification of main techniques/tools that are being used to collect, clean, analyze, and visualize data in this field; *(iii)* the primary data sources that are being used in data journalism projects and research; and *(iv)* the possible gaps in this field.

We believe the information gathered in this systematic review can be helpful to researchers, developers, and designers that are interested in data journalism, considering that different types of users may benefit from such visualizations, either as journalists or the general public.

Regarding future works in the field as mentioned earlier, we envision this text to be capable of becoming a methodological blueprint for updates in coming years. As the area of data-driven journalism continues to evolve, aligned with the academic and professional research on the field, we foresee the need to update the literature review in a few years time. Other possible implementations, as the use of our methodology to focus specifically on texts from a country or a timeframe, suggest future fields of development. Also, since the academic datasets were already collected, the creation of interactive and responsive visualizations could help scholars and students to perform a visual analysis and search of the works.

Acknowledgements. This work was partially supported by PUCRS (Edital N 05/2017 - Programa de Apoio a Integração entre Áreas/PRAIAS). The authors also acknowledge the support of CAPES (Coordenação de Aperfeioamento de Pessoal de Nvel Superior) in this work.

Appendix A - List of selected studies

S1 - Parsons P., Johnson R.B.: ProfNet: A Computer-Assisted Reporting Bridge to Academia. In: Newspaper Research Journal, Vol 17, Issue 3-4, pp. 29–38 (1996)
S2 - Garisson B.: Tools Daily Newspapers Use in Computer-Assisted Reporting. In: Newspaper Research Journal, Vol 17, Issue 1-2, pp. 1131–126 (1996)
S3 - Williams P., Nicholas D.: Journalists, news librarians and the Internet. In: New Library World, Vol 98, Issue 6, pp. 217–1223 (1997)
S4 - Garrison B.: Newspaper size as factor in use of computers for newsgathering. In: Newspaper Research Journal, Vol 20, Issue 3, pp. 72–185 (1999)
S5 - Martin H.: The changing information environment in the media: Case study the Guardian/Observer. In: Aslib Proceedings, Vol 51, Issue 3, pp. 91–196 (1999)
S6 - Mayo J., Leshner G.: Assessing the credibility of computer-assisted reporting. In: Newspaper Research Journal, Vol 21, Issue 4, pp. 68–182 (2000)
S7 - Maier S.R.: Digital diffusion in newsrooms: The uneven advance of computer-assisted reporting. In: Newspaper Research Journal Vol 21, Issue 2, pp. 95–110 (2000)
S8 - Deuze M.: Online journalism: Modelling the first generation of news media on the World Wide Web. In: First Monday, Vol 6, Issue 10 (2001)
S9 - Diakopoulos N., Goldenberg S., Essa I.: Videolyzer: Quality Analysis of Online Informational Video for Bloggers and Journalists. In: Proceedings of the SIGCHI Conference on Human Factors in Computing Systems (CHI'09), pp. 799–808 (2009)
S10 - Diakopoulos N., Naaman M., Kivran-Swaine F.: Diamonds in the rough: Social media visual analytics for journalistic inquiry. In: IEEE Symposium on Visual Analytics Science and Technology, pp. 115–122 (2010)
S11 - Diakopoulos N.: Game-y information graphics. In: Extended Abstracts on Human Factors in Computing Systems (CHI EA'10), pp. 3595–3600 (2010)
S12 - Cohen S., Hamilton J. T., Turner F.: Computational journalism. In: Communications of the

ACM, Vol 54, Issue 10, pp. 66–71 (2011)

S13 - Cohen S., Li C., Yang J., Yu C.: Computational journalism: A call to arms to database researchers. In: 5th Biennial Conference on Innovative Data Systems Research, Conference Proceedings (CIDR 2011), pp. 148–151 (2011)

S14 - Pulimood S. M., Shaw D., Lounsberry E.: Gumshoe: A model for undergraduate computational journalism education. In: Proceedings of the 42nd ACM technical symposium on Computer science education (SIGCSE'11), pp. 529–534 (2011)

S15 - Wagner E. J., Lin J.: In-depth Accounts and Passing Mentions in the News: Connecting Readers to the Context of a News Event. In: Proceedings of the 2011 iConference, pp. 790–791 (2011)

S16 - Hollings J.: The informed commitment model: Best practice for journalists engaging with reluctant, vulnerable sources and whistle-blowers. In: Pacific Journalism Review: Te Koaboa, Vol 17, Issue 1, pp. 67–89 (2011)

S17 - Francisco-Revilla L., Figueira A.: Adaptive Spatial Hypermedia in Computational Journalism. In: Proceedings of the 23rd ACM conference on Hypertext and social media (HT'12), pp. 313–314 (2012)

S18 - Weber W., Rall H.: Data visualization in online journalism and its implications for the production process. In: 16th International Conference on Information Visualisation, pp. 349–356 (2012)

S19 - Francisco-Revilla L.: Digital Libraries for Computational Journalism. In: Proceedings of the 12th ACM/IEEE-CS joint conference on Digital Libraries (JCDL'12), pp. 365–366 (2012)

S20 - Diakopoulos N., De Choudhury M., Naaman M.: Finding and assessing social media information sources in the context of journalism. In: Proceedings of the SIGCHI Conference on Human Factors in Computing Systems (CHI'12), pp. 2451–2460 (2012)

S21 - Pellegrini, T.: Integrating linked data into the content value chain: a review of news-related standards, methodologies and licensing requirements. In: Proceedings of the 8th International Conference on Semantic Systems, pp. 94–102 (2012)

S22 - Wu Y., Agarwal P.K., Li C., Yang J., Yu C.: On"one of the few" objects. In: 8th ACM SIGKDD International Conference on Knowledge Discovery and Data Mining (KDD'12), pp. 1487–1495 (2012)

S23 - Pellegrini, T.: Semantic Metadata in the News Production Process: Achievements and Challenges. In: Proceeding of the 16th International Academic MindTrek Conference (MindTrek'12), pp. 125–133 (2012)

S24 - Lesage F., Hackett R. A.: Between objectivity and opennessthe mediality of data for journalism. In: Media and Communication, Vol 2, Issue 2, pp. 42–54 (2013)

S25 - Parasie S., Dagiral E.: Data-driven journalism and the public good: "Computer-assisted-reporters" and "programmer-journalists" in Chicago. In: New Media Society, Vol 15, Issue 6, pp. 853–871 (2013)

S26 - Goasdou F., Leblay J., Karanasos K., Manolescu I., Katsis Y., Zampetakis S.: Fact checking and analyzing the Web. In: Proceedings of the 2013 ACM SIGMOD International Conference on Management of Data (SIGMOD'13), pp. 997–1000 (2013)

S27 - Lewis S.C., Usher N.: Open source and journalism: Toward new frameworks for imagining news innovation. In: Media, Culture Society, Vol 35, Issue 5, pp. 602–619 (2013)

S28 - Thelwall M., Buckley K.: Topic-based sentiment analysis for the social web: The role of mood and issue-related words. In: Journal of the Association for Information Science and Technology, Vol 64, Issue 8, pp. 1608–1617 (2013)

S29 - Anderson C. W.: Towards a sociology of computational and algorithmic journalism. In: New Media Society, Vol 15, Issue 7, pp. 1005–1021 (2013)

S30 - Lin Y.-R., Margolin D., Keegan B., Lazer D.: Voices of victory: A computational focus group framework for tracking opinion shift in real time. In: In: Proceedings of the International Conference on World Wide Web (WWW'13), pp. 737–748 (2013)

S31 - Gall M., Renders J.-M., Karstens E.: Who Broke the News?: An Analysis on First Reports of News Events. In: Proceedings of the 22nd International Conference on World Wide Web (WWW'13), pp. 855–862 (2013)

S32 - Lesage F., Hackett R.A.: Between objectivity and opennessThe mediality of data for journalism. In: Media and Communication, Vol 2, pp. 42–54 (2014)

S33 - Griffin P.: Big news in a small country-developing independent public interest journalism in NZ. In: Pacific Journalism Review, Vol 20, Issue 1, pp. 11–34 (2014)

S34 - Lewis S.C., Usher N.: Code, Collaboration, And The Future Of Journalism: A case study of the Hacks/Hackers global network. In: Journalism and Mass Communication, Vol 2, Issue 3, pp. 383–393 (2014)

S35 - Hassan N., Sultana A., Wu Y., Zhang G., Li C., Yang J., Yu C.: Data in, fact out: Automated monitoring of facts by FactWatcher. In: 40th International Conference on Very Large Data Bases (VLDB 2014), pp. 155–1560 (2014)

S36 - Appelgren E., Nygren G.: Data Journalism in Sweden: Introducing new methods and genres of journalism into old organizations. In: Digital Journalism, Vol 2, Issue 3, pp. 394–405 (2014)

S37 - Wu Y., Walenz B., Li P., Shim A., Sonmez E., Agarwal P. K., Li C., Yang J. Yu C.: iCheck: Computationally Combating Lies, D–ned Lies, and Statistics. In: Proceedings of the 2014 ACM SIGMOD International Conference on Management of Data (SIGMOD'14), pp. 1063–1066 (2014)

S38 - Sultana A., Hassan N., Li C., Yang J., Yu C.: Incremental discovery of prominent situational facts. In: 2014 IEEE 30th International Conference on Data Engineering (ICDE), pp. 112–123 (2014)

S39 - Gynnild A.: Journalism innovation leads to innovation journalism: The impact of computational exploration on changing mindsets. In: Journalism, Vol 15, Issue 6, pp. 713–730 (2014)

S40 - Regattieri L.L., Rockwell G., Chartier R., Windsor J.: TweetViz: Following Twitter hashtags to support storytelling. In: HT (Doctoral Consortium/Late-breaking Results/Workshops), Vol 1210 (2014)

S41 - Corby T.: Visualizing the news: Mutant barcodes and geographies of conflict. In: Leonardo, Vol 47, Issue, pp. 84–85 (2014)

S42 - Diakopoulos N.: Algorithmic Accountability: Journalistic investigation of computational power structures. In: Digital Journalism, Vol 3, Issue 3, pp. 398–415 (2015)

S43 - Broussard M.: Artificial Intelligence for Investigative Reporting: Using an expert system to enhance journalists ability to discover original public affairs stories. In: Digital Journalism, Vol 3, Issue 6, pp. 814–831 (2015)

S44 - Anderson C. W.: Between the Unique and the Pattern: Historical tensions in our understanding of quantitative journalism. In: Digital Journalism, Vol 3, Issue 3, pp. 349–363 (2015)

S45 - Lewis S.C., Westlund O.: Big Data and Journalism: Epistemology, expertise, economics, and ethics. In: Digital Journalism, Vol 3, Issue 3, pp. 447–466 (2015)

S46 - Graham C.: By the Numbers: Data Journalism Projects as a Means of Teaching Political Investigative Reporting. In: Asia Pacific Media Educator, Vol 25, Issue 2, pp. 247–261(2015)

S47 - Coddington M.: Clarifying Journalisms Quantitative Turn: A typology for evaluating data journalism, computational journalism, and computer-assisted reporting. In: Digital Journalism, Vol 3, Issue 3, pp. 331–348 (2015)

S48 - Hullman J., Diakopoulos N., Momeni E., Adar E.: Content, context, and critique: Commenting on a data visualization Blog. In: Proceedings of the 18th ACM Conference on Computer Supported Cooperative Work Social Computing (CSCW'15), pp. 1170–1175 (2015)

S49 - Knight M.: Data journalism in the UK: A preliminary analysis of form and content. In: Journal of Media Practice, Vol 16, Issue 1, pp. 55–72 (2015)

S50 - Plaue C., Cook L.R.: Data journalism: Lessons learned while designing an interdisciplinary service course. In: Proceedings of the 46th ACM Technical Symposium on Computer Science Education (SIGCSE'15), pp. 126–131 (2015)

S51 - Parasie S.: Data-Driven Revelation?: Epistemological tensions in investigative journalism in the age of big data. In: Digital Journalism, Vol 3, Issue 3, pp. 364–380 (2015)

S52 - Hassan N., Li C., Tremayne M.: Detecting Check-worthy Factual Claims in Presidential Debates. In: Proceedings of the 24th ACM International on Conference on Information and Knowledge Management (CIKM'15), pp. 1835–1838 (2015)

S53 - Young M. L., Hermida A.: From Mr. and Mrs. Outlier To Central Tendencies: Computational journalism and crime reporting at the Los Angeles Times. In: Digital Journalism, Vol 3, Issue 3, pp. 381–397 (2015)

S54 - Bonaque R., Cao T.D., Cautis B., Goasdou F., Letelier J., Manolescu I., Mendoza O., Ribeiro S., Tannier X., Thomazo M.: Mixed-instance querying: A lightweight integration architecture for data journalism. In: Proceedings of the VLDB Endowment, Vol 9, Issue 13, pp. 1513–1516 (2015)

S55 - Berendt B.: Power to the agents?! in the #WebWeWant, people will critically engage with data - And data journalism can help them want to do this. In: Joint Proceedings of the 5th International Workshop on Using the Web in the Age of Data (USEWOD '15) and the 2nd International Workshop on Dataset PROFIling and fEderated Search for Linked Data (PROFILES '15) co-located with the 12th European Semantic Web Conference (ESWC 2015), pp. 29–31 (2015)

S56 - Broussard M.: Preserving news apps present huge challenges. In: Newspaper Research Journal, Vol 36, Issue 3, pp. 299–313 (2015)

S57 - Rodrguez M.T., Nunes S., Devezas T.: Telling stories with data visualization. In: Proceedings of the 2015 Workshop on Narrative Hypertext (NHT'15), pp. 7–11 (2015)

S58 - Diakopoulos N.: The editor's eye: Curation and comment relevance on the New York Times. In: Proceedings of the 18th ACM Conference on Computer Supported Cooperative Work Social Computing (CSCW'15), pp. 1153–1157 (2015)

S59 - Birnbaum L., Boon M., Bradley S., Wilson J.: The news context project. In: Proceedings of the 20th International Conference on Intelligent User Interfaces Companion (UIU Companion'15), pp. 5–8 (2015)

S60 - De Maeyer J., Libert M., Domingo D., Heinderyckx F., Le Cam F.: Waiting for Data Journalism: A qualitative assessment of the anecdotal take-up of data journalism in French-speaking Belgium. In: Digital Journalism, Vol 3, Issue 3, pp. 432–446 (2015)

S61 - Kim S., Oh J., Lee J.: Automated news generation for TV program ratings. In: Proceedings of the ACM International Conference on Interactive Experiences for TV and Online Video (TVX'16), pp. 141–145 (2016)

S62 - Broussard M.: Big Data in Practice: Enabling computational journalism through code-sharing and reproducible research methods. In: Digital Journalism, Vol 4, Issue 2, pp. 266–279 (2016)

S63 - Griffin R.J., Dunwoody S.: Chair support, faculty entrepreneurship, and the teaching of statistical reasoning to journalism undergraduates in the United States. In: Journalism, Vol 17, Issue 1, pp. 97–118 (2016)

S64 - Hu Y., Lin Y.-R., Luo J.: Collective sensemaking via social sensors: Extracting, profiling, analyzing, and predicting real-world events. In: Proceedings of the 22nd ACM SIGKDD International

Conference on Knowledge Discovery and Data Mining (KDD'16), pp. 2127–2128 (2016)

S65 - Davies K., Cullen T.: Data Journalism Classes in Australian Universities: Educators Describe Progress to Date. In: Asia Pacific Media Educator, Vol 26, Issue 2, pp. 132–147 (2016)

S66 - Tabary C., Provost A.-M., Trottier A.: Data journalism's actors, practices and skills: A case study from Quebec. In: Journalism, Vol 17, Issue 1, pp. 66–84 (2016)

S67 - Appelgren E.: Data Journalists Using Facebook: A Study of a Resource Group Created by Journalists, for Journalists. In: Nordicom Review, Vol 37, Issue 1, pp. 1–14 (2016)

S68 - Zhu H., Lee Y. W., Rosenthal A. S.: Data Standards Challenges for Interoperable and Quality Data. In: Journal of Data and Information Quality (JDIQ) - Challenge Papers, Regular Papers and Experience Paper, Vol 7, Issue 1-2, Article No. 4 (2016)

S69 - Felle T.: Digital watchdogs? Data reporting and the news media's traditional 'fourth estate' function. In: Journalism, Vol 17, Issue 1, pp. 85–96 (2016)

S70 - Splendore S., Di Salvo P., Eberwein T., Groenhart H., Kus M., Porlezza C.: Educational strategies in data journalism: A comparative study of six European countries. In: Journalism, Vol 17, Issue 1, pp. 138–152 (2016)

S71 - Shehu V., Mijushkovic A., Besimi A.: Empowering Data Driven Journalism in Macedonia. In: Proceedings of the The 3rd Multidisciplinary International Social Networks Conference on SocialInformatics, Article No. 49 (2016)

S72 - Sidiropoulos E.A., Konstantinidis E. I., Veglis A. A.: Framework of a collaborative audio analysis and visualization tool for data journalists. In: 11th International Workshop on Semantic and Social Media Adaptation and Personalization (SMAP), pp. 156–160 (2016)

S73 - Zwinger S., Zeiller M.: Interactive infographics in German online newspapers. In: Proceedings of the 9th Forum Media Technology, pp. 54–64 (2016)

S74 - Hewett J.: Learning to teach data journalism: Innovation, influence and constraints. In: Journalism, Vol 17, Issue 1, pp. 119–137 (2016)

S75 - Lokot T., Diakopoulos N.: News Bots: Automating news and information dissemination on Twitter. In: Digital Journalism, Vol 4, Issue 6, pp. 682–699 (2016)

S76 - Le Borgne Y.-A., Homolova A., Bontempi G.: OpenTED browser: Insights into European Public Spendings. In: 1st Workshop on Data Science for Social Good (2016)

S77 - Salovaara I.: Participatory Maps: Digital cartographies and the new ecology of journalism. In: Digital Journalism, Vol 4, Issue 7, pp. 827–837 (2016)

S78 - Gertrudis-Casado M.-C., Grtrudix-Barrio M., lvarez-Garca S.: Professional information skills and open data. Challenges for citizen empowerment and social change. In: Comunicar, Vol 24, Issue 47, p. 39 (2016)

S79 - Orellana-Rodriguez C., Greene D., Keane M. T.: Spreading the news: How can journalists gain more engagement for their tweets?. In: Proceedings of the 8th ACM Conference on Web Science (WebSci'16), pp. 107–116 (2016)

S80 - Yang F., Du Y. R.: Storytelling in the Age of Big Data: Hong Kong Students Readiness and Attitude towards Data Journalism. In: Asia Pacific Media Educator, Vol 26, Issue 2, pp. 148–162 (2016)

S81 - Park D., Sachar S., Diakopoulos N., Elmqvist N.: Supporting comment moderators in identifying high quality online news comments. In: Proceedings of the 2016 CHI Conference on Human Factors in Computing Systems (CHI'16), pp. 1114–1125 (2016)

S82 - La-Rosa L., Sandoval-Martn T.: The transparency laws insufficiency for data journalisms practices in Spain. In: Revista Latina de Comunicacin Social, Vol 71, pp. 1208–1229 (2016)

S83 - Heravi B.R., Harrower N.: Twitter journalism in Ireland: sourcing and trust in the age of social media. In: Information, Communication Society, Vol 19, Issue 9, pp. 1194–1213 (2016)

S84 - Borges-Rey E.: Unravelling Data Journalism: A study of data journalism practice in British newsrooms. In: Journalism Practice, Vol 10, Issue 7, pp. 833–843 (2016)

S85 - Bucher T.: Machines dont have instincts: Articulating the computational in journalism. In: New Media Society, Vol 19, Issue 6, pp. 918–933 (2017)

S86 - S. Zwinger; J. Langer; M. Zeiller: Acceptance and Usability of Interactive Infographics in Online Newspapers. In: 21st International Conference Information Visualisation (IV), pp. 176–181 (2017)

S87 - Diakopoulos N., Koliska M.: Algorithmic Transparency in the News Media. In: Digital Journalism, Vol 5, Issue 7, pp. 809–828 (2017)

S88 - Boon M.: Augmenting media literacy with automatic characterization of news along pragmatic dimensions. In: Companion of the 2017 ACM Conference on Computer Supported Cooperative Work and Social Computing (CSCW'17), pp. 49–52 (2017)

S89 - Wu Y., Agarwal P.K., Li C., Yang J., Yu C.: Computational fact checking through query perturbations. In: ACM Transactions on Database Systems (TODS) - Invited Paper from ICDT 2014, Invited Paper from EDBT 2015, Regular Papers and Technical Correspondence, Vol 42, Issue 1, Article No. 4 (2017)

S90 - Cushion S., Lewis J., Callaghan R.: Data Journalism, Impartiality And Statistical Claims: Towards more independent scrutiny in news reporting. In: Journalism Practice, Vol 11, Issue 10, pp. 1198–1215 (2017)

S91 - Fan Q., Li Y., Zhang D., Tan K.-L.: Discovering newsworthy themes from sequenced data: A step towards computational journalism. In: EEE Transactions on Knowledge and Data Engineering,

Vol 29, Issue 7, pp. 1398–1411 (2017)

S92 - Langer J., Zeiller M.: Evaluation of the user experience of interactive infographics in online newspapers. In: 10th Forum Media Technology (2017)

S93 - Brolchin N.., Porwol L., Ojo A., Wagner T., Lopez E.T., Karstens E.: Extending open data platforms with storytelling features. In: Proceedings of the 18th Annual International Conference on Digital Government Research (dg.o'17), pp. 48–53 (2017)

S94 - Wihbey J.: Journalists Use of Knowledge in an Online World: Examining reporting habits, sourcing practices and institutional norms. In: Journalism Practice, Vol 11, Issue 10, pp. 1267–1282 (2017)

S95 - Bounegru L., Venturini T., Gray J., Jacomy M.: Narrating Networks: Exploring the affordances of networks as storytelling devices in journalism. In: Digital Journalism, Vol 5, Issue 6, pp. 699–730 (2017)

S96 - Boyles J.L., Meyer E.: Newsrooms accommodate data-based news work. In: Newspaper Research Journal, Vol 38, Issue 4, pp. 428–438 (2017)

S97 - Tandoc E.C., Jr., Oh S.-K.: Small Departures, Big Continuities?: Norms, values, and routines in The Guardians big data journalism. In: Journalism Studies, Vol 18, Issue 8, pp. 997–1015 (2017)

S98 - Patwari A., Goldwasser D., Bagchi S.: TATHYA: A Multi-Classifier System for Detecting Check-Worthy Statements in Political Debates. In: Proceedings of the 2017 ACM on Conference on Information and Knowledge Management (CIKM'17), pp. 2259–2262 (2017)

S99 - Lpez-Garca X., Rodrguez-Vzquez A.-I., Pereira-Faria X.: Technological skills and new professional profiles: Present challenges for journalism. In: Comunicar, Vol 25, Issue 53, pp. 81–90 (2017)

S100 - Hassan N., Arslan F., Li C., Tremayne M.: Toward Automated Fact-Checking: Detecting Check-worthy Factual Claims by ClaimBuster. In: Proceedings of the 23rd ACM SIGKDD International Conference on Knowledge Discovery and Data Mining (KDD'17), pp. 1803–1812 (2017)

S101 - Thurman N., Drr K., Kunert J.: When Reporters Get Hands-on with Robo-Writing: Professionals consider automated journalisms capabilities and consequences. In: Digital Journalism, Vol 5, Issue 10, pp. 1240–1259 (2017)

References

1. Bradshaw, P.: The inverted pyramid of data journalism. Online Journalism Blog 7 (2011)

2. Cairo, A.: The Functional Art: An Introduction to Information Graphics and Visualization. New Riders (2012)

3. Coddington, M.: Clarifying journalisms quantitative turn: a typology for evaluating data journalism, computational journalism, and computer-assisted reporting. Digit. Journalism **3**(3), 331–348 (2015)

4. Cox, M.: The development of computer-assisted reporting. Informe presentado en Association for Education in Jornalism end Mass Comunication. EEUU: Universidad de Carolina del Norte, Chapel Hill (2000)

5. Diakopoulos, N.: A functional roadmap for innovation in computational journalism. Nick Diakopoulos (2011)

6. Gray, J., Chambers, L., Bounegru, L.: The Data Journalism Handbook: How Journalists Can Use Data to Improve the News. O'Reilly Media, Inc., Sebastopol (2012)

7. Kawamoto, K.: Digital Journalism: Emerging Media and the Changing Horizons of Journalism. Rowman & Littlefield Publishers, Lanham (2003)

8. Kitchenham, B.: Procedures for performing systematic reviews. Keele University, Keele, vol. 33, pp. 1–26 (2004)

9. Kitchenham, B., Charters, S.: Guidelines for performing systematic literature reviews in software engineering. Technical report, Ver. 2.3 EBSE Technical report. EBSE (2007)

10. Nielsen, J.: Inverted Pyramids in Cyberspace. Nielsen Norman Group (1996)

11. Nielsen, J.: Scrolling and Attention. Nielsen Norman Group (2010)

12. Royal, C.: The journalist as programmer: a case study of the New York times interactive news technology department. In: International Symposium on Online Journalism (2010)

Data Mining for Prevention of Crimes

Neetu Singh[(⊠)], Chengappa Bellathanda Kaverappa,
and Jehan D. Joshi

University of Illinois at Springfield, Springfield, IL 62703, USA
{nsing2, cbell36, jjosh2}@uis.edu

Abstract. Preemptive measures are of utmost importance for crime prevention. Law enforcement agencies need to have an agile approach to solve everchanging crimes. Data analytics has proven to be an effective deterrent in the field of crime data analysis. Various countries like the United States of America have benefitted by this approach. The Government of India has also taken an initiative to implement data analytics to facilitate crime prevention measures. In this research paper, we have used R Studio, an open source data mining tool to perform the data analysis on the crime dataset shared by the Gujarat Police Department. To develop predictive model and study crime patterns we used various supervised and unsupervised data mining techniques such as Multiple Linear Regression, K-Means Clustering and Association Rules Analysis. The scope of this research paper is to showcase the effectiveness of data mining in the domain of crime prevention. In addition, an effort has been put forth to help the Gujarat Police Department to analyze their crime records and provide meaningful insights for decision making to solve the cases recorded.

Keywords: Data mining · Predictive model · Process modelling
Crime analysis

1 Introduction

Technology is a double-edged sword. Criminals have been using technology for various destructive purposes [1]. Preemptive measures are of utmost importance for crime prevention. This makes it imperative that law enforcement agencies need advanced crime analysis tools which will aid them in the process of crime prevention [2–4]. Various countries like the United States of America have benefitted by this approach [5, 6].

One of the challenges faced by the police departments is to minimize threats to society by investigating large volumes of data [7]. Various initiative has been taken by researchers to analyze crime data using data mining techniques [1, 8–11]. The Government of India has also taken an initiative to implement data analytics to facilitate crime prevention measures [11]. The Gujarat Police Department shares the same vision. A proposal was submitted by one of the authors to the Gujarat Police Department to access confidential crime data for research purposes. They graciously granted access to monthly and annual crime records for the years 2012 to 2016, for the State of Gujarat in India. The Gujarat Police Department categorized crime into 9 different categories such as Home Break-In, Injuries, Kidnapping, Murder, Attempt to

S. Yamamoto and H. Mori (Eds.): HIMI 2018, LNCS 10904, pp. 705–717, 2018.
https://doi.org/10.1007/978-3-319-92043-6_55

Murder, Police Raids, Rioting, Robbery and Theft. The goal of this paper is to develop a predictive model to predict future solving rates of the police department making them more proactive in nature and identifying crime associations and crime hotspots. This will further help the Gujarat Police Department to analyze their crime records and provide meaningful insights for decision making to solve the cases recorded. A survey on crime analysis found that 10% of the criminals commit 50% of the crimes [10]. So, any research initiative taken will help the police department solve crimes at a faster rate.

This research paper is a blend of supervised and unsupervised data mining techniques to analyze crime data. Here we implemented Multiple Linear Regression, Association Rules Analysis and K-Means Clustering algorithms to conduct a comparative study of various crime patterns from the dataset. The above-mentioned methods will help us identify underlying crime patterns and generate valuable insights from the dataset.

The motivation for conducting this research is to aid Gujarat Police Department in crime prevention and further help them understand the benefits of data mining in this domain. In this paper, different types of data mining methods have been applied and their subsequent results have been discussed which can be used by multiple enforcement agencies and as a point of reference for future research initiatives in the domain of crime analysis.

The paper is organized into six different sections. Section 1 is comprised of introduction. The background and information for crime analysis and prevention is discussed in literature review (Sect. 2). The methodology is discussed in Sect. 3. The different phases of data analysis are presented in Sect. 4. In Sect. 5 we discuss the results of data analysis which includes data visualization, graphs, thresholds and performance evaluation. The conclusion of the paper is presented in Sect. 6 which also includes the limitations and future scope of the research.

2 Literature Review

India is second most populous country in the world with a population of 1.3 billion people [11, 12]. According to the "7th Schedule Article 246 of the Indian Constitution", Law and Order is a subject of the state [13]. This has created enormous pressure on state government to prevent crimes in their respective states. In this research paper, the authors have analyzed crime records for the state of Gujarat using various data mining techniques. The state of Gujarat has a population of 60.3 million people according to the latest census data [14]. The recorded number of actual strength of police personal in the year 2011 for the state of Gujarat was 72,838. This makes the task of maintaining law and order extremely challenging owing to the sparse police to population ratio which is 129.89 per 100,000 people [15].

Understanding the administrative structure of the Gujarat State Police Department was important for the authors to effectively collaborate with them. For a state jurisdiction the Head of the police force is Director General of Police (DGP). In a state there are many districts. For a Range that is a cluster of neighboring districts are under the control of Deputy Inspector General of Police (DIGP). There are some states where it has two or more ranges based on geographical or population figure such a state is headed by the Inspector General of Police (IGP). A district is headed by Senior Superintendent of Police (SSP)/Superintendent of Police (SP) under whom the Assistant Superintendent of Police (ASP) and Deputy Superintendents of Police (DSP) act [11].

Researchers used various data mining techniques to help law enforcement agencies prevent crimes [16–19]. During our literature review we observed that to conduct crime hotspot analysis and crime pattern analysis various researchers preferred Density Based Clustering and K-Means Clustering which resulted in clusters based on the number of crime incidents recorded [1, 10, 20]. Based on the crime type and data type the method used for analysis greatly varies. Method such as social network analysis is used to analyze data from various social media platform, which is extremely important to prevent crimes [4]. Certain classification techniques such as K-NN (K Nearest Neighbors) classification help law enforcement agencies to classify crime types based on independent attributes [2].

Studying the work done by various other researchers we learnt that most of the researchers used one of the following above-mentioned data mining techniques. We identified this as our research gap and aimed to develop a holistic research paper which uses multiple unsupervised and supervised data mining methods along with statistical tests to test their performance and accuracy.

Market Basket Analysis using the Apriori algorithm has been a widely used data mining technique to study consumer behavior in the retail sector [21]. We wanted to study if market basket analysis can be used in other domains to study the underlying relationships in the data. We came across research carried out in the domain of transportation safety using Market Basket analysis to study the associations between fatal car crashes and various under lying factors [22]. We extrapolated this idea in the domain of crime data mining and after initial data preparation we used Market Basket analysis to study the associations between different crime types based on the number of crime incidence recorded.

3 Methodology

To effectively analyze the data using multiple data mining techniques and avoid any ambiguities we followed the SEMMA data mining implementation steps developed by the SAS institute [23]. SEMMA stands for "Sample", "Explore", "Modify", "Model", "Assess" [24].

- Sample: Various sampling strategies are used when the data is too large or complex to be analyzed. These sampling strategies try to replicate the properties of the population. In our dataset we analyzed the complete dataset to derive important insights and inferences.
- Explore: We conducted exploratory analysis by visualizing the data in R Studio to study descriptive statistics and to identify relationships between the variables in the dataset which helped us understand the data in a concise manner.

- Modify: Feature engineering is an important element which enriches the data for effective analysis and model development [10]. We added variables such as "Population", "Crime Density" and "Crime Code" for calculating the Crime Density we wanted to know the population in each city as population is an attribute present in crime density formula. Secondly, the Crime Code was added to apply Multiple Linear Regression in which the different Crime Types were encoded to numerical data for gaining statistical parameters. These variables improved model accuracy.
- Model: We developed predictive models using Multiple Linear Regression and used unsupervised learning methods such as K-Means Clustering algorithm and Apriori algorithm to study underlying patterns and associations in the data.
- Assess: To evaluate the performance of the model it was tested on unseen data. Various statistical parameters such as 'p-value', 't-value', 'R-Squared' and ANNOVA tests were performed to select the best model.

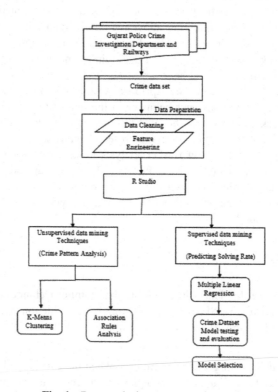

Fig. 1. Data analysis – conceptual model

A conceptual model is developed which would help us perform the data analysis activities in a structured manner. Figure 1 depicts this conceptual model. It illustrates the data analysis processes and techniques used for developing the predictive and classification model using SEMMA. First the data was obtained from the police department after which the annual crime data was extracted from the main dataset. This crime dataset is used for detailed analysis. To prepare the data various irrelevant fields and null values were omitted and the variable solving rate was checked for correctness and errors were modified. To further enrich the data, we added new features such as "Crime Density", "Crime Code" and "Population". Average imputation method was used to impute population values for districts whose populations weren't available [1]. The cleaned data was loaded to R Studio for further analysis.

We decided to use two approaches for our analysis "Crime pattern analysis" to study patterns and associations in the dataset using K-Means Clustering and Association Rules mining [4]. Predictive models were also developed using Multiple Linear Regression to predict the departments solving rate in subsequent years [25].

Various model testing techniques such as ANNOVA, variable importance tests and p-values were used for finalizing the variables and model selection. The model was validated by testing it on unseen test data. The model's stability was further tested by performing Ten-Fold cross validation [26].

4 Data Analysis

Data analysis uses various data mining techniques to analyze the data and uncover information previously unknown to us [26, 27]. Here selecting the best algorithms that best fits our research needs and help the Gujarat State Police Department was of utmost importance. Various steps were undertaken to successfully analyze the data and extract meaningful information from the data.

4.1 Data Preprocessing and Preparation

The dataset comprised of variables such as "Year", "City", "Crime type", "Public" (Cases recorded), "Found" (Cases resolved), and "Percentage" (Crime solving percentage). Data cleaning and transformation was carried out to derive better insights from the data. Three additional variables namely "Population", "Crime Density" and "Crime Code" were added. The population for the years 2012 to 2016 were derived based on the annual population growth rate of India and census data for the year of 2011 [1, 28]. In addition, the Crime density per 100,000 of the general population was calculated by mathematical formula:

$$Crime\ density = \left(\frac{Public(\text{Cases Recorded})}{Population} \right) * 100,000$$

The variable "Crime Code" was derived by encoding variable "Crime types" as sequential numbers from 1 to 9.

4.2   Data Mining Techniques

Data mining techniques are used to extract valuable information from the data [27]. These techniques are mainly of two types supervised and unsupervised.

Supervised learning: In this method the dataset is divided into two parts namely Training dataset and Test dataset. The models are developed using the training data and validated by comparing the model's prediction with that of the unseen test data [5, 27].

Unsupervised learning: In this class of data mining technique, the dataset is analyzed for underlying patterns and relations between different variables of a dataset [5, 27].

We have applied both of these data mining techniques to analyze the data, such as Multiple Linear Regression (supervised learning) to develop predictive models, K-Means Clustering and Association Rules mining (unsupervised learning) for studying the underlying patterns and the associations between different attributes in the dataset.

R programming provides us with various packages to perform analysis on dataset. We used R programming to clean the data, developed data mining models and test models. The following data mining methods were used to analyze our data.

Multiple Linear Regression: This method has more than one independent variables which predicts and establishes a relationship between our dependent variable. The difference between the predicted value P_i and actual value A_i is termed as error rate. The Regression can be expressed as Y_i [5, 27]: $Y_i = \beta_0 + \beta_1 x_1 + \ldots + \beta_n x_n + \in_i$
Where:

Y	denotes the measured value of dependent variable
X	denotes the value of the independent variables
β_0	denotes a constant
$\beta_1 \ldots \beta_n$	denotes estimated regression co-efficient
$\in_i$	denotes residual

K-Means Clustering: This method groups similar objects together to analyze crime patterns in the dataset. The algorithm uses distance computation functions to assign an observation to a cluster. An object is assigned to the cluster that it has the closest distance. The user decides the number of clusters in such a way that it minimizes the within sum of squares distance between the clusters [27, 29].

Association Rule Mining: The Association Rules analysis is an unsupervised learning technique which is used to extrapolate the rules associated with each other. In this case we have mined the associations of different types of crime committed with each other type [27, 30, 31]. The Association Rules are generated as a Transaction ID when applied. Each rule is represented by a transaction ID. The Association Rules are evaluated using the metrics - Support, Confidence and Lift.

Support is the measurement in which the crime type is repeated in a set of transactions given in the dataset. Confidence is the ratio of the support of two different crime types together to the support measure of individual crime committed regardless of associated crime is repeated in other transaction sets with other crime type. It also measures the accuracy of the associated rule [32]. Lift is measured as ratio of the support of two crimes together to the support of two crimes individually committed [30, 31].

The algorithm used for mining Association Rules is Apriori algorithm using R programming in R Studio. The different crime types are mined and crime types which are associated to each other based on the above three measures are extracted carefully. This shows the crime type that are committed by the criminals based on the previous committed crimes. The police department can use this prediction to analyze the next crime type of the suspect and be watchful for prevention of crime.

5 Results and Discussion

After this preliminary analysis of dataset, we would like to help the Gujarat Police Department to be more proactive and introduce counter measures in due course of time. We developed a predictive model using regression [5, 27] in R which would predict the cases resolved. These will be the cases which need to be solved on high priority to minimize the crime in Gujarat. The regression model provided an accuracy of 85.5%. We observed in Table 3 the "Crime Type Thieves" is not significant. This leads to an interesting finding that theft related cases are the least resolved ones which needs to be considered by Gujarat Police. Theft being the least resolved crime type might be one of the reasons that we found theft as one of the most happening crime in Gujarat as represented in Table 1.

Table 1. Regression model results

Model variables	Coefficient estimate	Std. error	t value	P value
(Intercept)	−18.216	3.677	−4.953	8.18e−07***
Crime type injuries	129.338	4.795	26.973	<2e−16***
Crime type kidnapping	41.597	4.820	8.630	<2e−16***
Crime type murder	21.021	4.818	4.363	1.38e−05***
Crime type murder attempt	18.607	4.741	3.924	9.11e−05***
Crime type raid	11.112	4.856	2.288	0.02226*
Crime type riot	28.269	4.842	5.838	6.54e−09***
Crime type robbery	14.714	4.734	3.108	0.00192**
Crime type thieves	−7.058	5.120	−1.378	0.16827
Cases recorded	0.405	0.006	61.996	<2e−16***
Population	7.718	0.957	8.066	1.53e−15***
Signif. codes:	***0.001	**0.01	*0.05	
Model statistics				
Adjusted R-squared: 0.8556 F-statistic: 853.4 p-value: <2.2e−16				

To test and validate the Multiple Linear Regression model we used the analysis of variance test and the model prediction function on unseen test data. Table 2 below displays the result of the analysis of variance test where the F-value of the variables of our choice are of significance.

Table 2. Analysis of variance

	Degrees of freedom	Sum square	Mean square	F- value	P-value
Crime type	8	5887627	735953	414.363	<2.2e−16***
Public	1	9153575	9153575	5153.733	<2.2e−16***
Population	1	115544	115544	65.055	1.528e−15***
Residuals	1429	2538055	1776		

Table 3 displays how well our predictive model performed on unseen test data we got an R-Squared value of 77% which signifies that our model did perform well on unseen test data as well.

Table 3. Model prediction statistics

RMSE	R squared	Mean absolute error
56.6043764	0.7701536	20.6575616

To check the impact and importance of the variable in our model we ran a variable importance test. Here a higher overall value signifies higher importance and relevance. From the Table 4 it is evidently clear that the variable 'Cases Recorded' has the highest significance towards predicting our crime solving rate.

Table 4. Variable importance table

Variable name	Overall
Crime type injuries	26.973354
Crime type kidnapping	8.630258
Crime type murder	4.362551
Crime type murder attempt	3.924314
Crime type raid	2.288467
Crime type riot	5.837583
Crime type robbery	3.108232
Crime type thieves	1.378487
Cases recorded	61.995915
Population	8.065636

We decided to perform K-Means Clustering and Association Rules Analysis to identify patterns and associations in the dataset which will help us to have deeper analysis of the crime data. To explore the underlying patterns in our data we used the K-Means clustering technique [9]. The Elbow method in Fig. 2 shows six clusters are optimal number of clusters as adding another doesn't improve the total within sum of squares. Using this as our K-value we performed K-Means clustering in R. Figure 3 represents the six homogenous clusters were created based on variables "Cases recorded" and "Cases resolved".

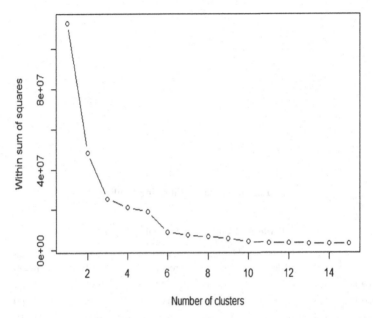

Fig. 2. A plot describing the within sum of squares (WSS) by Elbow method

As observed in Fig. 3, clusters can be segregated into different crime zones based on the number of cases recorded. In addition, Cluster 5 is a high-risk crime zone where the number of cases recorded is over 2500 crime instances which is extremely high as compared to the average.

After detailed analysis of cluster 5 it is evident that all the recorded instances of crime are related to theft (Table 5). Hence, it is recommended for the Gujarat Police Department to allocate more resources and prioritize theft related cases in Ahmedabad city. In Table 5 we found an interesting inference as the cases of theft are not properly resolved they do not have a high significance in helping us predict the cases resolved.

Our predictive model helped us to identify that theft is the most happening crime in Gujarat (Table 1). Crime pattern analysis using K-means clustering further confirms our results by providing the details that over 2500 recorded instances of crime are related to theft (Fig. 3, Table 5). So, planned to extrapolate different association between crime types using a novel approach of performing market basket analysis. To conduct this analysis using the annual crime dataset we split the dataset based on the crime type and grouped it based on cases recorded to create our baskets [32]. We used

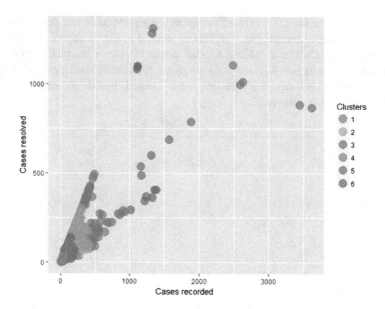

Fig. 3. K-means clustering result

Table 5. Contents of K-means cluster 5

City	Year	Crime code	Crime type	Cases recorded	Cases resolved	Cases resolved (%)
Ahmedabad	2012	6	Thieves	3622	871	24.04
Ahmedabad	2013	6	Thieves	3449	886	25.69
Ahmedabad	2014	6	Thieves	2589	999	38.59
Ahmedabad	2015	6	Thieves	2487	1108	44.55
Ahmedabad	2016	6	Thieves	2623	1014	38.66

different Support and Lift values to analyze these baskets and mine relevant Association Rules as displayed in Table 6.

Based on the lift values we can infer that suspects committing Murder and Riot are most likely to commit Kidnapping. Here a set of interesting rules that are relevant and will help the Gujarat State Police department are displayed in Table 6. Owing to these rules the department can narrow down their search to relevant suspects. In addition, the understanding of the association of crime types can help police department to focus on resolving those crime types simultaneously. For example, if police department strategize to resolve the crime types of murder, riot and robbery it can help them to resolve the kidnapping crime type.

Another interesting finding of the association rules analysis is that theft is usually not associated with any another crime type. Overall, we have observed theft is the most happening and recorded crime type in the state of Gujarat but is usually not associated with any other crime type. On the other hand, robbery is associated with most of the

Table 6. Association rules for crime types

Rules	Support	Confidence	Lift*	Count
[Murder, Riot] → [Kidnapping]	0.1	0.8	2.9	39
[Riot, Robbery] → [Kidnapping]	0.1	0.8	2.8	41
[Murder, Robbery] → Murder attempt]	0.1	0.8	5.8	38
[Kidnapping, Murder Attempt] → [Murder]	0.1	0.8	4.9	37
[Home Break-In, Murder Attempt, Robbery] → [Riot]	0.08	0.9	4.5	30
[Home Break-In, Murder Attempt, Raid] → [Robbery]	0.05	1.0	5.5	21

*Lift > 1 explains the significance of an association rule.

crime types. It would be interesting for us to further explore the difference between robbery and theft and focus on the cases reported and resolved for both crime types. We plan to extend our analysis on these lines in future.

6 Conclusion

As the population keeps growing with urbanization taking place new types of crime are committed making the cities vulnerable for the public to be safe. A smart and robust approach would help the law enforcement agencies to maintain public safety and peace in the cities by preventing crimes to be committed. The data mining techniques have been proven to be effective in analyzing the dataset and gather insights which are useful in many domains. This research paper used few data mining techniques to predict and identify the pattern of crime types committed in the state of Gujarat. This will help the Gujarat Police Department to identify the suspects and prevent them to further committing crime. The Association Rules suggest the Crime Types that are associated based on the cases recorded and the Multiple Linear Regression model has helped us to develop a predictive model which predicts the resolved cases with accuracy of 85.5%. The K-Means Clustering has shown the high crime risk zone city amongst the other cities.

In this research we have used yearly data (2012–2016) to predict the solving rate and to analyze the patterns of crime. The results of this research will help Gujarat police department to analyze their crime records and provide meaningful insights for decision making to solve the cases recorded. Currently we are working on monthly crime dataset to develop more robust models that could be implemented with the department. To successfully implement these methodologies, we are studying the As-Is crime solving process of the Gujarat Police Department and subsequently modelling a To-Be process which would integrate the data mining approach in the crime solving process.

Acknowledgement. The authors sincerely appreciate and are thankful for the cooperation of The Gujarat Police Department, India for providing us access to the dataset used in this study.

References

1. Malathi, A., Baboo, S.S.: An enhanced algorithm to predict a future crime using data mining. Int. J. Comput. Appl. **21**(1) (2011). ISSN 0975-8887
2. Hassani, H., et al.: A review of data mining applications in crime. Stat. Anal. Data Min.: ASA Data Sci. J. **9**(3), 139–154 (2016)
3. Bloomberg, J.: How the FBI Proves Agile Works for Government Agencies (2012). https://www.cio.com/article/2392970/agile-development/how-the-fbi-proves-agile-works-for-government-agencies.html
4. David, H., Suruliandi, A.: Survey on crime analysis and prediction using data mining techniques. ICTACT J. Soft Comput. **7**(3) (2017)
5. McClendon, L., Meghanathan, N.: Using machine learning algorithms to analyze crime data. Mach. Learn. Appl.: Int. J. (MLAIJ) **2**(1) (2015)
6. Li, X., et al.: GDP growth vs. criminal phenomena: data mining of Japan 1926–2013. AI Soc. **33**, 1–14 (2013)
7. Chen, H., et al.: Crime data mining: a general framework and some examples. Computer **37** (4), 50–56 (2004)
8. Tayal, D.K., et al.: Crime detection and criminal identification in India using data mining techniques. AI Soc. **30**(1), 117–127 (2015)
9. Thota, L.S., et al.: Cluster based zoning of crime info. In: Proceedings of the 2017 2nd International Conference on Anti-Cyber Crimes (ICACC), pp. 87–92. IEEE (2017)
10. Nath, S.V.: Crime data mining. In: Elleithy, K. (ed.) Advances and Innovations in Systems, Computing Sciences and Software Engineering, pp. 405–409. Springer, Dordrecht (2007). https://doi.org/10.1007/978-1-4020-6264-3_70
11. Gupta, M., et al.: Crime data mining for Indian police information system. In: Proceeding of the 2008 Computer Society of India (2008)
12. Wikipedia: World Population Prospects: The 2017 Revision. https://en.wikipedia.org/wiki/United_Nations_Department_of_Economic_and_Social_Affairs
13. Government of India: The Constitution of India (2015). http://lawmin.nic.in/olwing/coi/coi-english/coi-4March2016.pdf
14. Government of Gujarat: Official Gujarat State Portal (2011). http://www.gujaratindia.com/state-profile/demography.html
15. Bureau of Police Research and Development: Data on Police Organizations in India (2015). http://www.bprd.nic.in/WriteReadData/userfiles/file/201607121235174125303FinalDATABOOKSMALL2015.pdf
16. Nissan, E.: An overview of data mining for combating crime. Appl. Artif. Intell. **26**(8), 760–786 (2012)
17. Caplan, J.M., et al.: Joint utility of event-dependent and environmental crime analysis techniques for violent crime forecasting. Crime Delinq. **59**(2), 243–270 (2013)
18. Dos Santos, M.J., Kassouf, A.L.: A cointegration analysis of crime, economic activity, and police performance in São Paulo city. J. Appl. Stat. **40**(10), 2087–2109 (2013)
19. Yu, C.-H., et al.: Crime forecasting using data mining techniques. In: Proceedings of the 2011 IEEE 11th International Conference on Data Mining Workshops (ICDMW), pp. 779–786. IEEE (2011)
20. Birant, D., Kut, A.: ST-DBSCAN: an algorithm for clustering spatial–temporal data. Data Knowl. Eng. **60**(1), 208–221 (2007)
21. Chen, Y.-L., et al.: Market basket analysis in a multiple store environment. Decis. Support Syst. **40**(2), 339–354 (2005)

22. Pande, A., Abdel-Aty, M.: Market basket analysis of crash data from large jurisdictions and its potential as a decision support tool. Saf. Sci. **47**(1), 145–154 (2009)
23. Kurgan, L.A., Musilek, P.: A survey of knowledge discovery and data mining process models. Knowl. Eng. Rev. **21**(1), 1–24 (2006)
24. Azevedo, A.I., Santos, M.F.: KDD, SEMMA and CRISP-DM: a parallel overview. In: Proceedings of the IADIS European Conference Data Mining 2008, DM 2008 Proceeding, pp. 182–185 (2008)
25. João, P., et al.: Predictive model for criminality in lisbon (2010)
26. Adderley, R.: The use of data mining techniques in crime trend analysis and offender profiling. University of Wolverhampton, United Kingdom (2007)
27. Shmueli, G., et al.: Data Mining for Business Analytics: Concepts, Techniques, and Applications in R. Wiley, New York (2017)
28. Census of India: Ahmedabad City Census 2011 Data (2011). http://www.census2011.co.in/census/city/314-ahmedabad.html
29. Zubi, Z.S., Mahmmud, A.A.: Using data mining techniques to analyze crime patterns in the libyan national crime data. Recent Adv. Image Audio Sig. Process. **8**, 79–85 (2014)
30. Englin, R.: Indirect association rule mining for crime data analysis. Eastern Washington University, Cheney, Washington (2015)
31. Sevri, M., et al.: Crime analysis based on association rules using apriori algorithm. Int. J. Inf. Electron. Eng. **7**(3), 99 (2017)
32. Tan, P.-N., et al.: Selecting the right objective measure for association analysis. Inf. Syst. **29**(4), 293–313 (2004)

An Entity Based LDA for Generating Sentiment Enhanced Business and Customer Profiles from Online Reviews

Aniruddha Tamhane, Divyaa L. R., and Nargis Pervin[(✉)]

Indian Institute of Technology, Madras, Chennai, India
nargisp@iitm.ac.in

Abstract. The accelerated growth of the Web2.0 has led to an abundance of accessible information which has been successfully harnessed by many researchers for personalizing products and services. Many personalization algorithms are focused on analyzing only the explicitly provided information and this limits the scope for a deeper understanding of the individuals' preferences. However, analyzing the reviews posted by the users seeks to provide a better understanding of users' personal preferences and also aids in uncovering business' strengths and weaknesses as perceived by the users. Topic Modeling, a popular machine learning technique addresses this issue by extracting the underlying abstract topics in the textual data. In this study, we present entity-LDA (eLDA), a variation of Latent Dirichlet Allocation for topic modeling along with a dependency tree based aspect level sentiment analysis methodology for constructing user and business profiles. We conduct several experiments for evaluating the quantitative and qualitative performance of our proposed model compared to state-of-the-art methods. Experimental results demonstrate the efficacy of our proposed method both in terms topic quality and interpretability. Finally we develop a framework for constructing user and business profiles from the topic probabilities. Further we enhance the business profiles by extracting syntactic aspect level sentiments to indicate sentimental polarity for each aspects.

Keywords: Topic modeling · Latent Dirichlet Allocation · Profiling
Personalization · Sentiment analysis

1 Introduction

The digital revolution has led to an exponential increase in the information accessible online, which has found ample utility in personalization of products and services. By providing a more relevant and engaging experience to the users with personalized content, a significant increase in revenue and market share has been achieved. Many businesses like Yelp, Amazon, Netflix, etc. have their core business models centered around product personalization. Specifically, in the recommendation arena of movies, articles, books, etc. personalization is indispensable [23].

© Springer International Publishing AG, part of Springer Nature 2018
S. Yamamoto and H. Mori (Eds.): HIMI 2018, LNCS 10904, pp. 718–742, 2018.
https://doi.org/10.1007/978-3-319-92043-6_56

A vast majority of personalization techniques have focused on extracting explicit information on users and businesses through online ratings issued by a user. For example, Collaborative Filtering (CF) based algorithms personalize product recommendations based on the users' and items' historical ratings [16]. Social network based personalization algorithms further enhance the CF approaches by incorporating the user's social network related information. On the other hand, Content-based algorithms incorporate attributes extracted from the business itself to generate personalized recommendation. However, these approaches fail to capture user's inherent preferences and the ground-level facts about the businesses, which are better reflected in the reviews written by the users. Research on extracting information on user-business interactions through online reviews has focused on a variety of topic modeling and sentiment analysis approaches to gain insight into the user's perspective and opinions [4,5,37]. However, lack of generalization and difficulty in interpretability by humans have limited the scope of their practical applications.

In this work, we propose a topic modeling based sentiment enhanced entity (user/item) profiling methodology that generates user and item profiles as numeric vectors from the review text. The proposed framework consists of three stages: text preprocessing, topic modeling, and profile generation. In text preprocessing, we adapt a codeword insertion step to replace topic relevant terms with the domain specific category names based on the business' features. For example, in the context of restaurant reviews, the words "pizza" and "pepperoni" would be replaced by "American" if the restaurant serves North American cuisines and would be replaced by "Continental" if the restaurant serves Italian cuisines instead. We further parse the text to get Parts of Speech (POS) tags and a dependency tree in order to extract nouns for topic modeling and aspect level dependencies for sentiment classification.

In the topic modeling stage, we aggregate all the nouns and noun phrases from the preprocessed reviews at entity-level and generate two separate corpora, corresponding to user-level and business-level documents. We implement the entity Latent Dirichlet Allocation (eLDA) for extracting the topics, where "entity" refers to a user or item to be profiled. The derived topics are validated using several quantitative and qualitative (human judgment-based) metrics. The extracted topics are then mapped to the profile aspects in the profile generation stage. This ensures a domain-specific interpretability of the profiles. In addition, we use an aspect level sentiment classifier to classify the business aspects into strength/weaknesses as perceived by users. Finally, the entity profile vectors are generated as an aggregation of the probabilities of all the topics related to the aspects.

For this study, we have considered restaurant reviews posted by users in a publicly available dataset released by Yelp for Yelp dataset challenge to generate personalized profiles for users and restaurants. Our proposed method outperforms the benchmarked algorithms both in quantitative and qualitative metrics. This can be attributed to the following aspects of the proposed approach:

Noun Extraction: Extracting nouns and noun phrases retains the information critical to generating the profile, while ignoring the irrelevant information [22]. This in turn improves the topic quality and also reduces the computational time for topic modeling technique.

Codeword Insertion: Replacing multiple words referring to the same object with a codeword explicitly indicating the category reinforces domain-specific connections among words, that are not apparent from the corpus. Replacing redundant terms with codewords reduces the vocabulary size without loss of information. This enhances topic interpretability by concentrating the word-topic probability over a fewer set of words. It also improves discoverability of implicit topics that are not directly aggregated with a codeword.

The topic probabilities obtained from the proposed model are then used for generating entity profiles as a numeric vector of pre-determined dimension using the online reviews. Furthermore, we have enhanced the business profile by classifying the business aspects into user-perceived strengths and weaknesses using the parse tree based aspect level sentiment analysis.

The rest of the paper has been structured as follows: Sect. 2 presents literature related to this work. The solution overview has been described in detail in Sect. 3 followed by dataset description and experimental findings in Sect. 4 Sect. 5, respectively. Finally, conclusion and future work has been presented in Sect. 7.

2   Literature Review

In this section, we have reviewed the following topics relevant to our work: (1) Topic Modeling (2) Sentiment Analysis (3) Profiling and Personalization.

2.1   Topic Modeling

Information retrieval from massive text corpora has attracted attention from various research communities over the years. A plethora of work has focused on reducing the dimensionality of the word-frequency space by clustering various documents based on common underlying topics. The traditional Term Frequency-Inverse Document Frequency (TF-IDF) approach proposed by Salton and Buckley [28] reduces a document of arbitrary length to a real vector having a dimension equal to the total number of terms in the vocabulary of the entire corpus. Each vector element represents the term frequency (TF) weighted by the logarithm of inverse of the proportion of documents containing that term (IDF). Clustering algorithms are then applied to the TF-IDF matrix to cluster documents. This approach is appealing because of its simplicity and intuitive interpretability; however, it fails to capture underlying topics in the document, which can be modeled based on the co-occurrence of multiple terms. This problem has been addressed by Hoffman in the probabilistic Latent Semantic Indexing (pLSI) approach [12]. pLSI models a document as a probabilistic mixture of multiple topics, where a topic itself is a probabilistic distribution over the terms in the vocabulary. However, this method is not scalable since the number of parameters grow linearly with the number of documents in the corpus and lacks generalized application in other corpus.

The Latent Dirichlet Allocation (LDA) approach proposed by Blei et al. [5] addresses these issues by including a document level multinomial probability distribution over topics in the pLSI model. This ensures that the learnt parameters are not document specific, do not grow linearly with the number of documents and can be used to apply topics to another corpus built from the same vocabulary. LDA has been widely applied in a number of domains, successfully yielding topics of good quality. Variations of LDA such as local-LDA by Brody and Eldahad [6] and sentence LDA (sent-LDA) by Bao and Dutta [1] have been developed to further improve topic quality in specific domains. Another probabilistic topic modeling approach, Correlated Topic Models (CTM) has been developed by Blei and Lafferty [4] to model and extract topics having a high correlation. CTM supports the extraction of a greater number of topics from the same corpus with a higher log-likelihood; however, CTM is computationally more expensive as compared to LDA, and has a comparable performance for a lower number of topics.

The approaches discussed so far have the ability to soft cluster groups of words into various topics based on their co-occurrence in the corpus. However, they do not have any means of enforcing a domain-specific understanding of words that may not be captured in a bag of words approach because of the inherent richness of the domain's vocabulary. Bao et al proposed an LDA based topic modeling approach where ratings corresponding to the reviews were incorporated in the topic modeling algorithm in [2]. This approach improved recommendation quality, but failed to explicitly generate topics and entity profiles. A semantic approach to topic modeling has been proposed by Linshi [17], where words of similar sentiment polarity have been grouped to reveal the sentiment associated with the individual topics. This approach has demonstrated the potential for revealing additional information through the incorporation of semantic knowledge in topic modeling. In this study, we have adapted a semantic-LDA hybrid approach by incorporating domain-specific knowledge in topic modeling, resulting in better topic quality and interpretability.

2.2   Sentiment Analysis

Sentiment Analysis (SA) is the computational treatment of subjective opinions and emotions expressed by people in texts. SA has most commonly been approached as a classification problem [24], where entire documents or parts of documents are classified to be associated with various sentiments. SA is done on three levels: document-level, sentence-level and aspect-level. Document-level SA classifies a document into a multitude of sentiments such as positive, negative, neutral, etc. Sentence-level SA makes the same classification for individual sentences instead of documents and the aspect-level SA associates different sentiments to various aspects of the document [24].

Aspect-level SA has the potential to provide better insight into the distinct polarity associated with various themes in rich review texts compared to other techniques where an aggregated polarity is given. It primarily involves three major steps: aspect detection, sentiment classification and aggregation [29]. Aspect detection methods have generally been frequency based [13,18], syntactic [26,38] or machine learning based [14,19]. Frequency based and syntactic approaches are intuitive and have similar performance to machine learning techniques [29]. The sentiment classification, the second step, can occur separately or jointly [37] with aspect detection using either a lexicon-based bag of words approach for sentiment aggregation or implementing algorithms such as LDA, logistic regression, Hidden Markov Models, K-Nearest Neighbour, and Support Vector Machines for supervised and semi-supervised learning of the sentiments [24]. Aggregation of the sentiment associated with an aspect is the weighted average of all the individual sentiments [33,34]. In this study, we have employed a hybrid aspect-level SA approach to extract the sentiment level polarities associated with the aspects of business profiles. A parse-tree based syntactic approach has been used for aspect detection and a machine learning based supervised classification approach for sentiment classification. Finally, we have aggregated the sentiments associated with an aspect by taking a simple average of the classification probabilities.

2.3   Profiling and Personalization

The exponential growth of the Internet has led to abundance of information at the disposal of web users. Personalization has helped tackle information overload by limiting nonessential content. For catering personalized services, accurately profiling users and businesses has become paramount. This has opened a new area of research for many researchers to mine both implicit (visits, clicks, time spent, etc.) and explicit (reviews, stars, etc.) information for profiling and personalization. Collaborative Filtering (CF) is a technique used for personalizing content by grouping similar users and businesses based on historical ratings [16,27]. The underlying assumption is users who rated similarly in the past tend to rate similar in the future. One main limitation of similarity based CF is the time complexity does not grow linearly with number of users and businesses. Rather than exploiting the entire user space to find similar users, researchers have started using online social networks for identifying users' preferences. In [15,20,35] the latent features of friends from online social networks are assumed to be similar to target users' latent features and the influence from friends' networks are utilized for personalizing content to the target user. Although personalization can be facilitated by incorporating recommendations from friends in social network, there is a potential in incorporating individual preferences and perspectives about the businesses for quality profiling. In order to maintain individual preferences of experienced users along with social network influence Feng and Qian in [11] proposed a framework that fuses personal interest along with social influence. The personal interest factor captures the most desirable business categories for the target user by analyzing the historical ratings. The recent growth of GPS enabled gadgets and Web 2.0 Technologies have attracted users to update their location information via check-ins. Prior works in [7, 39–41] focused on only geographical influence for providing recommendations whereas recent works in [9, 36] uses both geographical and social influences for providing Point-Of-Interest (POI) recommendations. Though the previously stated works provide personalized information using explicit information (ratings or stars), they do not analyze the user preferences that are latent in the form of reviews, posts, blogs, etc.

In this study, we propose an elegant approach to generate profiles for users and items by incorporating latent features mined from the user reviews using codeword based entity-level LDA technique. Further, sentiment analysis technique has been employed to detect sentiments associated with the topics.

3 Solution Overview

3.1 Problem Formulation

In this study, we aim to construct M user and N item profiles as k-dimensional real-valued vectors $U_u \in \mathbb{R}^k \ \forall u \in \mathbb{U}$ and $I_i \in \mathbb{R}^k \ \forall i \in \mathbb{I}$ from review texts, where $\mathbb{U}$ is the set of users and $\mathbb{I}$ is the set of items, such that $|\mathbb{U}| = M$ and $|\mathbb{I}| = N$.

3.2 Solution Details

Herein, we propose a Latent Dirichlet Allocation (LDA) based entity profiling algorithm ($eLDA$) that ensures that the domain specific user interests are captured in user profile $U_u \forall u \in \mathbb{U}$. Item profiles $I_i \forall i \in \mathbb{I}$ captures the key business attributes along with the sentiment polarity which indicates the strengths and weaknesses as perceived by the users in general. The overall architecture of $eLDA$ has been presented in Fig. 1.

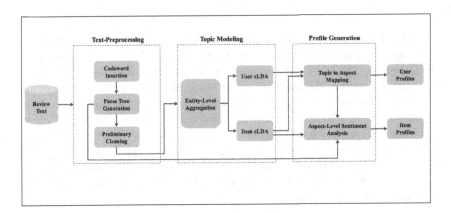

Fig. 1. Architecture for entity profile generation

Our algorithm consists of three major stages: text pre-processing, topic modeling, and entity profile generation. Each step has been explained in details below.

Text Pre-processing. Text pre-processing is customary in any Natural Language Processing technique. In eLDA this process consists of three stages, namely, codeword insertion, parse tree generation, and preliminary cleaning.

Codeword Insertion. We aggregate multiple words that represent the same category in the texts by replacing them with a common "codeword". For example, the words ice-cream, cake, gateau, pastry are replaced by the codeword "desserts". This work is focused on constructing restaurant profiles on a publicly available dataset collected from Yelp. To effectively tag the codewords, bag-of-words dictionary $\mathbb{B}$ corresponding to each codeword (category) has been generated. Instead of manually creating the codeword dictionary, crawling multiple recipe, cuisine and professional restaurant review websites with appropriately categorized recipe names would be a convenient route. As a result of this data collection process, our bag-of-words dictionary contains over 5000 words corresponding to 10 distinct cuisine categories, namely: American, Continental, Southern, Alcohol, Desserts, Asian, Indo-Arabic, Sandwiches and Breakfast-Brunch. We use the bag-of-words list to replace the words with the corresponding codeword, determined by the context or the business features F_i. In our case $|F| = 10$. For example, we have replaced the word "pizza" with "American" if the restaurant serves North American cuisine, and "Continental" if it serves Italian cuisines. Thus, we define the codeword insertion operator $C(v, \mathbb{B})$ for word v and dictionary $\mathbb{B}$ as:

$$C(v, \mathbb{B}) = \begin{cases} \mathbb{B}[v] & \mathbb{B}[v] \in F_i \\ v & \mathbb{B}[v] \notin F_i \end{cases} \tag{1}$$

Additionally, to detect if a restaurant enjoys location advantage, location specific keywords such as address of the restaurant has been added in a location codeword dictionary. Each dictionary category label is used as the respective codeword.

Parse Tree Generation. After codeword insertion, we parse the texts in order to generate the parse trees along with Parts Of Speech (POS) tags associated with the words. The POS tags are used to extract the nouns and noun phrases, which are later used in the topic modeling stage, to gain maximum critical information [22]. The parse trees are used for the aspect-level sentiment analysis of the business features in the profile generation stage. We have used the Compositional Vector Grammar based Parser [21,32] available as a part of the Stanford NLP package for generating parse trees in our dataset. The parse tree attaches a POS tag t and an index j denoting the tree-node position for each of the words obtained from the codeword transformation. Thus, the tree transformation converts each word v into a triplet:

$$T(v) = (v, t, j) \tag{2}$$

Here, $T(v)[i]$ corresponds to the word v, POS tag t and tree-node position j for $i = 1, 2, 3$, respectively.

Preliminary Cleaning. After generating parse trees, we remove the stop-words, punctuations and special characters in order to retain information critical to topic modeling and sentiment analysis. Finally, we lemmatize and stem the words to their root words in order to improve topic quality and interpretability [3].

Topic Modeling. We formulate the entity Latent Dirichlet Allocation (eLDA) algorithm as an extension of LDA [5] for extracting topics to construct entity-level profiles described as follows:

1. Extract nouns and noun phrases i.e., aggregate that all the $w = C(\mathbb{B}, v)$ s.t. $T(w)[2] = $ NN or NNS or NNP or NNPS after codeword insertion from each of the reviews.
2. Generate M user-level aggregated documents $E_u \; \forall \; u \in \mathbb{U}$ and N item-level aggregated documents $E_i \; \forall \; i \in \mathbb{I}$ s.t. E_e is the aggregation of all the pre-processed nouns present in the reviews pertaining to the entity $e \in \mathbb{U}$ or $e \in \mathbb{I}$.
3. For each entity aggregation E_e:
   (a) Choose $L \sim \text{Poisson}(\xi)$
   (b) Choose $\theta \sim \text{Dirichlet}(\alpha)$
   (c) For each of the L nouns in E_e:
       i. Choose a topic z_l from Multinomial (θ).
       ii. Choose a noun w_n from $p(w_n|z_{n,\beta})$, a multinomial probability distribution

We obtain the probability of $\mathbf{w}_u$, $\mathbf{w}_i$ (which are the term frequency forms of E_u and E_i), the user-level corpus D_u and item-level corpus D_i of entity-level aggregations as:

$$p(\mathbf{w}_u|\alpha, \beta) = \int p(\theta_u|\alpha) \left(\prod_{l=1}^{L_u} \sum_{z_{u_l}} p(z_{u_l}|\theta_u)p(w_{u_l}|z_{u_l}, \beta) \right) d\theta_u \qquad (3)$$

$$p(\mathbf{w}_i|\alpha, \beta) = \int p(\theta_i|\alpha) \left(\prod_{l=1}^{L_i} \sum_{z_{i_l}} p(z_{i_l}|\theta_i)p(w_{i_l}|z_{i_l}, \beta) \right) d\theta_i \qquad (4)$$

$$p(D_u|\alpha, \beta) = \prod_{u \in \mathbb{U}} \int p(\theta_u|\alpha) \left(\prod_{l=1}^{L_u} \sum_{z_{u_l}} p(z_{u_l}|\theta_u)p(w_{u_l}|z_{u_l}, \beta) \right) d\theta_u \qquad (5)$$

$$p(D_i|\alpha, \beta) = \prod_{i \in \mathbb{I}} \int p(\theta_i|\alpha) \left(\prod_{l=1}^{L_i} \sum_{z_{i_l}} p(z_{i_l}|\theta_i)p(w_{i_l}|z_{i_l}, \beta) \right) d\theta_i \qquad (6)$$

Finally, the joint probability of a topic allocation θ_e to an entity aggregation E_e is given by:

$$p(\theta_e, \mathbf{z}_e | \mathbf{w}_e, \alpha, \beta) = \frac{p(\theta_e, \mathbf{z}_e, \mathbf{w}_e | \alpha, \beta)}{p(\mathbf{w}_e, \alpha, \beta)} \tag{7}$$

We use Gibbs sampling for estimating the optimal parameters for inferring the topic distribution since the integral in Eq. 3 is generally hard to compute. Finally, we extract a set of topics T_u, T_i from document corpora D_u, D_i and the associated document-topic probability matrices P_u, P_i.

3.3 Entity Profile Generation

The entity profile generation stage is executed in three steps: topic to aspect mapping, aspect-level sentiment analysis, and profile generation.

Topic to Aspect Mapping. The user-level topics extracted from *eLDA* for the user-level document corpus D_u and business-level document corpus D_i are mapped to k domain-specific profile categories where the mapping is defined as:

$$\mathcal{M} : \{T_u, T_i\} \rightarrow \Pi \tag{8}$$

Here, Π refers to the domain specific set of categories. In general, $|\Pi| = k \leq max|T_u|, |T_i|$.

Aspect Level Sentiment Analysis. We use the parse tree transformation $T(v, t, j)$ generated in the pre-processing step for aspect identification and the corresponding sentiment classification. We use a supervised classifier $\mathbb{K}$ trained on a sentiment-labeled review dataset for a binary (positive/negative) classification of the sentiments. It is worthy to note that we do not apply sentiment classification for *User eLDA* as that would misrepresent the user profiles. We argue that if a user is concerned about a particular aspect of a restaurant, then only he/she mentions that in a review. Hence, the sentiment towards the identified aspects will always be positive.

The aspect level sentiment analysis proceeds as follows:

For every item level document $d_i \in D_i \ \forall i \in \mathbb{I}$:

1. Aspects $A_i \leftarrow \{v | t = \text{NN or NNS or NNP or NNPS} \ \forall T(v) = (v, t, j)\}$
2. $temp \leftarrow \{\}$
3. For each of the aspects v in A_i:
   (a) Generate the dependency text δ as the aggregation of the children of v, i.e. aggregate all u s.t. $j' = T(u, t', j')[3] \in children[j]$
   (b) After preliminary cleaning mentioned in Sect. 3.2, generate the Document-Term frequency matrix dtm_i for δ.
   (c) Use classifier $\mathbb{K}$ to obtain probability of positive sentiment $s_i = \mathbb{K}(dtm_i)$
   (d) Store s_i corresponding to v in $temp$ s.t. $temp[v] = s_i$
4. Use Eq. 8 to map individual aspects in A_i to the k profile categories and aggregate the sentiment probabilities for all aspects under each category as an arithmetic mean. Thus, $temp, \mathcal{M} \rightarrow S_i$
5. Store S_i in S.

Profile Generation. We aggregate the topic probabilities P_u and P_i using the mapping $\mathcal{M}$ given in Eq. 8 to obtain the k-dimensional profile category probabilities $\mathbb{P}_u$ and $\mathbb{P}_i$.

The User Profile U_u is defined as a k-dimensional numerical vector where the k elements of the vector denote the probability that the user u is interested in the each of the k categories in Π. Thus,

$$U_u = [\mathbb{P}_{ur}]_{k \times 1} \tag{9}$$

where $\mathbb{P}_{ur}$ is the probability that user u is interested in category Π_r.

The business profile I_i has been characterized in analogy to the user profile, with the additional incorporation of the normalized sentiment score S_i. The normalized sentiment score includes a sentiment polarity with the existing profile topics enabling to capture the positive and negative aspects. Thus, the business profile is defined as:

$$I_i = [\mathbb{P}_{is} \times f_{is}]_{k \times 1} \tag{10}$$

where $\mathbb{P}_{is}$ is the probability that business i is associated with category Π_s and S_{is} is the overall sentiment of business i with respect to category Π_s. Also, f_{is} is the sentiment polarity of item i with respect to category s defined as:

$$f_{is} = \begin{cases} -1 & S_{is} \in [0, l_1] \\ 1 & S_{is} \in [l_1, l_2] \\ 2 & S_{is} \in [l_2, 1] \end{cases} \tag{11}$$

Here, l_1 and l_2 are the experimentally decided cut-offs for the negative and neutral sentiments. Thus, a sentiment polarity of -1 denotes a negative sentiment, 1 denotes neutral sentiment and 2 denotes a positive sentiment.

4   Dataset Description

In this study, we use the publicly available dataset released by Yelp dataset[1]. The dataset contains user-item interactions in the form of ratings and textual user reviews along with business features. The dataset contains 144,072 businesses, located in United States of America (USA), United Kingdom, Canada and Germany. These business are categorized into 1191 categories such as restaurants, nightlife, religious organizations, etc. Out of 45,472 users, we consider only users who have reviewed at least five businesses globally which reduces the number of users to 11609 and number of businesses to 1983. With no loss of generality, we consider the reviews of businesses tagged as "restaurants", located in Phoenix city, USA. The basic dataset description is provided in Table 1.

5   Experimental Findings

In this section, we discuss the experiments conducted to evaluate our proposed topic modeling algorithm (*eLDA*) against benchmark models.

---

[1] https://www.yelp.com/dataset_challenge.

Table 1. Dataset description

Number of users	11,609
Number of businesses	1,983
Number of reviews	51,236
Business category	Restaurants
Location	Phoenix, AZ, USA

5.1 Benchmark Models

We have chosen three unsupervised models *TF-IDF*, *standard LDA* and *local-LDA* as benchmark models for evaluating the performance of our proposed model.

1. Term Frequency-Inverse Document Frequency (TF-IDF) [28]: This algorithm weighs the term frequency (TF) weighted by the logarithm of the inverse of the proportion of documents containing that term. A document-term matrix is then generated, with each document containing the TF-IDF score for all the terms in the vocabulary. K-means clustering technique is then employed for clustering the documents.
2. Standard LDA [5]: This algorithm treats each review as an individual document. Each document is further modeled as a probabilistic mixture of topics which themselves are distributed over the terms present in the vocabulary. This is a soft clustering technique that yields a probabilistic distribution of topics to a document.
3. Local LDA [6]: This algorithm is similar to *standard LDA* but treats each sentence as a document.

All the benchmark models are applied on review text which has been processed for removal of stopwords, numbers and punctuation marks. The corpus is then stemmed and lemmatized to obtain root words.

5.2 Intermediate Models

We introduce two intermediate models, *"LDA Noun-Codeword"* and *"LDA Noun"* to analyze the effect of extracting nouns and codeword insertion steps in isolation. *"LDA Noun-Codeword"* applies LDA on noun-codewords (some nouns replaced with noun-codewords) and each document corresponds to one review. *"LDA Noun"* applies LDA on nouns extracted from the reviews. In forthcoming sections, we compare the performance of our proposed models, *User eLDA* and *Item eLDA*, including the intermediate models against benchmark models.

5.3 Metrics for Evaluation

For evaluating the quality of the derived topics, we consider metrics such as perplexity, silhouette coefficient, word intrusion score and topic labeling. Quantitative metrics such as perplexity and silhouette coefficient are used to validate the model performance whereas word intrusion and topic labeling are qualitative measures used for evaluating the quality of the derived topics.

1. **Perplexity:** Perplexity measures the predictive performance of a model given a set of unobserved documents. Perplexity is a monotonically decreasing function of log likelihood of unseen documents which is defined as follows:

$$Perplexity(D_{test}) = exp\left(-\frac{\sum_{d\in\mathbb{T}} logp(w_d)}{\sum_{d\in\mathbb{T}} N_d}\right) \tag{12}$$

Where D_{test} is the test corpus containing documents belonging to test set $\mathbb{T}$ and N_d is the number of words in document d. Lower values of perplexity imply better performance of the model. Since *TF-IDF* does not have a log likelihood component, perplexity cannot be used to validate *TF-IDF*.

2. **Silhouette Coefficient:** Silhouette Coefficient measures how closely a document is assigned to its own cluster compared to other clusters. For a given document i, the silhouette coefficient is given as

$$s(i) = \frac{b(i) - a(i)}{\max\{a(i), b(i)\}} \tag{13}$$

Where $a(i)$ is the average distance between document i to all other documents in the same cluster and $b(i)$ is the lowest average distance between document i to all other documents from other clusters. Silhouette coefficient can range from -1 to 1, where 1 implies a document is well assigned to its own cluster than other clusters and -1 implies the opposite.

3. **Word Intrusion:** Word intrusion task aids in quantitively measuring the coherence of the topics [8]. In word intrusion task, a subject is presented with a set of most probable terms from a topic along with a randomly selected intruder term that does not belong to that topic. The task of the subject is to find the intruder term. For example, for a given set of words like $\{summer, winter, spring, autumn, dog\}$, a subject can easily identify dog as the intruder since all other words refer to seasons. When the terms in a topic lack such coherence, it becomes very difficult to identify the intruder term and subjects may randomly choose the intruder. Model precision is defined as follows:

$$MP_k^m = \frac{1}{S}\sum_s \mathbb{1}(i_{k,s}^m = w_k^m) \tag{14}$$

where MP_k^m is the word intrusion score generated for topic k and inferred from model m. Also, $i_{t,s}^m$ is the intruder word chosen by subject s and w_k^m is the actual intruder word. $\mathbb{1}(\cdot)$ is equal to 1 if the intruder word identified by the subject is same as actual intruder word w_k^m and 0 otherwise. Final word intrusion score for a model is the average word intrusion score for all the topics.

4. **Topic Labeling:** Topic labeling is the process of tagging domain-specific names to each of the topics. The degree to which subjects agree with the labeled terms for the topics can be quantified using Kappa statistics [31]. Cohen's Kappa is defined as follows.

$$\kappa = \frac{P_0 - P_c}{1 - P_c} \tag{15}$$

where P_0 is the proportion of mutual agreement and P_c is the proportion of agreement by chance. The values for Kappa can range from -1 to 1, where 1 represents perfect agreement, -1 perfect disagreement and 0 agreement by chance.

5.4   Model Performance

Predictive Power. In this section, we evaluate the performance of intermediate models against benchmark models. To validate the impact of using noun codewords to extract topics, we first compare the performance of intermediate models *LDA Noun* and *LDA Noun-Codeword* with benchmark models. Table 2 summarizes the perplexities for the held out documents for number of topics (K) varied from 5 to 100. Note that *TF-IDF* is not reported in Table 2 as it does not have a log likelihood component to measure perplexity.

Table 2. Performance comparison with model perplexity

Number of topics	Local LDA	Standard LDA	LDA noun	LDA Noun-Codeword
5	1013.11	1014.84	784.74	231.94
15	867.81	907.44	674.93	181.76
30	744.19	823.99	587.97	162.21
50	649.85	758.81	528.28	150.15
75	582.03	709.64	480.20	142.54
100	539.52	674.55	453.36	137.22

As we can see from Table 2, *LDA Noun-Codeword* significantly outperforms the benchmark models. This can be attributed to two reasons: firstly, by replacing most of the food terms in the nouns with codewords reduces the number of terms in the vocabulary. Secondly, it reduces the confusion of assigning codewords to specific cuisine topics. For example, "pizza" being a very popular food term can belong to multiple topics like "American", "Continental", etc. and assigning "pizza" to the relevant topic for a given document becomes challenging. However, information about the cuisines offered by each restaurant can be used to replace these food terms with cuisine terms. For example "pizza" can be replaced with "American" if the restaurant serves American food. The task is now simplified to assign these codewords to cuisine topics and therefore identifying each entity's cuisine preferences becomes easier. Also it helps to uncover other aspects of the restaurants like deals, hangout place etc. which remained hidden otherwise.

Cluster Quality. Table 3 summarizes the silhouette coefficients for all the models for varied number of topics. For LDA based models, each document is considered as one datapoint and the most probable topic as the cluster the document is assigned to. Each document's topic probabilities are used as distance vectors for computing distance. For *TF-IDF*, the document-term matrix with TF-IDF scores is used for measuring distance. We use Euclidean distance to compute the distance between the documents for all the models.

Table 3. Performance comparison with silhouette coefficient

Number of topics	TF-IDF	Local LDA	Standard LDA	LDA Noun-Codeword
5	0.016	0.281	0.234	0.312
15	0.005	0.290	0.114	0.111
30	−0.015	0.118	0.069	0.07
50	−0.013	0.092	0.051	0.046
75	−0.01	0.075	0.038	0.036
100	−0.029	0.067	0.029	0.031

As seen from Table 3, *Local LDA* has highest silhouette coefficient compared to other models. One should note that silhouette coefficient assumes hard clustering, where one document can be assigned to only one cluster. The silhouette coefficient for *Local LDA* is relatively high because *Local LDA* treats each sentence as one document and the number of topics discussed in a sentence is much lower than number of topics discussed in an entire review. Therefore, measurement with silhouette coefficient alone will not be a determinant factor for model performance. To compare the quality of the topics, we will use other metrics as well.

Performance of Entity Level Models. To determine the efficacy of the proposed approaches (*User eLDA, Item eLDA* and *LDA Noun-Codeword*), perplexity and silhouette coefficient are measured for different number of topics.

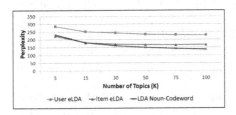

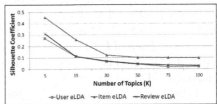

Fig. 2. Performance comparison with Model Perplexity

Fig. 3. Performance comparison with Silhouette Coefficient

Figure 2 demonstrates the superiority of *LDA Noun-Codeword*, for higher values of K with lowest perplexity. However, from Fig. 3, it can be observed that the silhouette coefficients for *Item eLDA* performs better compared to other models. Figure 4 compares the performance of Noun-Codeword models for 50 topics with respect to both perplexity and silhouette coefficient. Model performance is said to be better if the model perplexity is low and silhouette coefficient is high. To test the statistical significance of the performance of different models in terms of perplexity and silhouette coefficient, we perform a one-tailed two sample t-test for different number of topics. Table 4 summarizes the results of the statistical test.

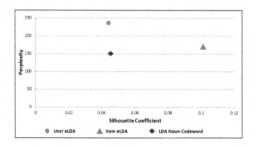

Fig. 4. Performance comparison of Noun-Codeword models (k = 50)

Table 4. Performance comparison with statistical significance

Metrics	User eLDA			Item eLDA			LDA Noun-Codeword		
	Mean	Std. dev.	p-value	Mean	Std. dev.	p-value	Mean	Std. dev.	p-value
Perplexity	246.43	21.31	0.0007***	179.97	22.58	0.25	167.64	35.29	-
Silhouette coefficient	0.088	0.095	0.41	0.189	0.144	0.87	0.101	0.108	-

Significance codes: p < 0.01*** p < 0.05** p < 0.1*.

The statistical test on *User eLDA* perplexity yields a p-value of 0.0007, which signifies the mean perplexity for *User eLDA* is statistically higher than *LDA Noun-Codeword* perplexity. The statistical test on *Item eLDA* perplexity yields a p-value of 0.25, and implies that mean perplexity for *Item eLDA* is statistically lower or not different from *LDA Noun-Codeword* perplexity. Although the perplexity of *User eLDA* is relatively higher than *LDA Noun-Codeword*, the performance of *User eLDA* is exceptionally exceeding compared to the traditional models (65.23% and 69.46% reduction in perplexity compared to *Local LDA* and *Standard LDA* respectively).

The statistical tests on silhouette coefficient yield p-values of 0.41 and 0.87 for *User eLDA* and *Item eLDA* respectively and therefore signify the mean silhouette coefficient for *User eLDA* and *Item eLDA* are statistically higher or not different from *LDA Noun-Codeword* silhouette coefficient.

The benefits of aggregating the noun-codewords at entity level (user/item) is multifaceted. Firstly, the size of the document-term matrix reduces drastically through aggregation and large number of terms can be accommodated in the document-term matrix. The number of documents in document-term matrix in this case have reduced by 77.34% and 96.12% for *User eLDA* and *Item eLDA* respectively, compared to *LDA Noun-Codeword*. Since the size of the document-term matrix is reduced, a reduction in the computational time can also be expected. Lastly, profiling the entities becomes much simpler as each document corresponds to an individual entity and the topic probabilities for each document can simply be used to represent entities' preference vectors.

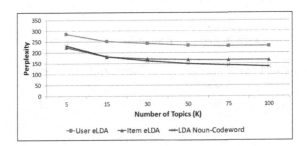

Fig. 5. Perplexity as a function of number of topics

Selecting the Number of Topics (K). To choose the optimal number of topics (K), held out perplexity is plotted as a function of number of topics for Noun-Codeword models. As seen from Fig. 5, the perplexity for the models is a decreasing function of number of topics, but remains stable after 50. Therefore, we pick 50 as the optimal topic number for our dataset.

5.5 Quality of Topics

The metrics discussed in Sect. 5.4 are crucial for evaluating model's performance. However, it is equally important to validate the quality of the derived topics. Herein, we present word clouds for each of the models, where the font size corresponds to the probability of the terms occurring in the topics. Later, we validate the quality of the topics using word intrusion and topic labeling technique.

Word Clouds. Due to space constraint, five random topics are selected from each of the models and are presented using word clouds. The size of each terms corresponds to the term probability for a given topic. The five topics include the following: cuisine (2), location (1), customer service (1) and order delivery (1). As seen from Fig. 6, for Noun-Codeword models, i.e *LDA Noun-Codeword*, *User eLDA* and *Item eLDA*, the term probabilities of noun codewords in cuisine topics ("Continental", "American") are exceptionally high. Similarly, the key terms for other topics such as location, customer service and order delivery have a relatively higher probabilities compared to other models. This visual representation depicts the relevance of terms in a topic for Noun-Codeword models, as the term probabilities are relatively higher compared to other models.

Topic Labeling. Topic labeling is a crucial step for understanding the context of the derived topics. Manually labeling topics ensures high labeling quality [8]. We selected top 15 topics (based on mean topic probabilities) from each model and employed a random user to manually label the topics. Two subjects were asked to rate these topic labels as "relevant" or "irrelevant". Cohen's Kappa is used to measure the inter-rater agreement for the labeled topics. Table 5 shows the Kappa statistics for different models.

Table 5. Cohen's Kappa for topic labeling

User eLDA	Item eLDA	LDA Noun-Codeword	Standard LDA	Local LDA	TF-IDF
0.42	0.66	0.63	0.19	0.24	0.21

It can be observed from Table 5 that Kappa statistic is high for Noun-Codeword models (*User eLDA, Item eLDA, LDA Noun-Codeword*). This indicates that our proposed model is able to generate of high quality topics and thereby reduces the chances of mislabeling the topics.

Word Intrusion. For evaluating the coherence of terms in topics, word intrusion task is performed. Top five terms based on high term probabilities are chosen from five topics, from all the models (shown in Fig. 6). For each of the topics, a low probable term from the same topic is chosen as intruder word. These set of terms are presented to 15 subjects. The task of the subjects is to find the intruder word from the given set of words. Since the probability of cuisine specific term in cuisine topics for Noun-Codeword models is extremely high, the cuisine term in isolation is sufficient for representing the entire topic. For example: in Fig. 6, *User eLDA* Topic 4 depicts that term *"Continental"* has a very high probability and is a representative of the entire topic. This eliminates the need for performing word intrusion task for cuisine topics for Noun-Codeword models and we argue that the model precision is always 1 for cuisine topics. For other benchmark models, we perform word intrusion on five topics whereas for Noun-Codeword models (*User eLDA, Item eLDA* and *Review Noun-Codeword*) we perform word intrusion task on only three non-cuisine topics (location, customer service and order delivery). Figure 7 presents model precision for all the models.

Figure 7 shows that Model Precision is high for Noun-Codeword models (*User eLDA, Item eLDA* and *Review Noun-Codeword*), which demonstrates better semantic coherence in the inferred topics.

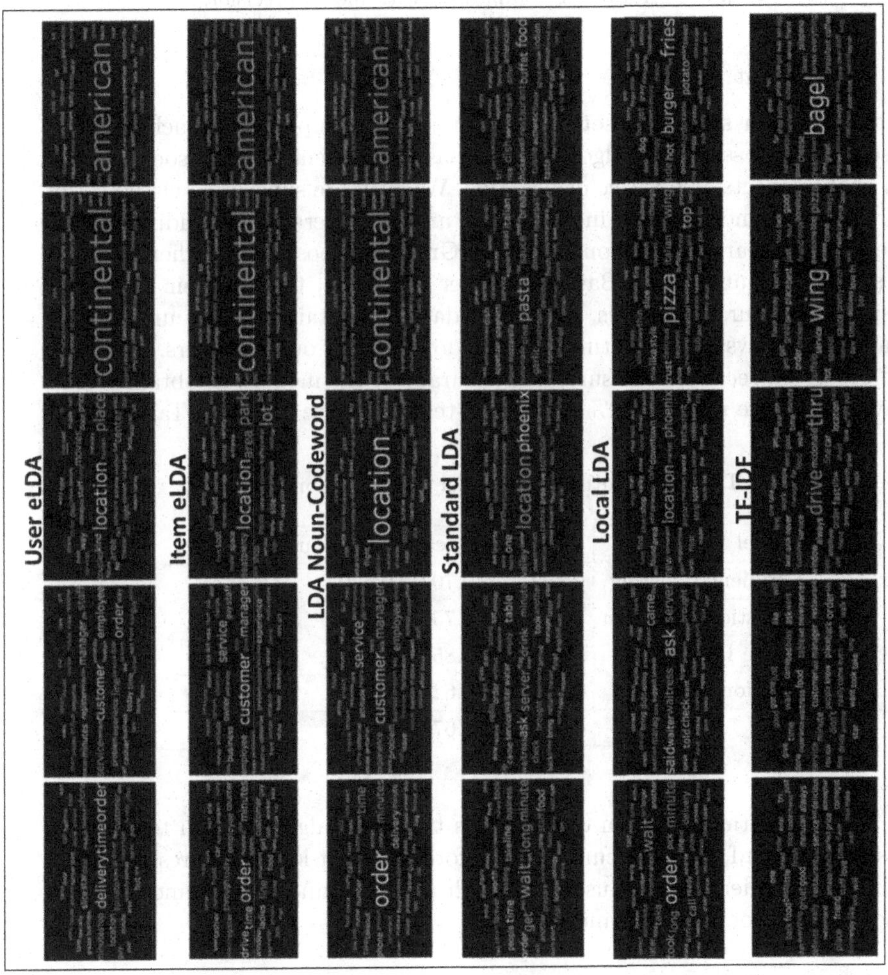

Fig. 6. Word cloud

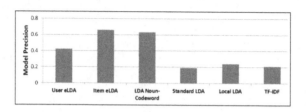

Fig. 7. Performance comparison with model precision

5.6   Sentiment Analysis

Sentiment Classifier Selection. We implement supervised machine learning sentiment classification algorithms to classify the sentiments associated with each of the aspects as positive or negative. We evaluate supervised classification algorithms commonly cited in sentiment analysis literature including logistic regression, K-Nearest neighbours (KNN), Gradient Boosting classifier, Random Forest classifier and Naive Bayes Classifier [24,25,29] to test their efficacy in classifying restaurant reviews. We use a dataset containing 1000 unique Yelp restaurant reviews[2] with sentiment labels for training our classifiers. The standard precision, recall, F-measure, and accuracy of the models, as obtained after an average of five random 80%–20% train-test splits are shown in Table 6.

Table 6. Sentiment classifier model performance

Model	Precision	Recall	F-measure	Accuracy
Gradient Boosting	0.752	0.707	0.729	0.737
Logistic Regression	0.806	0.757	0.780	0.793
Naive Bayes	0.656	0.856	0.742	0.700
Random Forests	0.863	0.620	0.721	0.755
KNN	0.689	0.675	0.680	0.687

Clearly, logistic regression outperforms the other algorithms in terms of F-measure score and overall accuracy. Therefore we select logistic regression as the sentiment classifier ($\mathbb{K}$) because of its high accuracy and well-balanced performance in terms of precision and recall.

6   Applications of Entity Profiles

6.1   Example of Profile Generation

The user and business profiles generated using our methodology have a multitude of potential applications. Our methodology generates the users and businesses

---

[2] https://github.com/Microsoft/microsoft-r/tree/master/microsoft-ml/Samples/
101/BinaryClassification/SimpleSentimentAnalysis.

profiles using review texts, thus reflecting the popular preferences and perspectives about the businesses. We demonstrate one example each for profiling users and businesses from a sample review taken from the Yelp dataset.

An Example of User Profiling. Consider a snippet of the user-level document for one user: *'Decent food, decent beer, fast service what more do you want? Price was fair too ! Some TVs to catch sports. Nothing special but gets the job done. Okay food, drinks, waitresses etc. The highlight is the karaoke...'*
We can intuitively say that this user is interested in food, drinks, service, prices and overall ambience based on the fact that he/she has mentioned these attributes. We demonstrate step-wise generation of the profile for this particular user:

1. After the pre-processing stage, the nouns and codewords are extracted. Thus, the above text is transformed into: "AMERICAN ALCOHOL service price tvs sports nothing job AMERICAN drinks waitresses highlight karaoke....". Thus, the text now contains only nouns and noun-phrase with the appropriate codewords, which are "AMERICAN" and "ALCOHOL" in this case, to denote the corresponding food category.
2. In the topic modelling stage, the pre-processed text is run through the *User eLDA* model to extract the corresponding topics T_u and the K-dimensional probability distribution vector p_u.
3. In the profile generation stage, a manually generated mapping $\mathcal{M}$ is used to map the $K = 50$ topics in T_u to the k user profile categories. In our case, we have $k = 15$ categories, corresponding to $\Pi = \{$american, continental, asian, indo-arabic, desserts, sandwiches, alcohol, nightlife, breakfast-brunch, southern, ambience, price, restaurant quality, service efficiency, location$\}$. Thus, the mapping $\mathcal{M}$ would map the topics "good waiter", "bad service", "bad waiter" to the category "Service quality". Also, the k-dimensional category probability vector P_u will be the aggregate of the K-dimensional topic probability vector p_u by taking the sum.
4. The user profile U_u is generated as: $[0.147, 0, 0, 0, 0, 0, 0.031, 0, 0, 0.039,$ $0, 0, 0.031, 0.112, 0.147]$, where the $U_u[i]$ corresponds to Π_i for all $k \in [1, 15]$. The topic probabilites for most of the terms in this example are very low and hence was rounded to zero. We can directly infer that the user has a very high interest in American cuisine (0.147), restaurant locations (0.147) and service efficiency (0.112), while being mindful of the restaurant quality (0.031), nightlife (0.031), and southern cuisine (0.039). This finding corroborates well with our initial intuition about the user.

An Example of Item Profiling. Consider a snippet of the item-level document for one restaurant: *'Brand new food business! They make you a pizza, you take it home and bake it. It's a create your own pizza place... But terrible service...'*

We can infer that this restaurant serves a take-at-home pizza, which is satis-
factory in terms of food, but poor in service. We can generate the item profile I_i for this restaurant following the steps similar to those mentioned for
generating user profiles. This item's profile obtained from *Item eLDA* model
is $[0.34, 0.04, 0, 0, 0, 0, 0.03, 0, 0, 0.04, 0, 0, 0.031, 0.23, 0]$. Further, we generate the
aspect level sentiment-polarities as follows:

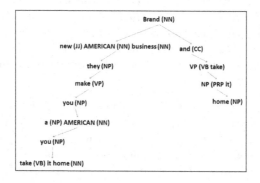

Fig. 8. Example of a parse tree for a single sentence

1. Firstly, we generate the parse tree from the text preprocessing stage as
   depicted in Fig. 8.
2. We then derive the aspect level sentiment score as explained in Sect. 3.3.
   Thus we obtain the following aspects and corresponding scores: {new american business: 0.83, AMERICAN: 0.69, home: 0.41 own pizza place: 0.63 and
   service: 0.14}.
3. We then aggregate the aspect level sentiments to category level sentiments
   using arithmetic mean: In this case, the sentiment for 'American' category
   would be an aggregate of sentiment scores of 'American', 'new american business' and 'own pizza place', which is 0.72. The aggregated sentiment scores
   for home and service correspond to service efficiency category and will be
   0.28.
4. Then the sentiment polarities are calculated as follows: American having a
   score of 0.59 will take a positive score of +2 and service with a score of 0.28
   will take a negative sentiment score of −1.
5. The topic probabilities for this business obtained from *Item eLDA* is
   $[0.34, 0.04, 0, 0, 0, 0, 0.03, 0, 0, 0.04, 0, 0, 0.031, 0.23, 0]$. Now the category sentiment polarities will be multiplied to the topic probabilities to generate
   the business profile I_i as: $I_i = [0.68, 0.04, 0, 0, 0, 0, 0.03, 0, 0, 0.04, 0, 0, 0.031,$
   $−0.23, 0]$. This vector will represent the business profile for the specified
   business.

We can infer from the generated business profile that the restaurant is perceived to have a very good american cuisine, a decent continental cuisine, alcohol

service, southern cuisines and overall quality, but bad service efficiency. Thus, the generated business profile closely reflects the inferred inutition about the restaurant.

6.2   Proposed Applications

The applications of these personalized profiles are multifaceted. We envision their direct applications in the following fields:

Recommender Systems. Recommender systems generate personalized recommendations by clustering similar users and items based on historical ratings [20,35]. Similarity measurements can be directly refined by incorporating the user and business profiles generated from text reviews. Further, our profiles can explain the recommendations in a more natural manner, by weighing in the user's interests and the business's strengths and weaknesses.

Web Personalization. Web personalization customize online content to suit individual users and effectively project the businesses [10]. Generated profiles can potentially make a twofold impact in this sector. Personalization of online content based on the generated profiles can potentially improve click-through and conversion rates.

Marketing. Traditionally, users and businesses are profiled using historical rating and purchase history to classify them into various market segments [30]. However, profiles generated using online reviews have the potential to unravel attributes such as price, quality, ambiance etc. which in-turn can improve the existing customer segmentation process.

7   Conclusion

In this paper, we propose an entity-level topic model - *eLDA*, which incorporates domain-specific noun-codewords aggregated at entity-levels (user and item levels) to derive underlying abstract topics hidden in textual reviews. Several experiments have been conducted on a large review dataset of restaurants to validate the predictive performance and quality of derived topics. Experimental results reveal that our proposed models, *User eLDA* and *item eLDA*, outperform other benchmark models with improved topic coherence. The topic probabilities obtained from *eLDA* models are then mapped to domain-specific aspects for building entity profiles. This pivotal step for personalization reveal the actual meaning of these latent features and hence improves the ease of interpretability, which was largely overlooked by the traditional CF-based personalized algorithms. The aggregation of documents to the entity levels reduces the size of large Document-Term Matrix (DTM) drastically which ensures a huge reduction in the computational complexity as well. Findings show that there is a gratifying reduction of 77.34% and 96.12% in DTM size for *User eLDA* and *item eLDA*, respectively compared to *LDA Noun-codeword*, with good predictive power and topic quality. In our future work, we plan to fuse the entity profiles to an existing recommender system algorithm and validate the quality of the recommendations.

References

1. Bao, Y., Datta, A.: Simultaneously discovering and quantifying risk types from textual risk disclosures. Manag. Sci. **60**(6), 1371–1391 (2014)
2. Bao, Y., Fang, H., Zhang, J.: TopicMF: simultaneously exploiting ratings and reviews for recommendation. In: AAAI, vol. 14, pp. 2–8 (2014)
3. Bird, S., Klein, E., Loper, E.: Natural Language Processing with Python: Analyzing Text with the Natural Language Toolkit. O'Reilly Media, Inc., Sebastopol (2009)
4. Blei, D.M., Lafferty, J.D.: Correlated topic models. In: Proceedings of the 18th International Conference on Neural Information Processing Systems, pp. 147–154. MIT Press (2005)
5. Blei, D.M., Ng, A.Y., Jordan, M.I.: Latent Dirichlet allocation. J. Mach. Learn. Res. **3**(Jan), 993–1022 (2003)
6. Brody, S., Elhadad, N.: An unsupervised aspect-sentiment model for online reviews. In: Human Language Technologies: The 2010 Annual Conference of the North American Chapter of the Association for Computational Linguistics, pp. 804–812. Association for Computational Linguistics (2010)
7. Cao, X., Cong, G., Jensen, C.S.: Mining significant semantic locations from GPS data. Proc. VLDB Endow. **3**(1–2), 1009–1020 (2010)
8. Chang, J., Gerrish, S., Wang, C., Boyd-Graber, J.L., Blei, D.M.: Reading tea leaves: how humans interpret topic models. In: Advances in Neural Information Processing Systems, pp. 288–296 (2009)
9. Cheng, C., Yang, H., King, I., Lyu, M.R.: Fused matrix factorization with geographical and social influence in location-based social networks. In: AAAI, vol. 12, pp. 17–23 (2012)
10. Eirinaki, M., Vazirgiannis, M.: Web mining for web personalization. ACM Trans. Internet Technol. (TOIT) **3**(1), 1–27 (2003)
11. Feng, H., Qian, X.: Recommendation via user's personality and social contextual. In: Proceedings of the 22nd ACM International Conference on Information & Knowledge Management, pp. 1521–1524. ACM (2013)
12. Hofmann, T.: Probabilistic latent semantic analysis. In: Proceedings of the Fifteenth Conference on Uncertainty in Artificial Intelligence, pp. 289–296. Morgan Kaufmann Publishers Inc. (1999)
13. Hu, M., Liu, B.: Mining opinion features in customer reviews. In: AAAI, vol. 4, pp. 755–760 (2004)
14. Jakob, N., Gurevych, I.: Extracting opinion targets in a single-and cross-domain setting with conditional random fields. In: Proceedings of the 2010 Conference on Empirical Methods in Natural Language Processing, pp. 1035–1045. Association for Computational Linguistics (2010)
15. Jamali, M., Ester, M.: A matrix factorization technique with trust propagation for recommendation in social networks. In: Proceedings of the Fourth ACM Conference on Recommender Systems, pp. 135–142. ACM (2010)
16. Konstan, J.A., Miller, B.N., Maltz, D., Herlocker, J.L., Gordon, L.R., Riedl, J.: GroupLens: applying collaborative filtering to usenet news. Commun. ACM **40**(3), 77–87 (1997)

17. Linshi, J.: Personalizing yelp star ratings: a semantic topic modeling approach. Yale University (2014)

18. Long, C., Zhang, J., Zhut, X.: A review selection approach for accurate feature rating estimation. In: Proceedings of the 23rd International Conference on Computational Linguistics: Posters, pp. 766–774. Association for Computational Linguistics (2010)

19. Lu, B., Ott, M., Cardie, C., Tsou, B.K.: Multi-aspect sentiment analysis with topic models. In: 2011 IEEE 11th International Conference on Data Mining Workshops (ICDMW), pp.81–88. IEEE (2011)

20. Ma, H., Zhou, D., Liu, C., Lyu, M.R., King, I.: Recommender systems with social regularization. In: Proceedings of the Fourth ACM International Conference on Web Search and Data Mining, pp. 287–296. ACM (2011)

21. Manning, C., Surdeanu, M., Bauer, J., Finkel, J., Bethard, S., McClosky, D.: The stanford coreNLP natural language processing toolkit. In: Proceedings of 52nd Annual Meeting of the Association For Computational Linguistics: System Demonstrations, pp. 55–60 (2014)

22. Martin, F., Johnson, M.: More efficient topic modelling through a noun only approach. In: 2015 Proceedings of the Australasian Language Technology Association Workshop, pp. 111–115 (2015)

23. Brusilovsky, P., Maybury, M.T.: From adaptive hypermedia to the adaptive web. Commun. ACM **45**(5), 30–33 (2002)

24. Medhat, W., Hassan, A., Korashy, H.: Sentiment analysis algorithms and applications: a survey. Ain Shams Eng. J. **5**(4), 1093–1113 (2014)

25. Pedregosa, F., Varoquaux, G., Gramfort, A., Michel, V., Thirion, B., Grisel, O., Blondel, M., Prettenhofer, P., Weiss, R., Dubourg, V., Vanderplas, J., Passos, A., Cournapeau, D., Brucher, M., Perrot, M., Duchesnay, E.: Scikit-learn: machine learning in Python. J. Mach. Learn. Res. **12**, 2825–2830 (2011)

26. Qiu, G., Liu, B., Bu, J., Chen, C.: Expanding domain sentiment lexicon through double propagation. In: IJCAI, vol. 9, pp. 1199–1204 (2009)

27. Resnick, P., Iacovou, N., Suchak, M., Bergstrom, P., Riedl, J.: GroupLens: an open architecture for collaborative filtering of netnews. In: Proceedings of the 1994 ACM Conference on Computer Supported Cooperative Work, pp. 175–186. ACM (1994)

28. Salton, G., Buckley, C.: Term-weighting approaches in automatic text retrieval. Inf. Process. Manag. **24**(5), 513–523 (1988)

29. Schouten, K., Frasincar, F.: Survey on aspect-level sentiment analysis. IEEE Trans. Knowl. Data Eng. **28**(3), 813–830 (2016)

30. Shaw, M.J., Subramaniam, C., Tan, G.W., Welge, M.E.: Knowledge management and data mining for marketing. Decis. Support Syst. **31**(1), 127–137 (2001)

31. Sim, J., Wright, C.C.: The Kappa statistic in reliability studies: use, interpretation, and sample size requirements. Phys. Ther. **85**(3), 257–268 (2005)

32. Socher, R., Bauer, J., Manning, C.D., et al.: Parsing with compositional vector grammars. In: Proceedings of the 51st Annual Meeting of the Association for Computational Linguistics, Long Papers, vol. 1, pp. 455–465 (2013)

33. Titov, I., McDonald, R.: A joint model of text and aspect ratings for sentiment summarization. In: Proceedings of ACL 2008: HLT, pp. 308–316 (2008)

34. Wang, H., Lu, Y., Zhai, C.: Latent aspect rating analysis on review text data: a rating regression approach. In: Proceedings of the 16th ACM SIGKDD International Conference on Knowledge Discovery and Data Mining, pp. 783–792. ACM (2010)

35. Yang, D., Zhang, D., Yu, Z., Wang, Z.: A sentiment-enhanced personalized location recommendation system. In: Proceedings of the 24th ACM Conference on Hypertext and Social Media, pp. 119–128. ACM (2013)

36. Ye, M., Yin, P., Lee, W.C., Lee, D.L.: Exploiting geographical influence for collaborative point-of-interest recommendation. In: Proceedings of the 34th International ACM SIGIR Conference on Research and Development in Information Retrieval, pp. 325–334. ACM (2011)

37. Zhao, W.X., Jiang, J., Yan, H., Li, X.: Jointly modeling aspects and opinions with a MaxEnt-LDA hybrid. In: Proceedings of the 2010 Conference on Empirical Methods in Natural Language Processing, pp. 56–65. Association for Computational Linguistics (2010)

38. Zhao, Y., Qin, B., Hu, S., Liu, T.: Generalizing syntactic structures for product attribute candidate extraction. In: Human Language Technologies: The 2010 Annual Conference of the North American Chapter of the Association for Computational Linguistics, pp. 377–380. Association for Computational Linguistics (2010)

39. Zheng, V.W., Zheng, Y., Xie, X., Yang, Q.: Collaborative location and activity recommendations with GPS history data. In: Proceedings of the 19th International Conference on World Wide Web, pp. 1029–1038. ACM (2010)

40. Zheng, V.W., Cao, B., Zheng, Y., Xie, X., Yang, Q.: Collaborative filtering meets mobile recommendation: a user-centered approach. In: AAAI, vol. 10, pp. 236–241 (2010)

41. Zheng, Y., Zhang, L., Xie, X., Ma, W.Y.: Mining interesting locations and travel sequences from GPS trajectories. In: Proceedings of the 18th International Conference on World Wide Web, pp. 791–800. ACM (2009)

Author Index